Engineered Rock Structures in Mining and Civil Construction

BALKEMA – Proceedings and Monographs
in Engineering, Water and Earth Sciences

Engineered Rock Structures in Mining and Civil Construction

R.N. Singh
Former Head, School of Civil, Mining and Environmental Engineering
Faculty of Engineering, University of Wollongong
Wollongong, NSW, Australia

A.K. Ghose
Former Director, Indian School of Mines
Dhanbad, India

CRC Press is an imprint of the
Taylor & Francis Group, an **informa** business
A TAYLOR & FRANCIS BOOK

Library of Congress Cataloging-in-Publication Data
Applied for

CRC Press
Taylor & Francis Group
6000 Broken Sound Parkway NW, Suite 300
Boca Raton, FL 33487-2742

First issued in paperback 2019

ISBN-13: 978-0-415-40013-8 (hbk)
ISBN-13: 978-0-367-39138-6 (pbk)

Visit the Taylor & Francis Web site at
http://www.taylorandfrancis.com

and the CRC Press Web site at
http://www.crcpress.com

Preface

The idea of this book was conceived some two decades ago for a variety of reasons: to bring together some international experience in hard rock mechanics and soft rock mechanics and to integrate it with own experience for the benefit of students through a lucid and readable text and to satisfy intellectual curiosity of the authors. A vast amount of published and unpublished research work has been collated, culled, sifted and synthesised to form the core material for this text which could help in developing an insight in rock mechanics, both theory and practice. The material that finally appears in this text has been honed in our undergraduate and postgraduate classes. The focus has essentially been on practical application of theory to extend the armoury of designers. It is hoped that the text should make significant contribution to the state-of-the-art practice providing guidance to the neophytes into the challenges of design for engineered rock structures. If the book could fill a general need by making accessible to students techniques that kindled the excitement, imagination and involvement of the present authors in this area, they will be more than delighted.

The contribution made by the students who sat through our courses in rock mechanics is appreciated. The first author acknowledges students at the University of Nottingham and University of Wollongong. The second author taught courses at Indian School of Mines, Dhanbad and Bengal Engineering College, Kolkata (both deemed University) and offered scores of short courses to Indian Mining Industry. The authors have worked closely with many researchers around the world whose work may have influenced the authors and perhaps inadvertently appeared in this text. Acknowledgement is made to the following colleagues for their close cooperation and valuable help: A.S. Atkins, S. Alameen, late H.A. Anireddy, N.I. Aziz, L. Armstrong, T. Aston, G.M. Bish, A. Barhkordarian, A. Birtle, S.K. Bordia, D.J. Brown, P. Buddery, V.B. Cassapi, S. Chakarvarty, Diamin Chen, I. Egretti, M. Eksi, R.J. Fawcett, C. Ghose, R.N. Gupta, A.S. Grant, C.R. Harvey, M. Jakeman, V.K. Joshi, G.R.L. Khanlari, A.N. Mehdi, S. Ngah, A.G. Pathan , I. Porter, S.M. Reed, D. Reddish, N.M. Raju, S. Rogers, B.H. Roberts, L. Sari, S.M.A. Shah, M.J. Sonter, N. Stuart, D. Stead, G. Sun, M. Sunu, B. Unver,

R. Varcoe, V. Venkateswaralu, G. Watson, B. Wells, D.N. Wright and A.M.H. Zadeh.

Special thanks go to Professor T. Atkinson, Dr. A. Isaac, late Professor E. L. J. Potts, late Professor B.N. Whittaker, Professor R.N. Chowdhury, Professor B. Smart, Professor I. Farmer, late Ivor Hawkes and V.S. Vutukuri. A debt of gratitude is owed to Professor E.T. Brown and Professor A.H. Wilson for the reproduction of some problems and drawings from their well known work in rock mechanics. Thanks are also due to J. Shonhardt, Peter Turner and many others who collectively and anonymously contributed towards the production of this book.

Finally, the authors acknowledge the support of their family and friends without which nothing worthwhile would get done.

Professor R.N. Singh
Professor A.K. Ghose

Contents

Introduction

Design and construction of structures in or on rock date back to antiquity. The concept of 'engineered rock structure', as opposed to structures sculpted by nature, refers to all man-made structures in or on rock mass which encompass a whole spectrum of structures – from underground to surface mines, to tunnels and caverns, and to dams, foundations and rock slopes. The fundamental divide that demarcates classical engineering design and that of 'engineered structure' in rock mass is that we are dealing with geological materials on 'as is where is basis', which are essentially discontinuous and where the load on the structure is often indeterminate. While precedent experience and in some cases heuristics guided the earlier practitioners for design of rock structures, the art of engineering rock structures rested basically in utilising the rock mass as the principal structural material and disturbing it as little as possible during the excavation process. One can see a neat and outstanding application of such design concept in extensive tunnels hewn in limestone in Roman times at Maastricht (Figure 1).

Heightened activities in construction in rock since the beginning of 20th century called for the emergence of the new discipline of rock mechanics that comprehends the study of mechanical responses of geological material. In fact, rock mechanics includes "all studies related to the physical and mechanical behaviour of rock and rock masses and the applications of this knowledge for the better understanding of geological processes in the fields of engineering" (Statutes of International Society for Rock Mechanics, 1963). While it would be difficult to pinpoint the origin of rock mechanics as an independent discipline of its own, mining geomechanics has been a basic mining science as reflected in the series of International Strata Control Conferences and the texts on rock pressure problems in mines. The publication of "Rock Mechanics and the Design of Structure in Rock" by Obert and Duvall in 1967 marked a major milestone in the evolution of rock mechanics as a scientific discipline. Significant advances have been made since then in the science of geomechanics and these have been reviewed exhaustively (Brady and Brown, 1995; Brown, 1998; Hudson, 1993).

The complexity of design in engineered structures in rock mass, however, is reflected in many publications (Ghose et al., 1973, Hoek and Brown, 1980;

Fig. 1

Hoek and Bray, 1981, Fairhurst, 1993; Hudson, 1999) because of multiplicity of variables, mechanisms and parameters which influence the design in rock mass. Despite multidisciplinary inputs into the science of rock mechanics from rock physics, geological engineering, applied geophysics, numerical techniques, problems remain in characterising the rock mass, predicting the effects of excavation disturbance and ensuring structural integrity over the design life of the structure. The strategy for design of engineered structures in rock masses has ranged over a whole host of disparate approaches – from closed form solutions, physical models, empirical approaches based on rock mass classification and a wide assortment of numerical models. Implicit in any design approach is the grey area that rock mass properties represent and the insight and understanding of practitioner to apply the principles to the field situation and circumstances. Design in rock mass calls therefore for culling as much of knowledge and science as available in the subject area, but most importantly in applying them with engineering judgment which resides in experience. In fact, engineered structures in rock mass pose the most intractable, if not challenging problem, as every design is an innovative one, uniquely tailored to a specific site and purpose!

This book is for any one involved in designing or implementing engineered structures in the rock mass addressing problems encountered in mining and civil construction. In particular, the book is designed to hone the skills of students in applying principles to real world design situation. There is currently an overflowing bookshelf in every library of introductory texts to thick volumes on rock mechanics, including the 5-volume compendium on *Comprehensive Rock Engineering* (Hudson, 1993). This book has no pretensions to such comprehensive works on rock mechanics. Instead, it is planned to introduce students and practitioners to design exercises, through graded problem solving. In mining and civil construction the concepts of factor of safety and what constitutes 'failure' of a structure may differ somewhat; otherwise there is much that is common ground. The chapters of the book follow a logical sequence – from testing techniques for intact rock, *in situ* testing techniques including stress measurement, to principles of design of underground structures, design of pillars and associated structures, subsidence engineering, stability of slopes and finally the over-riding importance of risk analysis in design of engineered structures. Each chapter has been designed so that the data presented are summarised in a tabular format for assisting the reader for a quick look-up, if not as an *aide-mémoire.*

References

Brady, B.H.G. and E.T. Brown (1993). *Rock Mechanics for Underground Mining*, 2nd Edition, Chapman and Hall, 571 pp.

Brown, E.T. (1998). *Ground Control for Underground Excavations–Achievements and Challenges'*, Proc. Int. Conf. On Geomechanics/Ground Control in Mining and Underground Construction, University of Wollongong, pp. 3-12

Ghose, A.K., B.R. Subramanyam and R.K. Abrol (eds) (1973). Rock Mechanics – Theory and Practice, The Institution of Engineers, Calcutta, 323 pp.

Fairhurst, C. (1993). *Analysis and Design in Rock Mechanics–The general Context, in Comprehensive Rock Engineering*, Vol. 2, Pergamon Press, pp. 1-29

Hoek, E. and E.T. Brown, (1980). *Underground Excavation in Rock*, Institution of Mining and Metallurgy, London

Hudson, J. (1999). *Technical Auditing of Rock Mechanics Modelling and Rock Engineering Design*, Rock Mechanics for Industry, Proc. of 37th U.S. Rock Mechanics Symposium, Vol. 1, Balkema, pp. 3-12

Hudson, J.A. (1993) (Editor). *Comprehensive Rock Engineering–Principles, Practice and Projects*, Vol. 1-5, Pergamon Press, Oxford, 4407 pp.

Obert, L., and Duvall, W.I., (1967). Rock Mechanics and the Design of Structures in Rock, John Wiley and Sons, New York, 650 pp.

Rock Characterisation for Rock Mechanics Design

1 Chapter

1.1 Introduction

The mechanical properties of rocks are important factors governing the behaviour of rock structures in response to an applied load. As the selection of appropriate criteria for the design of a rock structure and the evaluation of its stability by analytical methods depends upon the mechanical properties of the rocks, the determination of these parameters is an essential requirement of all rock mechanics investigations.

The mechanical properties of rocks depend primarily on their mineral composition, petrological structure and texture. It is, therefore, logical to attempt to correlate the strength properties with these physical parameters. This chapter outlines the types of tests which are carried out to determine strength and deformational parameters. A range of intrinsic properties and index properties must also be determined to correlate mineralogical composition to the tensor properties of the rocks for use in rock mechanics design. Rock characterisation data can be also used in the analysis of field results of a geotechnical monitoring programme and in analytical approaches such as the development of mathematical models.

The principal factors affecting the quality and integrity of the specimen are its size and shape, the condition of the rock material and, where relevant, the extent of any contamination brought about by the rock coring process. It is of importance, therefore, that the most common cylindrical-shaped specimen preparation should be compatible with the test concepts discussed below.

1.2 Classification of Rock Mechanics Tests

Generally, these tests fall into three categories as follows.

(1) Engineering design properties
(2) Intrinsic properties
(3) Index tests

1.2.1 Engineering Design Properties

The major engineering design parameters employed in a geotechnical analysis are uniaxial and triaxial strength properties and the deformation parameters of the rocks. Some of the more important tests are summarised in Table 1.1.

1.2.2 Intrinsic Properties

Table 1.2 shows various intrinsic properties of rocks which are determined in a routine rock mechanics test. These are density, porosity and permeability to water and so on, as shown in Table 1.2. Some argilliaceous rocks absorb water and become weak and may contribute to instability of any mine structure they are associated with. Therefore, swelling index is an important parameter in mines with a predominance of shale in the roof strata which are consequently prone to roof falls.

1.2.3 Index Tests

A large number of index tests has been devised. These tests can be carried out in the field with minimum preparation of samples and some can be correlated to uniaxial compressive strength values and other design properties. These include the following tests.

Point load index test

The point load index is given by the following equation:

$$I_s = \frac{P}{D^2} \tag{1.1}$$

where

I_s = point load index
P = applied load
D = dimetral distance between indentor at the time of failure (m)

Schmidt hammer hardness test

In this test, the rebound of a spring loaded piston on application on a prepared surface of rock indicates the rock hardness number. This number can be correlated to the uniaxial compressive strength and other properties of the rock.

Cone indentor hardness test

A prepared rock disc is tested on the cone indentor where a force applied by the indentor is measured against penetration of the rock

Table 1.1 Laboratory tests on intact rock

No.	*Test type*	*Loading geometry and stress path*	*Parameter*	*Core diameter (mm)*	*h/d ratio*	*Minimum number*
1.	Uniaxial compressive	Axial load on cylindrical specimen	$\sigma_c = \frac{P}{0.25\pi d^2}$	AX 20–75 mm	2.5–3.0	10–20
2.	Indirect tensile	Dimetral load on disc	$\sigma_t = \frac{2P}{\pi d t}$	NX or 50 mm	0.5–1.0	5
3.	Triaxial test 1. Single specimen 2. Knowledge of compressive and tensile strengths 3. Batch test 4. Multifailure state test 5. Strain controlled triaxial test	Triaxial test cylindrical specimens in a triaxial cell	1. Cohesive strength (C) 2. Angle of friction (ϕ) 3. Angle of fracture (α)	25–50 mm	1.5– 2.0	5–10
4.	Direct shear or punch test	Cylinder disc	$\tau = P/A$	25–50 mm cylinder	2.5–3.0	5–10
			A = area of shear	50–150 mm disc	0.25–0.5	5–10
5.	Joint testing	• Cut cylinders • Natural joint planes in core	C, ϕ	25–50 mm	2.5–3.0	5–10
6.	Young's modulus and Poisson's ratio	Axial loading of cylinder and measuring stress against strain	$E = \frac{\text{Stress}}{\text{Strain}}$ $\upsilon = \frac{\text{Lateral } \varepsilon}{\text{Axial } \varepsilon}$	25–50 mm	2.5–3.0	5

Table 1.2 Testing of intrinsic properties of rock

No.	*Test type*	*Method*	*Parameter*	*Core diameter (mm)*	*h/d ratio*	*Min nos*
1.	• Density	Wetting cylindrical specimen	Density = $\frac{W}{V}$	20–150 mm	2.5–3.0	10
	• Porosity • Water content • Absorption		Porosity = $\frac{\text{Pore V}}{V_{total}}$			
2.	Water permeability		$K = \frac{Q}{\rho}$	37 or 60 mm	2.5–3.0	5
	(i) Axial (ii) Radial	Directional controlled flow of water through a cylindrical specimen	Q = quantity ρ = pressure difference			
	(iii) Fissure		$K = \frac{Q}{2\pi LP} \log \frac{D_2}{D_1}$			
3.	Swelling index			25–50 mm	1.5–2.0	5–10
4.	*Cerchar* abrasivity index	Disc		50–150 mm disc	0.25–0.5	5–10
5.	Sonic testing	• Cut cylinders • Natural joint planes in core	C, ϕ	25–50 mm	2.5–3.0	5–10

Table 1.3 Index testing of rocks

No.	*Test type*	*Method*	*Parameter*	*Core diameter (mm)*	*h/d ratio*	*Min nos*
1.	Point load index	Loading between two conical indentors	Applied load and distance between two conical indentors	20–150 mm	2.5–3.0	10
2.	Schmidt hammer	Rebound on a prepared surface	Rebound number can be correlated to uniaxial compressive strength and other properties	37 or 60 mm	2.5–3.0	5
3.	Swelling index	Cylindrical specimen submerged in water	Measure pressure developed or expansion in volume	25–50 mm	1.5–2.0	5–10
4.	Cerchar abrasivity index	Disc-shaped sample	Scratch disc with stylus and measure wear flat on travelling microscope	25–50 mm cylinder	2.5–3.0	5–10
				50–150 mm disc	0.25–0.5	5–10
5.	Shore hardness	Rebound of diamond weight on a polished disc sample	Measure height of rebound on a scale	54 mm diameter specimen	0.5–1.00	15 readings
6	Cone indentor	Disc samples	Measure depth of indent by stylus against load	25–54 mm	2.5– 3.0	10

by the cone. The cone indentor hardness is given by the following relationship:

$$I_s = \frac{D}{P} \tag{1.2}$$

where

D = deflection of plate spring or force
P = penetration of the cone

The cone indentor hardness number can be correlated to the uniaxial compressive strength of the rock and other properties of the rock.

Abrasivity index

Some rocks are extremely abrasive and cause considerable damage to excavation tools during the mine development process. It, therefore, becomes imperative to use the drilling and blasting method to excavate such openings. Abrasivity index is therefore determined to classify rock, which can wear mining machinery and is amenable to drilling and blasting as a means of rock breaking (Pathak and Ghose, 1994).

1.3 Planning the Laboratory Rock Core Preparation

The most convenient way of testing rock in the laboratory is to conduct tests on cylindrical cores obtained by drilling. The flow chart in Figure 1.1 indicates the sequence of operational events taking place during the planning and production of rock core specimens.

1.3.1 Sample Selection

The importance of judicious sample selection in the overall success of the test programme cannot be overemphasised. It follows, therefore, that test specimens employed for the determination of strength or deformation properties should be of good-quality unweathered material, should be free from incipient fractures or damage and should be of regular shape and size. Similarly, rock core specimens used with porosity or permeability tests should be free from contamination, e.g. 'chemicals used with flush media'. In every case, where laboratory tests on rock core specimens are being conducted the test results should be representative of the average physical properties of the rock mass *in situ*. The timing of sample collection may also play an important role with regard to the end result of the tests, in that collection of samples should coincide with the availability of the facilities for specimen preparation, thus avoiding long delays and storage problems which, in turn, may result in deterioration and unnecessary additional costs.

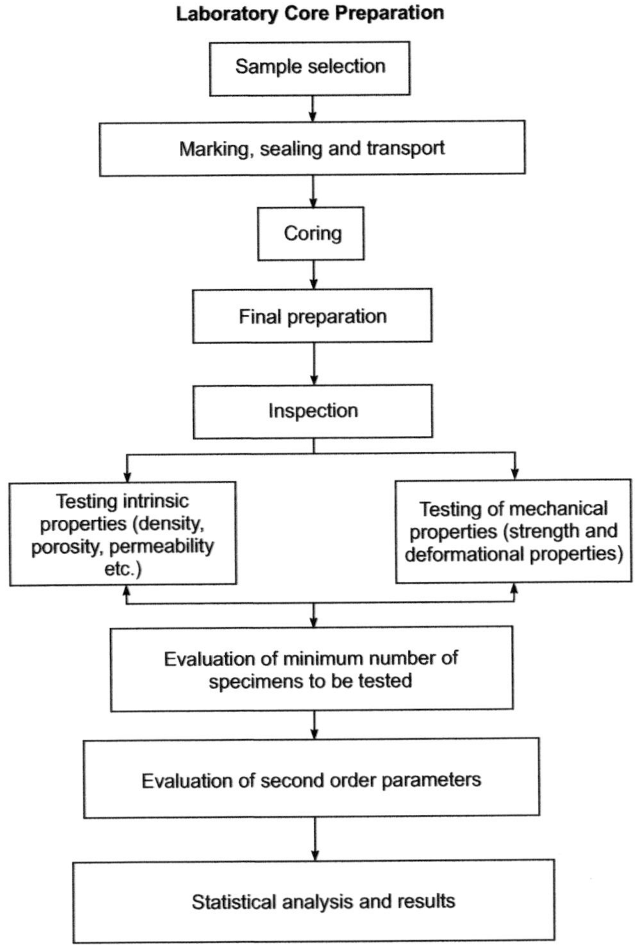

Figure 1.1 Laboratory rock core preparation and testing

1.3.2 Marking, Sealing and Transport

The practice of marking rock samples often serves two essential functions, the first to identify the site of origin, e.g. coalfield, mine or seam site and the second to permit accurate orientation of the sample in relation to the parent geology.

Sealing of the sample is a practice carried out at the site of origin or at the laboratory immediately upon delivery so as to retain the inherent moisture in the sample. The methods include the following.

(a) Hermetic sealing by polyeurethane varnish
(b) Polythene sheet, sheathing and bags

(c) Cling film
(d) Casting in foam (Resi-foam)

From 100 kg of material, it is possible to extract approximately 8–10 m of 38 mm diameter core and, with an intact core recovery rate of about 60% from friable rock material, this would yield about 40–60 testable core specimens.

Problems do occur, however, with certain sample materials, such as mudstone, siltstone etc., manifested by the development of hairline cracks along the bedding planes resulting in ultimate separation. Stress relaxation and dehydration have been found to promote this fracture process. The use of tipping gear when unloading heavy bulk sample material from lorries, may also contribute to damage to certain rock types and subsequent poor and uneconomic recovery rates.

1.3.3 Specimen Preparation

1.3.3.1 Problems of core recovery from friable rocks

The preparation of cored rock specimens from friable bulk samples often presents difficulties with regard to economic intact core recovery and is particularly relevant when attempting to core perpendicular to the bedding. Apart from bituminous coal, similar difficulties are also encountered with banded carboniferous siltstone, mudstones and seat-earths. In each of the abovementioned cases the problem appears to be due to weak stratification where separation may take place along the bedding. This may lead to foreshortening of the minimum core length requirements for useful laboratory tests. This chapter contains the findings of a research programme to develop various drilling techniques and equipment in order to maximise intact core recovery, thus increasing the cost-effectiveness of the laboratory core specimen preparation process (Cassapi, 1983).

1.3.3.2 Factors affecting intact core recovery

The quality of intact cores recovered from friable rock material is interdependent upon:

- machine characteristics,
- applied drilling parameters,
- selection of suitable coring bits
- condition of sample material

Details of the above factors are summarised below.

1.3.3.3 Criteria for efficient intact core recovery in soft rocks

(i) Drilling machine characteristics: The ideal characteristics of the drilling machine for core recovery in the laboratory are as follows.

(a) Stiffness of the drilling machine (Fig. 1.2)

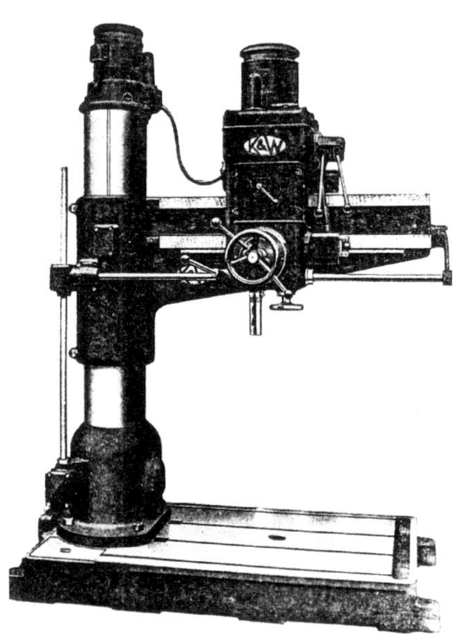

(a) soft drilling machine with gear box and one drilling and thrust motor

(b) Stiff drilling machine with indedependent drilling and thrust motors

Figure 1.2 Types of drilling machines for preparing cored specimens for laboratory testing (Atkinson and Cassapi, 1983)

(b) Minimum vibration
(c) Independent load application to the bit
(d) Adequate peripheral speed 120–180 m/min
(e) Rotation per minute of the drilling motor, 400–2000 rpm

(ii) Drilling parameters: The drilling parameters investigated were the effect of load on the bits, peripheral speed, flushing rate and torque on the drilling rate and core recovery as follows.

(a) Load on the bit: The effect of bit loading on the penetration rate is illustrated in Figure 1.3 which indicates that as the load increases the penetration rate of bit on the rock increases. However, above a certain value of load the drilling rate starts decreasing with the load and the drill starts to seize. Thus, there is an optimum load for a drill bit and a rock type upto which point drilling rate increases with the load on the bit.

(b) Peripheral speed: The effect of the peripheral speed of the bit on the penetration rate is illustrated in Figure 1.4 which indicates that as the peripheral speed of the drill bit increases, the penetration rate of the bit on the rock increases. However, above a certain value of peripheral speed the

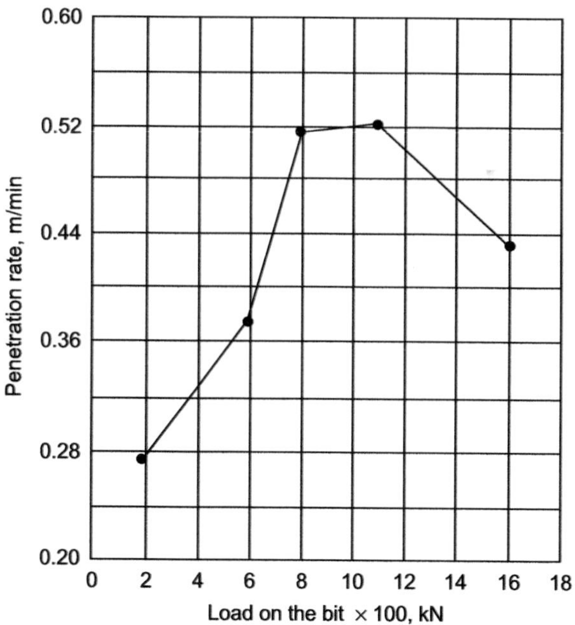

Figure 1.3 Effect of load on the bit on penetration rate

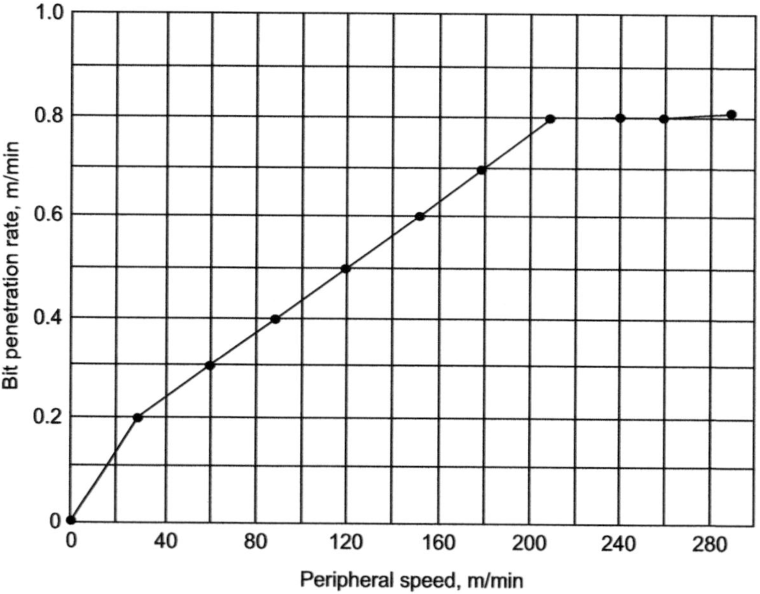

Figure 1.4 Effect of peripheral speed of the bit on penetration rate

drilling rate starts levelling off with an increase in the peripheral speed. Thus, there is an optimum peripheral speed for a drill bit and a rock type up to which the drilling rate increases with an increase in the bit peripheral speed.

(c) Effect of flushing rate on the rate of penetration: When the drilling bit cuts the rock, the resulting debris is removed by the flushing fluid or in the case of argillaceous rocks by flushing air. This exposes the new rock surface and drilling continues smoothly. As the flushing rate increases, the rate of penetration increases. In the absence of flushing, the debris will be continuously churned and considerable energy will be lost in grinding the rock cuttings resulting in an inhibition of the rate of drilling (Cassapi, 1983). Figure 1.5 shows the relationship between the rate of drilling and the flush rate. There is an incremental relationship between the rate of flushing and the rate of penetration up to a point above which the buoyancy of flushing fluid dramatically reduces the rate of penetration of the drill bit.

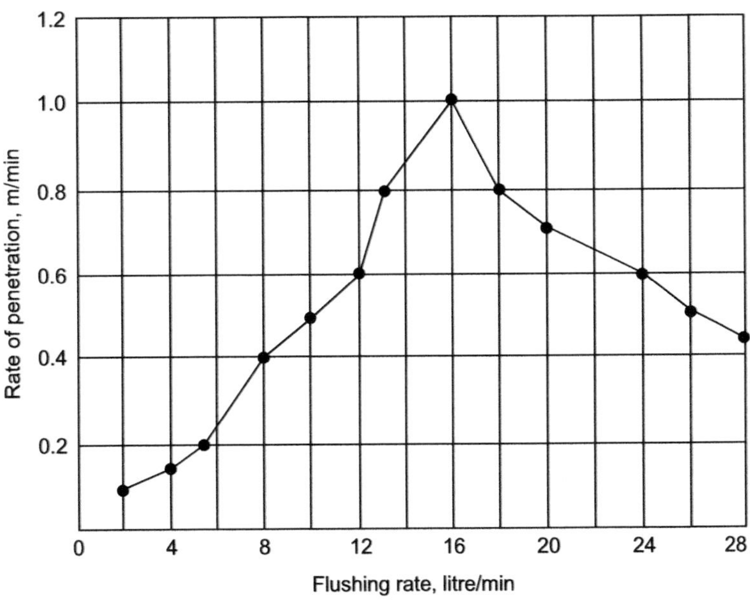

Figure 1.5 Effect of flushing rate on the rate of penetration of the bit

(d) Relation between torque and load on the bit: Figure 1.6 shows the performance of an impregnated bit versus surface set bits. It can be seen that the impregnated bit needs less torque to drill and the rate of penetration increases with the load on the bit and so does the torque. Torque requirements also increase with increased diameter of the bit. A step-wise increase in the torque requirement may be noticed with different types of bits.

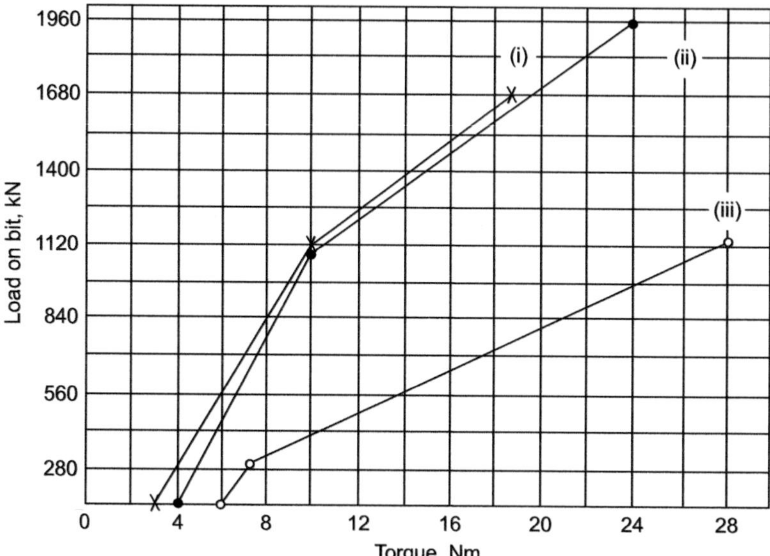

Figure 1.6 Correlation of torque as a function of load on the bit (1) 38 mm impregnated diamond bit, (2) 38 mm surface set and (3) 61 mm diamond impregnated bit

1.3.3.4 Core bit selection

A summary was compiled from information gathered from a series of tests with different types of core drills of similar size i.e. 38 mm (Figures 1.5 and 1.6). These were

(a) standard concrete type impregnated diamond bits,
(b) special purpose surface set (diatube) bits and
(c) special purpose extra thin wall impregnated bits.

The standard concrete core drill was found to be suitable for competent rock material and was amenable to both air and water flush systems. This drill, however, did not appear to perform as well with the more friable rock materials.

The surface set bit was designed for use with certain mudstone and clay materials and because of its marginally extra annular clearance and non-depreciating flush ways, was more amenable to non-aqueous flush media necessary with this type of material.

The extra thin wall diamond impregnated core bit was designed specifically for use with friable bituminous coal and seat-earth samples. This bit was composed of relatively larger grit sized stones in a soft matrix and a 35% reduction in kerf area permitted coring at much lower bit pressures than the standard concrete core drill.

Bit load is equal to 9 kg per cm^2 of kerf area per 100 diamond concentration: the kerf area of the extra thin wall bit is equal to 2.514 cm^2 compared with the standard concrete bit which has the kerf area of 3.860 cm^2. Therefore, with a diamond concentration of 50, the bit load difference is significant (114.3 kg) for the thin kerf bit in comparison to 175.6 kg for the standard bit (Cassapi, 1983).

1.3.3.5 Pre-coring preparation

The coring of friable rock samples of irregular shape and size on a powerful high speed machine can be a most hazardous practice. Therefore, proper care and attention to locating and securing the rock sample to the machine is essential. The concern for safety often leads to the practice of over-tightening the securing clamps or vice, thus imposing excessive stress upon the rock sample. While these imposed stresses may or may not affect the characteristics of the physical properties of the rock, it is clear that intact core recovery is reduced. This is caused by the weakening affects of 'overstressed clamping to the rock material' resulting in unsafe working conditions due to disintegration of the sample when core material is removed. It was apparent therefore that in the interest of safety and economic core recovery, new techniques would have to be evolved to eliminate the physical hazards and at the same time, increase the core recovery potential from each individual rock sample.

Coring trials carried out on seat-earth rock samples set in encapsulating polyurethane foam, indicated much improved core production on previously clamped or vice held samples. Three different materials were investigated for their strength and reliability in setting rock samples in readiness for machining (1) resin; (2) herculite plaster and (3) concrete.

Setting the sample in concrete was found to be the cheapest, strongest and most reliable means of captivating the rock sample for machining purposes. The single disadvantage being that coring cannot take place for at least 24 hours after setting, however, this is distinctly outweighed by its other merits. The proportion of aggregate and cement finally decided upon, was four parts aggregate to one part cement. Based on drillability tests, Table 1.4 sets out the specifications of speed, bit load, at flush rate for different rock types.

1.4 Final Preparation

1.4.1 Cutting to Length

This operation is normally carried out on a diamond wheel electric masonry saw. The core is held in a jig to facilitate safe handling and cut to a length allowing for the finish grinding of the faces. There are those special cases where the material is so friable that even the most careful machining techniques can result in failure of the specimen when attempting to cut to

Table 1.4 Drillability test on various rock types

Rock type	*UCS (MPa)*	*Drill bit type*	*Speed (rpm)*	*Bit load (kN)*	*Water flush rate (l/min)*
Medium grade sandstone	65	38 mm diamond impregnated	1250–1500	750–850	20
Fine grade sandstone	85	38 mm diamond impregnated	1000–1250	850–950	16
Pennant grit (Welsh)	180	38 mm diamond impregnated	750–1000	1250–1400	20
Limestone	90	38 mm diamond impregnated	1100–1250	1000–1200	16
Mudstone	45	38 mm diamond surface bit	750–1000	950–1400	Air
Shale	30	38 mm diamond surface bit	750–1000	950–1100	Air
Coal	<25	38 mm extra thin wall diamond impregnated	1000	650–850	12
Seat-earth	29	38 mm extra thin wall diamond impregnated	800–1000	580–650	Air
Siltstone	64	38 mm diamond impregnated	900–1000	1100–1200	16

length by diamond saw. In this case, the use of a standard high-speed steel hacksaw blade used with a hand-operated frame has yielded good results, providing great care and attention is given to holding the specimen when cutting.

1.4.2 Grinding of Face Ends

The face end grinding of rock core specimens can be carried out on a standard mechanical engineering surface grinder. An Adcock and Shipley surface grinder is a typical machine that could be exclusively used for this purpose.

1.4.3 Inspection

After the specimen has been finally ground to size, the specimen is submitted for inspection before testing. The inspection will take account of dimensional accuracy according to ISRM (1972) conditions and the conditional integrity of the rock relative to the test concepts. The specimen is then carefully wrapped in cling film to maintain the best possible condition before testing or during storage.

1.4.4 Tolerances on Dimensions of Cylindrical Specimens

Recommendations are made by the International Society of Rock Mechanics (1972) regarding the tolerances and specifications of cylindrical rock core specimens and these are listed below.

(1) The ends of the specimen shall be ground or lapped flat to within 0.02 mm.

(2) The ends of the specimen shall be square to the perpendicular axis and to be within 0.001 radian.
(3) The sides of the specimen shall be smooth and free from any surface irregularities, and be straight to within 0.3 mm (0.010") over its entire length.
(4) In preparation of the test specimen, the rock shall not be subjected to any treatment (such as chipping with a hammer) liable to induce incipient fracture.

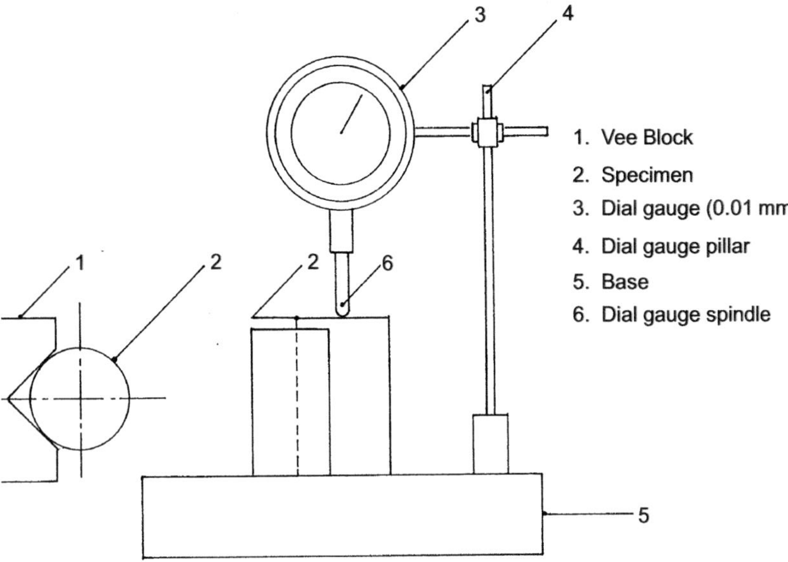

Figure 1.7 Typical specimen inspection jig

1.5 Minimum Number of Samples Required for a Test Series

Results obtained by testing individual specimens sometimes indicate a high degree of scatter and are subsequently, but not always, representative of the parent rock body. Therefore, it is often necessary to conduct a greater number of tests in order to limit costs without compromising accuracy or reliability. In the past, three methods have been used, namely (1) Method of varying degree of confidence, Protodyakanov (1969); (2) Sub-sampling technique, Yamaguchi (1970) and (3) Method of limitation of variance, Yang and Hardy (1967). The methods of limitation of variance and sub-sampling technique require tedious calculations and are not amenable to manual computation.

The methods depend upon the degree of scatter of results and need a suite of computer programs.

A relatively simple technique of calculating the minimum number of tests required for a test series using a method of varying degree of confidence is given in the following section.

1.5.1 Method of Varying Degree of Confidence

The method comprises testing the uniaxial compressive strength of rock on the preliminary number of tests 'N' on a particular rock type and carrying out the statistical analysis as follows.

$$X_m = \text{mean value} = \frac{\Sigma X_i}{N} \tag{1.3}$$

$$S = \text{standard deviation} = \sqrt[2]{\frac{S(X_i - X_m)^2}{N-1}} \tag{1.4}$$

V = coefficient of variation in the preliminary test of N specimen

$$= \frac{S}{X_m} \times 100 \tag{1.5}$$

The following relationship can be used for determination of number of specimens 'n' to be tested.

$$n = \left[\frac{(P+1)KV}{100\,(P-1)}\right]^2 \tag{1.6}$$

where,

$$P = \frac{100 + r}{100 - r} \tag{1.7}$$

P = a number depends upon permissible deviation of mean value and is equal to 1.5 for 20% permissible deviation

r = maximum permissible error

For example, P can be calculated for various values of r as follows:

If $r = 20\%$

$$P = \frac{100 + 20}{100 - 20} = 1.5$$

If $r = 10\%$, then $P = \dfrac{100 + 20}{100 - 20} = \dfrac{110}{90} = 1.22$

K = depends upon the desired confidence interval and the number of specimens tested (N) in a preliminary experiment as given in Table 1.5.

Table 1.5 Value of K

Degree of freedom	*Level of probability (confidence level)*		
N-1	90%	95%	99%
1	6.31	12.7	63.7
2	2.92	4.3	9.92
3	2.35	3.18	5.84
4	2.13	2.78	4.60
5	2.02	2.57	4.03
6	1.94	2.45	3.71
7	1.89	2.36	3.50
8	1.86	2.31	3.36
9	1.83	2.26	3.25
10	1.81	2.23	3.17
11	1.80	2.20	3.11
12	1.78	2.18	3.05
13	1.77	2.16	3.01
14	1.76	2.14	2.98
15	1.75	2.13	2.95
16	1.75	2.12	2.92
17	1.74	2.11	2.90
18	1.73	2.10	2.88
19	1.73	2.09	2.86
20	1.72	2.09	2.85
25	1.71	2.06	2.80
30	1.70	2.04	2.75
Known Stnd. D or infinity	1.64	1.96	2.53

Example 1 In a compressive strength test on coal, the results of preliminary tests are as follows: 29.5, 25.2, 18.2, 17.3 and 23.3 MPa respectively. If the permissible deviation from the mean is 15%, probability of distribution around mean is at 95% confidence level, calculate the minimum number of specimens required for the test.

Solution

$$X_m = \frac{SX_i}{N} = \frac{113.5}{5} = 22.7$$

$$S = \sqrt{\frac{S(X_i - X_m)^2}{N-1}} = [25.56]^{0.5} = 5.056$$

$$V = \frac{100\,S}{X_m} = \frac{100 \times 5.056}{22.7} = 22.27$$

$$P = \frac{100 + r}{100 - r} = 1.35$$

$$K = 2.78 \text{ from Table 1.5}$$

$$n = \left[\frac{(P + 1)K \cdot V}{100(P - 1)}\right]^2 \qquad i = \left[\frac{(1.35 + 1)2.78 \times 22.27}{100(1.35 - 1)}\right]^2$$

$$= 17$$

Example 2 In a tensile strength test, the results of preliminary tests are 3.80, 4.86, 5.66, 3.40, 4.07 and 3.08 MPa respectively. If the permissible deviation from the mean is 15% and degree of confidence of the estimation is 95%, calculate the minimum number of samples required to be tested.

Solution

$$X_m = \frac{\Sigma X_i}{N} = \frac{24.81}{6} = 4.13$$

$$S = \sqrt{\frac{\Sigma(X_i - X_m)^2}{N - 1}} = 0.96$$

$$V = \frac{100S}{X_m} = \frac{100 \times 0.96}{4.13} = 23.25;$$

$$P = \frac{100 + r}{100 - r} = \frac{115}{85} = 1.35$$

$$K = 2.57 \text{ from Table 1.5}$$

$$= 16$$

$$n = \left[\frac{(P + 1)K \cdot V}{100(P - 1)}\right]^2 \qquad n = \left[\frac{2.35 \times 2.57 \times 23.25}{0.35 \times 100}\right]^2$$

$$= 16$$

Laboratory experiments have been conducted on a variety of rock types and the minimum number of tests required are tabulated in Table 1.6

Table 1.6 Minimum number of samples in compressive strength test for various rock types

Rock type	*Minimum number of tests required (n)*
Homogeneous rock	2-3
Shale	5
Mudstone	6
Pennant sandstone	16
Red sandstone	25
Silty seat-earth	19
Limestone	14

1.6 The Economics of Laboratory Core Specimen Preparation

Indiscriminate selection of samples can significantly affect both quality and quantity of intact core recovery and by doing so affect the quantity of material required to produce sufficient numbers of core specimen needed. This may, in turn, incur further costs involved with selection and transport of additional material. These additional costs may in some cases place certain constraints upon the limitation of the project.

Rock core specimen costs can be calculated by the following equation.

$$C_T = \frac{C_1 + C_2 + C_3 + C_4}{N} \tag{1.8}$$

where

C_T = total unit cost
C_1 = sample collection and transport cost
C_2 = cost of sealing and pre-coring treatment
C_3 = manpower cost for core preparation
C_4 = cost of bit and grinding equipment

where

$$C_4 = \left\{\frac{W \times L_1}{L_2}\right\}$$

L_1 = bit life in metres
L_2 = length of core drilled (m)

***Example* 3** Cost of collecting sample includes wages cost for a technical person at the rate of A$ 22.50 per hour and overtime rate of A$ 33.75 per hour. Time spent in collecting and transport of sample from underground mine to pit head is eight hours and travelling time from the mine to the laboratory is three hours both ways. The cost of using a personal car to bring the sample from the mine to the laboratory is at the rate of A$ 0.51 / km and distance travelled for the round trip is 200 km.

Cost of sealing the specimen is A$ 10.50 and the cost of precasting the sample is A$ 6.00 material cost and A$22.50 is labour cost. If the time required to core 90 specimens is one day and sizing and lapping the specimen takes two days, calculate the cost of producing one specimen. Assume that the drill bit cost is A$ 0.16 per specimen, cost of sizing and lapping is A$ 0.35 per specimen.

Solution

Wages cost of collection of sample = $8 \times 22.50 + 33.77 \times 3$ = A$ 281.3
Travelling cost = 0.51×200 = A$ 102

Cost of sealing = A$ 10.50
Pre-coring treatment =A$ 26.50
Manpower cost for core preparation = $3 \times 8 \times 22.50$ = A$ 540.00
Cost of bit and grinding equipment = 0.51×90 = A$ 45.90

$$C_T = \frac{(382.30 + 37.00 + 540 + 45.90)}{90} = \text{A\$ } 11.17 \text{ per specimen}$$

1.7 Intrinsic Properties of Rock

Rock cores prepared for various laboratory tests can be used to determine the intrinsic properties of rocks.

Measurement of Rock Density

Dimensions of prepared cores are carefully measured at the ambient temperature and the volume (V) of the core is calculated. The accurate weight (W) of the sample is determined and the bulk density calculated as follows.

$$\text{Density } \gamma = \frac{W}{V} \tag{1.9}$$

γ = bulk density of rock

Other densities of rock can be:

γ_s = density of solid mineral matter
γ_{sat} = saturated density of sample
γ_d = dry density of sample

Porosity

The core is then oven-dried at 105°C for 24 hours and the dry weight is measured. The specimen is submerged in water for three days and the saturated weight is determined. The porosity is calculated as

$$n = (W/V) \tag{1.10}$$

Void ratio,

$$e = \frac{V_V}{V_S} = \frac{n}{(1-n)} \tag{1.11}$$

V_S = volume of solid materials
V_w = volume of water in void
V_V = volume of voids
n = porosity
e = void ratio

Moisture Content

Moisture content can be calculated as the weight of water divided by the weight of solid mineral content expressed as a percentage:

$$w = \frac{W_W}{W_S}\% \tag{1.12}$$

W_S = weight of solid mineral matter after oven drying to constant weight at 105°C

W_W = weight of water in the voids

Void Ratio

Saturated moisture content is calculated when the voids are fully saturated. It is called void index and is used extensively as primary characteristics of the rock material. Void ratio can be expressed as

$$i = \frac{W_W}{W_S} \tag{1.13}$$

where

i = void ratio

W_W = weight of water in the voids

W_S = weight of solid grains in specimen

Dry Apparent Specific Gravity

$$G_b = \frac{W_s}{V \times \gamma_w} = \frac{\gamma_d}{\gamma_w} = \frac{n}{i} \tag{1.14}$$

where

V_w = volume of water in void

γ_d = dry density of the sample

γ_w = density of water in void

G_b = dry apparent specific gravity

Saturated Apparent Specific Gravity

$$G_b^1 = \frac{(W_s + W_w)}{V_w \times \gamma_w} = \frac{\gamma_{sat}}{\gamma_w} = \frac{G_b}{(1-n)} \tag{1.15}$$

G_b^1 = saturated apparent specific gravity

1.8 Permeability to Water Flow in Rocks

Permeability tests incorporating water flow on cylindrical rock samples of 40 mm diameter and 60 mm length are usually performed under a water pressure of 26 MN/m^2, where a precise rate of percolation is measured. As it is difficult to measure flow rates lower than 0.01 cm^3 per hour, pressure gradients up to 7 MPa are used in the laboratory to test impervious samples to measure the desired rate of flow.

1.8.1 Types of Permeability Tests

The following types of permeability tests have been devised.

(i) Longitudinal permeability test

The cylindrical surface of the rock specimen is coated with a thick layer of resin and the specimen is held inside a permeability cell as shown in Figure 1.8(a). Water under pressure of 26 MPa is allowed to flow through the rock specimen and the amount of water flow is measured.

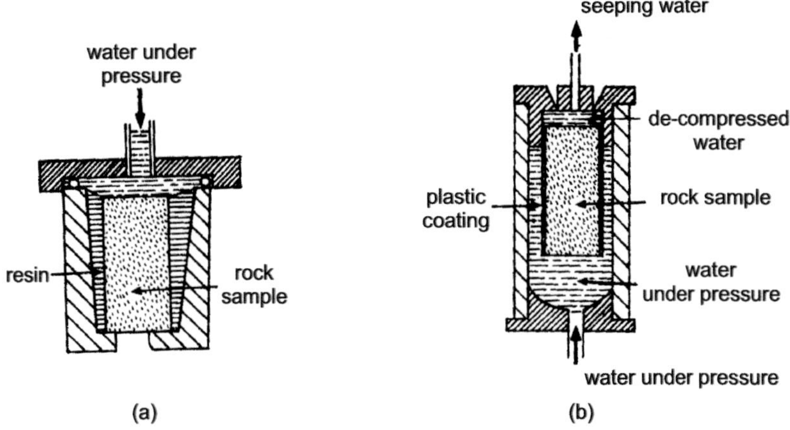

Figure 1.8 Test arrangements for longitudinal permeability test (Bernaix, 1966)

A sample is normally saturated in water before the test in order to exclude air entrapped in the pore space of the rock. The permeability factor K can be calculated using the following formula:

$$K = \frac{QL}{PA} \tag{1.16}$$

where

Q = rate of water percolating through the sample

L = length of the sample
P = pressure differential between two faces of the specimen

Figure 1.8(b) shows the longitudinal percolation test developed by Bernaix (1966). The cylindrical specimen is coated with plastic on its curved surface. The specimen is secured on a specimen holder with a water cavity at the atmospheric pressure. This assembly is fitted into a permeability cell where water is introduced at pressure. The rate of flow of percolating water is measured and the permeability coefficient calculated by using Equation 1.16.

(ii) Fissure permeability on artificial joint

The test arrangement in Figure 1.8 can be modified for a specimen containing an artificial longitudinal joint or fissure in which the flow of water is predominantly through the joint or a fissure flow. The flow rate is comparatively high and fissure permeability is calculated as per Equation 1.16.

(iii) Radial permeability test

A test specimen 60 mm diameter and 150 mm long is used through which a concentric hole of 12 mm diameter and 125 mm long is drilled. A tube of 12 mm diameter is fitted in the first 25 mm of the hole within the specimen. The specimen ends are coated with plastic or resin so that flow of water is predominantly radial in direction. Three alternative testing arrangements can be made:

1. Convergent flow
2. Divergent flow
3. Radial permeability test under varying strains.

Test arrangements are shown in Figure 1.9. The permeability factor in the case of the radial test can be calculated using the following formula:

$$K = \frac{Q}{2\pi LP} \log \frac{D_2}{D_1} \tag{1.17}$$

where D_2 = outer diameter of specimen
D_1 = outer diameter of specimen
Q = quantity of flow per second
L = length of test cavity (m)
P = pressure difference creating water flow (MPa)

Figure 1.9 shows the radial percolation test in a triaxial test arrangement under a confining pressure of $P_1 > P_0$ (water pressure). Figure 1.10 shows the test results of a percolation test on gneiss specimen.

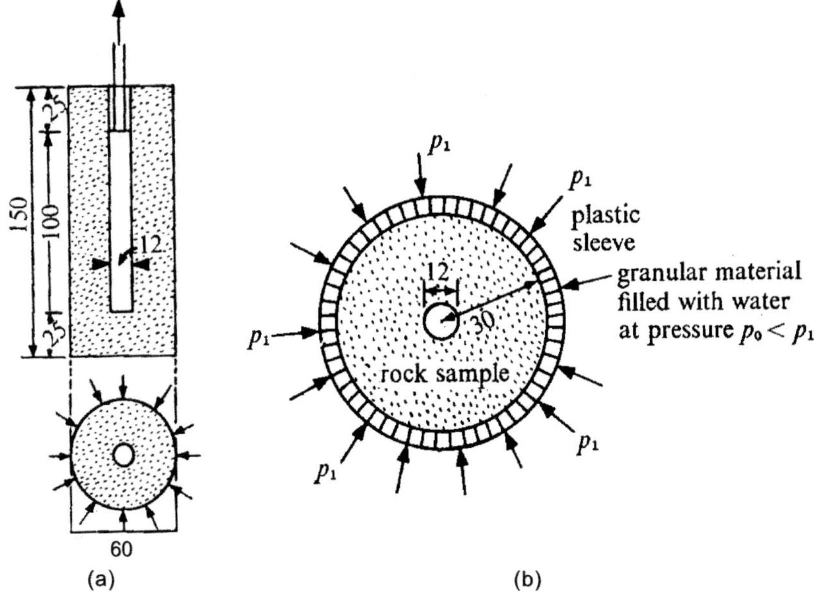

Figure 1.9 Radial permeability test arrangement under confining pressure

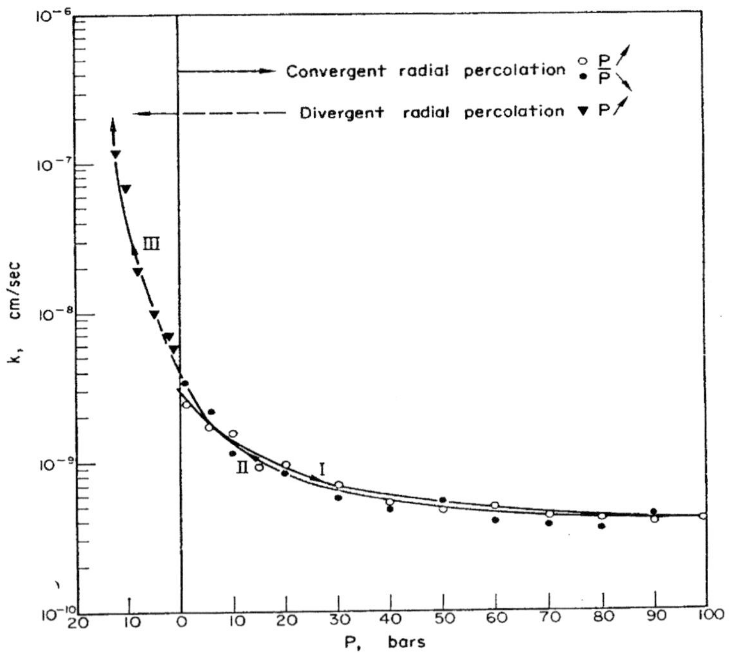

Figure 1.10 Results of radial percolation test on gneiss (Bernaix, 1966)

1.9 Discussions and Conclusions

The preparation of core specimens in friable rock is expensive. In order to obtain a maximum of useful data, economical use of specimens should be made. Therefore, non-destructive tests should be carried out before conducting destructive testes.

The drilling machine used for the extraction of rock cores from friable material should possess adequate frame stiffness and capacity. It should also possess an infinitely variable speed system and should be able to apply constant load on the bit during drilling of a specific rock.

Each of the above systems should be compatible with the sizes and types of core bits being used.

Considerations should also be given to the use of either water or air flushing systems in relation to rock mineralogy. Air flushing system should be used with argillaceous rocks.

Sufficient instrumentation should be used with the drill for recording the applied drilling parameters during drilling. These parameters are as follows.

(i) Speed
(ii) Load on bit
(iii) Rate of flush being used

The applied drilling parameters have been shown to possess an interdependence of complex nature. It has been established that increased penetration rates can be achieved by

(i) increased thrust and
(ii) increased peripheral cutting speed

In each case, there appears to be a critical point where increased performance stalls. This phenomenon does not take into account tool wear. Extensive studies in the economics of tool penetration as a function of tool wear could reveal a satisfactory solution.

Finally, it must be emphasised that speed of penetration is not the most important criteria, rather the quality and quantity of intact rock core recovery per unit sample is of significance. Weight and volume tests on the rocks should be conducted before determining the strength and deformational properties of the rock. The intrinsic properties of rock can give some insight into the design properties of the rock by correlation.

References

Atkinson T., and V.B. Cassapi (1983). 'The Preparation of Laboratory Cored Specimens from Friable Rocks', *Mining Engineer*, Vol. 142, No. 259.

Attwell, P.B. and I.W. Farmer (1976). *Principles of Engineering Geology*, Chapman and Hall, John Wiley and Sons, New York, ISBN 0 412 11400 3, 1035 pp.

Adler, L. and M.C. Sun (1976). *Ground Control in Bedded Formations*, Virginia Polytechnic Institute and State University, Bulletin 28, Research Division, Reprint March 1976, 263 pp.

Amedei, B. and O. Stephansson (1997). *Rock Stress and its Measurement*, Chapman and Hall, London.

Bernaix, J. (1966). 'New laboratory methods of studying the mechanical properties of rocks', *International Journal of Rock Mechanics and Mining Science and Geomechanics Abstracts*, Vol. 6, No. 1, pp. 43–90

Bernaix, J. (1974). 'Properties of rock and rock mass', General report, Third International Congress in Rock Mechanics, ISRM, Oslo, 1974, 11 pp.

Bienawiski, Z.T. (1984). *Rock Mechanics Design in Mining and Tunnelling*, A.A. Balkema, ISBN 90 6191 507 4, 272 pp.

Bienawiski, Z.T. (1989). *Engineering Rock Mass Classification*, McGraw Hill, New York.

Biron, C. and E. Arioglu, (1983) *Design of Support in Mines*, John Wiley and Sons, New York, 237 pp.

Brady, B.H.D. and E.T. Brown (1985). *Rock Mechanics for Underground Mining*, George Ellen and Unwin, London.

Cassapi, V. (1983), *Aspects of rock core specimens preparation with particular reference to machine tool design*, M.Phil. Thesis, University of Nottingham, 189 pp.

Coates, D.F. (1970). *Rock Mechanics–Principles*, Department of Energy, Mines and Resources, Mines Branch, Monograph, 8.74-8.95.

Farmer, I. (1985), *Coal Mine Structures*, Chapman and Hall Ltd., ISBN 0 412 250306, 310 pp.

Granholm, S. (1983). *Mining with Backfill*, A.A. Balkema, Rotterdam, ISBN 90 6191 509 0, 455 pp.

Gramberg, J. (1989). *A Non-conventional View on Rock Mechanics and Fracture Mechanics*, A.A. Balkema, Rotterdam, Netherlands, ISBN 90 6191 806 5, 250 pp.

Goodman, R. (1976). *Methods of Geological Engineering*, West Publishing Company, New York, ISBN 0-8299-0066-7.

Goodman, R. (1976). *Methods of Geological Engineering*, West Publishing Company, New York.

Hoek, E. and J.W. Bray (1977). *Rock Slope Engineering*, The Institution of Mining and Metallurgy, London.

Hoek, E. and E.T. Brown (1980). *Underground Excavations in Rocks*, The Institution of Mining and Metallurgy, London.

ISRM (1972). 'Committee of Laboratory Tests: Suggested methods for determining the uniaxial compressive strength tests of rock materials and the point load index', Document 1, Oct. 1972, 12 pp.

ISRM (1981). 'Rock characterisation, testing and monitoring: Suggested methods', E.T. Brown (Ed), Pergamon Press, UK.

Jaeger C. (1972). *Rock Mechanics and Engineering*, Cambridge University Press.

Jaeger, J.C. and N.G.W. Cook (1969). *Fundamentals of Rock Mechanics*, Methuen.

Kidybiniski, A. and M. Kwasniewski (1984). *Application of Rock Mechanics to Planning and Design Prior to Mining*, A.A. Balkema, 279 pp.

Obert, L. and W.I. Duvall (1967). *Rock Mechanics and Design of Structure in Rock*, John Wiley and Sons, New York, 650 pp.

Pathak, K., and Ghose, A.K. (1994). 'Abrasivity of Indian Coalmeasure rocks and wear of drill bits', Proc. of Int. Synp. on Mine Planning and Equipment Selection, A.A. Balkema, pp. 477–482.

Peng, S.D. (1978). *Coal Mine Ground Control*, John Wiley and Sons, 450 pp.

Protodyakanov, M.M. (1961). 'Methods of studying the strengths of rocks, used in the USSR', *Proc. of the International Symposium of Mining Research*, Rolla, Missouri, Vol. 2, pp. 649–668.

Protodyakanov, M.M. (1969). 'Method of determining the strengths of rock under uniaxial compression', In *Mechanical Properties of Rocks* N. Protodyakanov and Koffman, translated from Russian, Jerusalem, Israel Program of Scientific Translation, 1969, pp. 1–12

Rowlands, D. (1967). 'Preparation of specimens for rock mechanics research, *Mining and Minerals Engineering*, Vol. 3, No. 11, November, pp. 428–430

Vutukuri, V.S. and K. Katsuyama (1994). *Introduction to Rock Mechanics*, Industrial Publishing and Consulting Ltd., Tokyo, 275 pp.

Vutukuri, V.S., R.D. Lama, and S.S. Saluja (1974). *Handbook on Mechanical Properties of Rock*, 4 volumes, Trans. Tech. Publications Clausthal, Germany, 280 pp.

Yang, Y.J. and H.R. Hardy Jr. (1967). 'Techniques of evaluating number of tests required to provide meaningful strength data', Pennsylvania State University, Interim report. 1967.

Yamaguchi, U. (1970). 'A number of test pieces required to determine strength of rock', *International Journal of Rock Mechanics and Mining Science and Geomechanics. Abstracts* Vol. 7, No. 2. pp. 209–227.

Assignments

Question 1 Describe the types of tests to be carried out in any rock mechanics investigation for rock characterisation. What are the requirements of these tests in relation to specimen preparation?

Question 2 Describe with the aid of a flow diagram the planning of specimen preparation for laboratory testing. Describe various steps to be taken for preparation of specimens for laboratory compressive strength tests.

Question 3 Describe the radial flow permeability test to be undertaken in a triaxial cell. Describe with the aid of a diagram the test results for gneiss for various confining stresses.

Question 4 What is the tolerance of dimensions for preparation of cylindrical test specimens as recommended by ISRM (1972)?

Question 5 Describe the criteria for efficient core recovery in soft and friable rocks with respect to machine characteristics, drilling parameters and types of bits. Illustrate your answer with diagrams.

Question 6 Why is it necessary to control the number of tests to be carried out in a laboratory investigation to obtain meaningful data? Describe the method of varying degree of confidence in calculating for the minimum number of tests to be carried out for uniaxial compressive strength.

Question 7 (a) Discuss the problems of core recovery in friable rock in the laboratory testing of intact rock material. Outline the various factors influencing the rate of core recovery during laboratory core preparation.

(b) In a compressive strength test on coal, the results of preliminary tests are as follows: 12.5, 15.2, 10.2, 17.3 and 16.2 MPa, respectively. If the permissible deviation from the mean is ±10% and the probability of distribution around the mean is at 90% confidence level, calculate with the aid of the Table 1.5 the minimum number of specimens required for the test.

Question 8 In a compressive strength tests on coal, the results of preliminary tests are as follows: 22.5, 25.2, 22.2, 17.3, 23.3 and 26.2 MPa, respectively. The permissible deviation from the mean is ± 10% and probability of distribution around mean is at 95% confidence level. Calculate with the help of Table 1.5 minimum number of specimens required for the test.

Uniaxial Testing of Rock in Compression

2 Chapter

2.1 Introduction

The uniaxial compression test consists of loading a rock specimen in a compression testing machine and continually recording the strains induced in the rock specimens at various stress levels. The tests are usually carried out on small cylindrical specimens of 20 to 75 mm in diameter with the length 2.5–3 times its diameter. The ultimate strength of rock material is characterised by the stress value at failure given by the following relationship:

$$\sigma_c = \frac{P}{A} = \left[\frac{P}{\pi/4\, D^2}\right] \tag{2.1}$$

where P = failure load

A = initial cross-sectional area at mid-height of the specimen $= \frac{\pi}{4} D^2$

D = diameter at mid-height of the specimen

σ_c = uniaxial compressive strength of rock

The stress-strain curves obtained during the uniaxial compression tests yield the following information.

- Young' s modulus
- Poisson's ratio
- Uniaxial compressive strength
- Internal structure and behaviour of rocks in cyclic loading
- Second order parameters indicating internal structure and behaviour of rock

2.2 Uniaxial Compressive Strength of Rock

In concept, compressive strength determination is extremely simple, but in practice it is very difficult to create a uniform uniaxial stress field on a rock specimen due to the factors discussed elsewhere. The test results obtained are not a unique property of rock alone but dependent upon the characteristics of both rock and testing machines. Two types of testing machines are normally used.

1. Conventional soft testing machines
2. Stiff testing machines

2.2.1 Conventional Soft Testing Machine

The test arrangement in a conventional (soft) testing machine is shown in Figure 2.1. A cylindrical specimen is loaded between the movable and fixed heads of a loading machine. Two cylindrical platens incorporating a spherical seating and having the same diameter as the specimen are placed between the machine and the specimen. The purpose of the spherical seat is to provide a uniform axial stress distribution on the specimen even when the ends of the specimen are exactly parallel.

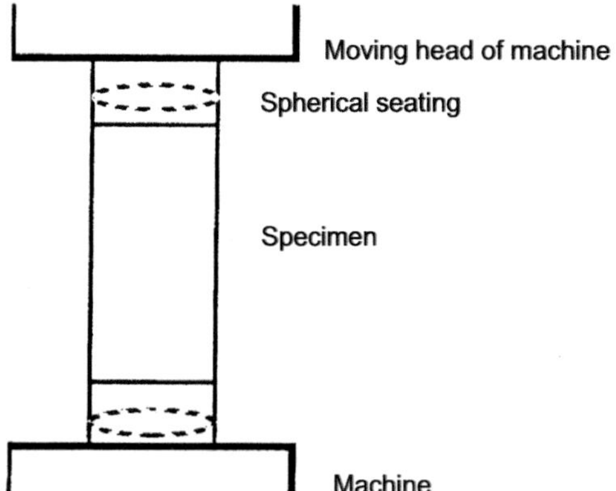

Figure 2.1 Conventional testing system

The load-displacement graph of a rock in this loading system is shown in Figure 2.2 where the load deformation characteristics of the machine are plotted on the left-hand side of the graph while that of the specimen on the right. At an initial stage of loading the deformation of the specimen and the machine are linear. When the specimen is further loaded, the load/deformation characteristics of the specimen become non-linear; more deformation being observed for a given incremental load until the specimen fails violently. Thus, the test only gives stress-strain characteristics of rock in the pre-failure state.

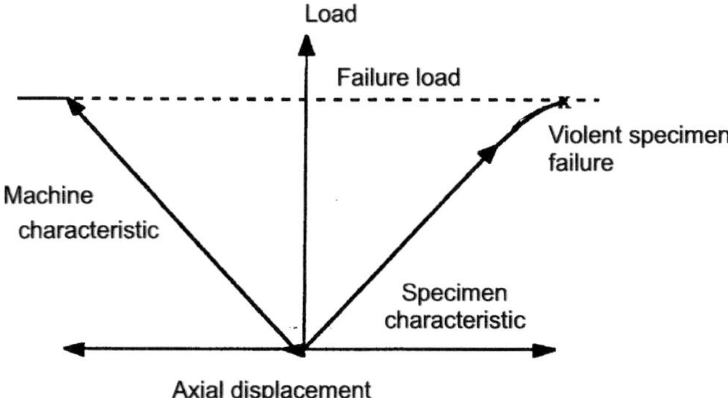

Figure 2.2 Load displacement response in a sample in conventional test.

The uniaxial compression test is normally carried out in one of the following types of testing machines.

(1) Hydraulic or screw driven testing machine characterised by a constant rate of loading
(2) Stiff testing machine capable of controlling the rate of deformation of the specimen
(3) Electro-hydraulic servo-control feed back machine controlling the rate of displacement of the specimen

During a compression test in a conventional hydraulic or screw driven testing machine, a constant rate of loading is applied to a rock specimen. Thus, under this testing condition the load (independent variable) on a rock specimen increases or decreases linearly with time in both pre-failure and post-failure regime respectively. The brittle explosive nature of rock specimens failure is attributed to the release of stored elastic strain energy from the test machine on to the specimen. As the specimen is loaded both the machine and specimen deform. The testing machine will deform in a completely elastic manner while stored strain energy will be released as an elastic recovery of the machine when the test specimen is at the verge of failure. At this point rock becomes unstable upon reaching its ultimate strength and fails uncontrollably as shown in Figure 2.3(a).

2.2.2 Loading in Stiff or Servo-controlled Testing Machines

Conversely, when the displacement of the specimens is controlled at a constant rate, the dependent variable load will rise to the specimens ultimate strength and then diminish in the post-failure regions. Under this test condition it is possible to obtain a complete stress-strain relationship of the specimen (Figure 2.3[b]). The concept of controlled failure testing dictates that in order to obtain a complete stress-strain relationship for rock, total

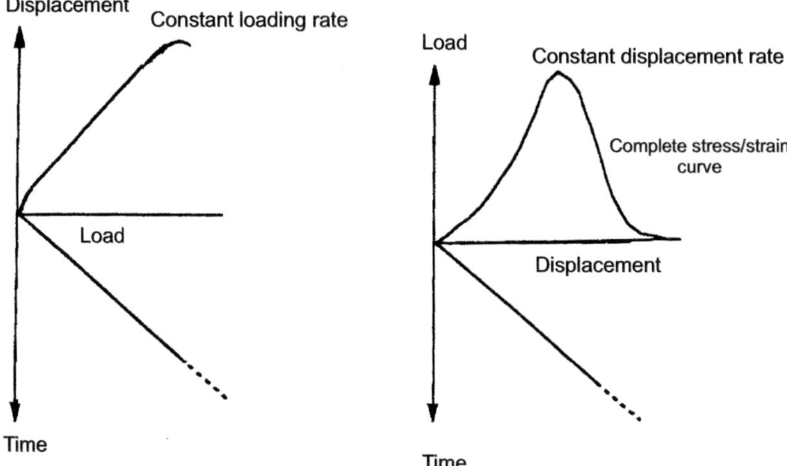

(a) Constant rate of loading (conventional) (b) Constant displacement rate (stiff testing)

Figure 2.3 Comparison between loading characteristics of conventional and stiff testing machines

control must be achieved over the release of stored strain energy from the machine and the rate of displacement applied to the specimen. This can be achieved by two basic types of testing machines.

(1) Soft machine with uncontrolled displacement
(2) Stiff machine controlling displacement

Stiff testing machines have high longitudinal stiffness where load is applied to the specimen by moving the machine cross-heads together at a constant displacement rate.

By using a testing machine which incorporates a closed-loop electro-hydraulic servo-controlled system. A constant rate of displacement is achieved by the closed-loop operation of the servo-system adjusting continuously the actual displacement to that of the required programmed displacement.

The effect of the variation in machine stiffness on the control of rock failure is shown in Figure 2.4. The testing machine is loaded to a sample's ultimate load along the line OL (soft machine) and OA in stiff machine. The specimen is loaded along the path OA, assuming the idealised force-displacement curve during the failure process OABC. From the peak load, a stiff machine elastically unloads along the line AH′. The incremental axial strain applied to the specimen, due to the release of stored energy is noted sufficient to promote rapid failure. Conversely, the soft machine unloads along line AL′ and supplies a large incremental axial strain on the rock. This will cause rapid displacement with disintegration of the test specimen.

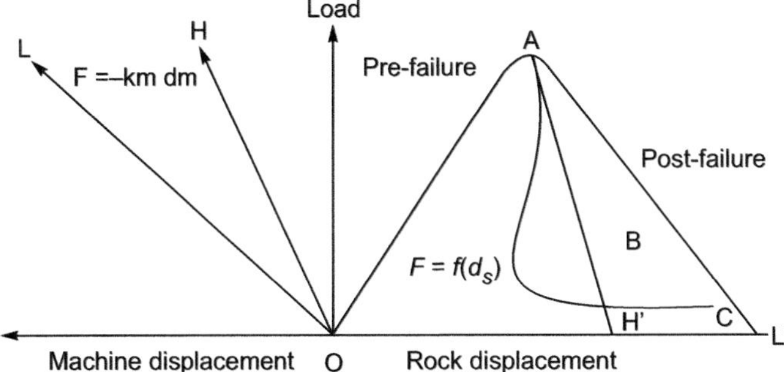

Figure 2.4 Effect of variation of machine stiffness on the failure of rock

2.3 Factors Effecting Uniaxial Compressive Strength of Rock

As mentioned, the uniaxial compressive strength is a simple test concept in theory but it is very difficult to obtain the representative material values in practice. This is due to the fact the stress distribution on the specimen by the loading machine is a complex process. It is very difficult to realise uniaxial compressive stress on the sample during testing. Factors on which the uniaxial compressive strength depends are as follows.

(1) Length of the specimen on stress distribution
(2) Effect of friction between platens and specimen end surfaces
(3) Effect of specimen geometry
  - Shape
  - Size
  - Effect of h/D ratio on strength
(4) Rate of loading the specimen
(5) Rate of strain
(6) Effect of moisture content
(7) Effect of intrinsic properties on strength
  - Effect of porosity
  - Effect of density

2.3.1 Stress Distribution on Cylindrical Specimen

2.3.1.1 Stress distribution on the end of specimen due to end effect

When the cylindrical specimen is loaded on a compression testing machine, the stress induced on the specimen is not always the same as the applied stress due elastic mismatch of the machine and the specimen. This introduces horizontal frictional forces on the end surfaces of the specimen. Many researchers have calculated the stress distribution on the end surface of the specimen and expressed in terms σ_1 of Figure 2.5(a). The general consensus is that the stress at the centre of the specimen is less than or near 1 and decreases slightly up to r/R of 0.6 and then increases exponentially. The stress at the outer boundary of the specimen is 1.6 times σ_1 as shown in Figure 2.5(b).

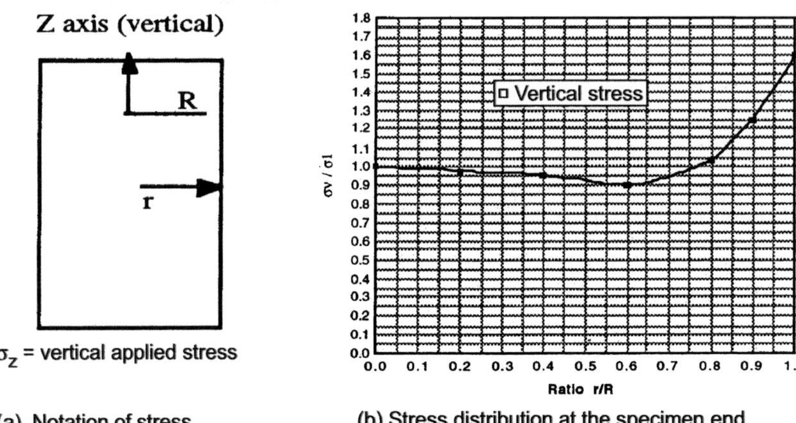

(a) Notation of stress (b) Stress distribution at the specimen end

Figure 2.5 Stress distribution on a cylindrical sample (D'Appolonia and Denmark, 1951)

End effect σ_z is vertical stress calculated by the finite element method at a distance R from the centre of the specimen.

2.3.1.2 *Stress distribution on mid-section of the specimen*

Figure 2.6 shows the stress distribution at the mid-section of the specimen from the centre of the specimen to the outer boundary of the specimen. It can be seen that the stress at the centre is slightly higher than the applied vertical stress and at the circumference slightly lower than applied vertical stress. However, in the average stress at the mid-section of the specimen is equal to applied vertical stress.

2.3.2 Friction between Machine Platens and the End Surface of Specimen

When a specimen is compressed between the platens of the testing machine, it tends to expand laterally as it shortens, while the frictional constraints at the plane of contact tend to prevent the expansion. As a result the specimen

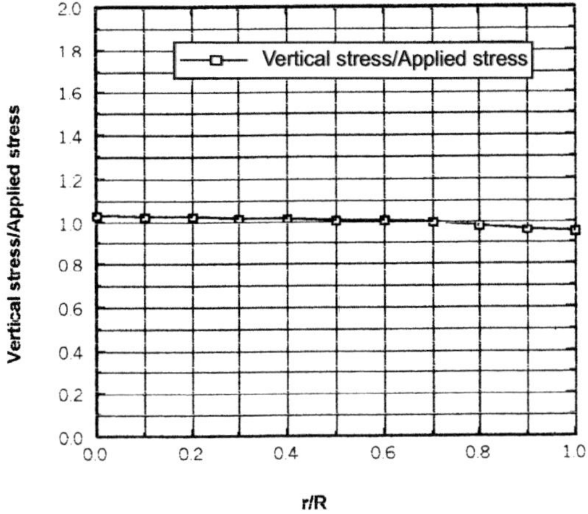

Figure 2.6 Stress distribution on the mid-section of a cylindrical specimen (after Girijavallabhan, 1970)

is not in the state of uniaxial compression at the specimen's ends. As the specimen length increases with reference to its diameter, the region in the mid-length of specimen has uniform compressive stress. Figure 2.7 shows that the specimen with the h/d ratio below 2 is clearly in triaxial state of stress while with the h/d ratio greater than 2 have regions at the mid-length of the specimen which are in the state of uniaxial compression.

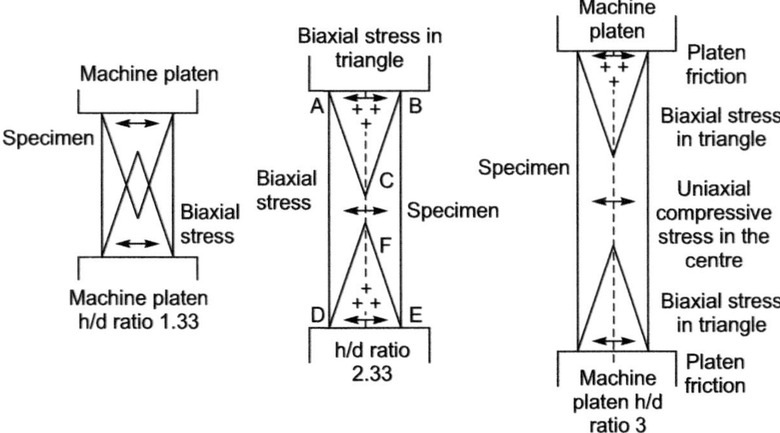

Figure 2.7 Effect of length of specimen on stress distribution on a cylindrical specimen (based on Bordia, 1971)

2.3.4 Effect of Specimen Geometry

2.3.4.1 *Effect of shape*

It is customary to determine the cubic strength of concrete in civil engineering with an h/d ratio of 1 because of ease in casting of samples. It is, however, preferable to use cylindrical specimen for rocks for ease in drilling core samples and subsequent preparation of specimen's ends. It may also be noted that the stress distribution in cylindrical specimens with an h/d ratio > 2.5 is symmetrical around the axis in case of rock testing.

2.3.4.2 *Variation of strength with height to diameter ratio (h/d ratio)*

The compressive strength of rock varies with the change in height to diameter ratio of the specimen. In specimens with h/d ratios less than 2, the stress distribution is predominantly in triaxial state and as a consequence the specimen exhibits very high compressive strength. Slender specimens with very high h/d ratios between 3.5 and 4.0 fail in tension due to elastic instability. The variation of strength of rock with change in h/d ratio has been determined by many research workers (Mogi, 1966; Bunting, 1911; Grosvenor, 1963; Hobbs, 1964). There is general agreement that the strength of rapidly decreases from h/d ratio of 1 to 2.5 (see also Figure 2.8). At the h/ d ratio 3, the compressive strength of rock stabilises. Some of relationships are:

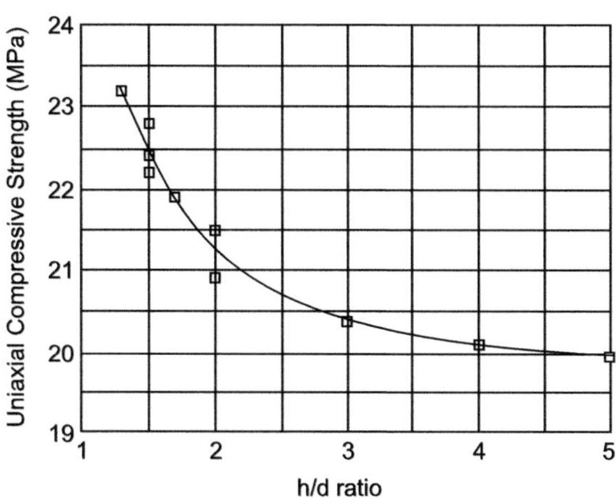

Figure 2.8 Variation of strength with h/ d ratio for Dunham Dolomite (after Mogi 1966)

(1) Compressive strength test on coal by Meikle and Holland (1965)

$$\sigma_c = C \times \left[\frac{h}{d}\right]^n \tag{2.2}$$

where

h = height of the sample (m)
d = diameter of the specimen (m)
σ_c = uniaxial compressive strength (MPa)
C = a variable for different rocks (22.5 to 35.8 MPa for coal)
n = a variable between 0.54 – 0.66 for various rocks

(2) Strength of anthracite (Bunting, 1911)

$$\sigma_c = 4.04 + 2.07\,\frac{h}{d} \tag{2.3}$$

σ_c = uniaxial compressive strength (MPa)
h = height of the sample (m)
d = diameter of the specimen (m)

(3) Calculation of uniaxial compressive strength for rock of height to diameter ratio 1 (Obert, Windes and Duvall, 1946)

The ASTM standard for compressive strength on natural building stones suggests using the following equation for calculating compressive strength for stone or rock specimens with h/d ratio of 1:

$$\sigma_{c1} = \frac{\sigma_c}{0.778 + 0.222 \times \left(\frac{d}{h}\right)} \tag{2.4}$$

σ_c = uniaxial compressive strength (MPa) (as tested in the laboratory for h/d ratio >1)
h = height of the sample (m)
d = diameter of the specimen (m)
σ_{c1} = calculated uniaxial compressive strength for h/d ratio of 1

(4) Calculation of uniaxial compressive strength for rock specimen with height to diameter ratio 2 (Protodynakanov, 1969)

It was suggested that the following formula for calculating the compressive strength corresponding to h/d ratio of 2 be used:

$$\sigma_{c2} = \frac{8\sigma_c}{7 + 2\frac{d}{h}} \tag{2.5}$$

where

σ_c = uniaxial compressive strength (MPa) (as tested in the laboratory for h/d ratio >1)
h = height of the sample (m)
d = diameter of the specimen (m)
σ_{c2} = calculated uniaxial compressive strength for h/d ratio of 2

2.3.4.3 *Effect of size on strength*

The strength of specimens usually decreases with increase in their size. Figure 2.9 shows the relationship between cube strength of coal as determined by Binwienski (1969). The curve indicates that cube strength of 1.5 m wide cube of coal can be taken as the *in-situ* strength of coal.

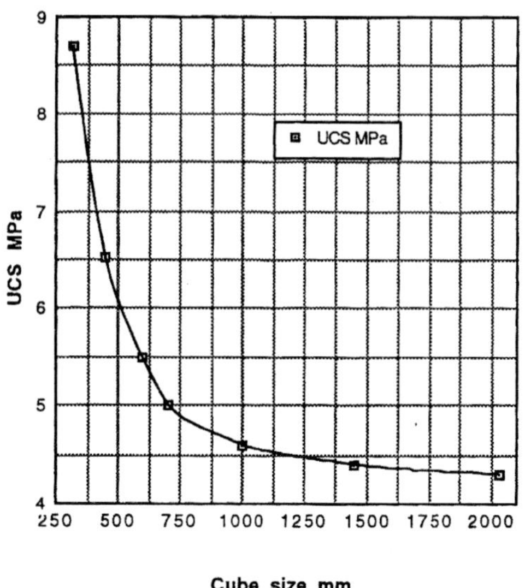

Figure 2.9 Variation of strength of coal with the size of cube

2.3.5 Variation of Strength with Friction and d/h Ratio

Meikle and Holland (1965) determined the effect of coefficient of friction between coal and steel platen varying between 0.04 and 0.3 for d/h ratio of 3 to 9. The result corresponded to the following form:

$$\sigma_c = C\left[\frac{h}{d}\right]^{-n} \tag{2.6}$$

It can be seen in Figure 2.10 that the value of C varied between 23.6 and 35.8 MPa and that of n between 0.04 and 0.3 for three levels of the coefficient of friction. Graph 2 represents the average curve for variation of strength with d/h ratio between 3 and 9.

2.3.6 Effect Rate of Strain and Rate of Loading on Compressive Strength

Figure 2.11 shows the effect of rate of strain on uniaxial compressive strength of cannel coal. It is reported that at lower rate of strain indicated by

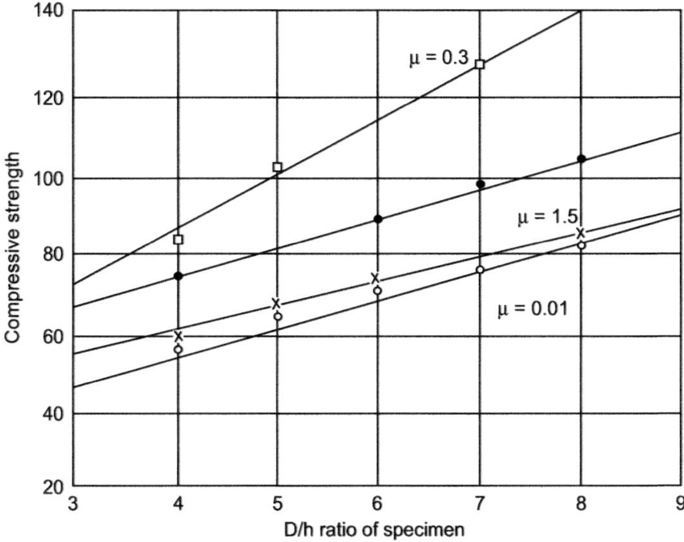

Figure 2.10 Relationship between d/h ratio, μ on compressive strength (Meikle and Holland, 1965)

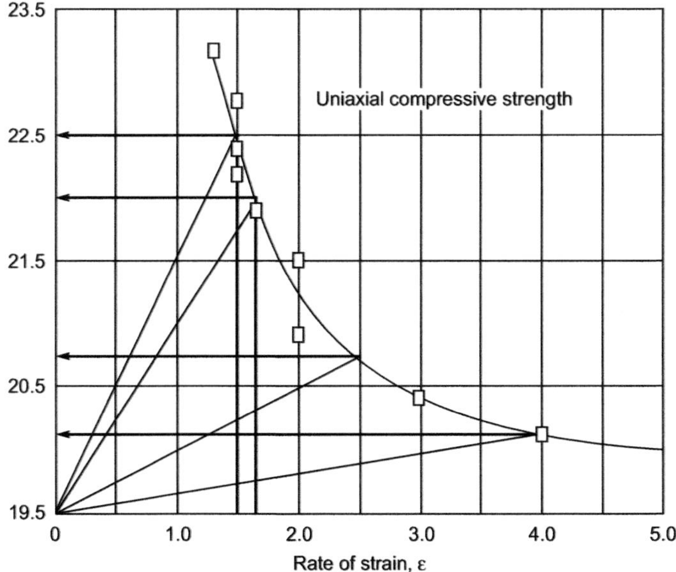

Figure 2.11 Effect of strain rate on compressive strength

line 4-4, the compressive strength was 20.1 MPa and stress/strain curve was 0-4. At a higher rate of strain as indicated by line 2.5–3, the stress/strain curve was 0–3 and uniaxial compressive strength was 20.65 MPa. Thus, the compressive strength increased with higher rate of straining and the specimen failed abruptly and violently.

Houpert (1970) also investigated the effect of rate of loading on the compressive strength of rock. Figure 2.12 indicates the increase in the strength with increase in loading rate.

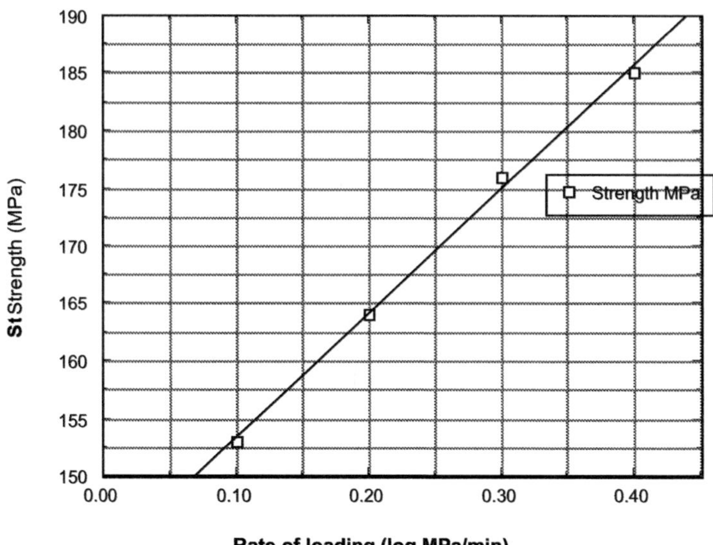

Figure 2.12 Variation of compressive strength σ_c with loading rate (Houpert, 1970)

Figure 2.13 shows the relationship between the compressive strength of rock and the strain rate of the testing machine. It has been shown that the increase in strain rate from 10^{-7} cm/s to 10 cm/s corresponds with increase in the compressive strength of marble from 7.5 to 18 MPa, granite (11.8 to 27.8 MPa) and andesite (19 to 38 MPa).

2.3.7 Environment

(a) Moisture

Colback and Wiid (1965) investigated the effect of moisture on the compressive strength of quartzitic sandstone (Figure 2.14). The results indicate that there is negative exponential reduction in strength as relative humidity and moisture content of the specimen increases.

(b) Mineralogy

The rocks containing quartz as a binding material are the strongest, followed by those with calcite and ferrous minerals. Rock grains bound by

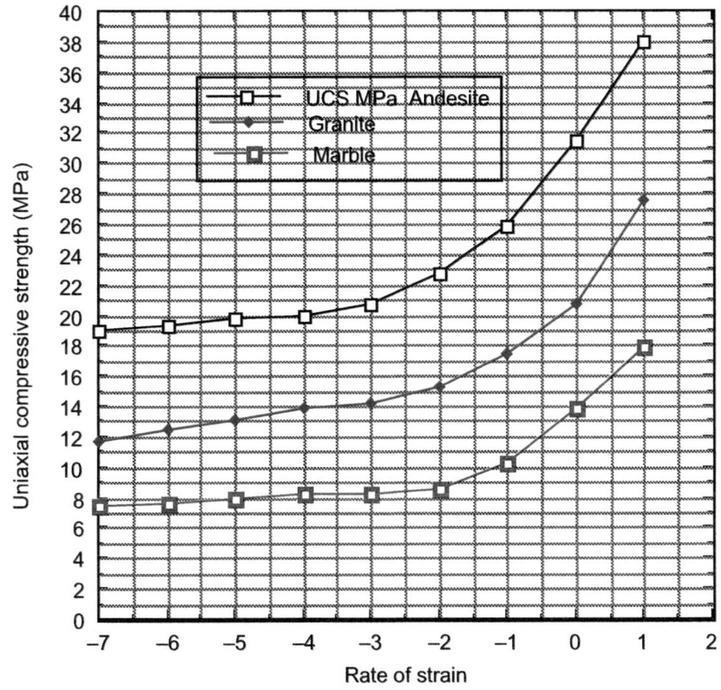

Figure 2.13 Variation of compressive strength with strain rate after Kobayashi (1970)

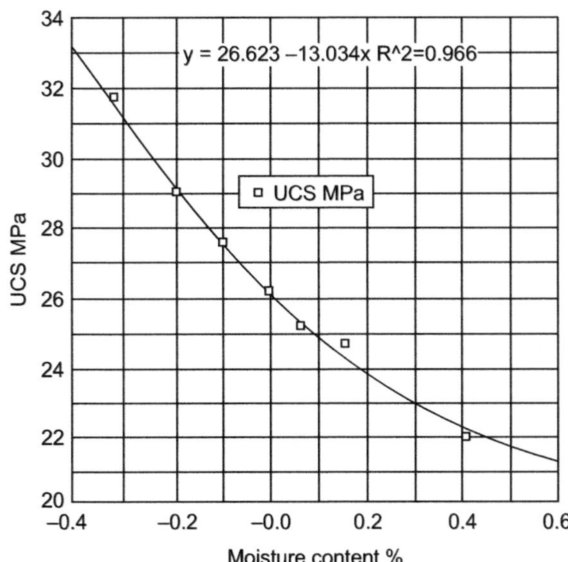

Figure 2.14 Relationship between compressive strength and moisture content (Colback and Wiid, 1965)

clay as a binding material are the weakest. In general, the higher the quartz content, the greater the strength (Price, 1960).

(c) Grain size

Brace (1961) reported that the strength of rock is greater for the fine grain size to a medium to coarse grained rock.

(d) Density

It has been observed that the denser the rock, the higher is its compressive strength (Figure 2.15).

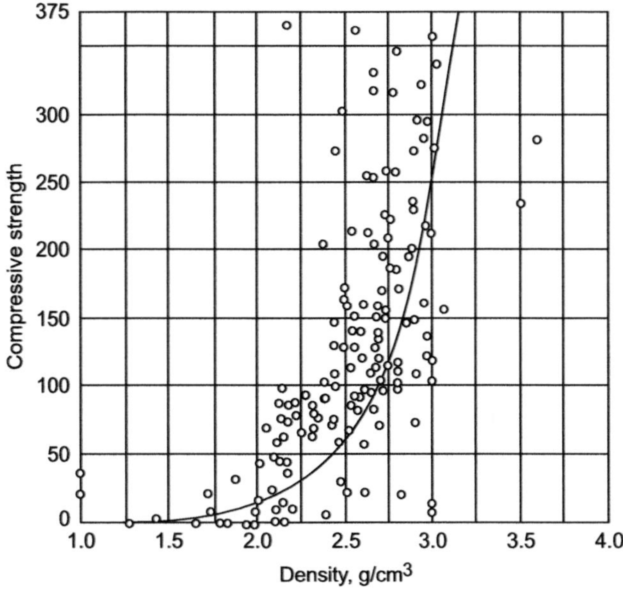

Figure 2.15 Relationship between uniaxial compressive strength and density

(e) Porosity

In general, compressive strength decreases with increase in porosity of rock. Figure 2.16 shows the relationship between uniaxial compressive strength and porosity (Price, 1960; Kowalski, 1966).

2.4 Mode of Failure on Compressive Strength Test

It has been observed in laboratory testing that test specimens commonly fail in the following modes as shown in Figure 2.17.

- General crumbling with conical ends

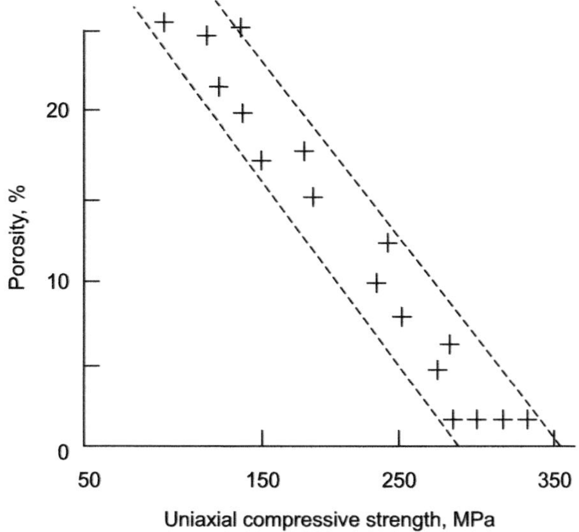

Figure 2.16 Relationship between uniaxial compressive strength and porosity (Price, 1960)

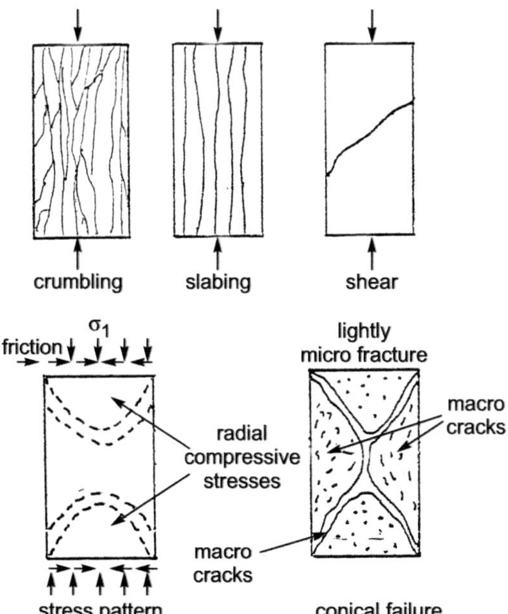

Figure 2.17 Mode of failure during the compressive strength test

- Development of one or two major cracks resulting in slabbing type failure
- Shear failure along a single oblique plane

The first mode of failure is the most common, where the conical wedge-shaped end segments are due to the end constraints of the loading platens and due to shear failure. This is common when the h/d ratio is too small. The second mode of failure is most common when the friction effect of the platen is eliminated. The third type of failure is observed extensively, most likely due to rotation of platen or lateral translation of platens relative to each other. Thus, this type of failure is the function of the loading system (Hawkes and Mellor, 1970).

2.5 Standardised Test Conditions for Conducting Uniaxial Compressive Strength of Rock

The foregoing discussions conclude that it is necessary to deploy standardised test procedure to induce uniform uniaxial compressive stress on the test specimen to carry out uniaxial compressive strength test. These standardised conditions are as follows.

(1) The specimen should be between 20 and 60 mm in diameter in size and the h/d ratio should be between 2.5 and 3.0. The test results should be reduced to h/d ratio of 1 or 2 according to Equation 2.4 and Equation 2.5 so as to facilitate comparison of results between rocks tested in different laboratories.

(2) End of the cylindrical specimens should be prepared in accordance to standards set in chapter 1 (1.4.4).

(3) The testing machine should have at least one spherical platen. It should be lubricated with light inert oil. The platens should be ground and the specimen should be placed centrally between the platens.

(4) Steel inserts of the same diameters as the specimens and overall length at least equal to diameter is recommended.

(5) Friction between the platens and end surfaces of the specimen has significant effect on induction of compressive stress in the specimen. It is, therefore, recommended not to use friction reducers during the test.

(6) Increased rate of loading increases the rock strength. A rate of loading 0.5 to 1.0 MPa/s is recommended by the ISRM commission.

(7) Effect of humidity is to decrease the compressive strength of rock. It is recommended by ISRM that the specimen be stored for five to six days in an environment of $20 \pm 2°C$ and 50% humidity.

2.6 Determination of Young's Modulus and Poisson's Ratio for Rock Specimen

The static Young's modulus of rock, E and the associated Poisson's ratio, υ are measured on the rock specimen during conventional compression tests (Figure 2.18). For accurate determination of these parameters, it is necessary to measure strain at the centre of the specimen with h/d ratio > 2.5.

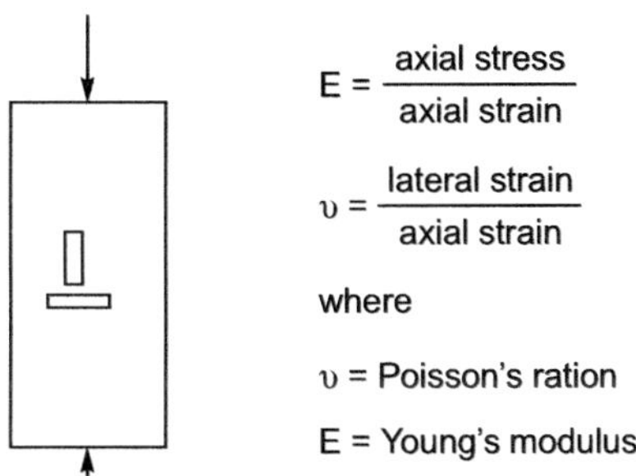

Figure 2.18 Determination of Young' s modulus and Poisson' s ratio in the uniaxial compression test

The deformation measurement device can be selected from one of the following techniques.

- High-grade micrometer dial gauge reading up to .002 mm
- Linear variable differential transducers (LVDT)
- Electric resistance strain gauges

Electrical resistance strain gauges are applied directly to the rock specimen in axial as well as lateral directions. This permits observation of data necessary for the calculation of Young's modulus as well as Poisson's ratio. Generally three axial gauges and three lateral gauges are equally spaced around the specimen at mid-height.

When tracing the load strain diagram, for deformation in the parallel direction to load, it can be seen that strain seldom varies linearly with load (Fig. 2.19). Four distinct modulus of elasticity can be calculated.

1 Tangent modulus at mid-stress
2 Secant modulus
3 Chord modulus
4 Initial tangent modulus

These moduli are defined in Figure 2.20. When reporting the Young's modulus, usually tangent modulus at 50% of the compressive strength is reported. The Secant modulus of rock at any stress level is sometimes referred to as the deformation modulus.

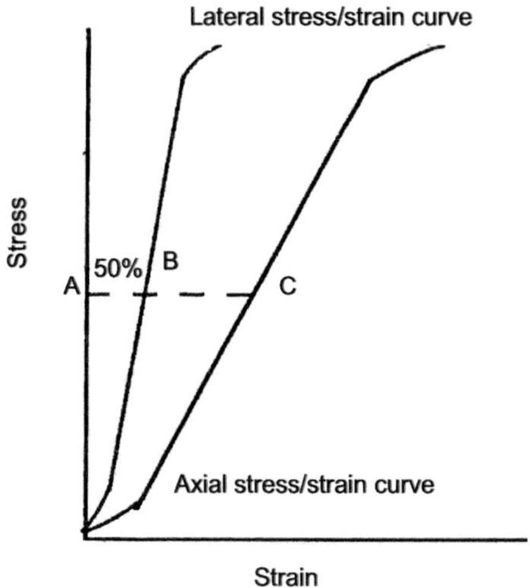

Line ABC is at 50% of ultimate failure load, $\upsilon = \frac{AB}{AC}$ and E = slope of stress/strain line

Figure 2.19 Axial and lateral stress/strain curves for a rock specimen

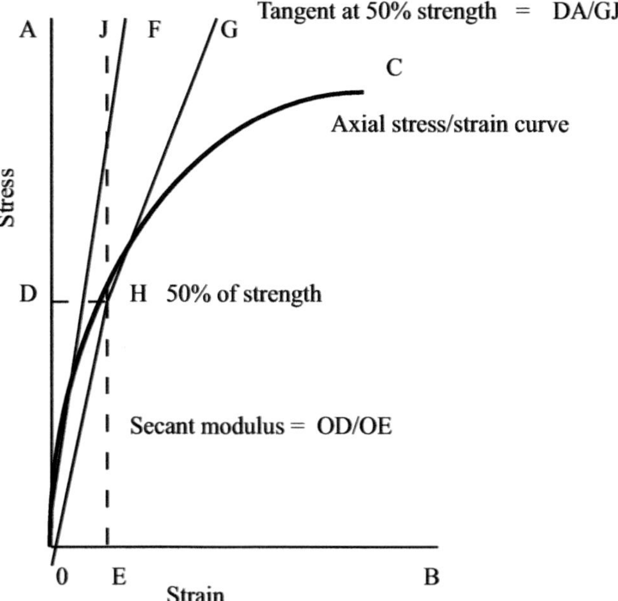

Figure 2.20 Representation of three types of modulus of elasticity

Figure 2.21 shows the stress/strain curve for a porous rock during successive loading and unloading. It may be noted that the loading and unloading curves follow a different path and for each unloading cycle there is residual strain. These residual strain values increase with each cycle of loading. It is possible to separate out permanent strain from elastic strain in each cycle of loading. Inclination of permanent deformation curve toward abscissa indicates the onset of instability.

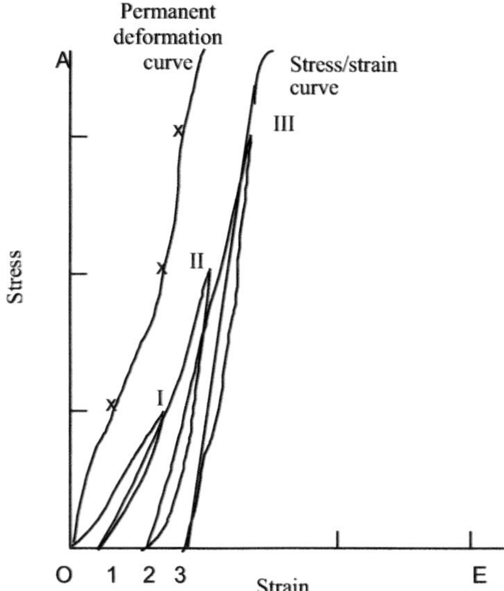

Figure 2.21 Stress strain curves on cyclic loading

2.6.1 Types of Stress-Strain Curves

The stress-strain curves for rocks vary profoundly based on the rock characteristics, method of testing, rate of loading, size and shape of the specimen and type of the loading machine. The stress/strain curves obtained for different rocks in conventional testing machine can be classified into the following categories as shown in Figure 2.22.

Category A shows a linear elastic behaviour indicating one value of modulus of elasticity up to the point of failure. This is exemplified by most igneous rocks, basalt, diabase, gabbro, fine grain sandstone, limestone, quartz and quartzite.

Category B showing strain softening behaviour where the curve is concave towards abscissa and shows pronounced strain for slight increase in load. This indicates breakdown of the rock as stress increases. The elastic modulus of rock is maximum at the early stage of loading followed by

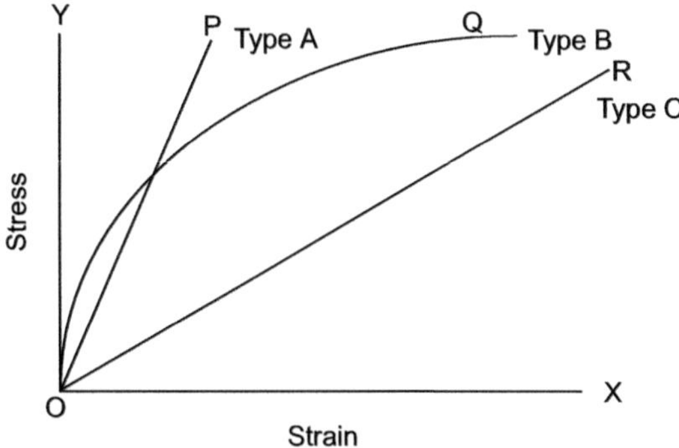

Figure 2.22 Classification of stress/strain curves

continual reduction at higher stage of loading. Examples of such rocks are shale, silt stone, tuff, weaker limestone, well cleated and bedded coal.

Category C curves are strain hardening curves that are concave towards ordinates. In such curves, the modulus of elasticity is the lowest at the initial stage of loading and continuously increases with increase in stress level. This behaviour is caused by closing of pore spaces and micro-cracks in rocks with increase in stress, thus, giving higher value of modulus of elasticity. Such behaviour is exemplified by sandstones, coal, rock salt and certain metamorphic rock. Some practical examples of stress stress/strain curves for various rocks are shown in Figure 2.23.

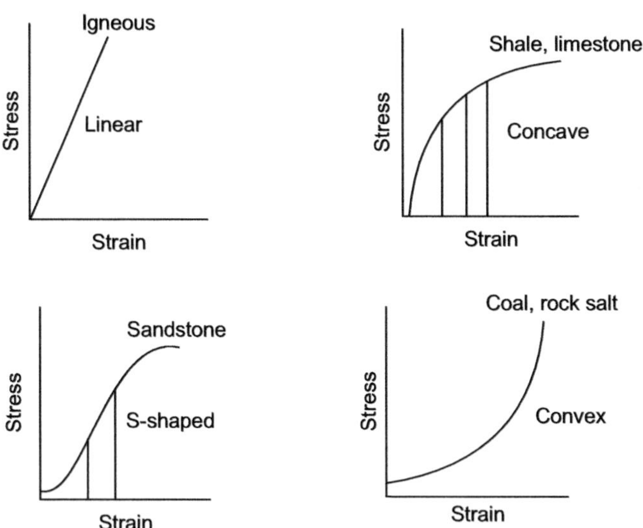

Figure 2.23 Practical example of stress/strain curves for various rocks

2.6.2 Second Order Parameters of Rock

(1) Stress-strain diagrams of rock on cyclic loading can be used to assess the structure of rock and predict how the rock will behave *in situ*. For example in cyclic loading, a fine grained, well cemented or compact rock will display very little residual strain for each unloading cycle than an unconsolidated rock. In case of compact rock, the loading and unloading curves are very close to each other while for uncompacted rock these curves are further apart and there are high residual strains. There is a tentative relationship between uniaxial compressive strength and the modulus of elasticity (Deere and Miller, 1966).

$$E = 350\ \sigma_c$$

where

E = modulus of elasticity, MPa

σ_c = uniaxial compressive strength of rock

(2) Attempts have been made to classify rock according to stiffness as presented in Table 2.1.

Table 2.1 Classification of rock according to stiffness

Classification	*Elastic modulus*
Quasi elastic (very stiff)	> 80 MPa
Quasi elastic (stiff)	40–80 GPa
Medium stiffness	20–40 GPa
Non-elastic (soft)	10–20
Non-elastic (very soft)	<10 GPa

It has been suggested by Deere and Miller (1966) that modulus ratio E/σ_c, combining the strength with deformation may be used to classify rocks. The following classification has been suggested.

High modulus rock (E/σ_c)	> 500 (highly elastic rock)
Medium modulus (E/σ_c)	200-500 (most igneous rocks, dolomite, limestone, diabase, basalt and other flow rocks)
Low modulus (E/σ_c)	<200 (some shale, mudstone, claystone and seat-earth have low modulus and distinct tendency to non-elastic deformation)

(3) Classification of rocks on the basis of their axial deformation characteristics as well as strength is of considerable importance in rock design.

The stiffness of rocks has been subdivided into very stiff, stiff, medium stiffness, low stiffness and yielding categories. Strength of rocks has been classified from very weak, weak, medium strength, strong and very strong. There is a tentative correlation between the uniaxial compressive strength and stiffness of rocks dividing the rock in low modulus, medium modulus and high modulus rocks as indicated in Figure 2.24.

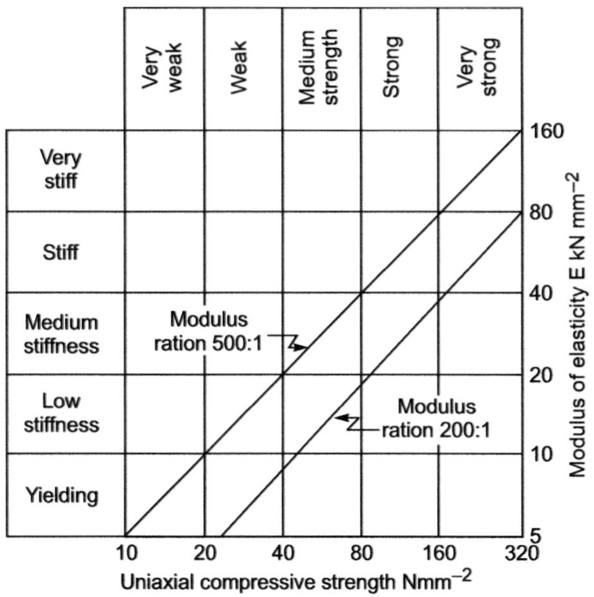

Figure 2.24 Classification of rock on the basis of their uniaxial compressive strength and tangent modulus at 50% strength (Deere and Miller, 1966)

(4) Ratio of elastic modulus in loading and unloading
It has been noted that the modulus of the loading curve and unloading curve for rock can give considerable information regarding its internal structure. For compact rocks, the loading and unloading curves are almost parallel and the E_L/E_d ratio is equal to 1. In weaker and uncompacted rocks, there is considerable increase in E_L/E_d ratio. This property of rock can be expressed as the ratio of the elastic modulus of rock as determined from loading and unloading curves. The classification based on ratio of loading (E_L) and unloading (E_d) is given in Table 2.2.

Table 2.2 Classification based on the ratio of loading and unloading Modulus

Classification	E_L/E_d
Strong rock	1
Medium strength rock	2–5
Weak rock	5–10

5) A compact and high-strength rock gives more consistent results and has a very low standard deviation of the test results. Thus, the variation of the ratio of the standard deviation to mean compressive strength is indicative of fissuration of rock *in situ*. Table 2.3 presents the classification of rock based on the standard deviation of the uniaxial test results.

Table 2.3 Fissures and the compressive strength of rock

Fissures	*Rock type*	S_d/*Mean* σ_c *ratio*
No microfissures	Compact limestone	0.005
Average microfissures	Biotite gneiss	0.22
Few microfissures and intense macrofissures	Jurassic limestone	0.25
Microfissures, microfractures and intense macrofractures	Poor gneiss	0.3
Very intense microfissures and microfractures	Very poor gneiss	0.37

(6) Bernaix (1969) contended that large diameter rock specimens contain more microfissures and microfractures and thus, display lower strength than the smaller diameter cylindrical specimens. The ratio of uniaxial strength of 10 cm diameter specimens (R_{10}) was compared with those of 60 cm diameter (R_{60}). The ratio of R_{10} and R_{60} was used as a second order parameter to classify the fissuration (Table 2.4).

Table 2.4 Fissures and the compressive strength of rock

Fissures	*Rock type*	R_{10} *and* R_{60} *ratio*
No microfissures	Compact limestone	1.0
Average microfissures	Biotite gneiss	1.25
Few microfissures and intense macrofissures	Jurassic limestone	1.40
Microfissures, microfractures and intense macrofractures	Poor gneiss	1.9
Very intense microfissures and microfractures	Very poor gneiss	2.9

2.7 Post-failure Behaviour of Rock in Compression

If a rock specimen is loaded in a stiff testing machine or in a servo-controlled machine incorporating the strain control, a complete stress and stress curve as shown in Figure 2.25 is obtained.

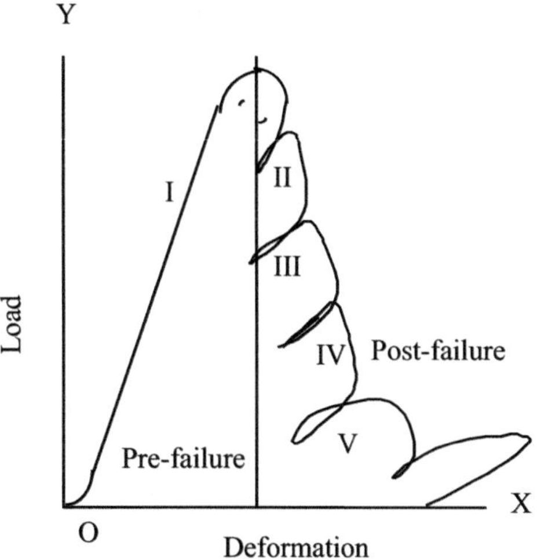

Figure 2.25 Complete load deformation curve for a rock in cyclic loading at post-failure stage

Wawersik (1968) carried out post-failure tests on a number of rocks and found that although the pre-failure behaviour was similar in many rocks, post-failure behaviour could be classified into two main categories as follows.

1. Class I stable fracture propagation where rocks retain their partial strength even if their maximum load carrying capacity has exceeded and more work is required for further lowering the load bearing capacity of the specimen. Highly porous clastic sedimentary rocks like sandstone, limestone, granite and Tennessee marbles are examples of this class.
2. Class II unstable fracture propagation occurs so that the amount of energy stored in the specimen at the time of reaching its peak strength must be used to further deform the specimen. In such case, fracture propagation can be arrested only when the strain energy in the specimen is fully consumed. Examples are basalt, granite and Solenhofen limestone. The dividing line between Class I and Class II is AD as indicated in Figure 2.26.

2.7.1 Effect of Size and Shape on Post-failure Stiffness

The complete stress and strain curve of rock depends upon the specimen size while keeping the volume of the specimen constant. The following sub-section discusses the effect of size and shape on the test results.

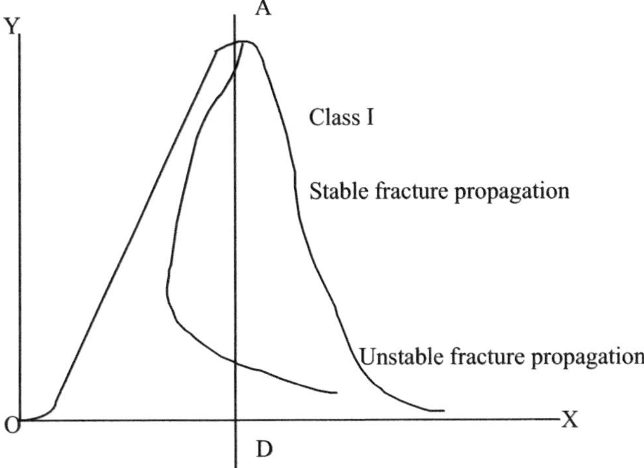

Figure 2.26 Classification of complete stress/strain curves

2.7.1.1 *Size effect*

Figure 2.27 shows the effect of size on the uniaxial complete stress-strain curve for rocks. It may be noted that the specimen size has no effect on the pre-failure modulus of elasticity of rock while both the uniaxial compressive strength and the post-failure stiffness vary with specimen size. The post-

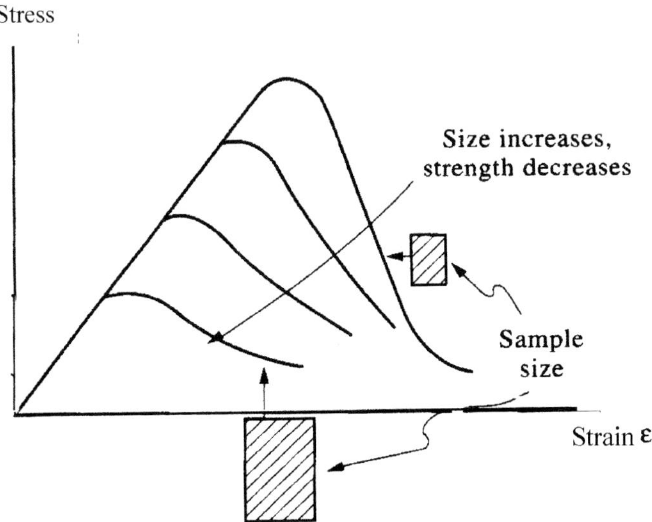

Figure 2.27 Influence of size of specimen on complete stress-strain curve (Harrison and Hudson, 2000)

failure curve for smaller specimens with high h/d ratio are steeper and, therefore, have high stiffness and the high uniaxial compressive strength. This behaviour can be attributed to the fact that the micro-crack population in a rock specimen is proportional to the size of the specimen. Therefore, larger specimens have more likelihood of containing a serious flaw culminating in failure of the specimen at a lower stress.

2.7.1.2 *Shape effect*

Figure 2.28 shows the effect of shape on the complete stress-strain curve of rock. It can be seen that as the slenderness of the specimen increases the uniaxial compressive strength decreases and the stiffness increases. For a 100-mm diameter specimen, there is a dramatic reduction in the stiffness of the specimen as the h/d ratio changes from 3 : 1 to $\frac{1}{3}$: 1. On the other hand, a squat specimen will have a high stiffness and a high uniaxial compressive strength.

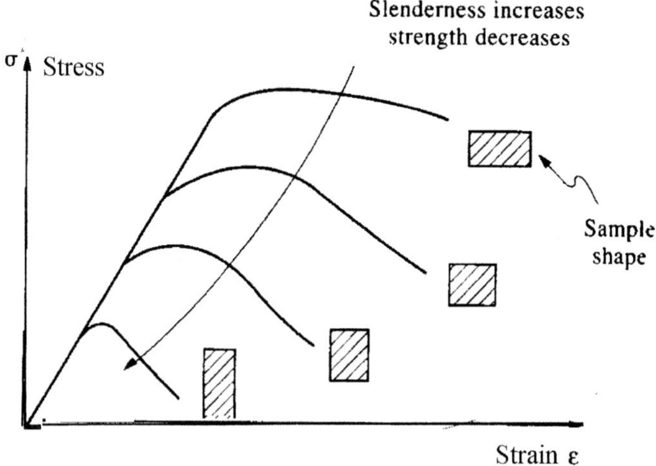

Figure 2.28 Effect of shape on the complete stress/strain curve (Harrison and Hudson, 2000)

Thus from the study of the post-failure behaviour of rocks, the following conclusions can be made (Wawersik and Brace, 1971; Hudson *et al.*, 1971a and b):

- Post-failure curves with high h/d ratio are steeper and have high stiffness.
- Post-failure curves with low h/d ratios have low stiffness.
- Post-failure curves for the same h/d ratio–larger specimens have steeper stiffness.

- For 100-mm diameter specimen, as h/d ratio changes from 3 : 1 to 1 : 1, the stiffness lowers dramatically.
- For 50-mm diameter post-failure stiffness lowers slightly and changes dramatically as h/d ratio reduces from 3 : 1 to 1.
- For 19-mm diameter specimens, stiffness lowers slightly and decreases as h/d ratio changes from 3 : 1 to1 : 1.

Figure 2.29 is a summary of the effect of size of specimen as well as increase in h/d ratio from 1 : 1 to 1 : 3 on the post-failure stiffness of rock.

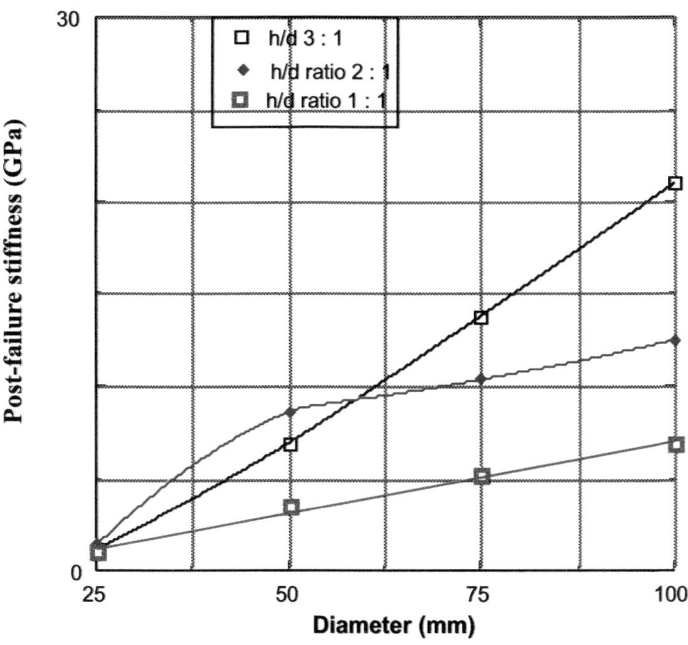

Figure 2.29 Variation of the post-failure stiffness as a function of h/d ratio and specimen diameter

2.8 Conclusions

The following are the main conclusions of this chapter.

(1) The uniaxial compressive strength of rock is a simple concept that can be used for classification of rocks according to strength. However, it is very difficult to induce uniform uniaxial stress on a cylindrical specimen due to elastic mismatch between the properties of the platens of the testing machine and that of the specimen.

(2) A number of factors affect the uniaxial compressive strength of rock. The type of testing machine is one of the most important factors. A conventional testing machine can only control rate of loading and induces elastic energy into the specimen. This allows a brittle specimen to fail abruptly on reaching peak strength. A stiff testing machine or a servo-controlled machine can control the rate of deformation of the specimen. It uses the internal strain energy of the specimen on reaching the peak strength to continue the deformation of the specimen. Thus, it allows a complete stress-strain curve to be obtained. These complete stress-strain curves give the post-failure stiffness of rock so necessary for pillar design in underground mines.

(3) The height-to-diameter ratio of the specimen is one of the most important factors in controlling the uniaxial compressive strength of rock. Civil engineers usually report the test results for h/d ratio of 1, while the mining engineers for the h/d ratio of 2. Usually the compressive strength is determined on specimens with h/d ratio of 2 to 3 and reduced to h/d ratio of 1 or 2 according to the industry for which the test is conducted.

(4) The effect of the size of the cube of coal on the compressive strength is the most important factor in determining the strength of the coal pillar. The larger the cube, more structural defects and weaknesses it will contain. Thus, the larger cube will have lower compressive strength until it reaches an asymptotic value. It has been found by Bienwiaski (1968) that a cube of a coal with sides 1.5 m gives the *in situ* strength of coal.

(5) Uniaxial stress-strain curves in compression are used to calculate Young's modulus and Poisson's ratio of rocks. A steep stress-strain curve indicates high modulus while the shallow curve indicates lower elasticity of the rock.

(6) The stress-strain curve for loading and unloading for a perfectly elastic material is a straight line. The stress-strain curve for non-elastic materials like rock may show hysteresis during unloading. The modulus of the loading curve may be much steeper than that of the unloading curve.

(7) Second order parameters of rock indicate how the rock will behave *in situ*.

(8) If a rock is loaded on a stiff testing machine or on a servo-controlled machine, complete stress-strain curve is obtained. There are two types of post-failure curves, one showing stable fracture propagation, the other displaying unstable fracture propagation.

(9) The post-failure behaviour of rocks also vary with h/d ratio and the width of the specimen.

References

Bernaix, J. (1969), 'New laboratory, methods of studying the mechanical properties of rock, *International Journal of Rock Mechanics and Mining Science,* 6. pp. 43-90.

Bordia, S.K. (1971). 'The Effect of Size and Stress Concentration on Dilatancy and Fracture of Rocks', *International Journal of Rock Mechanics and Mining Science and Geomechanics Abstracts,* Vol. 8, No. 6, pp. 629–640.

Burstein L.S. (1969), 'Effect of Moisture on the Strength and Deformity of Sandstone', *Sov. Mining Science,* No. 5, pp. 573–576.

Bieniawski, Z.T. (1968), 'The effect of specimen size on strength of coal', *Int. Jour. of Rock Mechanics and Mining Science,* Vol. 5, No. 4, pp. 325–335.

Bieniawski, Z.T. (1969), 'Deformational Behavior of the Fractured Rock Under Multiaxial Compression', *Proc. Int. Conference Structure, Solid Mechanics and Engineering Materials,* Southampton, pp. 589–598.

Brace, W.F. (1961), 'Dependence of fracture strength of rock on grain size', *Proceedings 4th Symp. of Rock Mechanics,* University Park, Penn. USA, pp. 99–103.

Bunting, D. (1911), 'Chamber Pillars in Deep Anthracite Mines', *Trans. AIME,* Vol. 42, pp. 236–245.

Colback P.S.B., and Wiid, B.L. (1965), 'The Influence of Moisture Content on the Compressive Strength of Rocks', *Third Canadian Rock Mechanics Symposium,* Toronto, pp. 65–83.

D'Appolonia, E. and N.M. Denmark (1951), 'A Method for the Solution of the Restrained Cylinders under Compression', *Ist US National Congress on Applied Mechanics,* Chicago, Illinois, pp. 217–226.

Deere, D.U., and Miller, R.P. (1966), Engineering classification and index properties of intact rock, Air Force Weapons Laboratory, Technical Report No. AFNL-TR-65-116, New Mexico.

Girirajabhalabhan, C.V. (1970), 'Stresses in Restrained Cylinder under Uniaxial Compression', *Jour. Soil Mech. and Found. Div., ASCE,* Vol. 96, No. SM2, pp. 783–787.

Harrison, J.P. and J.A. Hudson (2000), *Engineering Rock Mechanics,* Pergamon Press, 2 Vol. 5, pp. 499.

Hawkes, I., and Mellor, M. (1970), '*Uniaxial Testing in Rock Mechanics Laboratories*', Engg. Geol., Vol. 4, pp. 177–285.

Hobbs, D.W. (1964). 'A simple method for assessing the uniaxial compressive strength of rock', Int. J. *Rock Mechanics and Mining Science,* 1, pp. 5–15.

Houpert, R. (1970), 'The uniaxial compressive strength of rocks', *Proc. Second Congress of International Society of Rock Mechanics,* Belgrade, Vol. 2, pp. 49–55.

Hudson, J.A., E.T. Brown and C. Fairhurst (1971a), 'Shape of the Complete Stress-Strain Curves for Rock', *13th Symposium in Rock Mechanics,* Urbana, Illinois, pp. 773–795.

Hudson, J.A., S.L. Crouch and C. Fairhurst (1971b), 'Soft, Stiff and Servo-controlled Testing Machines–A Review with Reference to Rock Failure', *Eng. Geo.,* Vol. 6, pp. 155–189.

Kowalski, W.C. (1966), 'The interdependence between the strength and void ratio of limestones and marls in connection with their water saturating and anisotropy', *Proc. Ist International Congress on Rock Mechanics,* Lisbon, Vol. 1, pp. 143–144.

Kobayashi, R. (1970), 'On mechanical behaviour of rocks under various loading rates', *Rock Mechanics in Japan,* Vol. 1, pp. 56–58.

Lundborg, N. (1967), 'Strength of Rock Like Materials', *Int. Jour. Rock Mechanics and Mining Science,* Vol. 8, No. 5, pp. 269–272.

Meikle, P.G. and C.T. Holland (1965), 'The effect of friction on the strength of model coal pillar'. *Tran. Soc. of Mining Engineers,* AIME, Vol. 232, No. 4, pp. 322–327.

Mogi, K. (1966). 'Some precise measurements of fracture strength of rocks under uniform compressive stress, *Felsmechanik und Ingineargeology*, 4, pp. 41–55.

Obert, L., S.L. Windes and Duvall, W.I. (1946), 'Standard tests for determining the physical properties of mine rock', USBM RI 3891, 61 pp.

Protodynakanov, M.M. (1969), 'Methods of Determining the Strength of Rock under Uniaxial Compression', In: Protodynakanov, M.M., M.I. Koifman *et al.*, (eds.) (Translated from Russian), Israel Program of Scientific Translation, pp. 1–8.

Paul, B. and M. Gangal (1966), 'Initial and Subsequent Fracture Curves for Biaxial Compression of Brittle Material', *8th Symposium in Rock Mechanics*, Minneapolis, Minn., pp. 113–141.

Peng, S.D. (1971), 'Stress between Elastic Circular Cylinders Loaded Uniaxially and Triaxially', *Int. Jour. Rock Mechanics and Mining Science*, Vol. 8, No. 5, pp. 399–432.

Price, N.J. (1960), 'The compressive strength of coal measures rock', *Colliery Engineering*, Vol. 37, pp. 283–292.

Wawersik, W.R., (1968). Experimental study of the fundamental mechanisms of rock failure in static uniaxial and triaxial compression and uniaxial tension. Thesis University of Minnesota.

Wawersik, W.R. and W.F. Brace (1971), 'Post-failure Behavior of Granite and Diabase', *Rock Mechanics*, Vol. 3, pp. 61–85.

Assignments

Question 1 In an uniaxial compressive strength test for coal, the dimensions of the specimen are as follows:

Diameter at mid-span at four positions – 37.52, 37.53, 37.54, 37.55 mm

Length measured at three places – 118 mm, 117.9 mm, 118 mm

The load at failure is 86.5 kN. Calculate the uniaxial compressive strength of rock and estimate the strength for (1) h/d ratio of 1, (2) h/d ratio of 2, (3) Strength of coal pillar, where W_p = 14.5 m and height of the pillar h = 3.6 m and the strength of pillar is given by:

$$\sigma_p = 7.18\, W_p^{0.46} \times h^{-0.66}$$

where,

W_p = width of the pillar (m)

h = height of the pillar (m)

σ_p = strength of pillar MPa

Question 2 Discuss briefly the effects of the machine characteristics on the uniaxial stress and strain curves of rock.

Give an example of stress and strain curves you would expect from a hard and soft rock tested on a servo-controlled testing machine.

Question 3 Discuss briefly the various factors which affect the results of the uniaxial compressive strength of rocks in the laboratory.

Tensile, Triaxial and Shear Strength of Rock

3
Chapter

3.1 Introduction

This chapter describes the indirect tensile test, direct shear test and triaxial shear test of rocks. As most rock structures and underground openings fail in a tensile stress field or in shear mode, these tests are important parameters in rock mechanics design. While the indirect tensile strength test is an important parameter which can be used for rock strength classification, the shear strength test yields rock parameters of cohesive strength 'c' and internal angle of friction 'ϕ' which may be used for the design of structures in rocks. Therefore these tests are normally carried out in any routine rock mechanics programme.

3.2 Brazilian Disc (Indirect Tensile) Test (Figure 3.1)

In this test, a solid disc of rock having a diameter 'D' and thickness 't' is placed between the platens of the testing machine and a compressive load is applied across the diameter. Rupture usually occurs along the line of

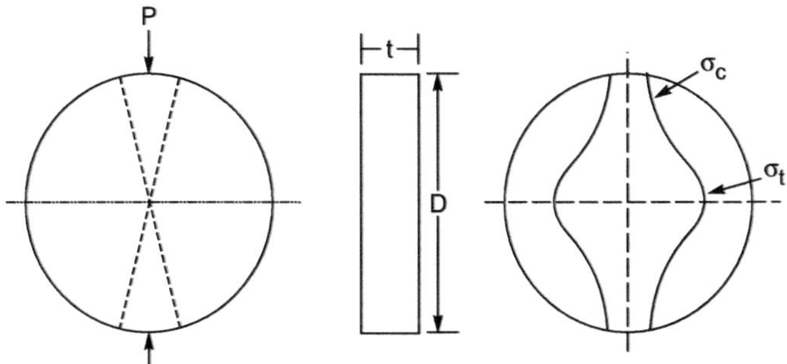

Figure 3.1 Brazilian disc test

loading with a complex stress distribution on the specimen. In the centre of the disc the tensile stresses are given by

$$\sigma_t = \frac{2P}{\pi D t} \tag{3.1}$$

where,

P = applied load at failure, MN
D = diameter of the disc (m)
t = thickness of the disc (m)
σ_t = tensile strength of rock

In spite of the apparent simplicity of this test, failure should be observed along the centre of the disc in order for the specimen to fail in tension. If, however, the specimen failure takes place near the platens, the test should be considered invalid.

In order to ensure that the failure originates at the centre, it is suggested that the following points be implemented.

(1) The load distribution on the specimen should be obtained through a cardboard or wood platen cushion 0.64 mm in thickness (Colback, and Wiid 1966). Thickness of cushion should be $0.01 \times D$ (Hawkes and Mellor, 1971).
(2) Minimum diameter of the specimen should be NX core (54 mm diameter) with h/d ratio of 0.5 : 1 (Hawkes and Mellor, 1971).
(3) The specimen should be loaded through a curved jig with spherical seating giving an angle of contact of $\text{Tan}^{-1}\frac{1}{12}$.
(4) A rate of deformation of 64 mm/minute is recommended.

Table 3.1 gives the range of tensile strength of some selected rocks.

3.3 Direct Shear Test

The direct shear strength is defined as the shear stress required to fail the rock when the stress normal to the plane of failure is zero. The shear strength of the rock is given by:

$$\tau_1 = \frac{\tau}{A} \tag{3.2}$$

where,

τ_1 = shear strength
τ = shear force at failure along the shear plane
A = area of cross-section of the shear plane

The following methods can be used for direct shear tests.

Table 3.1 Example of tensile strength of some selected rocks (Vutukuri et al, 1974)

Rock	σ_C	σ_T	ρ
Marlstone	2.78	0.33	1.86
Monazite	75.7	10.2	2.67
Mudstone	76.8	35.6	2.72
Mudstone	23.85	11.79	2.55
Limestone	142.7	18.6	2.65
Marble	66.9	1.79	2.71
Phyllite	313.71	22.75	3.30
Quartzite	293.00	35.2	4.04
Rock salt	21.4	0.83	2.16
Sandstone	71.7	3.45	2.06
Shale	62.9	10.70	1.82
Siltstone	118.5	34.4	2.66
Slate	180.64	25.5	2.93
Syenite	215.12	15.86	2.75

Single Shear Test (Figure 3.2)

A cylindrical sample is clamped on the specimen holder and a shear force is applied perpendicular to the curved surface through a sharp edged platen. The shear force at failure divided by the area of cross-section of the failure surface will give shear strength.

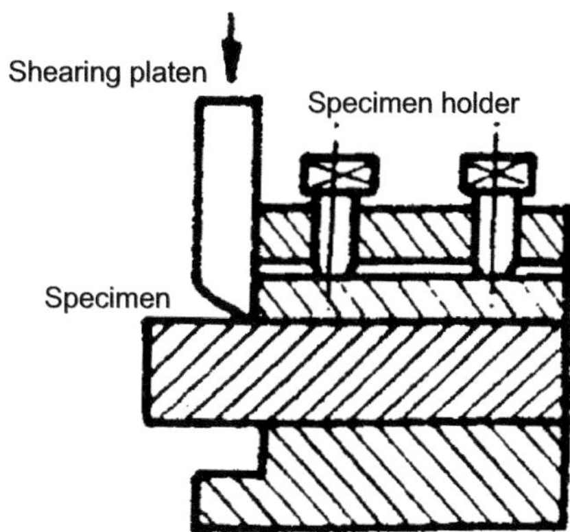

Figure 3.2 Arrangement for single shear test (after Vutukuri, et al, 1974)

Double Shear Test (Figure 3.3)

A cylindrical specimen is clamped in a double shear box and a shear load is applied through a double-edged platen with a cylindrical groove at the bottom of the platen fitting the specimen.

The specimen is sheared along two parallel planes. The shear strength is calculated by dividing the load at failure by the sum of the two areas of cross-section at failure.

$$\tau_1 = \frac{\tau}{2A} \tag{3.3}$$

where

τ_1 = shear strength
τ = shear force at failure along the shear plane
A = area of cross-section of shear plane

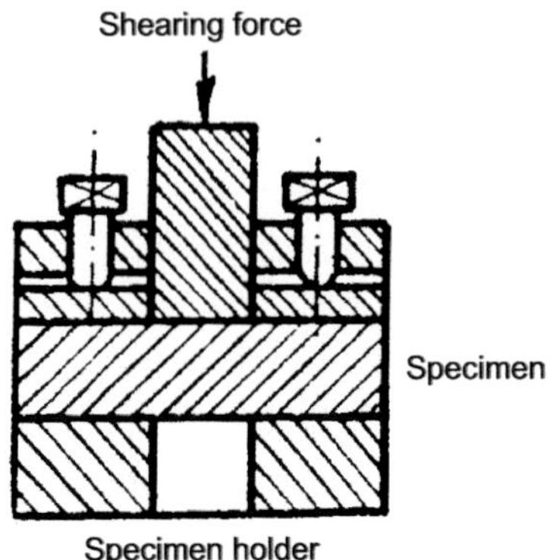

Figure 3.3 Test arrangement for double shear test (Vutukuri et al, 1974)

Punch Test (Figure 3.4)

This test is carried out on a disc-shaped specimen held at the bottom of a shear box fitted with a hollow slot of the same diameter as the punch. A disc-shaped specimen is loaded by a circular punch. The load at failure divided by curved area of failure surface will give the shear strength of the rock.

$$\tau_1 = \frac{\tau}{\pi DT} \tag{3.4}$$

where,

τ_1 = shear strength
τ = shear force at failure along the shear plane
T = thickness of disc specimen
D = diameter of the punch

Limitations of Direct Shear Test

(1) In these tests, stress concentration is caused by curved edge and induced tensile stresses due to bending of specimen cannot be

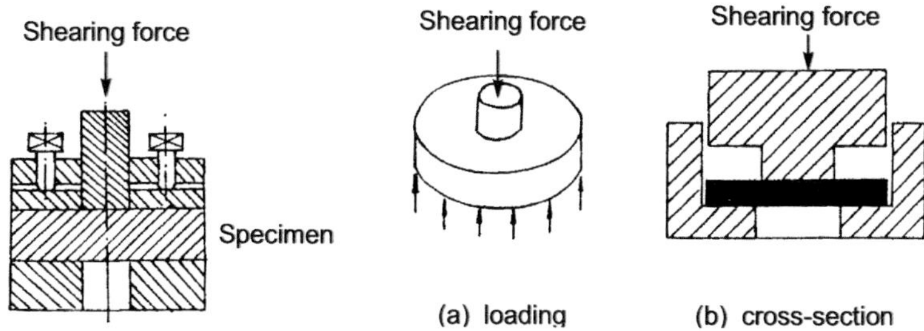

Figure 3.4 Test arrangements for the punch test (Vutukuri et al, 1974)

quantified. Thus, the results are unable to represent the true shear strength of the rock.

(2) The test results are surface sensitive, affected by the interaction of the friction between the platen and the rock.

(3) These tests are recommended when only relative values of shear strength of different rocks are required.

3.4 Triaxial Testing of Rocks

In the triaxial test, the major principle stress is applied along the axis of the cylindrical rock specimen by the testing machine and a confining stress is applied to the curved surface of the specimen through an impervious metal or rubber jacket by the confining fluid as shown in Figure 3.5. In general, the

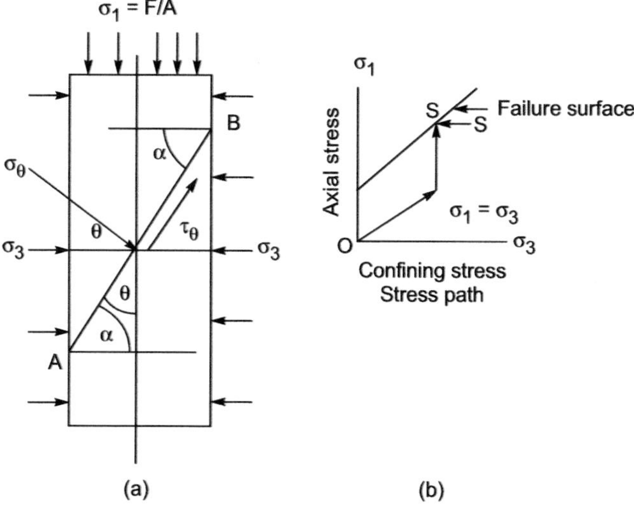

Figure 3.5 Principal stresses in a triaxial test specimen

load is applied to the specimen in such a way that the confining pressure increases proportionately with the axial stress σ_1 until the confining stress σ_3 reaches p. The confining stress is then kept constant while axial stress σ_1 is increased until the specimen fails.

The stress at failure as defined by point S on the failure envelope is determined. Figure 3.5(b) shows the stress path during the test. At the time of failure, the load at failure divided by the area of cross-section 'A' gives the principal stress σ_1 at failure. The plane AB in Figure 3.5(a) shows the plane of failure of the specimen. It can be seen that σ_θ is normal stress at right angles to the plane of failure and τ_θ is the shear stress parallel to the plane of failure AB. These results can be conveniently interpreted in Mohr's circle of stress in Figure 3.6.

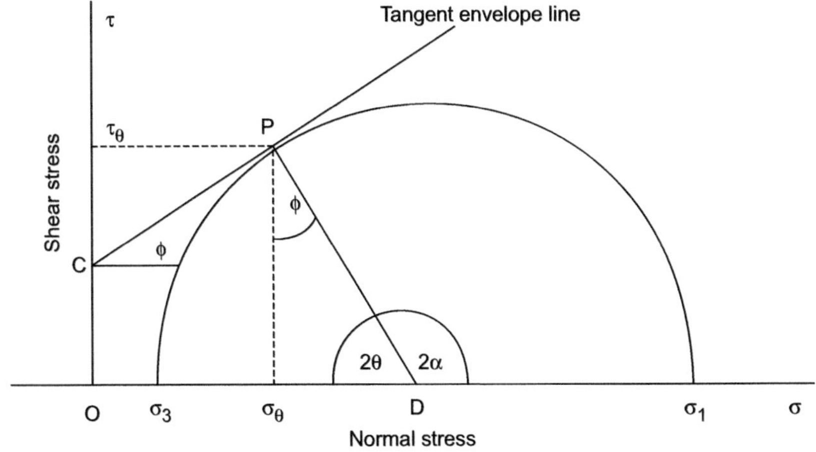

Figure 3.6 Mohr's circle for triaxial test results

3.4.1 Interpretation of Results from a Single Test

During the triaxial test on a single specimen, the parameters noted are the principal stress at failure, confining stress and angle of fracture to an accuracy of 1°. The Mohr's circle is started on the abscissa of the plot of shear stress vs normal stress between points σ_1 and σ_3 with the diameter, D as a centre and taking $\left(\dfrac{\sigma_1 - \sigma_3}{2}\right)$ as radius. In Figure 3.6, the following notations have been used:

$$\phi = \text{internal angle of friction}$$
$$\sigma_\theta = \text{normal stress acting on the failure plane}$$
$$c = \text{cohesive strength of rock (OC)}$$
$$\tan 2\theta = \tan \phi$$
$$\alpha = \text{angle of fracture}$$

3.4.2 Interpretation of Results between Axial Stress σ_1 and Confining Stress σ_3

A more accurate and representative triaxial test is carried out on the batch of specimens and the results are interpreted by drawing a relationship between the various axial stresses and the corresponding confining stresses. The intercept of the axial stress and confining stress curve with the ordinate will be the uniaxial compressive strength σ_c of the rock and the tangent of the slope 'β' is the material property of the rock in triaxial stress (Figure 3.7).

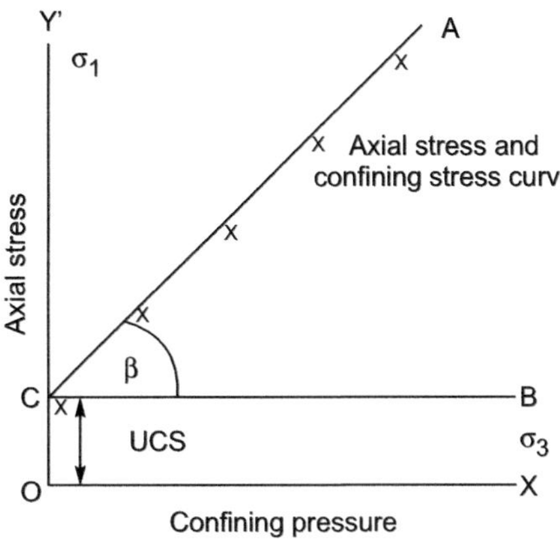

Figure 3.7 Plot of axial stress to confining stress in triaxial test

where,

$$Y = \sigma_c + \tan\beta \times \sigma_3$$
$$\beta = \text{slope of } \sigma_1 \text{ and } \sigma_3 \text{ curve}$$
$$\sigma_c = \text{uniaxial compressive strength of rock}$$
$$\tan\beta = K = \text{triaxial stress factor}$$

It can be shown that ϕ and C in Figure 3.6 can be related to β and σ_c in Figure 3.7 as follows:

$$\tan\phi = \frac{(\tan\beta - 1)}{2\tan^{1/2}\beta} \tag{3.5}$$

$$\tan\beta = \frac{(\sin\phi + 1)}{(\sin\phi - 1)} \tag{3.6}$$

and for a linear Mohr's envelope, the relationship between σ_c and 'C' can be represented by

$$\sigma_c = \frac{(2C \times \cos\phi)}{1 - \sin\phi} \tag{3.7}$$

where,

C = cohesive strength of rock
ϕ = internal angle of friction
σ_c = uniaxial compressive strength of rock

3.4.3 Determination of Cohesive Strength from a Series of Triaxial Tests

For determination of the shear strength of rock from triaxial tests, a series of tests are conducted and the strength values σ_1 are plotted for the different values of the confining pressures σ_3. The values of σ_1 will always vary for the different values of σ_3. The data are plotted in the 'σ' and 'τ' planes by plotting a group of Mohr's circles and then a common tangent is drawn to represent a Mohr's envelope (Figure 3.8). The intersection of Mohr's envelope with the ordinate represents the cohesive strength and the slope of the tangent gives the internal angle of friction for the rock.

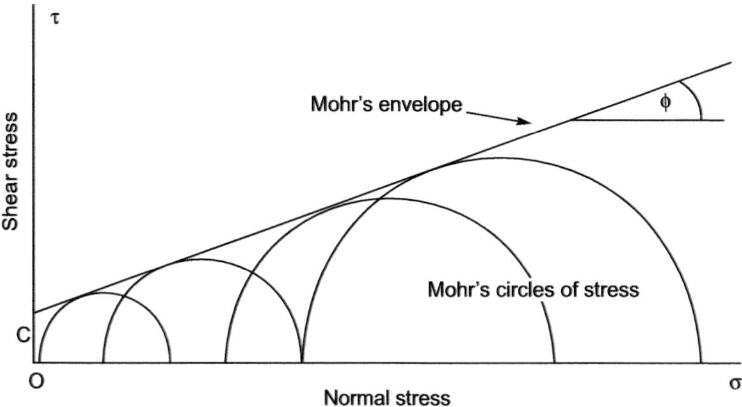

Figure 3.8 Mohr's envelope determined from the failure stresses in a group of triaxial tests

3.4.4 Extrapolating Shear Strength from Tensile Strength and Uniaxial Compressive Strengths

Wuerker (1959) extrapolated the shear strength of rocks by drawing Mohr's circles of tensile strength and uniaxial compressive strength as indicated in Figure 3.9. An interception of the common tangent on the ordinate represents cohesive strength C and the slope of the tangent of the internal angle of friction 'ϕ' of the rock.

Horibe (1970) utilised Griffith's hypothesis for the interpretation of triaxial test data as shown in Figure 3.10. A circle with diameter equivalent

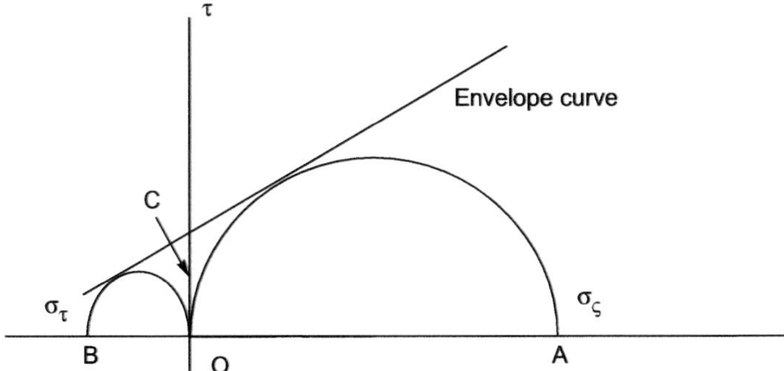

Figure 3.9 Mohr' s representation of tensile, compressive and shear (cohesive) strength (Wuerker, 1959)

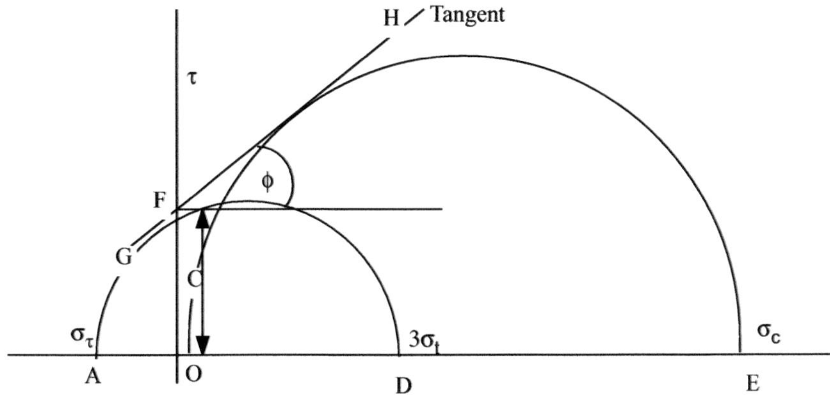

Figure 3.10 Calculating C and ϕ from Griffith's hypothesis

to 4 σ_t passing through points $-\sigma_t$ and $+3\sigma_t$ is drawn. A Mohr's circle with a diameter equal to σ_c is drawn through O. A common tangent through both the circles intercepts the τ axis at the cohesive strength.

The shear strength is given by the following equation:

$$\tau_t = \frac{(\sigma_c - \sigma_t)}{2x\sqrt{\sigma_t}(\sigma_c - 3\sigma_t)} \tag{3.8}$$

where,

τ_t = shear strength of rock
σ_t = uniaxial tensile strength of rock
σ_c = uniaxial compressive strength of rock

3.4.5 Triaxial Test Specimens

The test specimens are right cylinders within the tolerances described below.

- The sides of the specimen should be generally smooth and free of abrupt irregularities with all elements straight to within 0.1 mm over the full length of the specimen.
- The ends of the specimens should be parallel to each other and at right angles to the longitudinal axis.
- The end surface should be ground or lapped flat to 0.02 mm and should not depart from preperdicularity to the axis of the specimen by more than 0.001 radians.
- The specimen should have an h/d ratio between 2 and 2.5.
- The specimen's diameter should not be less than NX core size, 54 mm.

3.4.6 Triaxial Test Apparatus

The test apparatus required for carrying out the triaxial tests is as follows.

(1) A universal compression testing machine of suitable capacity of between 150 and 250 MN.

(2) A suitable triaxial cell (Hoek and Franklin type cell [Figure 3.11] is most popular).

(3) A hydraulic pressure pump together with a pressure intensifier for maintaining constant desired confining pressure 's_3'.

(4) Suitable deformation measurement devices:
   - (i) A high-grade dial micrometer with least count of 0.002 mm
   - (ii) A linear variable differential transformer securely attached to the cylinder
   - (iii) Electrical resistance strain gauges applied directly to the rock specimen, three in the axial direction and three equally spaced in lateral direction

(5) Flexible membrane of suitable material to exclude confining fluid from the specimen. (Neoprene rubber tubing of 1.6 mm wall thickness and 40–60 Duro hardness, Shore type A, has been found suitable for this purpose).

In the triaxial cell, the specimen, enclosed in an impregnable flexible membrane, is placed between two hardened spherical seated platens. A constant confining pressure is applied to the specimen through the membrane. The platens are made of tool steel hardened to minimum of Rockwell 58 Rockwell C and ground to a flatness of 0.0127 mm. The cell itself consists of a high-pressure cylinder with an overflow valve, a base and suitable entry ports for injecting the hydraulic fluid and applying the

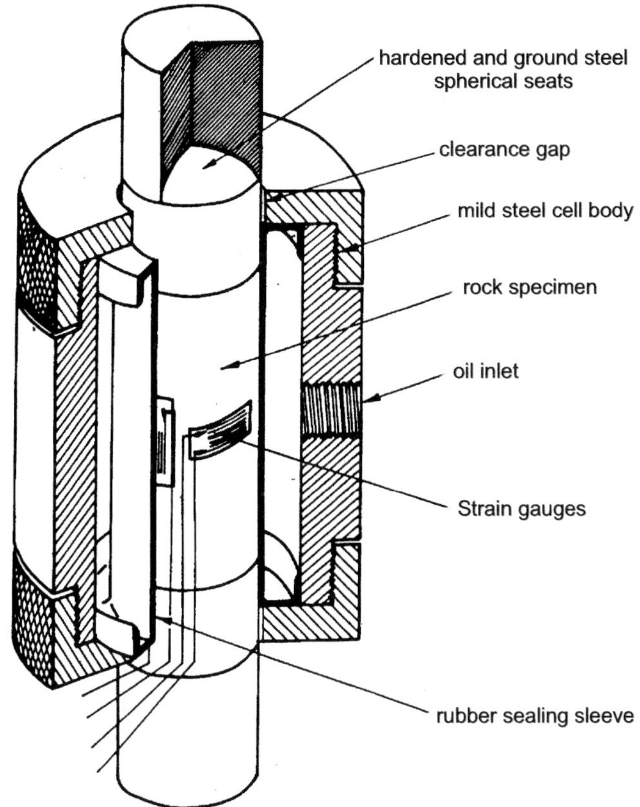

Figure 3.11 Hoek and Franklin triaxial cell

confining pressure. In addition, suitable hoses, pressure gauges and valves are also required.

A strain gauged specimen is placed inside a Neoprene flexible membrane and set up inside a triaxial cell. After setting up two sets of spherical seating at either end of the specimen, a confining pressure is applied to the specimen. A slight axial load (110 N) is applied to the specimen while the confining pressure is set to a desired level. A typical stress strain curve for a rock is shown in Figure 3.12. In an initial stage OA, there is more strain for slight increase in stress indicating the compaction of the specimen. After the initial period, stress is proportional to strain as indicated by line AA′. In stage A′ B, the stress/strain curve becomes non-linear and it reaches the peak strength at point B. Thus, at point A′ the specimen starts to yield and reaches the fracture strength at point B. In many rocks the yield point and

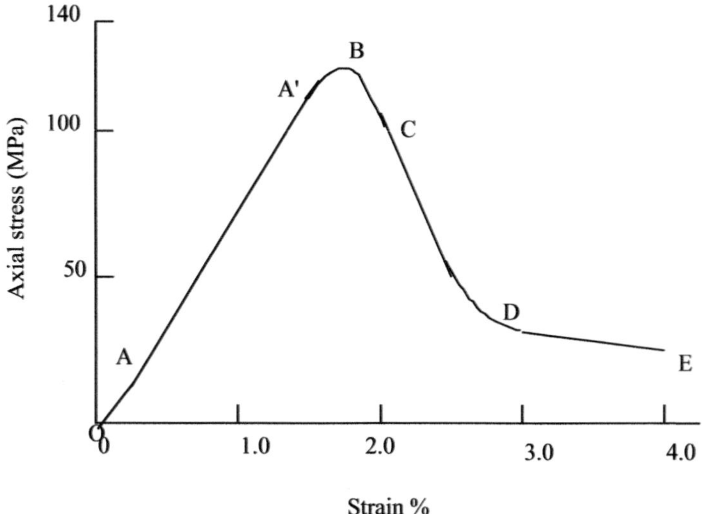

Figure 3.12 Stress-strain curve in triaxial loading

fracture stress are very close to each other. This fact is further illustrated in the following section.

3.5 Examples of Triaxial Strength of Rocks

Evans and Pomroy (1966) carried out triaxial tests on anthracite coal from Wales. Results in Figure 3.13 show the yield strength as well as fracture strength of anthracite for various confining pressures. Figure 3.14 shows the triaxial test results on Deep Duffryn coal, which indicates that axial stress and confining pressure results are non-linear. The uniaxial compressive strength of coal is 5 MPa. The fracture strength as well as yield strength of coal increases with the increases in confining pressure. It may be recalled that the yield strength indicates the onset of non-linearity of axial stress and confining pressure curve. It may also be noted that during the test when yield stress is reached, the fracture stress is closely approaching. Implications of these characteristics are discussed in the next section. It may also be noted that the axial stress and confining stress graphs for most rock types are non-linear.

3.6 Multiple Failure State Triaxial Test

In a single stage triaxial test, a number of specimens is required to obtain peak stress and residual stress envelopes and consequently shear strength

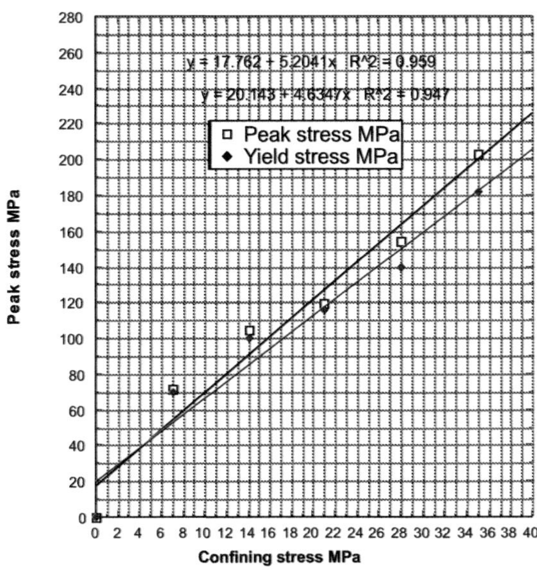

Figure 3.13 Variation of axial yield stress and failure stress with confining pressures for Welsh anthracite (Evans and Pomroy, 1966)

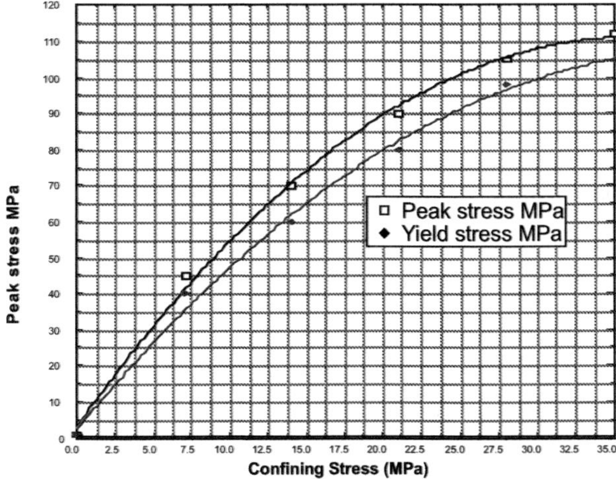

Figure 3.14 Variation of axial stress and failure stress with confining pressure for Deep Duffryn coal (Evans and Pomroy, 1966)

parameters. This requires four to five specimens to define the Mohr-Coulomb failure envelope if the rock is fairly uniform. Otherwise, even greater numbers of costly cores are required to determine an average envelope.

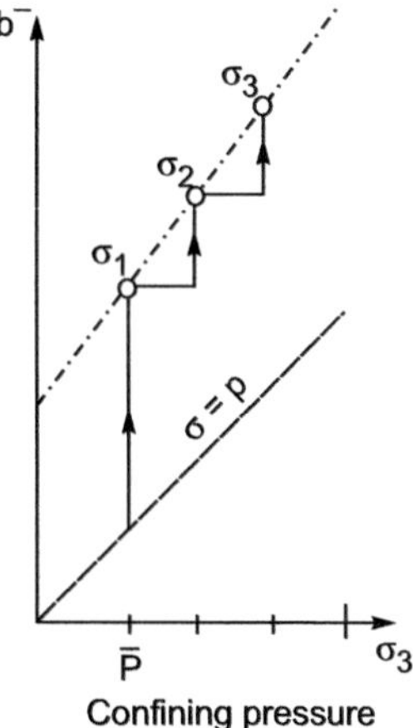

Figure 3.15 Multiple failure test loading path

The multiple failure state triaxial test has been developed by Kovari and Tisa (1975). If properly conducted, the test can produce Mohr-Coulomb failure envelope by testing just one specimen. Thus, the strength parameters 'c' and 'ϕ' can be determined for each core sample. Hence, it is possible to study the variables of strength parameters by conducting a multiple failure state triaxial test.

The test consists of loading a specimen in a triaxial cell by imposing a small amount of axial load and confining stress simultaneously until a predetermined confining stress is reached. Then, the axial stress is increased so that the yield stress is reached. At this point the confining pressure is abruptly increased, which increases the strength of the rock (Fig. 3.15). In this way, three to four points on the peak stress envelope are obtained. Once the specimen has failed, the axial load will drop from D to F on the residual stress envelope. Then by step-wise reduction of confining stress, a number of points on the residual stress envelope can also be obtained. A multiple failure state triaxial test loading path is shown in Figure 3.16.

The success in conducting this test strongly depends upon

- careful control of loading process,
- ductility of the specimen and
- stiffness of the machine

3.7 Continuous Failure State Triaxial Test

The continuous failure state triaxial test is the logical extension of the multiple failure state test, when the specimen is loaded and unloaded in

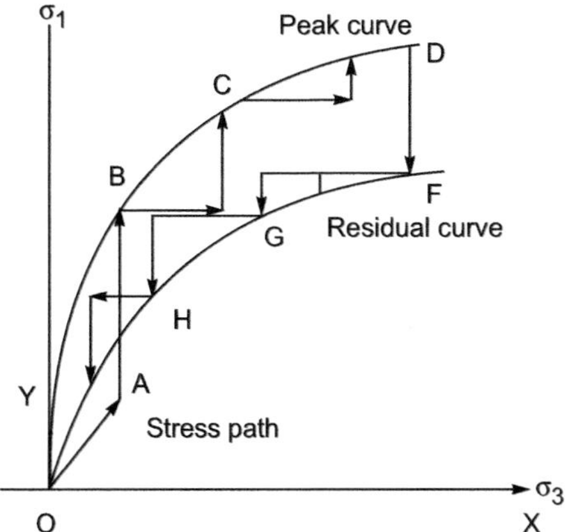

Figure 3.16 Multiple failure state triaxial test

strain-controlled mode. The specimen is set up in the triaxial cell in the same manner as in the multiple failure state triaxial test. After reaching the first point on the peak strength curve, the confining pressures as well as axial stress are continuously increased so that the stress path follows the peak envelope. This is carried out until the specimen fails and residual stress is reached. The confining load is then reduced at the same rate as the rate of unloading and the stress path of the residual curve is followed as the confining pressure is reduced to zero. Figure 3.17 shows the continuous failure state triaxial test.

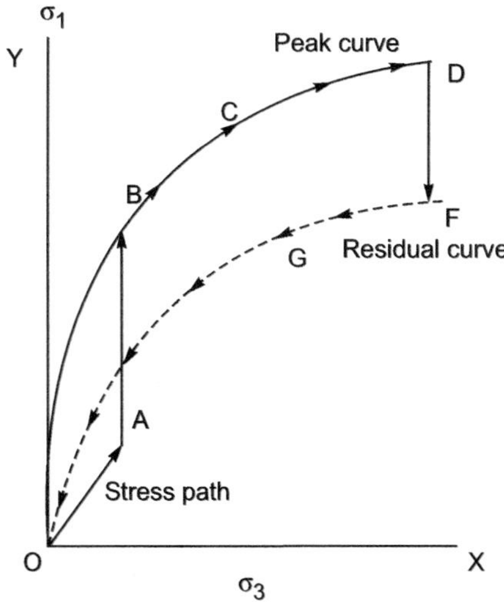

Figure 3.17 Continuous failure state triaxial test

3.8 Triaxial Tests for Various Confining Pressures

Wawersik and Fairhurst (1970) carried out complete stress and strain curves for marble in a confined as well as an unconfined state (Fig. 3.18). As usual there is an increase in the axial strength with increased confining pressure. It can be observed that the onset of the pronounced part of the stress-strain curve is insensitive to the confining pressure. The post-failure curve displays a sudden drop in strength followed by gentle slopes when tested under low confining pressures.

At high confining pressures, the brittle–ductile transition takes place and the rock starts showing flow characteristics. This behaviour is presented in Figure 3.18. Figure 3.19 shows the peak, residual and consolidated stress envelopes of the rock as derived from the stress/strain diagram.

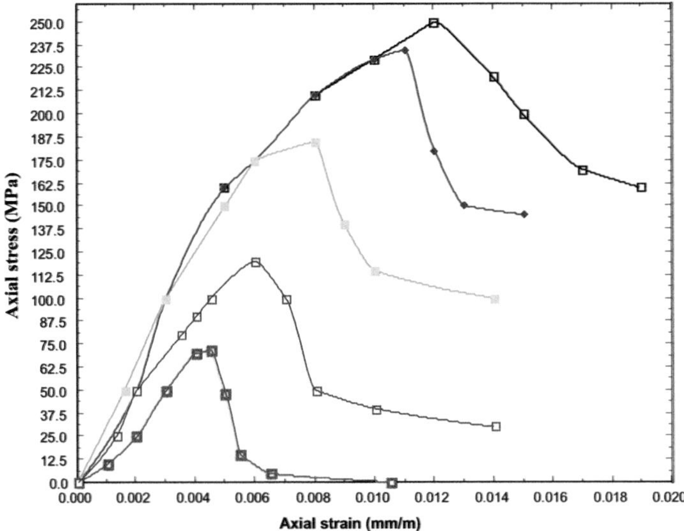

Figure 3.18 Complete stress/strain diagrams for various confining pressures

3.9 Example of Triaxial Tests

The triaxial tests on rock materials show that under high confining pressure, the brittle rocks undergo an increase in strength. The more brittle the rock, the stronger is the strength increase under confining stress. Figure 3.20 illustrates the triaxial strength of rock under increasing confining pressures from four to five times the uniaxial compressive strength. It may be noted that the effect of confining pressure on strength increase varies from rock to rock.

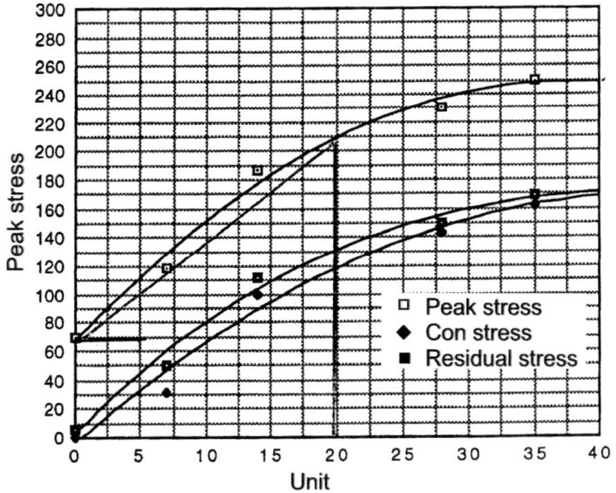

Figure 3.19 Triaxial test results in axial and confining stress, peak, residual and consolidated envelopes

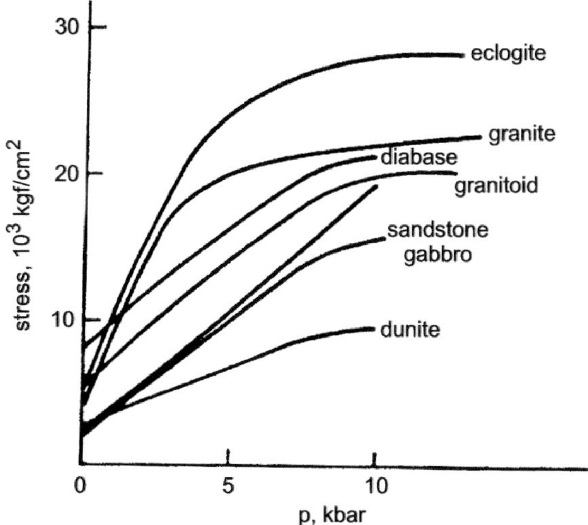

Figure 3.20 Triaxial strength of rock under increasing confining pressure (Ryabinin *et al.*, 1971)

The rate of strength increase depends upon the rock type as well as the level of confining pressure.

3.10 Factors Affecting Triaxial Test Results

The following factors affect the triaxial test results.

(1) *h/d ratio* At lower confining pressures, h/d ratio has considerable end-effects. At high confining stresses, the failure strength becomes independent of h/d ratio.

(2) *Shape* Tests on double bell-shaped specimens have indicated that there is no the appreciable effect on the strength of rock due to differences in a specimen's configuration.

(3) *Loading path* A triaxial specimen on single stage loading can take different loading paths. Tests have indicated that change in loading path does not change the strength of rock.

(4) *Temperature* Various rocks show different effects of temperature on the triaxial strength, but generally, the triaxial strength of a rock increases with a moderate increase in temperature. At very high temperature, the triaxial strength decreases with increased temperature.

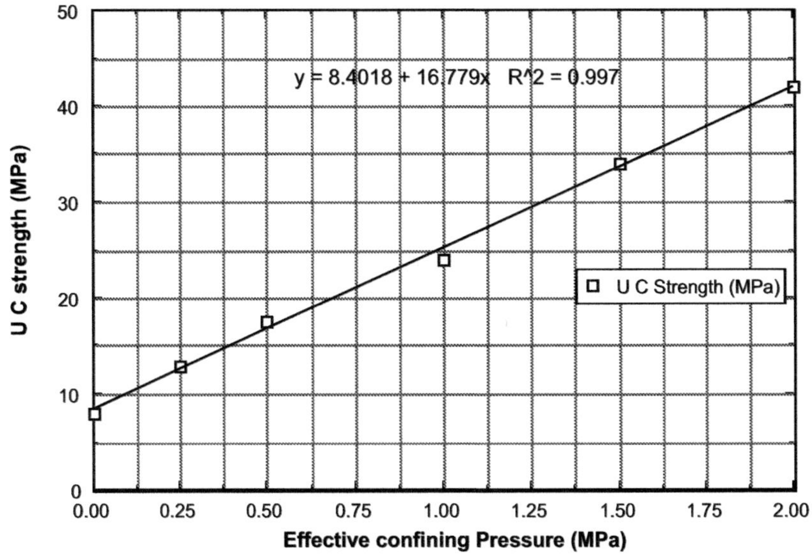

Figure 3.21 Ultimate strength of Berea sandstone as a function of effective confining pressure (Handin *et al.*, 1963)

(5) *Pore pressure* The triaxial strength of porous rocks decreases with increases in the effective confining pressure (confining pressure–pore pressure). However, the triaxial strength of impervious rocks has no correlation with effective confining pressures.

(6) *Chemical nature of pore fluid* Chemically active fluids like water reduce the strength of rock considerably under triaxial loading.

(7) *Strain rate* The effect of high strain rates and high rates of loading increases the strength of rocks.

(8) *Combined effect of confining pressure, temperature (30°C) and normal formation pore pressure.*

The UCS of water saturated rocks increases as a function of depth below the surface (Fig. 3.21) (Handin *et al.*, 1963).

3.11 Shear Strength of Rock Joints

In rock slope stability analysis, it is important to determine the shear strength of the potential failure surface which may consist of a single discontinuity. The test arrangement for the shear strength of joint is given in Figures 3.22 and 3.23.

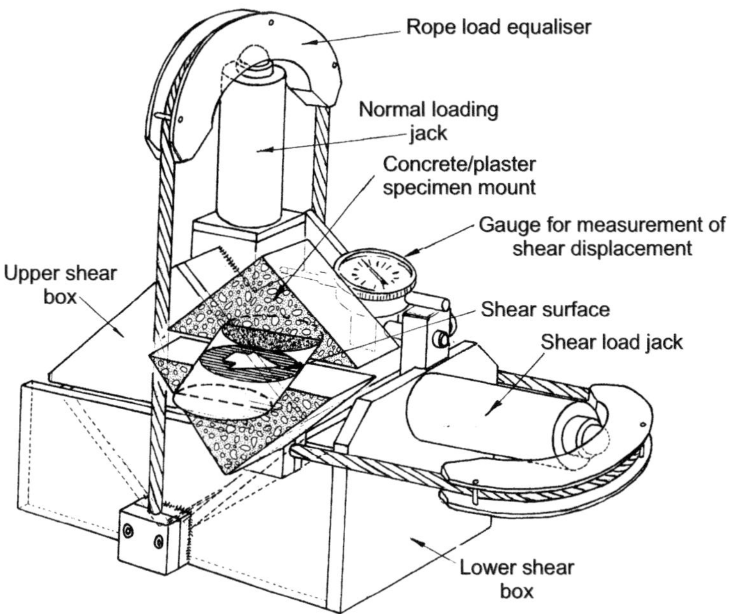

Figure 3.22 Robertson's portable shear box (Hock and Bray, 1981)

3.11.1 The Shear Box

The shear testing of joints is carried out in a Robertson's portable shear box (Figure 3.22) which is 51 cm long, 48 cm high and weighs about 40 kg. The machine is fitted with two jacks, one for providing normal load to the specimen and another for providing shear load. Some apparatus are fitted with a third jack for reversing the direction of shearing while maintaining the normal load constant. This feature is useful in determining the residual shear strength of certain type of discontinuities.

3.11.2 Test Specimen

Test specimen is prepared by drilling a large diameter rock core perpendicular to the joint surface. The end surfaces of the core are cut parallel to the joint plane. The portable shear box comprises two shear boxes resting on each other and carrying specimen mounts. Two parts of the specimen are tied together with a piece of wire and cast in plaster into upper and lower specimen mounts. When the plaster is dry, the plaster mounts are placed into the upper and lower shear boxes. The upper shear box carries a normal jack located in an upright position by a rope load equiliser. The lower shear box carries a horizontal load jack put into position by a rope stirrup. Two precise dial gauges are provided for measuring shear and normal displacements. Two hand-operated hydraulic pumps together with the pressure gauges are provided to monitor normal and shear loads.

3.11.3 Test Procedure

Each specimen is subjected to a normal stress σ_n applied across the discontinuity surface and a shear force required to cause displacement 'u' is measured (Figure 3.23a). A plot of shear stress 'τ_θ' against displacement 'u' at a constant normal stress level is given in Figure 3.23(b). At very small displacements, the specimen behaves elastically and the shear stress increases linearly with displacement. As the forces resisting movement are overcome the shear load / shear displacement curve becomes non-linear and then reaches a peak. Thereafter, the shear stress required to cause further

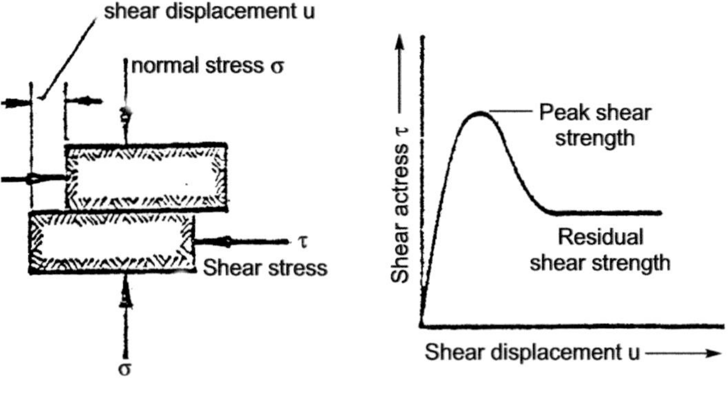

(a) Shear tests on joints (b) Shear stress/displacement curve

Figure 3.23 Shear test on a single rock joint

displacement drops rapidly to a constant value characterised as the residual shear strength.

Figure 3.24(a) shows the peak strength values obtained from the tests carried out at different normal stress levels. This relationship is linear with a slope equal to the peak friction angle ϕ_p and an intercept on the shear load axis 'c_p' the cohesive strength of the cementing material which is independent of normal stress. The peak shear strength is defined by the following equation:

$$\tau = c_p + \sigma_n \tan \phi_p \tag{3.9}$$

Plotting the residual shear strength against the normal stress gives a linear relationship defined by the following equation:

$$\tau = \sigma_n \tan \phi_r \tag{3.10}$$

If the discontinuity surface is inclined at an angle 'i' to the shear stress direction, the relationship between applied shear stress and normal stress is given by:

$$\tau = c_p + \sigma_n \tan (\phi_p + i) \tag{3.11}$$

This equation is also valid for rough joints having regular surface projections inclined at an angle 'i'.

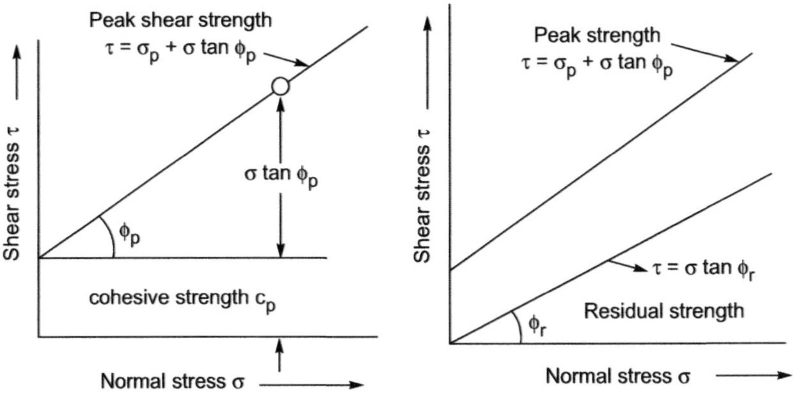

(a) Peak stress envelope of joint (b) Peak and residual envelope of joint

Figure 3.24 Peak shear strength and residual strength for different normal stress values.

3.12 Conclusions

(1) As most underground openings either fail in tension or shear, it is necessary to determine the uniaxial tensile strength and shear strength of rock in the laboratory.

(2) As it is difficult to grip a specimen in the specimen holder in the direct tensile test and also to drill a sufficiently long rock sample, an indirect tensile test or Brazillian test has been devised.
(3) Table 3.1 indicates that the uniaxial compressive strength is approximately six times the indirect tensile strength.
(4) Direct shear tests on the intact rock material indicate that either tensile or bending forces are introduced on the specimen. So the force required to fail the specimen is lower than the average value. Hence, these tests usually give lower shear strength for the rock materials. In addition, the process of preparation of the specimens affects the surface property of the rock and will also affect the strength values.
(5) The most common method of determining shear strength of rock for the design of underground structures is the triaxial test. This test determines the cohesive strength and internal angle of friction of the rock.
(6) By plotting the axial stress at failure against the confining stress, the slope of the curve gives the triaxial stress factor and uniaxial compressive strength of rock. These parameters can be related to the cohesive strength and internal angle of friction of the rock.
(7) The triaxial stress factor (K or tan β) is a material property, which can be used for design of structures in rock such as pillars or for assessing the stability status of an underground roadway.

References

Balmer, G. (1952). 'A general analytical solution for Mohr's envelope', *ASTM. Proc.*, Vol. 52, pp. 1260–1271.

Evans, I. and Pomroy C.D., (1966). 'Strength, Fracture and Workability of Coal', Pergamon Press, London, 277 pp.

Griffith, A.A. (1921). 'The phenomena of rupture and flow in solid', *Phil. Transaction Royal Society*, London, Series A, Vol. 221, pp. 163–198.

Handin, J. *et al.* (1963). 'Experimental deformation of sedimentary rocks under confining pressures: pore pressure tests', *Bull. Am. Assoc. Petroleum Geologists*, Vol. 47, No. 5, pp. 717–755.

Handin J. *et al.* (1967). 'Effect of intermediate prinicipal stress on the failure of limestone, dolomite and glass at different temperature and strain rates'. *Journal of Geophysical Research*, Vol. 72, No. 2, pp. 611–640.

Hobbs, D.W. (1964). 'The strength and the stress-strain characteristics of coal in triaxial compression'. *J. Geol.*, Vol. 72, No. 2, pp. 212–231.

Hoek, E. and J.A. Franklin (1968). 'A simple triaxial cell for field or laboratory testing of rocks', *Transaction of the Institution of Mining and Metallurgy*, Vol. 77, sec. A., pp. A22–A26.

Hoek, E and J. Bray, (1981) Rock slope engineering 3rd Edition, London, Inst. of Mining and Metallurgy.

Hawkes, I. and M. Mellor (1971). 'Measurement of Tensile Strength by Diametral Compression of Disc and Annuli', *Eng. Geol.*, Vol. 5, pp. 173–225.

Kovari, K. and A. Tisa (1975). 'Multiple Failure State Stress and Strain Controlled Triaxial Tests'. *International Journal of Rock Mechanics and Mining Science,* Vol. 7, No. 1.

Murrell, S.A.F. (1962). 'A criterion of brittle fracture of rocks and concrete under triaxial stress and the effect of pore pressure on the criterion'. *Proceedings of the Fifth Symposium on Rock Mechanics,* Minneapolis, pp. 563–577.

Murrell, S.A.F. (1958). 'The strength of coal under triaxial compression'. *Proceedings Confererence of Mech. Properties of the Non-metallic Brittle Materials'*, London, pp. 123–145.

Ryabinin, Y.N. *et al.* (1971). 'Mechanical properties and processes in solids under high pressure'. *J. Geophysics Res.*, Vol. 76, No. 5, pp. 1370–1375.

Von Karman, Th., (1911). 'Strength tests with triaxial compression'. *Z. Ver. deut. Ing.*, Vol. **55**, pp. 1749–1757.

Vutukuri, V.S., Lama, R.D., and Saluja, S.S. (1974). Handbook on Mechanical Properties of Rocks, Vol. 1, Trans Tech. Publications, Clausthal.

Wuerker, R.G. (1959). 'The shear strength of rocks'. *Mining Engineering,* Vol. 11, No. 10, pp.1022–1026.

Wawersik, W.R. and C. Fairhurst (1970). 'A study of brittle rock fracture in laboratory compression experiments'. *International Journal of Rock Mechanics and Mining Sciences,* Vol. 7, No. 5, pp. 561–575.

Assignments

1. In a laboratory test, the uniaxial tensile strength of sandstone was found to be 10 MN/m^2 and uniaxial compressive strength 45 MN/m^2. Draw the Mohr's diagram of strength and evaluate graphically the internal angle of friction and cohesive strength of the rock.
2. A particular rock failed under a major stress of 123 MN/m^2 with a corresponding minor principal stress of 30 MN/m^2. The angle of fracture was 55°. Evaluate graphically the normal stress and shear stress at failure and the internal angle of friction.
3. A series of shear tests were performed on a weak rock. Each test was carried out until the sample sheared, and the principal stresses for each test were:

No.	σ_3 (kN/m^2)	σ_1 (kN/m^2)
1	200	570
2	300	875
3	400	1162

Plot the Mohr's stress circle and hence determine the strength envelope and the angle of internal friction of the rock. If the confining pressure was 600 kN/m^2, extrapolate the failure stresses of the rock sample.

4. In example 3, plot the σ_1 and σ_3 curve for the samples and estimate the uniaxial compressive strength and the triaxial stress factor K or ($\tan\beta$) for the rock.
5. A tunnel is driven in a competent rock mass at a depth of 1000 m. The average vertical rock load is 0.027 MPa per metre depth and the ratio of horizontal stress to vertical stress (σ_h/σ_v) is 0.3. The maximum compressive stress that can be sustained by the tunnel walls is 2.7 times the maximum principal stress. The cohesive strength of the rock is 14 MPa and the angle of internal friction is 42°.

   Draw Mohr's circle of the state of stress before and after the excavation, the envelope and estimate the radial stress required to be applied to the tunnel walls to stabilise the excavation.
6. A soil failed under major principal stress of 300 kN/m^2 with a corresponding confining stress of 100 kN/m^2. If, for the same soil the minor principal stress is 200 kN/m^2, determine graphically what the major principal stress at failure would be if (a) $\phi = 0°$ and (b) $\phi = 45°$.

Time-dependent Behaviour of Rocks

4 Chapter

4.1 Introduction

From the previous chapters it may be inferred that rock is not a perfectly elastic material and there is some time lag in the recovery of strain after the stress is removed. It has been shown that during the cyclic loading of some rock materials that the loading curve and unloading curves of some types of rocks exhibited delay in elastic recovery of strain (known as hysteresis). There was also some permanent deformation in the specimen when stress was completely removed during each cycle of unloading. Further, it has also been shown that some strain controlled uniaxial tests incorporating high rates of strain, such as sonic and impact tests, resulted in high uniaxial strengths. Thus, the elastic response of many rocks to stress is time-dependent.

This chapter deals with the time-dependent deformational behaviour as well as long-term strength of rocks.

4.2 Time-dependent Deformation Behaviour

When stress is applied to rock it can deform in a variety of ways. A perfectly elastic material deforms instantaneously on the application of load and undergoes instant recovery of strain on the removal of stress. Thus, rock can be divided into the following categories depending upon its time-related behaviour.

1 *Perfectly Elastic Material*

A perfectly elastic material undergoes elastic deformation on the application of stress and instantaneous recovery on removal of load.

2 *Real material*

In a real-life material, the deformation is accompanied by a time lag during the application of the stress. Both reversible and irreversible time-dependent deformations are exhibited even though the stress level may remain below the elastic limits.

4.3 Rheology

Rheology is the study of the flow of material under stress. Models are used to define the fundamental types of behaviour of various materials. Descriptions of some of the fundamental rheological models are given below.

Hookean Materials

This type of model displays the elastic deformation of the material as a spring. On application of load on a spring with stiffness k, the spring undergoes deformation and sustains strain ε, as indicated by Equation 4.1 (and Figure 4.1) and recovers any strain on the removal of σ.

$$\sigma = k\varepsilon \tag{4.1}$$

where

k = spring stiffness
σ = stress
ε = strain

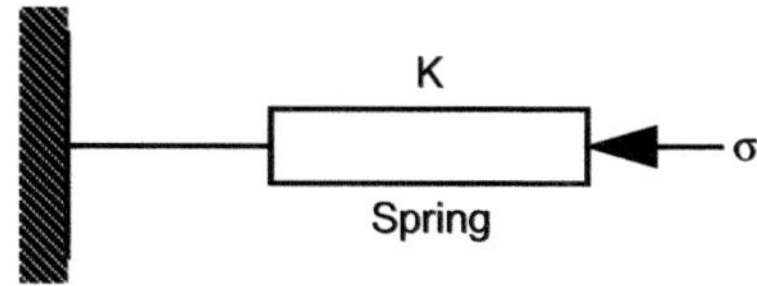

Figure 4.1 Spring model

Newtonian or Viscous Material

Newtonian material is defined as a dashpot and represents a Newtonian liquid. The fundamental equation for this is that on the application of stress σ, the stress is proportional to the rate of strain and η represents the time dependent viscosity coefficient of the liquid (Figure 4.2).

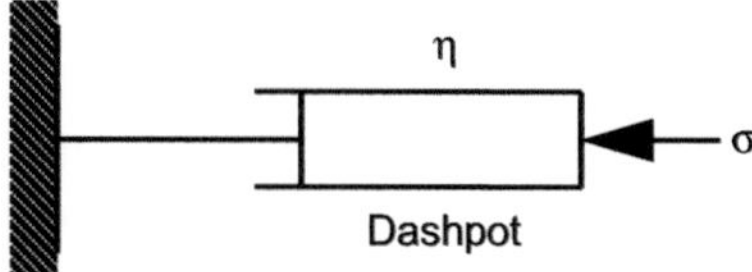

Figure 4.2 Dashpot model

$\sigma = \eta\dot{\varepsilon}$ if, $\sigma = 0$ and $t = 0$ (Dashpot)

$$\dot{\varepsilon} = \frac{\sigma}{\eta}$$

$$\dot{\varepsilon} = \frac{\sigma \times t}{\eta} \quad \text{(by integration)} \tag{4.2}$$

St Venant Model Showing Frictional Material

This rheological model is considered to be a plastic frictional contact containing a St Venant element and its inclusion in a model represents the visco-plastic behaviour of the material (Figure 4.3).

Figure 4.3 Frictional contact

The three models described above are basic models representing the behaviour of different materials.

Maxwell Material

A perfectly visco-elastic material (liquid) is represented by a combination of a spring and a dashpot in series. The spring represents the elastic behaviour of the material while the dashpot incorporates the time-dependent behaviour of the material. Figure 4.4 shows a graphical presentation of the model and the time strain behaviour of the material.

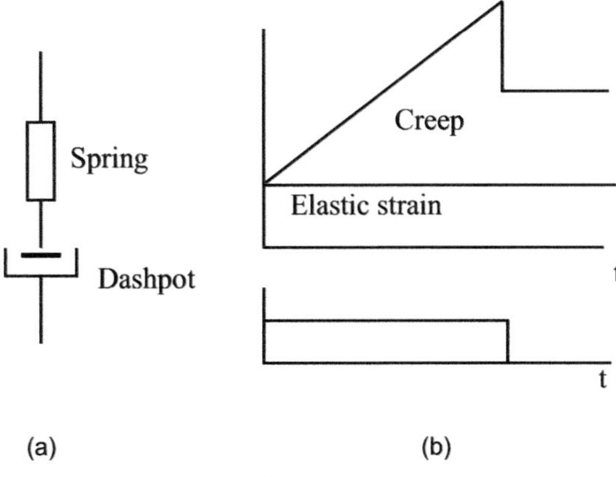

Figure 4.4 Maxwell model

Total strain is given by the summation of the elasatic strain represented by the spring and the time dependent strain represented by the dashpot.

$$\varepsilon = \frac{\sigma}{K} + \frac{\sigma t}{\eta} \tag{4.3}$$

If a constant stress σ is applied to the model from time $t = 0$ to time $= t$, the spring undergoes an instant elastic strain depending upon the magnitude of stress (Figure 4.4b). A time-dependent deformation is represented by the dashpot in series.

Kelvin Model (Visco-elastic Solid)

The Kelvin model (Figure 4.5) is represented by a spring and dashpot in parallel showing limited delayed strains. The basic relationship is given in Equation 4.4a.

$$\sigma = \varepsilon K + \eta \frac{de}{dt} \tag{4.4a}$$

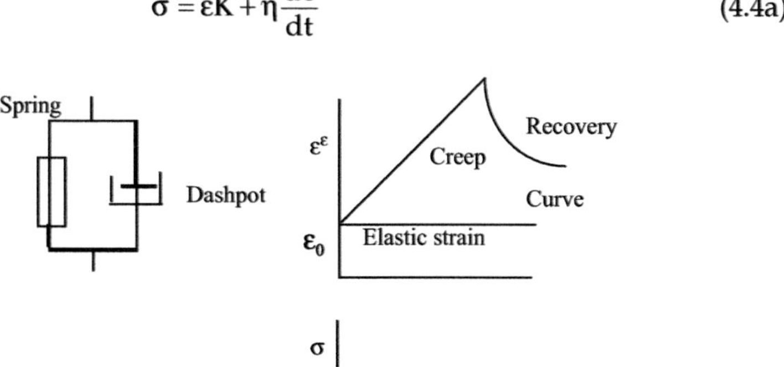

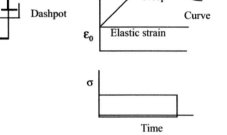

Figure 4.5 Kelvin model

Rearranging, we have $dt = \eta \dfrac{\delta\varepsilon}{(\sigma - \varepsilon K)}$ (4.4b)

Integrating, we get $t = \dfrac{\eta}{K} \log(\sigma - \varepsilon K) + C$ (4.4c)

when $t = 0$ and $\varepsilon K = 0$; $0 = \dfrac{-\eta}{K} \log(\sigma) + C$; $C = \log \sigma$

$$t = \frac{-\eta}{K} \log(\sigma - \varepsilon K) + \frac{\eta}{K} \log \sigma$$

$$\frac{-tK}{\eta} = -\log \sigma + \log(\sigma - \varepsilon K) = \log \frac{(\sigma - \varepsilon K)}{\sigma}$$

$$\frac{(\sigma - \varepsilon K)}{\sigma} = e^{-tK/\eta}; \qquad \varepsilon = \frac{\sigma}{K}(1 - e^{-tK/\eta}) \tag{4.4d}$$

Now assume the stress σ is removed at $\varepsilon = \varepsilon_m$

$$\eta \frac{d\varepsilon}{dt} + K\varepsilon = 0; \qquad dt = \frac{-\eta \, d\varepsilon}{K \varepsilon}$$

$$t = \frac{-\eta}{K} \log \varepsilon + C$$

when $\quad t = 0$, and $\varepsilon = \varepsilon_m \quad$ gives $\quad C = \log \varepsilon_m$

$$-\frac{tK}{\eta} = \log \varepsilon - \log \varepsilon_m$$

$$e^{-tK/\eta} = \varepsilon/\varepsilon_m; \quad \varepsilon = \varepsilon_m \times e^{-tK/\eta} \tag{4.4e}$$

Equation 4.4e is known as strain recovery equation and η/K is called relaxation time.

Visco-plastic Model

A visco-plastic model can incorporate the characteristics of instantaneous elasticity followed by viscous plasticity as used in the Bingham model. The generalised expression for strain is represented by:

$$\dot{\varepsilon}_{vp} = \frac{1}{\eta(\gamma)} < F(\sigma,\gamma) > \eta \frac{\partial G(\sigma,\gamma)}{\partial \sigma} \tag{4.5}$$

where

γ = scalar quantity equal to viscoplastic strain
$F(\sigma,\gamma)$ = visco-plastic failure criteria
$G(\sigma,\gamma)$ = plastic potential of volumetric plastic deformation (dilation)
$\eta(\gamma)$ = viscosity constant or a variable of gamma representing tertiary creep
$\dot{\varepsilon}_e$ = elastic strain rate
$\dot{\varepsilon}_{vp}$ = visco-plastic strain rate

The variation of F with γ represents the strain softening behaviour of the rock mass.

4.4 Flow in Rock

Figure 4.6 shows a generalised constant load creep curve representing the strain–time relationship for a typical rock at different magnitude of stress. On application of stress, the rock undergoes instantaneous elastic deformation followed by three stages of creep.

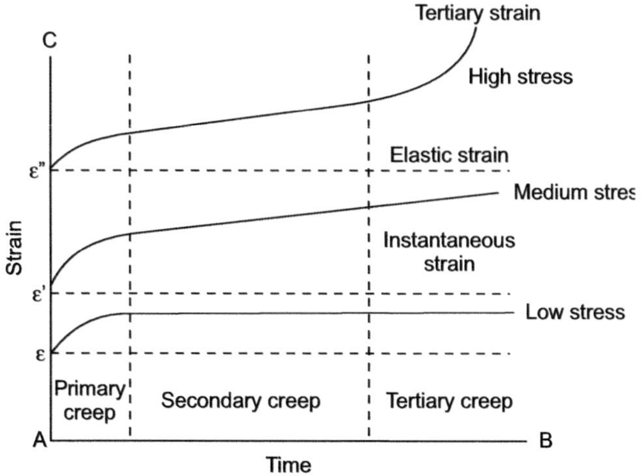

Figure 4.6 Generalised creep curve for rocks

(1) Primary creep is characterised by gradually decreasing creep rates that last for a short period not exceeding 24 hours.
(2) The second stage of creep where the rate of creep is stationary over a prolonged period and the specimen generally does not fail.
(3) The third stage of creep is tertiary creep in which the strain rate increases with time eventually leading to failure of the specimen. If the magnitude of stress is high with respect to that leading to instantaneous failure of the specimen, the tertiary creep period is too short to permit measurements of strains.

Therefore, at low stress levels with respect to the uniaxial compressive strength of the rock only transient creep is present in the rock sample. In a purely volumetric stress field, the time-dependant behaviour is confined to transient or primary creep only. For higher stress levels, secondary and tertiary creep are also manifested in the specimen as shown in Figure 4.6.

4.5 Factors Affecting Creep

Creep rate is dependent upon a multitude of factors as given below.

Stress Level

It may be noted that the strain rates increase with the level of applied stress as indicated in Figure 4.7. Laboratory tests carried out on various rock types.

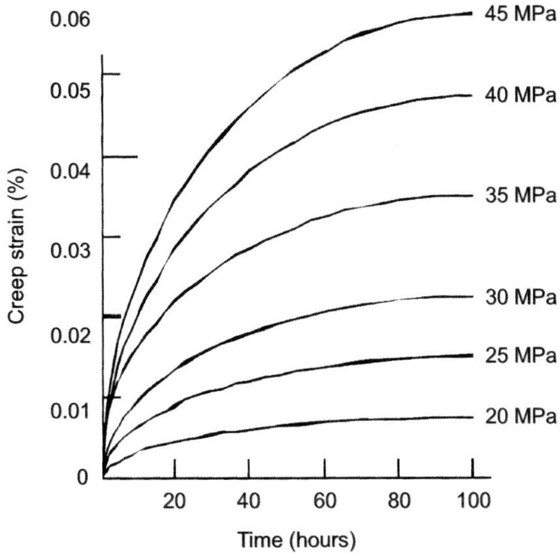

Figure 4.7 Creep versus time for various values of stress

Robertson (1963) has indicated that at a given time 't', the transient creep is expressed by the following equation:

$$\dot{\varepsilon} = D\sigma^n$$

$$\varepsilon = Dt\sigma^n \qquad \text{where } D = \text{constant for rock} \qquad (4.6)$$

Robertson (1963) compiled the values of 'n' for a number of rocks tested in a uniaxial creep test in the laboratory at room temperature (Table 4.1). This creep equation is valid for the primary state of creep and not necessarily valid for steady-state creep.

Table 4.1 Value of creep constants evaluated in uniaxial creep test at room temperature in the laboratory

Rock type	*Maximum strain x 10^{-3}*	*Maximum stress MPa*	*Constant n*
Alabaster (saturated)	5	0.40	2.0
Gabbro	0.01	0.20	1.0
Granite 1	1	7.25	3.3
Granite 2	3	2.03	3.0
Grainodiorite	0.2	0.2	1.0
Shale	3	0.2	2.7
Slate	1	6.9	1.8
Limestone	7	2.90	1.7
Halite	20	0.61	1.9

Simulation of Three Stages of Creep

Connecting visco-elastic and visco-plastic models in series one can carry out numerical simulation of three stages of creep at constant compressive stress. It can be demonstrated that by considering the material properties as constant, primary and secondary creep can be simulated as shown in Figure 4.8. A tertiary stage of creep indicates a decrease in viscosity and or shear strength with increasing creep strains.

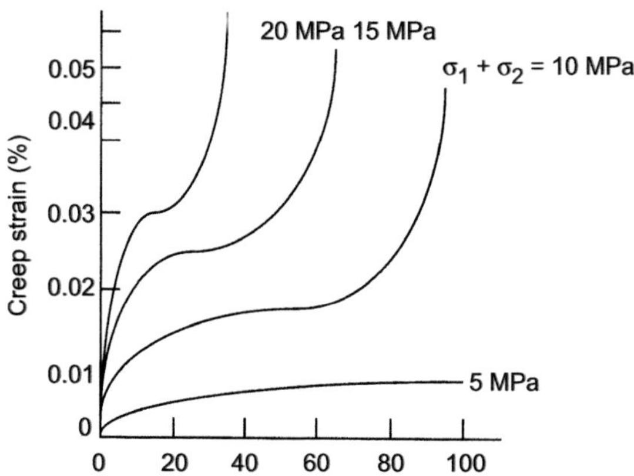

Figure 4.8 Numerical simulation of primary, secondary and tertiary stages of creep

Nature of Stress

The nature of stress is important in the creep testing of rock in the laboratory. It has been shown that on the application of the following stress regimes, creep rates increase appreciably.

(1) When torsion stress is applied to a cylindrical rock sample at low stress levels, high rates of creep strains are observed in the specimen.
(2) In bending and tensile stress fields at low stress levels, high creep strain rates are observed. It is also observed that the creep rate is six times in tension than it is in a compressive stress field.
(3) It is also observed that in a compressive stress field the tertiary creep takes place at relatively high stress levels giving advance warning of the impending failure of the specimen.

Deviator Stress

It is observed that the application of a high deviator stress to the specimen results in an appreciable increase in creep rate as shown in Figure 4.9.

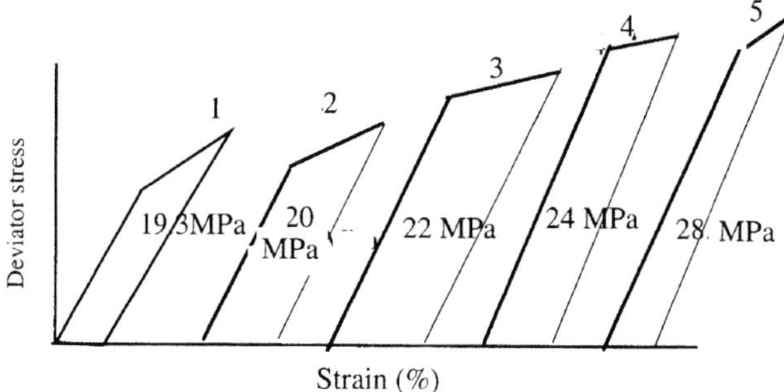

Figure 4.9 Effect of deviator stress on creep

Confining Pressure

It has been observed that an increase in the confining pressure on the rock specimen results in an increase in the strength of the rock.

Figure 4.9 indicates an increase in the creep rate with increase in confining pressures on various cycles of loading on a limestone specimen. It may also be observed that yield points also increase with an increase in confining stress.

Effect of Temperature

The study of creep rates at elevated temperature is of interest to geologists studying the behaviour of rock material at higher than normal temperature.

Figure 4.10 shows the creep results of rock salt at elevated temperatures indicating that the creep is a temperature-dependent phenomenon.

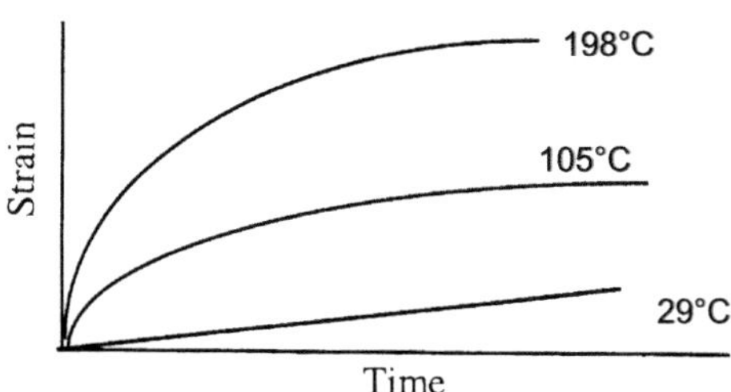

Figure 4.10 Effect of temperature on creep of rock salt

Effect of Water and Humidity

Water can affect the measurement of creep properties of rock in three possible ways.

1. Saturation of the specimen

2. Development of pore pressure affecting its shear strength
3. Effect of change in humidity on measurement of strains in electric strain gauges.

Figure 4.11 shows the effect of water saturation on the creep properties of slate and porphyrite. It can be seen that both the primary creep rate and secondary creep rates of the saturated slate specimen is twice that of the dry slate specimen. In the case of porphyrite, the creep rates of saturated specimens are 2.5 to 3 times higher than that of the dry rocks.

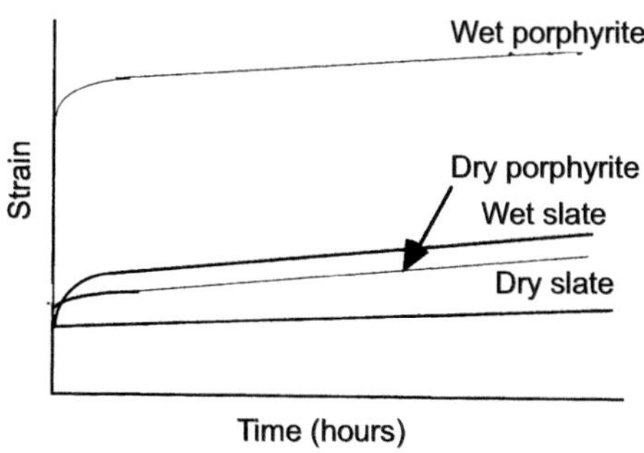

Figure 4.11 Effect of saturation on creep properties of rocks

In the case of a drained triaxial test, the consequence of pore pressure is to reduce the effective stress on the specimen and as a consequence reduce the shear strength of the rock. Therefore, the creep rate will increase during these tests.

Price (1963, 1964) studied the effect of humidity on the drift of strain readings in the electric strain gauges. It can be seen in Figure 4.12 that when the specimens and the electric strain gauges were uncoated with polyeurathane the specimen indicated considerable variations in strain. It is therefore customary to spray the specimens and electric strain gauges with

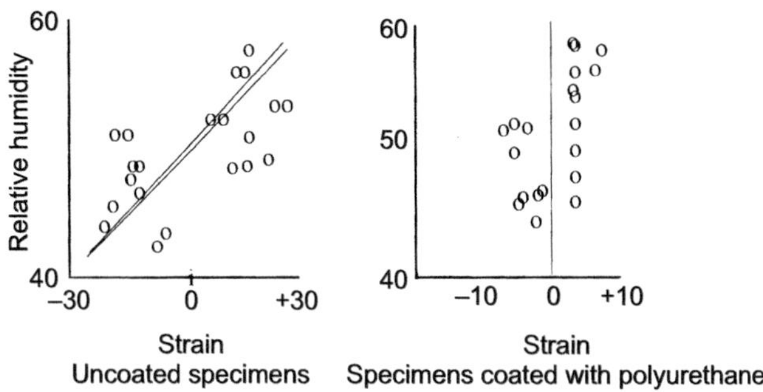

Figure 4.12 Effect of humidity on creep

water proof coatings in a creep test to alleviate the effect of humidity on the measurement of strains.

Effect of Grain Size on the Creep Properties of Rock

Grain size in a rock is one of the major factors affecting the porosity of the rock. Porous material has lower density than the compact rocks and increased stress concentration on the grain boundaries peripheral to the pore spaces results in higher creep rates. Figure 4.13 shows that the rock with large grain sizes shows higher primary and secondary rates of creep.

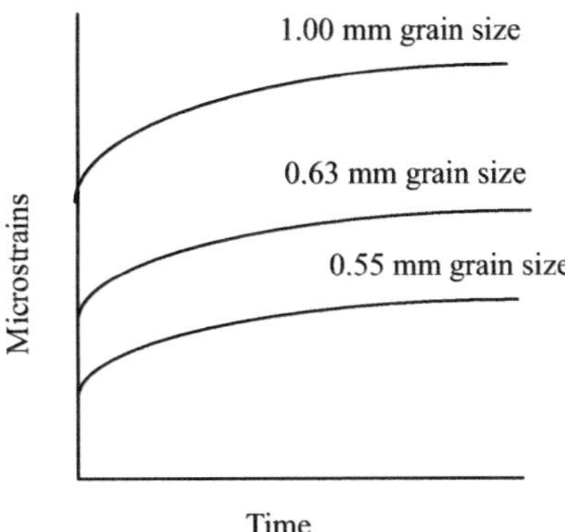

Figure 4.13 Effect of grain size on creep

4.6 Need for Controlling Various Factors in Creep Tests

Creep tests on rocks are carried out over a long period during which test conditions have to be controlled in order to obtain acceptable test results. The following factors influence the creep test results.

(1) Testing machine should have constant load over a long period.
(2) Strain measuring instruments should be sensitive and drift free.
(3) Temperature and humidity must be controlled.
(4) Test specimen should conform to the standards for preparation of samples.
(5) Test stress should be standardised as a fixed percentage of uniaxial compressive strength.
(6) Laboratory stress conditions should conform to *in situ* stress conditions, i.e. compressive, tensile, shear or triaxial as the case may be.

In general, hydraulic machines are not capable of maintaining uniform loading over a prolonged period of time and, therefore, resort is made to using a lever type of loading machine to carry out the creep test.

Figure 4.14 shows various loading configurations used in creep tests based on a dead weight and lever arrangement. Table 4.2 summarises the types of creep tests used, their loading geometry, method of applying load and measuring corresponding creep rates. Since the variation of temperature and humidity affect the test results, it is important that the creep testing apparatus is housed inside a temperature and humidity controlled chamber.

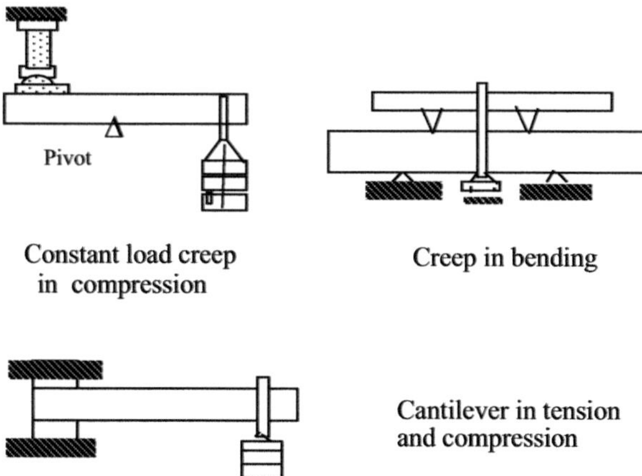

Figure 4.14 Constant loading arrangements for creep tests

4.7 Creep Test Results

Figure 4.15 shows the results of some creep tests on rocks in the laboratory. The following three rock types have been considered.

1. Granite
2. Millstone grit
3. Slate

All these rock types show relatively low instantaneous strains followed by high rates of primary creep. The primary creep is non-linear with decreasing rate of strain. On removing the stress the strain recovery is also non-linear.

Creep curve of rock salt, shown in Figure 4.16, has the following characteristics.

- Instantaneous strains are low
- Primary creep period is short
- The duration of the steady state creep curve is prolonged depending upon the stress level
- Tertiary creep curve is too short to permit observation

Table 4.2 Creep testing equipment of intact rock

Loading machine	*Type of apparatus*	*Loading geometry*	*Method of applying load*	*Method of measuring strain*	*Salient features*	*References*	*Comments*
1. Bending	Mechanical beam testing apparatus	Four-point loading apparatus	Dead weight	Optical or mechanical gauges	• Three-point loading system • Periodic measurement of strains	Phillips (1931) Williams and Elizzi (1976)	• Simple and stable system • Sample size 30 mm
2. Torsion testing	Mechanical	Torsion couple at lower end of specimen through a pulley	Dead weight	Optical record of mechanical gauge (10–12 magnification)	Photographic record of oscillograph	Lomnitz (1956)	• Specimen 450 mm • Minimum shear strain 1.4×10^{-6} radians
3. Compression tension	Mechanical load reversal type		Weight through two levers with overall ratio 1:100	Martin lever, LVDT or strain gauges	• Load reversal by spring and jack assembly • Creep test at high temperature	Misra and Murrel (1965), Singh (1975)	Frame tends to become heavy for large diameter specimen
4. Uniaxial compression or tension	Hydraulic loading		220 t capacity jack at 22 MPa pressure + pressure intensifier and controller	LVDT, Martin mirror or end-to-end displacement or strain gauges	• Solenoid valve to release pressure • Pressure maintained by accumulator	Wawersik (1972)	• High pressure gas • Specimen size 25.4 mm × 70 mm long
5. Uniaxial creep to oil shale	Spring-loaded hydraulic frame	Loading frame through spring	Hydraulic jack and spring to keep constant load	0.00025 mm dial gauge	Relaxation behaviour of oil shale studied	Chong *et al.* (1980)	2.54 × 2.54 × 5.1 cm specimen

(contd.)

Table 4.2 (contd.)

Loading machine	*Type of apparatus*	*Loading geometry*	*Method of applying load*	*Method of measuring strain*	*Salient features*	*References*	*Comments*
6. Triaxial creep	Hydraulic pressure by actuator membrane	• Loading between two pistons • Polyurethane • Confining pressure by hand pump or screw driven regulator	Confining pressure by servo-controlled regulator and feedback	• End-to-end displacement • Integrated radial deformation technique (Crouch, 1970)	Volume monitoring is affected by servo-controlled pressure regulator	Wawersik (1972)	Specimen 25.4 mm × 50 mm
7. Triaxial creep	Pneumatic hydraulic axial pressure through plunger		• Pneumatic/ hydraulic creep cell • High pressure upto 35 MPa	• LVDT for axial strain • Load measured by strain gauge load cells	Independent of external features	Williams and Ellizzi (1977); Cogan (1976)	• Specimen 25 mm in diameter and 75 mm long • PVC jacket
8. Biaxial creep to joint	Biaxially loaded frame with constant load test	Sliding surface 45° to principal stress	• Two flat jacks 22 MPa and 17.5 MPa • 35 MPa intensifier and accumulator and hydraulic pump	• Semiconductor strain gauges • LVDT continuously recording pressure and shear	• Constant major principal stress • Lowering minor principal stress	Dieterich and Conard (1967)	• Specimen 1500 mm × 1500 mm × 400 mm • Sliding surface 45° to principal stress

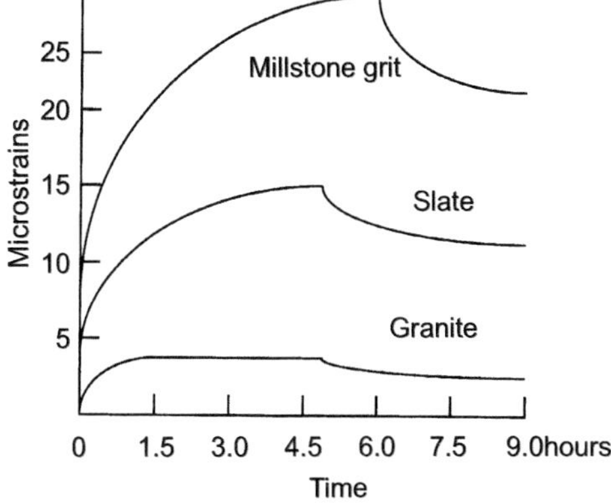

Figure 4.15 Creep test results of some rocks

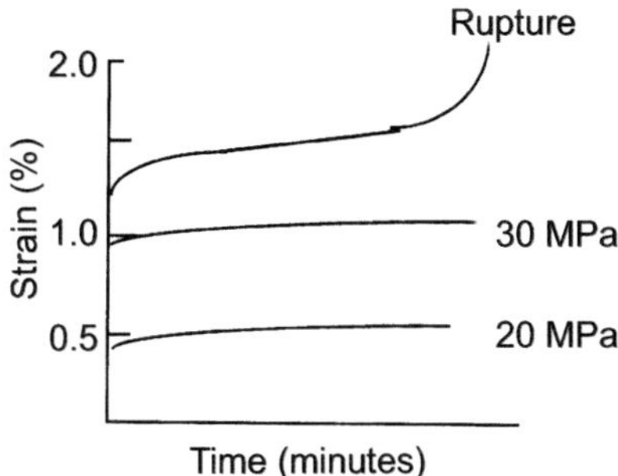

Figure 4.16 Creep testing results for rock salt in compression

When rock fails in secondary creep, the rock may rupture at any time when the strain rate is either constant or decreasing. A decreasing rate of strain does not necessarily ensure stability.

4.8 Creep Testing *in situ*

Primary Creep *in situ*

In relation to the design of structures in rock, the time-dependent roof sag is measured in rock salt and potash mines. Normally transient creep or primary creep is too short to be measurable. Immediately following an underground blast there is usually evidence of the transient creep as manifested by a small fall, development of roof fracture or by audible and micro-seismic noises.

Steady state creep has been observed over long periods in roof and pillars. Roof sag rate in an experimental room in an oil shale mine was constant over an 800-day period showing a prolonged secondary creep period without failure as shown in Figure 4.17.

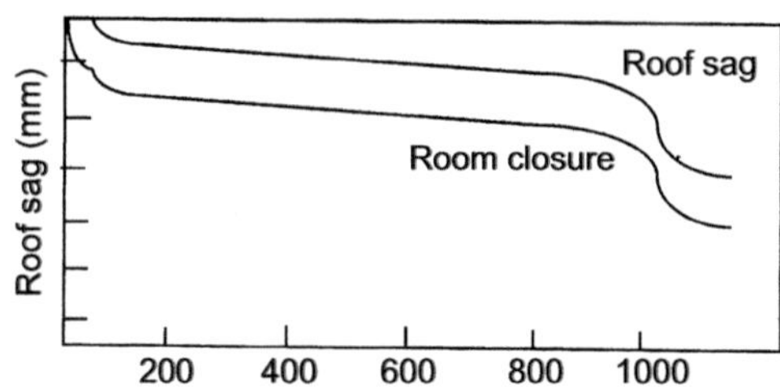

Figure 4.17 Roof sag measurements in an oil shale mine

Lateral Deformation of Pillars

This has been measured in German potash and sylvanite mines. The deformation rate was constant from 63 days to 10 years and the strain rate up to 4% per year (Figure 4.18).

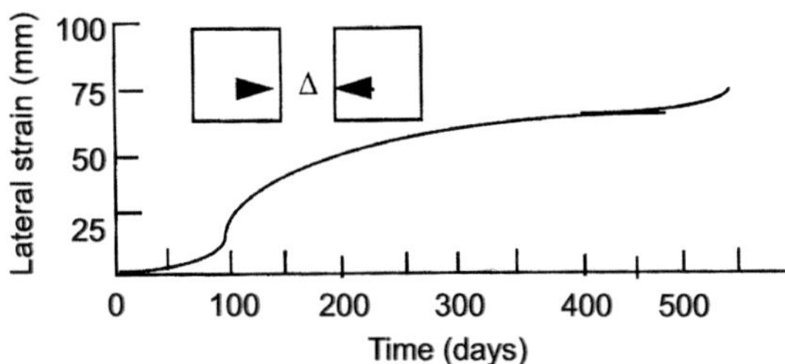

Figure 4.18 Lateral deformation of pillar

Measurement of *in situ* creep has exhibited a varying period of transient, secondary and tertiary creep. Micro-seismic studies of rock masses has indicated that tertiary creep period involving rock up to several tonnes varies from a few hours to several weeks for rock masses of several 100 tonnes. A method of estimating time to failure of tertiary creep of a pillar, roof or a slope in an open pit mine is presented in the next section.

4.9 Estimation of Time of Failure of a Slope

Rock *in situ* exhibits a varying degree of transient steady state and tertiary creep. In open pit mining it is usual to observe a tension crack on the overburden benches of the pit. Sometimes these instabilities involve millions of tonnes of overburden material.

It is therefore customary to carry out borehole extensometer measurements across the tension cracks between two anchor points in the borehole and take time-dependent strain readings. A typical strain time diagram is shown in Figure 4.19. In this diagram, primary, secondary and tertiary periods of creep are isolated. In the tertiary creep graph, two points A_1 and A_3 are chosen and total tertiary creep strain between these points 2Δ is noted. A point A_2 is chosen in the tertiary creep graph so that strain between A_1 and A_2 and between A_2 and A_3 is Δ. Times t_1, t_2 and t_3 corresponding to reach strains A_1, A_2 and A_3 are noted in the abscissa in Figure 4.19. The time to rupture of the slope can be calculated using the following equation.

$$T_r = \frac{t_2^2 - t_1 \times t_2}{2t_2 - t_1 - t_3} \tag{4.7}$$

where,

T_r = time to rupture
t_1 = time to reach the initial strain point in tertiary creep curve
t_2 = time to reach the middle strain point in the tertiary creep curve
t_3 = time reach the third strain point in the tertiary creep curve

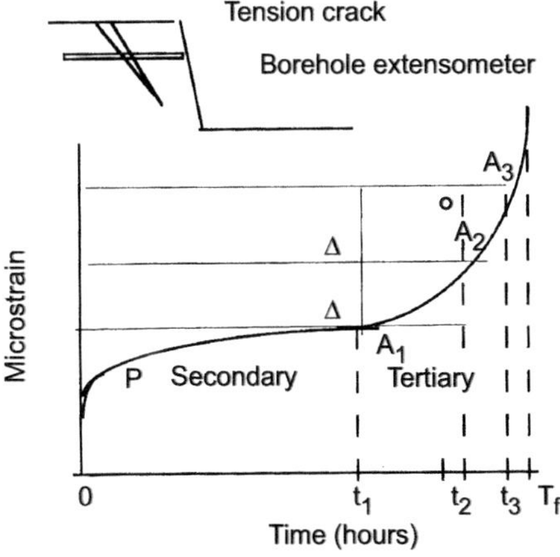

Figure 4.19 Estimation of time of failure of a slope

Figure 4.20 shows a graphical method of projecting the rupture life of a slope in the creep range as developed by Saito (1969). The method comprises plotting the enlarged portion of the tertiary section of the creep curve and selecting points A_1, A_2 and A_3 as indicated earlier. Select a point M midpoint between A'_1 and A'_2 and draw a quarter circle with A_2 as a centre and A_2 M

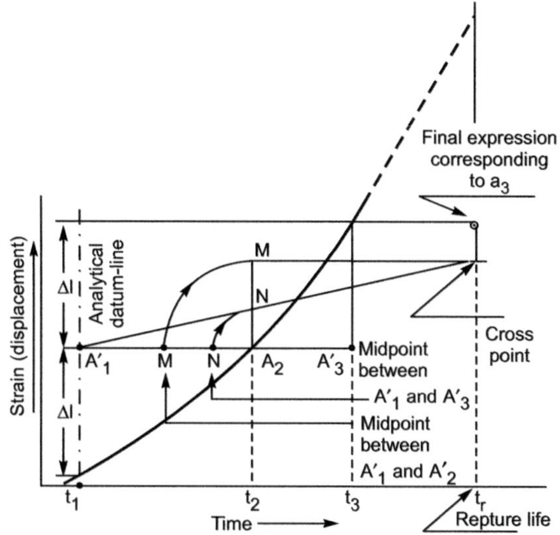

Figure 4.20 Graphical analysis for rupture life in the tertiary creep range (after Saito, 1969)

as radius. Similarly, by selecting a point N mid-distance between points A'_1 and A'_3, draw an arc with A_2N as radius and A_2 as a centre. The inter-section between the tangential line A'N' and horizontal line M'X represents the point of rupture of the slope. The projection of point X on the abscissa as 't_r' gives the rupture life of the slope and a point A_4 on the tertiary creep curve will represent the total displacement before the rupture of the slope. This method has been refined in a recent publication by Federico *et al.* (2002).

4.10 Long-term Strength of Rock

Construction engineers are interested in finding the long-term strength of rock with respect to a particular stress level below which the rock structure will not fail. It has been noted that the strength of rock depends upon the test environment and loading conditions. Rate of loading of the specimen and strain rates both affect the uniaxial strength. It has been also noted that dynamic tests incorporating sonic and impact loads on the specimen give very high apparent strengths of the rock material. It has also been observed that the time to fail at sustained load is inversely proportional to the magnitude of the stress. Thus, the long-term strength of rock is defined as the stress below which rock will not fail.

The main difficulty in determining long-term strength is that only in extremely homogeneous rocks consistent results can be obtained. A large number of tests over a prolonged period are required in order to determine the long-term strength of rock (Griggs, 1936; Phillips, 1948; Wiid, 1966).

4.10.1 Direct Methods

The direct method of determining long-term strength comprises the application of different constant stresses on the rock specimens and the highest value of stress under which no failure takes place is termed the long-term strength of the rock. It is customary to express the long-term strength as a percentage of the uniaxial compressive strength of rocks.

Figure 4.21 shows the strength-time relationship for Solenhofen limestone in confining stress that indicates that the long term strength is some 65% of the uniaxial compressive strength.

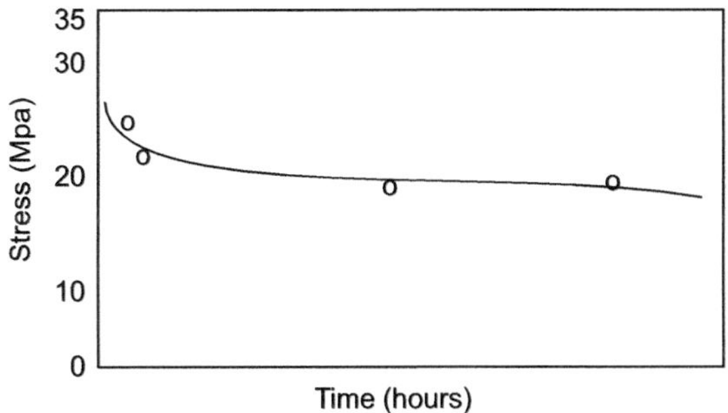

Figure 4.21 Strength-time behaviour of limestone (Griggs, 1936)

Similarly, Figure 4.22 indicates the strength time relationship for saturated alabaster as determined by Griggs (1936) showing that the long-term strength determination for 1000 hours of testing is some 30% of the uniaxial compressive strength.

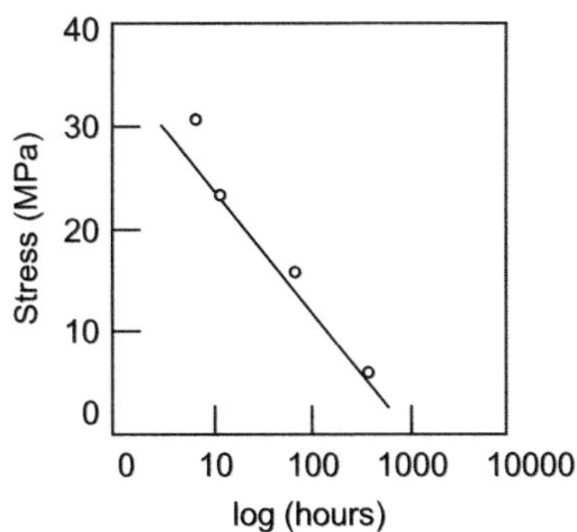

Figure 4.22 Strength-time behaviour of Alabaster

Singh (1977) determined a strength time graph for sandstone which indicated that the asymptotic value of stress with time represents the long-term strength (Figure 4.23).

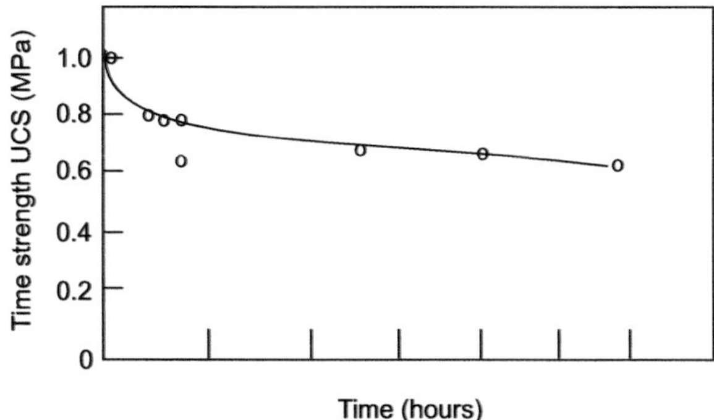

Figure 4.23 Strength-time behaviour of sandstone

The direct method of determining long-term strength is tedious, repetitive and time-consuming and therefore some indirect methods, as discussed in the following sections, have been developed.

4.10.2 Indirect Methods

Indirect methods of evaluating long-term strength require relatively short period tests together with knowledge of deformational behaviour of rock to interpret the results to evaluate long-term strength as follows:

1. Creep test method
2. Volumetric strain method
3. Log stress-log strain method
4. Stress-strain rate method
5. Loading rate method
6. Micro-seismic method

Creep test method

Stress at which the steady state creep rate is zero represents the long-term strength of the rock. Plotting compressive stress and the secondary state creep velocity curve and extrapolating the necessary results obtain this strength (Figure 4.24).

The strain-time relationship as obtained during the test is given in the following equation:

$$\varepsilon = A \sinh B(\sigma - S) \tag{4.8}$$

where ε = strain rate

σ = stress

A, B and S are empirical constants derived by curve fitting

σ = S in Equation 4.8 represents that at $\varepsilon = 0$ onset of micro-fracture will take place.

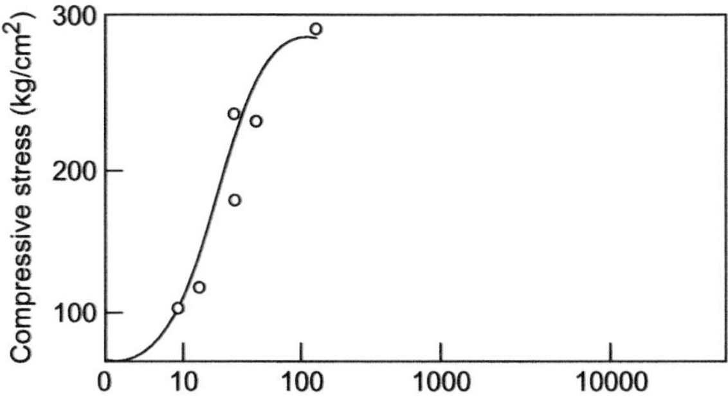

Figure 4.24 Steady state creep velocity in, micro-m/day

Figure 4.25 shows the applied load as a percentage of the uniaxial compressive strength of sandstone and the secondary creep rate in mm/day indicating linear relationship (Price, 1966). The result indicates that the long-term strength of a Pennant sandstone is 20% of the uniaxial compressive strength of the rock.

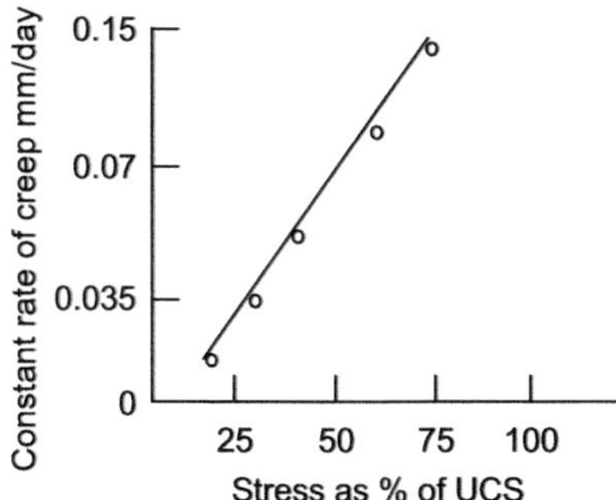

Figure 4.25 Creep rate method

Volumetric strain method

There are two types of fracture propagation in the rock specimen during testing under a uniaxial compressive load, namely:

1. stable fractures and
2. unstable fracture

The stable fracture process is a function of loading and can be controlled during the uniaxial compressive strength test in the laboratory. During unstable loading conditions, extensive micro-fractures occur in the specimen that cannot be controlled. The stress at which unstable fracture starts can be taken as the long-term strength of rock.

$$\varepsilon_V = \varepsilon_A + 2\varepsilon_L \tag{4.9}$$

$$\varepsilon_V = \text{volumetric strain}$$
$$\varepsilon_A = \text{axial strain}$$
$$\varepsilon_L = \text{lateral strain}$$

Figure 4.26 shows the stress-strain curve of rock in a uniaxial stress field indicating the stress-axial strain and lateral strain curves. In the stress and volumetric strain curve in the left-hand side of the diagram, the flexure point of the volumetric strain indicates the long-term strength of the rock (Singh, 1977; Wiid, 1966).

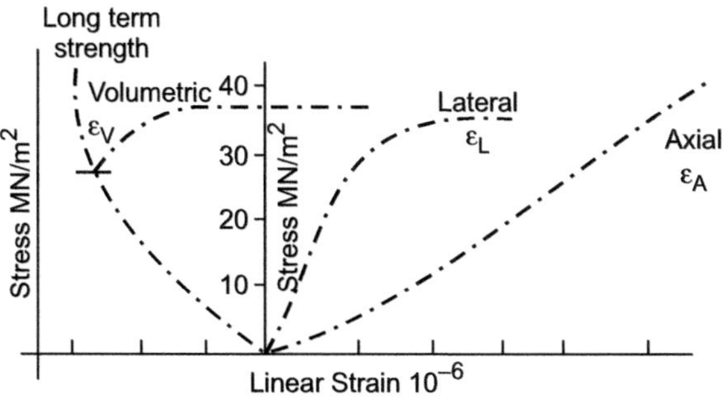

Figure 4.26 Long-term strain from the stress-strain diagram (Singh, 1977)

Log stress-log strain method

The stress-strain curves of many rocks like marble and limestone are concave towards the abscissa. Figure 4.27 depicts the log stress/strain curve method of determining long-term strength. The concave stress strain curve indicates that at the lower stress level for a small increase in stress there is more deformation of the rock resulting in compaction of voids within the rock. In the medium stress field, the stress-strain relationship is linear. At a higher stress field the curve again tilts towards abscissa showing the onset of failure of the specimen. This tendency of the stress-strain curve is more pronounced in the log stress and log strain curve using the same data. The

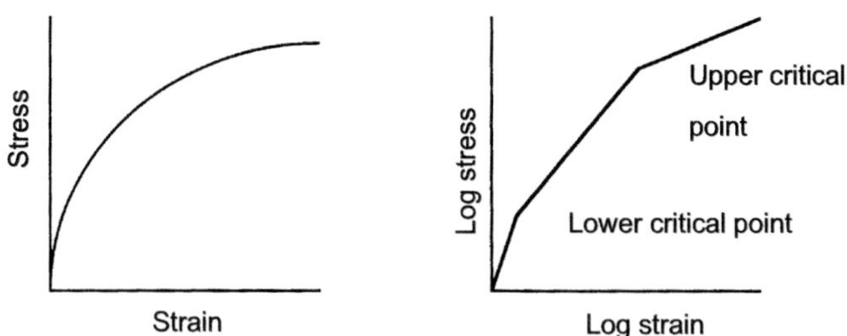

Figure 4.27 Log-log plot of the concave stress-strain curve (Singh, 1977)

stress-strain curve displays two pronounced kinks showing lower and upper critical points of the stress-strain curve corresponding to 15–25% and 70–90% stress of the compressive strength. The upper critical point indicates crack initiation in the sample which should be taken as the criterion for estimating the long-term strength of the rock.

Figure 4.28 shows the long-term strength of Sicilian marble, Wombayan marble and North Broken Hill ore as determined by the log stress and log strain method (Singh, 1977).

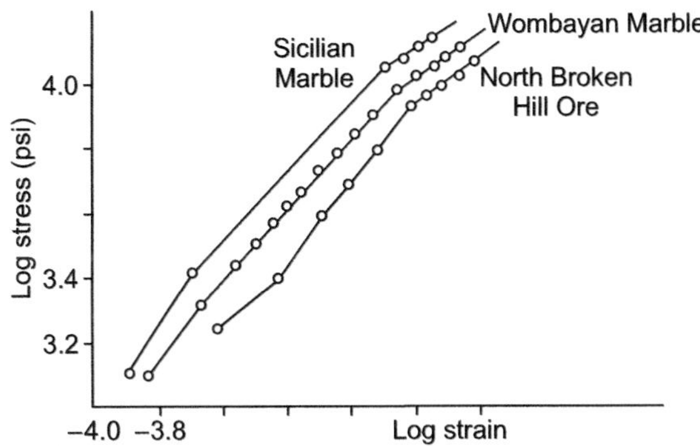

Figure 4.28 Long-term strength of some hard rocks (Singh, 1977)

Stress-strain rate method

The stress/axial strain curve loaded at a constant rate of loading has three distinct features.

1. Initially, strain increases rapidly with stress due to closing of micro-cracks and fissures.
2. In the second stage, the curve displays elastic behaviour if the sress rate is constant.
3. Rapid increase in the strain rate with a constant rate of stress causes unstable fracture.

The point of unstable fracture can be located on the strain rate/stress curve. This point is independent of the rate of loading (0.7 to 0.28 MPa sec).

Loading rate method

A specimen is loaded at a constant loading rate and a record of the axial and transverse strains are kept (Figure 4.29). The flexural point on the load time curve and where the lateral strain and the axial strain curves become non-linear is the long-term strength of the rock.

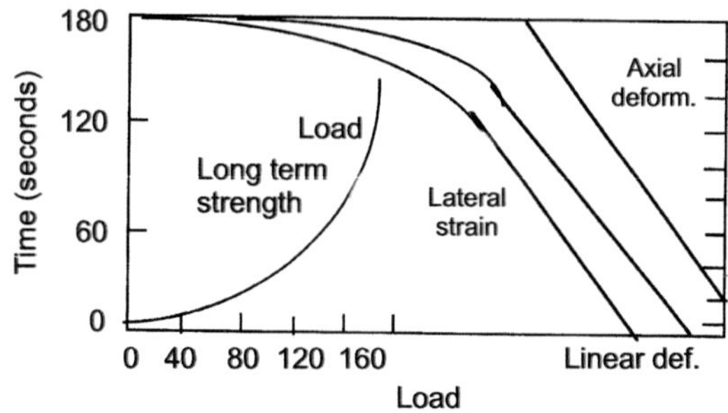

Figure 4.29 Loading rate method of determining long-term strength (Singh and Bamford, 1971)

Micro-seismic method

Micro-fracturing of rock during compression can be detected by measuring frequency of radiated elastic waves.

- At low stress, the micro-seismic activity increases due to closing of cracks.
- At intermediate stress, the micro-seismic activity is low.
- When the specimen is at the verge of failure, the micro-seismic activity increases again corresponding to the beginning of a rapid rate of development of micro-fractures (Figure 4.30).

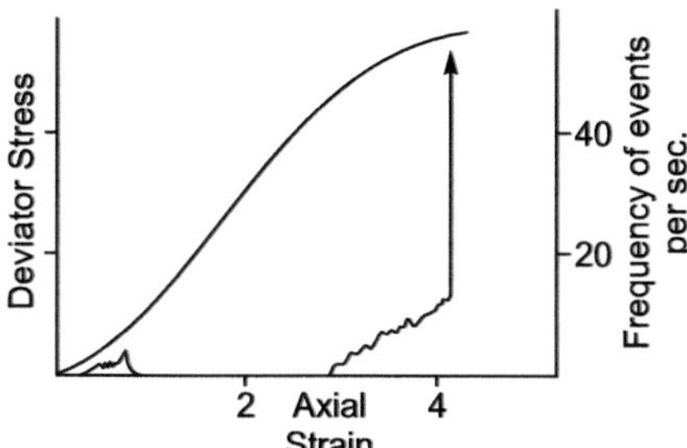

Figure 4.30 Micro-seismic method of determining long-term strength of rock

The upper critical point where micro-seismic activity increases is taken as an estimate of the long-term strength of the rock.

4.11 Conclusions

Many types of rocks exhibit time-dependent strain behaviour when loaded under constant state of stress. A typical rock type under constant compressive stress would show three stages of creep. The primary stage of the strain time curve is characterised by decreasing rate of creep, the secondary stage of creep shows a constant creep rate while the tertiary state of creep curve shows successive increasing creep rates. At the time of specimen failure, the tertiary creep rate becomes infinite.

It has been shown that the creep behaviour of various rocks can be simulated using rheological models.

Excavations in many rock types like halite, oil shale, potash and syenite display various stages of time-dependent strain behaviour. The primary stage of creep is observed during the excavation of a development roadway by the growth of a fracture or by creation of audible or micro-seismic noises. The secondary stage of creep has been observed in some potash mines extending over a period of 10 years showing a constant rate of creep. The tertiary state of creep in some underground or surface mining excavations involve from hundreds to ten of thousands of tonnes of rock material. The time for an impending failure to take place can be estimated from the analysis of the strain-time graph.

The determination of long-term strength of rock is based on the time-dependent properties of rocks. Direct methods of testing the long-term strength of rocks are protracted and tedious. Various indirect methods of evaluating long-term strength of rock have been devised that involve the understanding of the stress and deformation behaviour of rocks.

References

Attwell, P. and I.W. Farmer (1973). 'Fatigue Behaviour of Rock'. *International Journal of Rock Mechanics and Mining Sciences and Geomechanics Abstracts*, Vol. 10, No. 1, January, pp. 1–9.

Afrouz, A. and J. Harvey (1974). 'Rheology of rocks with soft to medium strength range'. *International Journal of Rock Mechanics and Mining Sciences and Geomechanics Abstracts*, Vol. 1, pp. 280–290.

Chong, K.P., P.M. Hoyt and J.W. Smith (1980). 'Effects of Strain rate on oil shale fracturing', *International Journal of Rock Mechanics and Mining Science*, 17, pp. 35–43.

Cogan J. (1976). 'An approach to creep behavior of failed rocks'. *US Rock Mechanics Symposium*, pp. 400–411.

Dorn, J.E. (1954). 'Some fundamental experiments in high temperature creep'. *Jour. Mech. Phys. Solids*, Vol. 19, pp. 17–42.

Dieterich, J.H. and G. Conrad (1967). 'Pre-seismic slip in a large scale friction experiment'. *US Geological Survey*, pp. 110–117.

Dusseault, M.A. and C.J. Fordham (1993). 'Time dependent behavior of rocks'. *Comprehensive Rock Engineering*, Vol. 3, pp. 119–147.

Federico, A., C. Fidelibus and G. Interno (2002). 'The prediction of landslide time to failure—A state of the art'. *Third International Conference on Landslides, Slope Stability and the Safety of Infrastructures*, 11–12 July 2002, Singapore, pp. 167–180.

Griggs, D.T. (1936). 'Deformation of rocks under high confining pressure'. *Journal of Geology*, Vol. 44, No. 5, pp. 541–577.

Griggs, D.T. (1940). 'Experimental flow of rocks under conditions favouring recrystallisation'. *Bulletin of Geological Society of America*, Vol. 51, pp. 1001–1022.

Hardy, H.R. Jr. (1967). 'Determination of inelastic parameters of geological materials from incremental creep experimentations'. American Institute of Mining, Metallurgical and Petroleum Engineers, Paper No. SPE 1707.

Lomntz, C. (1956). 'Creep measurements in igneous rocks'. *J. Geol.*, Vol. 64, pp. 473–479.

Misra, A.K. and S.A.F. Murrel (1965). 'An experimental study of the effect of temperature and stress on creep of rock'. *Geophysics, J. Abs. Royal Society*, Vol. 9, No. 5, July, pp. 509–535.

Munday, J.L.J., A.E. Mohammed and R.K. Dhir (1997). 'A criterion for predicting long-term strength of rock'. *Proceedings of the Conference on Rock Engineering*, Core U.K., 4–7 April, Newcastle Upon Tyne, pp. 127–135.

Murrel S.A.F. and A.K. Misra (1962). 'Time dependent strain or creep in rocks and similar non-metallic materials'. *Transaction of the Institution of Mining and Metallurgy*, Vol. 71, pp. 353–378.

Obert, L. (1965). 'Creep in model pillars, salt, trona and potash'. USBM RI 6703, pp. 23.

Phillips, D.W. (1931). 'The nature and physical properties of coal measures strata'. *Trans. Inst. Min. Engr.*, Vol. 80, Part 4, January, pp. 212–239.

Phillips, D.W. (1948). 'Tectonics in mining'—Part 8. *Colliery Engineering*, Vol. 25, Part 4, January, pp. 312–316.

Price, N.J. (1963). *A study of time strain behaviour of coal measures rock*. National Coal Board, MRE, Report No. 2332/1963.

Price, N.J. (1964). 'A study of time strain behaviour of coal measures rock'. *International Journal of Rock Mechanics and Mining Sciences and Geomechanics Abstracts*, Vol. 1, No. 2, March, pp. 277–303.

Robertson, E.C. (1964). 'Viscoelasticity of Rocks', in State of Stress in the Earth's Crust, W.R. Judd. (Ed), Elsevier, pp. 181–233.

Rummel, F. (1969). 'Studies of time dependent deformation of some granite and eclogite rock samples under uniaxial, constant compressive stress and temperatures upto 400°C'. *J. Geophysics*, 35, 17–42.

Saito, M. (1969). 'Forecasting the time occurrence of a slope failure'. *Sixth International Congress on Soil Mechanics*, pp. 537–541.

Singh, D.P. and W.E. Bamford (1971). 'The prediction and measurement of long-term strength of rocks'. *Proceedings of First Geomechanics Conference*, Australia, August, pp. 37–44.

Singh, D.P. (1975). 'A study of creep of rock'. *International Journal of Rock Mechanics and Mining Sciences and Geomechanics Abstracts*, Vol. 12, No. 9, pp. 271–275.

Singh, D.P. (1977). 'Long term strength of rocks'. *Colliery Guardian*, November, pp. 861–881.

Wiid, B.L. (1966). *The time dependent behavior of rock; considerations with regard to research program*, Report number Meg 24 of CSIR p. 71.

Williams, F.T. and M.A. Elizzi (1977). 'Creep properties of Sherburn Gypsum under Triaxial Loading'. *Proceedings of the Conference on Rock Engineering*, Core U.K., 4–7 April, Newcastle Upon Tyne, pp. 71–83.

Wawersik, W.R. (1972). 'Time dependent rock behavior in uniaxial compression'. *14th Rock Mechanics Symposium*, University of Pennsylvania, pp. 85–106.

Wawersik, W.R. (1974). 'Time dependent behavior of rock in compression'. *Third Congress of ISRM*, pp. 357–363.

Williams, F.T. and M.A. Elizzi (1976). 'An apparatus for determination of the Time-dependent Behavior of Rocks'. *International Journal of Rock Mechanics and Mining Sciences and Geomechanics Abstracts*, Vol. 13, No. 8, pp. 245–248.

Williams, F.T. and M.A. Elizzi (1976). 'Bending Creep Tests in Gypsum'. *Journal of Iraq Engr. Society*, Vol. 21, No. 1, Ser. 58, pp. 3–14.

Vutukuri, V.S. and R.D. Lama (1974). 'Mechanical properties of rocks'. Vol. 2, Chapter 9, *Time-dependent Properties of Rocks*, Trans Publisher, Clausthal, pp. 219–322.

Assignments

Question 1 (a) Draw an idealised creep curve for rock indicating three stages of creep.

(b) Describe the factors affecting creep results in the laboratory. Illustrate your answers with diagrams.

Question 2 The results of a time-strain test for a rock in uniaxial compressive stress field are given below:

Time (hours)	*Micro-strain 10^{-6} (mm/mm)*	*Time (hours)*	*Micro-strain 10^{-6} (mm/mm)*
0	1000	100	1360
10	1120	200	1520
25	1200	225	1600
40	1240	275	2000
50	1250	300	2400
75	1300		

Plot the strain-time graph and extrapolate the time at which the rock specimen will rupture.

Question 3 Give examples of creep curves for some practical materials as follows:

(i) Creep curve of slate
(ii) Creep curve for rock salt
(iii) Roof convergence measurements in an oil shale mine
(iv) Lateral deformation of pillars in potash mine.

Question 4 Define the long-term strength of rock. Why it is difficult to determine this parameter in laboratory by direct method? Describe two indirect methods of determining this property of rock.

Index Properties of Rocks

5 Chapter

5.1 Introduction

This chapter describes the tests carried out by engineers or geologists by portable or handheld equipment at a construction or mining site to assess the rock mass index properties. These index properties can be correlated to the strength and deformation properties of the rock material to provide design information.

5.1.1 Reasons for Developing Index Tests

The index tests are devised to overcome some of the difficulties encountered in the laboratory tests as follows.

(i) Laboratory testing of rock material is elaborate, time consuming and therefore, expensive.

(ii) Delay in assessment

(iii) Index tests are essentially field tests devised to obtain test results

- (a) without much specimen preparation
- (b) with portable equipment
- (c) correlated to strength and deformation properties for design calculations
- (d) some tests give representative of properties of rock-mass or *in situ* properties
- (e) sometimes open boreholes tests for lithological classification and structural mapping can also be correlated to the index properties.

5.2 Index Properties of Rock

(i) *Brazilian test* can be correlated to tensile strength used for classification of rocks by strength for *drillability, rock breakage* and *crushing* classifications.

(ii) *Point load index* for strength classification of rocks which can be used by geotechnical engineers for predicting strength and deformation properties of rock and applicable to the *design of mining excavations.*

(iii) *Dynamic impact strength test* (Pomeroy, 1955; Ghose et al, 1964) to estimate resistance of coal to degradation and applicable to *rock workability studies.*

(iv) *Cone indentor hardness test* developed by British Coal, Mining Research and Development Establishment for assessing strength of rock in underground coal mines.

(v) *Slake durability test* gives resistance of rock to weathering and, therefore, applicable to assess the durability of rock for near surface excavations or swelling of roof and floor strata in underground excavations.

(vi) *Schmidt hammer test*, a rebound test devised for field conditions and can be used for estimating strength and deformation characteristics of rock.

(vii) *Shore Scleroscope test*, a non-destructive test for classifying hardness and, therefore, selecting excavation machinery or blasting.

(viii) *Neutron-neutron logs* are carried out in the open borehole which can be correlated to rock density and point load index test. Table 5.1 summarises the index properties of rocks.

5.3 Brazilian Test

At a borehole site, a cylindrical sample of rock is roughly cut and selected in such a way that the thickness of the core approximately is half of the diameter. The load at failure can be related to the uniaxial tensile strength as follows:

$$\sigma_t = \frac{2P}{\pi D t} \tag{5.1}$$

where σ_t = uniaxial tensile strength
P = load at failure on a portable machine,
D = diameter of the core (m)
t = thickness of core (m)

- Diameter of core=minimum 54 mm
- Diameter-to-thickness ratio = 2 to 1
- Minimum number of tests=5
- Loading rate=6.5 mm/minute

Table 5.1 Index testing techniques for intact rock

	Loading geometry	*Parameters*	*Sample: Diameter*	*Sample: h/d ratio*	*Sample: No sample*	*Loading rate*	*Recommended loading geometry*	*Limitations*
Brazilian test	p, d	$\sigma_t = \frac{2P}{\pi dt}$	NX core 54 mm	0.5 – 1	5	0.64 cm/min	• Loading through curved jig with spherical seating • Angle of contact $= \frac{1}{6}$	Failure in biaxial stress field
Point load index	p, d	$I_s = \frac{P}{d^2}$ $\sigma_c = 24\ I_s$	50 mm	1 – 1.5	10 – 15	No standard loading rate		
Impact strength index	1.8 Kg, 20 blows, 4.45cm	$\sigma_{c2} = 1.46\ I - 60.45$ MPa	3 – 10 mm		6	Impact rate should not be faster than 1 per 2 sec		Results should be used to estimate resistance to degradation and workability of rock
Cone indenter hardness		D = deflection of strip I = depth of indentation $\sigma_c = K\frac{D}{I}$	• 12.5 mm × 12.5 mm × 6 mm or • 27 mm dia, and 12 mm thick	0.1:1	6		Minimum depth of indentation should be 0.2 – 0.25 mm	• Should not be tested near the edge of specimen • Generally applicable to rock with UCS < 150 MPa

contd.

Table 5.1 (contd.)

	Loading geometry	*Parameters*	*Sample*			*Loading rate*	*Recommended loading geometry*	*Limitations*
			Diameter	*h/d ratio*	*No sample*			
Schmidt hammer hardness		Advisable to use > 40 mm diameter					Core size < 40 mm can be applied to unshaped joint surface	Not advisable to use small dia. core size
Slake durability		SI (d) = $\frac{\text{Final dry weight}}{\text{Initial dry weight}}$	40–60 mm		10 lumps		Can be applied to unshaped lumps	Errors large for small values of slake durab. index
Shore hardness	Model C-2	Shore hardness scale 0–140	12–50 mm	Min. thick = 1 cm Surface = 1 cm^2	20 determinations	Indentation spacing 5 mm	Sticking surface circular = 0.1 to 0.4 D	Flat surface using No 1800 A10 abrasive powder

- Recommended loading geometry—through curved jig with spherical seating so that the loading area of contact is 9.4°
- Failure of specimen across loading platen

5.4 Point Load Index Test

A cylindrical core obtained from a borehole is approximately cut to the length to diameter ratio 1.5 to 1 and diametrically loaded against conical platens on a portable loading machine. The load at failure can be related to the point load index of rock as follows:

$$I_s = \frac{P}{d^2} \tag{5.2}$$

$$\sigma_c = 24\, I_s$$

where

P = load at failure

d = diametrical distance between conical platens at failure (m)

I_s = point load index of rock

σ_c = uniaxial compressive strength of rock (MPa)

I_{s50} = point load index for 50 mm diameter core (Fig. 5.1 - size correction)

Idealised conditions for this test are as follows.

- Portable loading machine
- Use of calibration chart for size correction (Figure 5.1)
- Minimum core diameter 50 mm
- Length-to-diameter ratio 1.5 to 1
- Number of samples – 10 to 15
- No standard rate of loading
- Loading platen 60° conical platen with 5 mm curvature tip

The point load index for a given test specimen (I_s) can be correlated to I_{s50} as per the following equation (Brook, 1985; Hassani, et al., 1981)

$$\log_{10} I_{s50} = 0.256 + \log_{10} I_s - 1.008e^{-0.027D} \tag{5.3}$$

5.4.1 Relationship between Point Load Index and Strength Parameters of Coal Measures Rocks

The point load index test can be conveniently used for quick estimation of rock strengths such as

- uniaxial compressive strength
- uniaxial tensile strength
- triaxial parameters

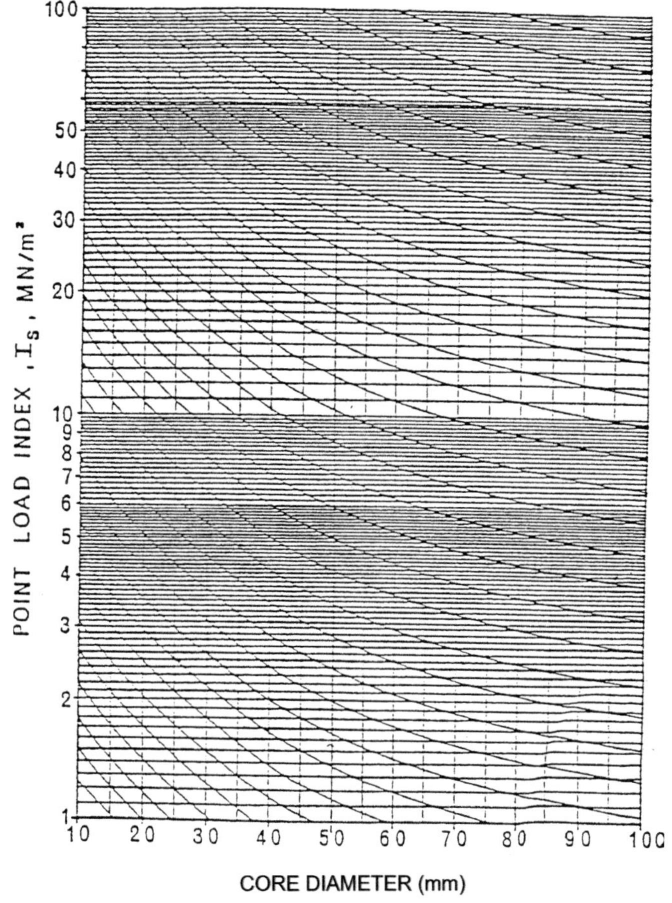

Figure 5.1 Size correction chart for point load index test (Hassani et al., 1981)

The correlation of the point load index number with various strength and deformation parameters is given below.

(a) Uniaxial compressive strength

$$\sigma_c = 29\ I_{s50} \tag{5.4a}$$

where

σ_c = uniaxial compressive strength (MPa)

I_{s50} = point load index for 50 mm diameter specimen

(b) Tensile strength

$$\sigma_t = 0.095\ \sigma_c = 2.76\ I_{s50} \tag{5.4b}$$

where

σ_t = uniaxial tensile strength (MPa)

σ_c = uniaxial compressive strength (MPa)
I_{s50} = point load index for 50 mm diameter specimen

(c) Triaxial strength

$$\sigma_1 = \sigma_c\left[\frac{\sigma_m}{\sigma_c} + \frac{\tau_m}{\sigma_c}\right] = 29I_{s50}\left[\frac{\sigma_m}{\sigma_c} + \frac{\tau_m}{\sigma_c}\right] \quad (5.5)$$

$$\sigma_3 = \sigma_c\left[\frac{\sigma_m}{\sigma_c} - \frac{\tau_m}{\sigma_c}\right] = 29I_{s50}\left[\frac{\sigma_m}{\sigma_c} - \frac{\tau_m}{\sigma_c}\right]$$

Triaxial strength may now be obtained by the following functional relationship:

$$\tan\beta = \frac{(K-1)}{2\sqrt{K}}$$

$$c = \frac{\sigma_c(1-\sin\phi)}{2\cos\phi} \quad (5.6)$$

where

c = cohesive strength
K = triaxial stress factor = $\tan\beta = \dfrac{\sigma_1}{\sigma_3}$
ϕ = internal angle of friction
σ_m = maximum principal stress
τ_m = maximum shear stress

Thus, point load index tests have the following advantages.

(i) Cost-effectiveness
(ii) Time saving
(iii) Avoidance of sample deterioration in prolonged laboratory tests.

5.5 Impact Test by MRDE

This test is devised by Pomeroy (1955) to estimate the natural strength of coal after it has been removed from the coal seam in order to assess the degradability of coal that controls its subsequent breakage. This may occur in loading coal on to the face conveyor, during transfer from one conveyor to another or into storage bunkers, during loading into trams and skips and during screening and washing.

A standard steel plunger of 1.8 kg and 42 mm in diameter is dropped on the coal fragments larger than 10 mm from a standard distance in a 45 mm diameter cylindrical vessel closed at the bottom by a screwed cap

(Figure 5.2). A steel cap, through which the plunger handle passes, is fitted top the cylinders. It has two purposes.

- Prevents dust from escaping, and
- Allows the plunger to drop for a distance of 300 mm.

Figure 5.2 MRDE impact strength apparatus (Pomeroy, 1955)

At least 20 blows are given to 100 gm of coal sample which is 3 mm to 10 mm in size range at a rate not faster than 1 impact every two seconds. The percentage of coal that is broken down to pass through the 5 mm sieve is used to calculate the coefficient of strength. The weight of coal in grams remaining on the 3 mm sieve, including the material trapped in the sieve, is the impact strength index of the coal.

The uniaxial compressive strength of coal can be estimated as follows:

$$\sigma_{c2} = 1.46 \times I - 60.45 \text{ MPa} \tag{5.7}$$

Figure 5.3 shows the correlation between the compressive strength of coal and the impact strength index. Results should be used to estimate resistance to degradation and workability of rock.

Standard test conditions:

- Specimen size: 3 – 10 mm
- Number of samples: 5 or 6
- Impact rates: not faster than 1 or 2 per second

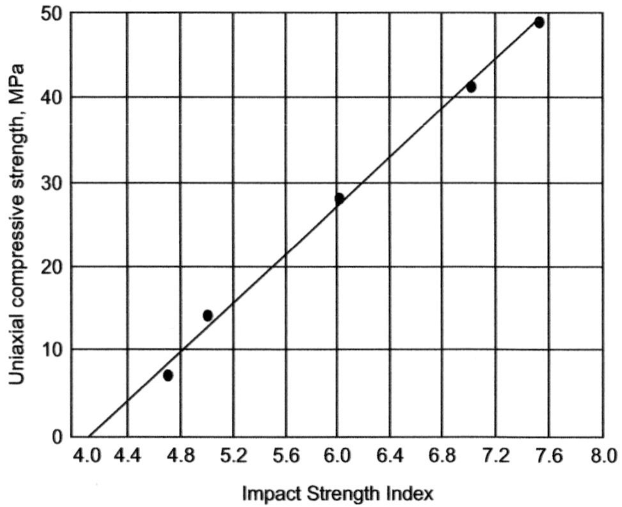

Figure 5.3 Relationship between compressive strength of coal and impact strength index number

5.6 Schmidt Hammer Rebound Test

Schmidt hammer is a portable and inexpensive device which can be used to obtain rapid results on intact rock and rocks joints in the rock masses surrounding the mining excavations and also on prepared samples in the laboratory. The test consists of determining the hardness of rock material by measuring the degree to which a steel hammer will rebound from a prepared rock surface. A rebound number is indicated which has been shown to correlate with uniaxial compressive strength taking dry density of rock into considerations (Hucka, 1965; Deere and Miller, 1966).

$$I_{SH} = 0.5\,\sigma_c \tag{5.8}$$

I_{SH} = Schmidt rebound number

σ_c = uniaxial compressive strength (MPa)

Coefficient of correlation = 0.86

5.6.1 Method of Operation

The Schmidt hammer, a spring-loaded impact plunger, is placed against a smooth rock surface or the prepared surface of a laboratory sample and depressed against the surface. This compresses the spring, thereby automatically releasing the energy stored in the spring and causing an impact of the hammer against the impact plunger. The height of rebound of the hammer mass is shown by the rider on a scale and its value is taken as the measure of hardness. A minimum of 20 rebound readings is taken and the lowest and highest readings are discarded. Mean value of the rebound number is taken as Schmidt hardness number.

Schmidt hammer models are available in different levels of impact energy; the most popular instrument for sedimentary rock is the L-type hammer with an impact energy of 0.74 Nm. Figure 5.4 shows the relationship between Schmidt's hardness number and the uniaxial compressive strength.

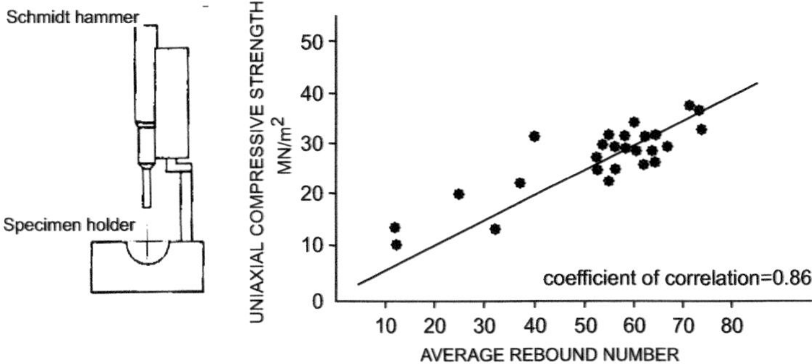

Figure 5.4 Relationship between Schmidt hammer rebound number and uniaxial compressive strength (Singh and Cassapi, 1984)

The following conclusions can be drawn from the Schmidt hammer index tests:

- The Schmidt hammer tests give an acceptably accurate assessment of rock strength.
- Some difficulties have been experienced in comparing the strength of the same rock type at different field locations.
- The Schmidt hammer tests conducted in the laboratory on rectangular specimens and borehole cores predicted much lower rebound numbers than those obtained in rock mass *in situ.*

5.6.2 Correlation of Strength and Deformation Parameters From Schmidt Hammer Tests

The following empirical equations (developed by Singh and Cassapi, 1984) can be used to estimate the strength and deformation parameters of the coal measure rocks as follows:

$$\sigma_c = 2\, I_{SH}$$
$$\sigma_t = 0.23\, I_{SH} - 0.81$$
$$c = 0.305\, I_{SH} + 8.17$$
$$K = 0.031\, I_{SH} + 2.98 \qquad (5.9)$$
$$\tan\beta = \frac{\sigma_1}{\sigma_3}$$

$$\sigma = \sigma_c + K\,\sigma_3$$
$$= 2\,I_{SH} + (\,0.031\,I_{SH} + 2.98)\,\sigma_3$$

or

$$\tau = c + \sigma_n \tan\phi$$
$$\tau = (\,0.305\,I_{SH} + 8.17)) + \sigma_n \tan\phi$$
$$\tan\phi = \frac{[1.98 + 0.03 I_{SH}]}{\sqrt{(2.98 + 0.031 I_{SH})}}$$
$$E = (25.37 + 0.5\,I_{SH})\,(GPa)$$
$$\frac{1+\upsilon}{E} = (\,159.28 - 3.36 I_{SH}) \times 10^{-12} \tag{5.10}$$

where

σ_c = uniaxial compressive strength (MPa)
I_{SH} = Schmidt hammer hardness index
c = cohesive strength (MPa)
τ = shear strength (MPa)
σ_n = normal stress (MPa)
ϕ = internal angle of friction
K = triaxial stress factor
υ = Poisson's ratio
E = Young's modulus (GPa)

5.7 Slake-Durability Test

The purpose of this test is to determine the weather resistance and degradation of argillaceous rocks such as shale, mudstone, seat-earth and siltstone under impact. The test can also be extended to other rock types as an aid to rock classification, for selection and quality control of materials for rock fills, road and concrete aggregates, in selecting plant, equipment and technique for rock excavations and predicting the problems of excavation stability and selection of support.

5.7.1 Method of Test

In essence, the slake durability test is conducted on oven dry rock lumps which are rotated in a test drum of a standard sieve mesh. The drum is half-immersed in a water bath at 20°C. During the test, the finer product of the slaking passes through the mesh into the water bath. The slake durability index (I_d) is the percentage ratio of the final to the initial dry weights of the rock in the drum.

The slake durability index values range from 30 to 100 for coal measure rocks and percentage errors should be within 3% for most cases. The results

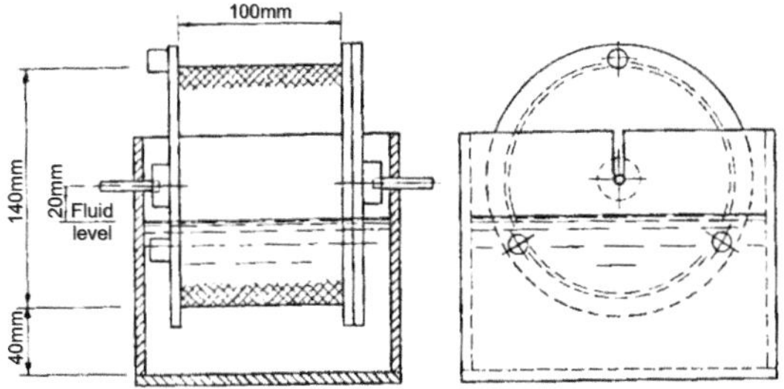

Figure 5.5 Slake durability apparatus (Franklin and Chandra, 1972)

of slake durability tests carried out in the laboratory for coal measures rocks are shown in Figure 5.6 which indicates non-linear variation of slake-durability index with uniaxial compressive strength.

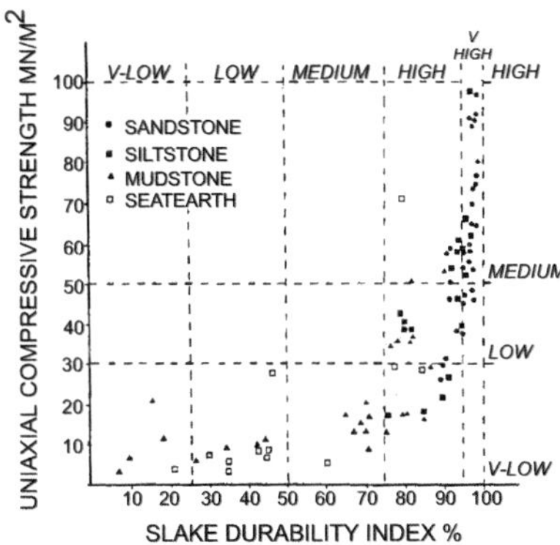

Figure 5.6 Variation of uniaxial compressive strength with slake durability index of coal measure rocks (Singh *et al.*, 1983)

5.8 Borehole Logging and Correlation of Strength and Deformation Parameters

It is often necessary to drill open boreholes for carrying out investigations and for assessing rock behaviour at an exploration site in order to obtain full range of borehole log information as follows:

(i) Density log for lithological classification of strata
(ii) S-wave and P-wave velocity logs

These logs are necessary to derive dynamic moduli of rocks for excavation modelling or via empirical correlation to predict strength parameters. It may be noted that precise continuous S-wave logging has proved to be difficult in near surface mud rocks, which are not particularly supportive of S-waves. It is often necessary to make assumptions, about P-wave to S-wave velocity so that elastic modulus graph can be derived.

Full wave train velocity logging techniques can also be used to obtain acoustic amplitude and attenuation logs to be computed. In this way acoustic discontinuities can be readily identified which are associated with physical discontinuities in the rock mass.

5.8.1 Relationship between Openhole Wireline Logs, Elastic Parameters of Rock

From the knowlege of the P-wave and S-wave velocities obtained from the open borehole log, the elastic parameters of rock can be obtained as per following equations:

$$\upsilon = 0.5\left[\frac{\left\{\left(\frac{V_P}{V_S}\right)^2 - 1\right\}}{\left\{\left(\frac{V_P}{V_S}\right)^2 + 1\right\}}\right] \tag{5.11}$$

$$E = 2\rho V_S^2(1+\upsilon)$$

$$\mu = \rho V_S^2$$

$$G = \rho\left[\frac{V_S}{V_P}\right]^2 - \frac{4}{3}\mu \tag{5.12}$$

where

μ = shear modulus
G = bulk modulus
ρ = bulk density
υ = Poisson's ratio
E = Young's modulus
V_P = P-wave velocity
V_S = S-wave velocity

Establishment of empirical correlation between geotechnical and geophysical database is important for quick site investigation using open borehole logs. It is possible to estimate intact rock strength by determining point load index test at close intervals on the borehole cores and correlate it with the neutron-neutron logs and discontinuity log in the borehole (Figures 5.7 and 5.8).

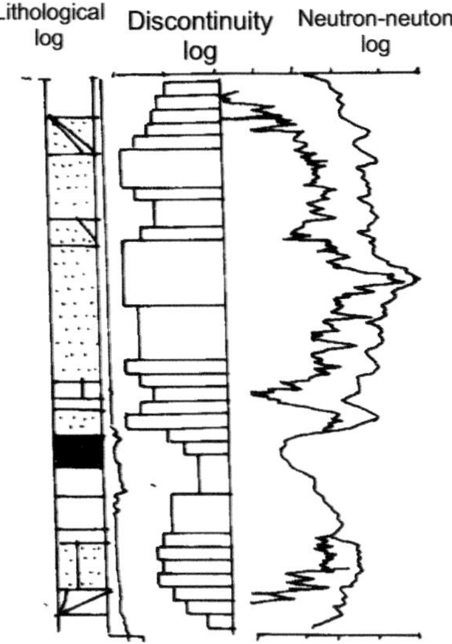

Figure 5.7 Section of raw point load index data with neutron-neutron log response and also discontinuity spacing data

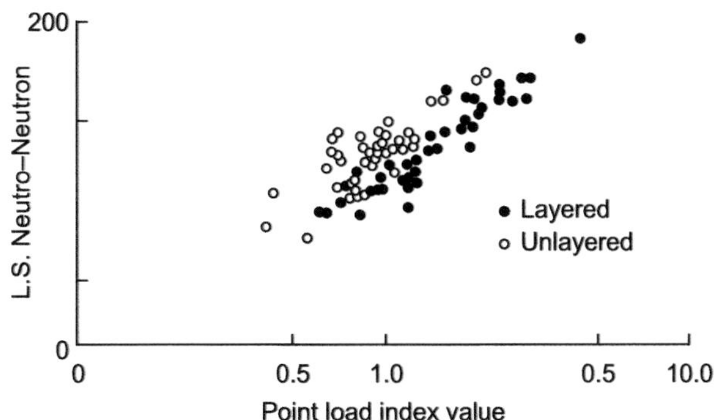

Figure 5.8 Correlation between LS neutron-neutron log and I_s index test (Ecklington *et al.*, 1982)

5.9 Rock Hardness and Abrasiveness

This is a function of hard mineral composition and the competency, strength and bonding capacity of the matrix material. It can be shown that the harder the rock, the greater is the resistance to penetration by the cutting tool.

5.9.1 Abrasiveness

The abrasive wear of the cutting tools due to rock/pick interaction is important as the cost and delays incurred for the replacement of the worn out parts reflect upon overall machine performance.

Various factors affect the abrasiveness of rock, particularly the following:

(a) Mineral composition
(b) Hardness of mineral constituents
(c) Grain size and shape
(d) Type of the matrix material
(e) Physical properties of rocks including strength, hardness and toughness

5.9.2 Assessment of Hardness

Some of the following methods have been used in the determination of the hardness of geological materials.

- Moh's hardness
- Roswell Scale
- CERCHAR index
- Schmidt hammer hardness
- Shore scleroscope hardness
- Cone indentor hardness
- Schimazek index

5.10 The Shore Scleroscope

The Shore scleroscope was originally used to determine the hardness of metals and has been improved to measure the rebound height of a small-indentor diamond which falls from a fixed height and rebounds freely from a specially prepared surface of the rock specimen (Figure 5.9).

At least 20 determinations should be made with at least 5 mm space between each indentations and only one test performed on the same spot. A mean hardness number is determined which can be correlated to the uniaxial compressive strength of rock. Figure 5.10 presents the correlation between Shore hardness index and uniaxial compressive strength.

5.11 National Coal Board Cone Indenter

The NCB cone indenter (Figure 5.11) was developed at MRDE in 1972 to determine rock hardness.

Figure 5.9 Shore scleroscope

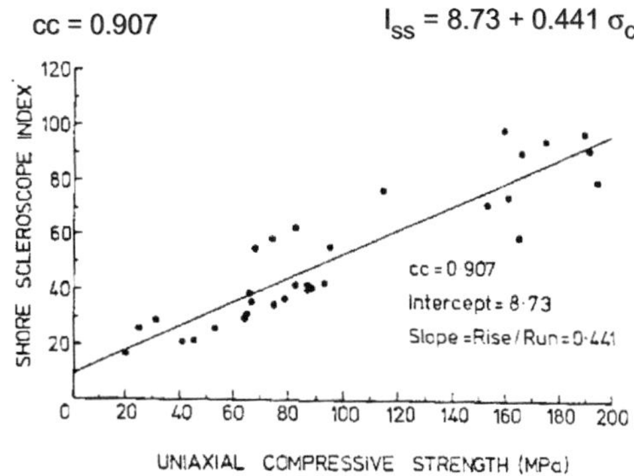

Figure 5.10 Correlation between Shore scleroscope hardness and uniaxial compressive strength

Figure 5.11 NCB cone indentor

It measures the penetration of a tungsten-carbide cone on a prepared core specimen 38 mm in diameter and 10 mm thick, under a normal force as follows:

Weak material	— 14 N
Strong	— 40 N
Very strong	— 140 N

The tungsten carbide cone is applied with a micrometer screw and thimble and a dial gauge is used to measure the deflection of a spring leaf from which normal force could be determined.

Figure 5.12 shows the correlation between the cone indentor hardness and uniaxial compressive strength.

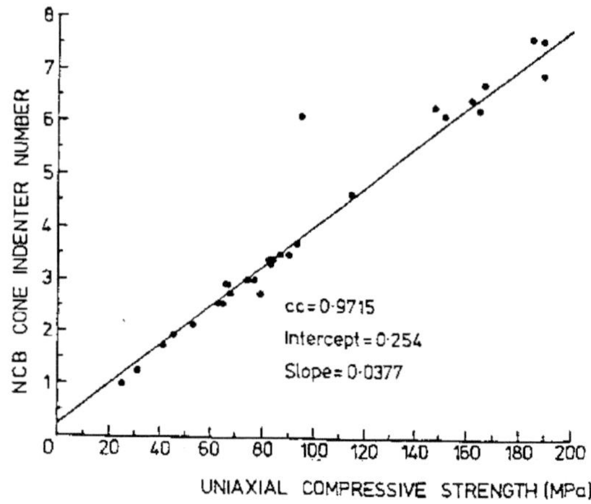

Figure 5.12 Correlation between the cone indenter hardness and uniaxial compressive strength

5.12 CERCHAR Index Test

This test is extensively used by the French coal mining industry and mining machinery manufacturers for defining abrasiveness potential of rocks.

The test consists of scratching the surface of the rock specimen for a distance of 10 mm with a sharp, 90° conical, steel stylus made of En24 and En25 steel and accurately heat treated to 610 Vickers, under a normal load of 70N.

The abrasiveness of the rock is determined by measuring the resultant wear flat worn on the stylus. The wear flat is measured by a travelling microscope to which is attached a dial micrometer graduated in 0.01 mm. The value of 0.425 mm diameter wear flat indicates the CERCHAR index of

4.25, indicating a highly abrasive rock. The accuracy of the test relies on accurate measurement of the wear flat and the accuracy of heat-treatment of the stylus. Table 5.2 presents abrasiveness classification and Table 5.3 gives typical abrasiveness index for some selected rocks.

Table 5.2 Abrasiveness classification of rocks (Atkinson and Cassapi, 1984)

Classification	*CERCHAR Index*	*Rock Type*
Extremely abrasive	> 4.5	Hornblende gneiss, pegmatite, Pennant sandstone
Highly abrasive	4.25–4.5	Amphibolite, granite
Abrasive	4.0–4.25	Granite, gneiss, schist, Pyroxene Darleydale sandstone
Moderately abrasive	3.5–4.0	Sandstone and siltstone
Medium abrasiveness	2.5–3.5	Gneiss, Californian granite
Low abrasiveness	1.25–2.5	Portland sandstone
Very low abrasiveness	<1.25	Limestone

Table 5.3 Typical examples of abrasive rocks (Atkinson and Cassapi, 1986)

Rock type	*Index*
Limestone	0.2–1.2
Portland sandstone	1.5
Diabase and granite	3.0–4.2
Californian granite	3.5
Dolerite	3.5
Marble	4.16
Darleydale sandstone	>4.5
Marble	4.16
Bio-cordite gneiss	4.56
Amphilobite	4.48
Pegmatite granite	4.95

5.13 Deduction of the Fracture Toughness Indices from Rock Hardness

Figure 5.13 presents a graph of fracture toughness indices and the cone indenter hardness for four different rocks, i.e. Welsh limestone, Newhurst granite, fine and coarse grained sandstone. A statistical regression analysis showed that the linear relationships between fracture toughness and NCB cone indenter number were reliable because high correlation coefficients up to 0.96 to 0.97 for K_{IC}-I and K_{IIC}-I curves respectively were obtained. These

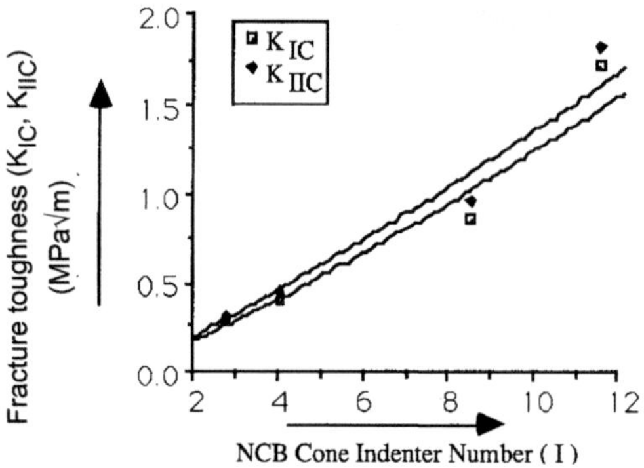

Figure 5.13 Correlation between the cone indenter hardness and fracture toughness

relationships can be expressed by the following equations (Singh and Sun, 1989):

$$K_{IC} = -0.24 + 0.155 \times I$$
$$K_{IIC} = -0.023 + 0.165 \times I \quad (5.13)$$

where

K_{IC} = Mode I fracture toughness (MPa√m)
K_{IIC} = Mode II fracture toughness (MPa√m)
I = Cone indenter hardness number

This is a significant result when consideration is given to the wide range of grain size of the rocks investigated.

Similarly, the Shore scleroscope hardness is also closely related to the fracture toughness (K_{IC} and K_{IIC}) of rock (Singh and Sun, 1989) as indicated by the correlation coefficients of 0.98 in the following relationships:

$$K_{IC} = -0.071 + 0.022 \times SH \quad (5.14)$$
$$K_{IIC} = -0.046 + 0.023 \times SH$$

where SH is the Shore hardness number.

Figure 5.14 presents the empirical relationship between the fracture toughness and the Shore scleroscope hardness number.

5.14 Conclusions

Laboratory tests discussed in previous chapters are elaborate, protracted and time-consuming and, therefore, take considerable time from taking the sample to obtaining the results. Index tests are designed to be carried out at field sites or on small specimens collected on the field site to obtain rapid information to assist in decision-making. Index tests, therefore, are very important in any rock mechanics investigation.

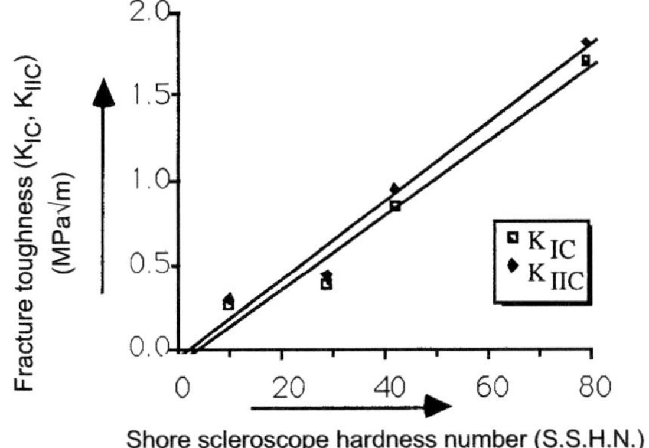

Figure 5.14 Correlation between Shore scleroscope hardness and fracture toughness

References

Al-Ameen, S.I. and M.D. Waller (1992). 'Dynamic impact abrasion index for rocks'. *International Journal of Rock Mechanics and Mining Sciences and Geomechanics Abstracts,* Vol. 29, No. 6, pp. 555–560.

Alpan, I. (1957). 'An apparatus for measuring swelling pressure in expansive soils'. 4th Int. Soil. Mechanics Conference, Filn Engi. 1, 3, 5

Atkinson, T., V. Cassapi, and R.N. Singh (1986). 'Assessment of Abrasive Wear Resistance Potential of Rock Excavation Machinery'. *Int J. of Mining and Geological Engineering,* July, Vol. 4, No. 3, pp. 151–163.

Atkinson, T. and V. Cassapi (1984). 'The prediction and reduction of abrasive wear in mine excavation machinery'. International Conference on Tribilogy, in Mineral Excavations, War on Wear, Institute of Mechanical Engineering, Nottingham University, pp. 165–174

Bieniawski, Z.T. (1974). 'Estimating the strength of rock materials'. *J. South African I. Min. Met.* pp. 312–320.

Bieniawski, Z.T. (1975). 'The point load index test in geotechnical practices'. *Engineering Geology,* Vol. 9, pp. 1–11.

Broch, E., and J.A. Franklin (1972). 'The point load index test'. *International Journal of Rock Mechanics and Mining Sciences and Geomechanics Abstracts,* Vol. 9, pp. 669–697.

Brook, N. (1977). 'A method of overcoming both size and shape correction in point load testing'. *Proceedings of Rock Engineering,* Newcastle-upon-Tyne, 4 to 7 April, pp. 53–70.

Brook, N. (1985). 'The equivalent core diameter method of size and shape correction in point load testing'. *International Journal of Rock Mechanics and Mining Sciences,* Vol. 22, pp. 61–70.

Brown, E.T. (1981). 'Suggested methods for determining hardness and abrasivness of rocks, rock characterisation, testing and monitoring'. *ISRM Suggested Methods,* Pergamon Press, Oxford, pp. 95–105.

Cassapi, V.B. (1987). 'Application of rock hardness and abrasive indexing to rock excavation equipment selection'. Ph. D. Thesis, University of Nottingham.

Carter, P.G. and M. Sheddon (1977). 'Comparison of Schmidt hammer, point load index and unconfined compressive test in carboniferous strata'. *Proceedings of Rock Engineering*, 4–7 April, Newcastle Upon Tyne, pp. 197–207.

Das, B. (1974). 'Vicker hardness concept in the light of Vicker's Impressions'. *International Journal of Rock Mechanics and Mining Sciences*, Vol. 11, pp. 85–89.

Deere, D.V. and R.P. Miller (1966). 'Engineering classification and index properties for intact rocks'. University of Illinois, U.S. Department of Commerce, National Technical Information Service.

Duncan, N., M.H. Dunne and S. Pitty (1969). 'Swelling characteristics of rocks'. *Water Power*, May, pp. 185–192.

Ecklington, P.A.S., P. Stouthammer and J.R. Brown (1982). 'Rock strength predictions from wireline logs'. *International Journal of Rock Mechanics and Mining Sciences*, Vol. 19, pp. 91–97.

Evans, I. and C.D. Pomroy (1966). *Strength, Fracture and Workability of Coal*, Pergamon Press, Oxford, p. 132 and p. 277.

Franklin, J.A. and R. Chandra (1972). 'The slake durability test'. *International Journal of Rock Mechanics and Mining Sciences*, Vol. 9, p. 325.

Franklin, J.A., U.W. Vogler, J. Szlavin, J.M. Edmond and Z.T. Bieniawski (1979). 'Suggested methods for determining water content, porosity, density, absorption, relative properties, swelling and slake durability'. ISRM Commission on Standardisation of Laboratory and Field Tests, *International Journal of Rock Mechanics and Mining Sciences*, Vol. 16, pp. 141–156.

Ghose, A.K., D. Barat, and S. Bagchi (1964)., 'Some preliminary studies on the Impact Strength Index of Indian Coals', Jl. of Mines, Metals and Fuels, 12, pp. 153-155, 164.

Ginenski, S.F. and L.J. Lee (1965). 'Comparison of laboratory swell tests to small scale field tests'. *Proc. Int. Res: Eng. Conf. on Expansive Clay Soils*, Texas. pp. 108–119.

Greminger, M. (1982). 'Experimental Studies of the Influence of Rock Anisotropy on Size and Shape Effect in Point Load Testing'. *International Journal of Rock Mechanics and Mining Sciences*, Vol. 19, pp. 241–246.

Hucka, V.A. (1965). 'A rapid method of determining the strength of rock in situ'. *International Journal of Rock Mechanics and Mining Sciences*, Vol. 2, pp. 127–134.

Haramy, K.Y. and M.J. De Marco (1985). 'Use of Schmidt hammer for rock and coal testing'. *26th US Rock Mechanics Symposium*, Rapid City, South Dakota, pp. 549–555.

Hassani, F.P., M. Scoble and B.N. Whittaker (1981). 'Application of Point Load Index Test to Strength Determination and Proposals for a New Size Correction Chart'. *21st U.S. Rock Mechanics Symposium*, Rolla.

ISRM (1985). 'Suggested methods for determining point load strength'. *International Journal of Rock Mechanics and Mining Sciences*, Vol. 22, pp. 51–62.

Janach, W. and A. Merminod (1982). 'Rock abrasivity test with a modified Schmidt hammer'. *International Journal of Rock Mechanics and Mining Sciences and Geomechanics Abstracts*, Vol. 19, pp. 43–45.

Kolek, J. (1958). 'An appreciation of Schmidt hammer'. *Mag. Conc. Res.*, Vol. 10, No. 28, pp. 27–36.

Nascimento, U., R. Oliveira and R. Gracia (1968). 'Rock swelling test'. *Proceedings of Int. Symposium in Rock Mechanics*, Madrid.

National Coal Board (1972). 'The cone indenter'. *Mining Research and Development Establishment*. Handbook No. 5, Stanhope Bretby, Burton on Trent.

Pells, P.J.N. (1975). 'The use of point load index test in predicting the compressive strength of rock materials'. *Australian Geomechanics Journal,* Vol. 95, No. 1, pp. 54–56.

Pomeroy, C.D. (1955). Simple methods for the assessment of Coal Strength, MRE Report No. 2022.

Poole, R.W. and I.W. Farmer (1980). 'Consistency and repeatability of Schmidt hammer rebound data during field testing'. *International Journal of Rock Mechanics and Mining Sciences,* Vol. 17, pp. 167–171.

Protodyakanov, E.I. (1950). Ugol. Vol. 25, p. 20.

Rabia, H. and N. Brook (1977) 'The Shore hardness of rock'. *International Journal of Rock Mechanics and Mining Sciences,* Vol. 11, pp. 335–336.

Read, J.R.L., P.N. Thornton and W.M. Regan (1980). 'A rational approach to the Point Load Test'. *3rd Australian New Zealand Conference on Geomechanics,* Wellington, Pt 2, 11–16 May, pp. 2–35 to 2.39.

Reichmuth, D.R. (1968), 'Point load testing of brittle materials to determine tensile strength and relative brittleness. *Proceedings 9th International Symposium in Rock Mechanics,* University of Colorado, Denver, pp. 134–159.

Sauna, M. and T.J. Peters (1982). 'The CERCHAR abrasivity index and its relation to rock mineralogy and petrography'. *Rock Mechanics,* Vol. 15, pp. 1–7.

Schmidt Hammer Type L operating instructions (1960). *Proc., S.A.,* P.O. Box 158, Zurich, Switzerland.

Schmidt, R.L. (1972). *Derivability studies—percussive drilling in the field,* US Bureau of Mines, RI 8684.

Sheorey, P.R., D. Bharat, M.M. Das, K.P. Mukherjee and B. Singh (1987). 'Schmidt hammer rebound data for estimation of large scale in situ strength'. *International Journal of Rock Mechanics and Mining Sciences,* Vol. 21, pp. 39–42.

Shore Instrument and Manufacturing Company, Inc., 80 Commercial Street, Freeport, New York, 11520, USA.

Singh, R.N. (1983). Testing of rock samples from under water trenching operations off Folkstone for the Central Electricity Generating Board, December, pp. 1–101.

Singh, R.N., F.P. Hassani and P.A.S. Elkington (1983). 'The application of strength and deformation index testing to the stability assessment of coal Measures excavations'. *24th U.S. Symposium on Rock Mechanics,* June, pp. 599–609.

Singh, R.N. and V.B. Cassapi (1984). 'Application of Predevelopment Mining Data to Mine Design in Remote Mining Areas'. *J. of Mines, Metals & Fuels,* April–May, pp. 165–172.

Singh, R.N. and G.X. Sun (1989). 'The relationship between fracture toughness, hardness indices and mechanical properties of rock. Nottingham University, *Mining Department Magazine,* Vol. 39, pp. 49–62.

Singh, R.N. (1988), 'Index and toughness testing of some selected rocks'. Special report to Geoffery Walton Practice. Consulting Mining Engineers and Geologists, Charlesbury, Oxford, September, p. 12.

Stacey, T.R. and C.R. Page (1986). 'Practical handbook for Underground Rock Mechanics'. *Trans. Clausthal,* pp. 77–78.

Sundae, L.S. (1974), 'Effect of specimen volume on apparent strength of three igneous rocks'. US Bureau of Mines RI 7846, Washington.

Szlavin, J. (1974), 'Relationship between some physical properties of rock determined by laboratory tests'. *International Journal of Rock Mechanics and Mining Sciences,* Vol. 11, pp. 57–66.

West, G. (1989), 'A review of rock abrasiveness testing for tunneling'. *International Journal of Rock Mechanics and Mining Sciences*, Vol. 26, pp. 151–160.

Wijk, J. (1980), 'Sclerograph measurements on rock materials'. *Geotechnical. Testing Journal*, Vol. 3, pp. 55–65.

Wright, D.N. (1996), 'The development of rock classification system for use with diamond tool. PhD thesis, University of Nottingham, May, 200 pp.

Assignments

Question 1 In a point load index test, the load at failure is 0.015 N and the distance 'D' between the two platens at the time of the failure is 0.0536 m. Calculate the point load index and estimate (i) I_{s50}, (ii) the uniaxial compressive strength, (iii) the tensile strength, (iv) the cohesive strength and (v) the triaxial stress factor K, if the internal angle of friction of rock is 37°.

Question 2 In a Schmidt hammer hardness test on coal measures rocks, the rebound values were as follows:

32, 39, 37, 41, 36, 35, 39, 37, 37, 32, 39, 38, 33, 40, 38, 39

Estimate (i) the uniaxial compressive strength of the rock, (ii) the cohesive strength, (iii) the internal angle of friction, (iv) the triaxial stress factor and (v) the equation of the peak envelope.

Large-scale *in situ* Testing of Rock Mass

6 Chapter

6.1 Introduction

The mechanical properties of rock mass differ greatly from that of the rock material due to the presence of micro fractures, geological features such as faults, stratification, joints, presence of water, scale effect and regional variations in lithology. *In situ* measurements are, therefore, mandatory in rock mechanics investigations for one of the following purposes.

(i) *In situ* state of stress and rock mass properties are determined to provide basic information for the design of structures in and on rock.

(ii) Continuous monitoring of a rock structure during construction and operational stages is necessary to determine the safety status of the structure and to check with the design calculations.

6.2 Types of *in situ* Tests

(a) Large scale *in situ* tests are undertaken to evaluate the following properties.

  (i) Modulus of deformation of rock mass, together with its regional and directional variations

  (ii) *In situ* compressive, triaxial compressive and shear strength of rock

  (iii) Residual stress in the vicinity of rock structures

(b) Borehole instrumentation for rock mechanics tests

  (i) Stress measuring instruments

  (ii) Modulus of deformation instruments

  (iii) Displacement measuring instruments

6.3 Pressure Chamber Tests

6.3.1 Cylindrical Pressure Chamber Tests

In these tests, an adit or an underground roadway obtains the access to the rock mass where a chamber of specific shape is constructed for conducting load/deformation tests. In general, three types of pressure chamber tests have been devised.

1. Cylindrical pressure chamber tests
2. Steel-lined cylindrical pressure chamber tests
3. Spherical pressure chamber tests

(a) Unlined cylindrical pressure chamber test

Concrete dams plug a limited length of a circular tunnel, and the unlined space behind the dams is filled with water under pressure P. The deformation of the tunnel radius relative to the tunnel axis is measured in different directions. Stress/deformation curves showing unit rock strain ε versus P are drawn for various directions. The average value of the modulus of deformation 'E' of the rock mass *in situ* is calculated using Jaeger's formula.

$$E = \frac{Pa^2(1+\upsilon)}{r\,\Delta r} \tag{6.1}$$

where

P = internal pressure (MPa)
a = radius of the tunnel (m)
r = radial distance of a point where deflection is measured with respect to tunnel axis (m)
Δr = change of radius (m)
υ = Poisson's ratio

The test arrangement for the unlined pressure tunnel test is given in Figure 6.1. For this test to be valid it is necessary that the length of pressurized section of the tunnel be at least 10 times 'd' and the deflection is measured at mid-length.

(b) Steel-lined cylindrical pressure chamber test

This test was devised for Essen Pressure tunnel in Switzerland for the measurement of the modulus of elasticity, total modulus of deformation of the rock mass and stresses developing within the thin steel lining. Detailed theories have been developed for estimating the proportion of pressure loss in steel lining, taking into account the modulus of elasticity and Poisson's ratio of both rock (E_1,υ_1) and steel (E_2, υ_2).

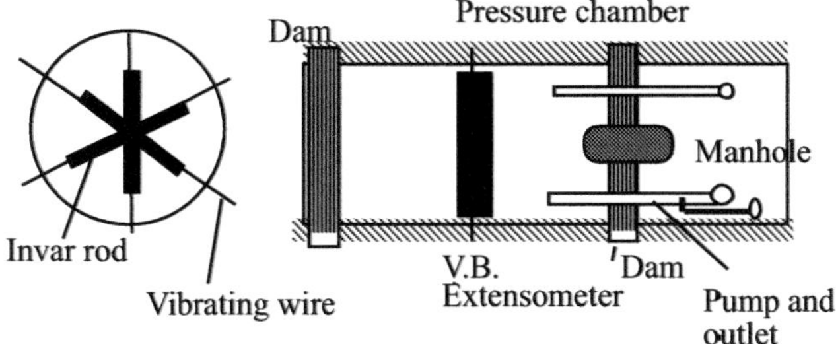

Figure 6.1 Cylindrical pressure chamber test

The radial displacement δ_R at radius C in rock is given by

$$\delta_R = \frac{P_c \times C(1+\upsilon_2)}{E_2} \tag{6.2}$$

where

C = external diameter of tunnel (m)
P_c = pressure transmitted by steel liner to rock (MPa)
p = hydrostatic pressure in the pressure tunnel
b = inner diameter of tunnel (m)
A = a variable

$$P_c = \frac{2'p'b^2}{E_1(C^2 - b^2)A} \tag{6.2a}$$

$$A = \frac{(1+\upsilon_2)}{E_2} + \frac{(1+\upsilon_1)C^2 + (1+\upsilon_2)}{E_1(C^2 - b^2)} \tag{6.2b}$$

6.3.2 Spherical Pressure Chamber Test

In the Kemano project, the effective modulus of elasticity of rock was measured by a steel sphere (37 mm thick plate), 3 m in diameter, concreted in the rock and placed up to varying pressures of 250 KN/m^2. A spherical chamber was used instead of a cylinder, because a cylinder would have been short and end conditions would have made the evaluation of results difficult. This method also permits one to evaluate three-dimensional anisotropy of rock mass.

The test arrangement for spherical pressure chamber test is given in Figure 6.2. The water pressure within the liner will be resisted partly by the liner and partly by the rock, the proportion being resisted by each being determined by the condition that the deflection of the liner must equal the radial deflection of rock (Bleifuss, 1955).

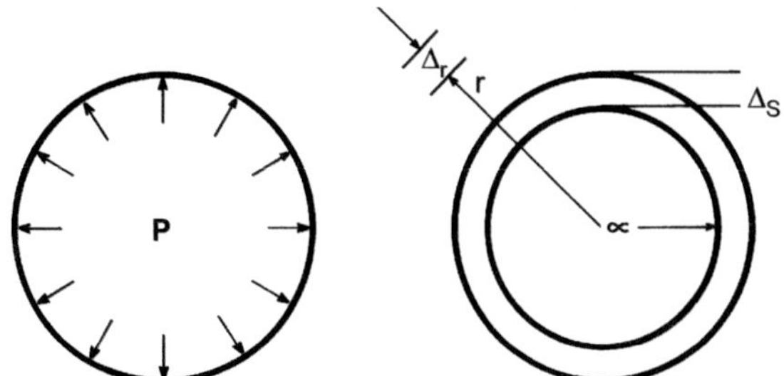

Figure 6.2 Spherical pressure chamber test

Brewer (1952) gives the modulus of elasticity E by the following equation:

$$E = \frac{1}{\Delta_S}\left(7.2\,\gamma P_r + R(P - P_s)^{12a - r/r}\right) \quad (6.3)$$

Where,

P = pressure in the steel sphere (measured in test)
Δ_S = deflection of the steel liner
P_s = pressure absorbed in the steel liner
a = radius of the steel sphere (measured in the test)
r = radius of the yield zone
t = thickness of the steel liner
E_s = Young's modulus of steel
υ_s = Poisson's ratio of steel

$$P_s = \frac{E_s t(1 - \upsilon_s)\Delta_S}{a^2} \quad (6.3a)$$

$$r^2 = \frac{a^2(P - P_s)}{82.65\,h} \quad (6.3b)$$

P_r = Pressure absorbed into rock

$$P_t = \frac{(P - P_s)a^2}{r^2} \quad (6.3c)$$

Pressure chamber tests are very expensive and can be justified for major projects only.

The pressure range for this test is between 210 kN/m^2 and 245 kN/m^2.

6.3.3 Radial Press (Austrian Method)

This method is a modification of the pressure chamber test, used in Austria, Switzerland (Seeber, 1961) and former Yugoslavia (Kujundzic, 1965). In the Austrian method, through a radial press load is applied to a limited length of the tunnel, 2 m in diameter, by a radial cylinder. The Yugoslav method uses flat jacks for applying radial loads to the rock through the tunnel lining.

The test rig (Figure 6.3) consisted of a 5 m long cylindrical frame which can be put into compression by radial wedges (Austrian Method) or flat jacks (Yugoslavian method) to load rock up to 75 kN/m^2. The radial displacements are measured in rock mass 150 mm away from the concrete lining. The deformations are measured from the fixed axis of the tunnel. The deformation caused by the cylindrical jacks or curved jacks differ greatly from those obtained by lined cylindrical pressure chamber tests. The deformation varies along the cylinder, the maximum deflection being at the centre of the loaded section.

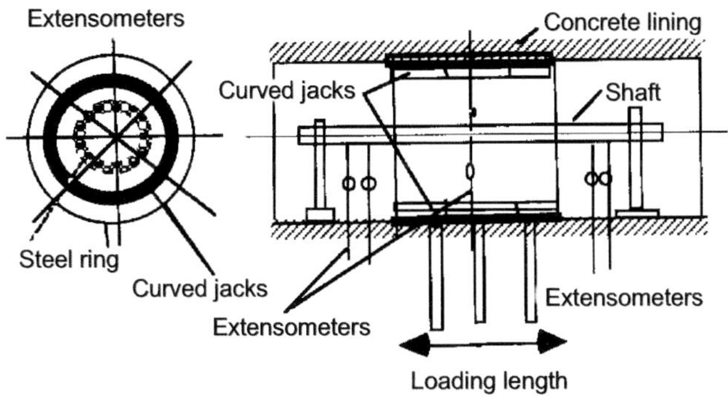

Figure 6.3 Radial press

The results can be interpreted as follows:

$$E = \frac{k\,Pr(1+\upsilon)}{\Delta r} \tag{6.4}$$

k = shape factor, 0.7 in outer section and 0.79 in central section
υ = Poisson's ratio (0.2)
P = load (MN)
r = radius of the pressure chamber (m)
Δr = radial deformation (m)

6.4 Plate Bearing Test

In this method load is applied to a flat surface of the rock mass and the resulting surface deformation or deformation at a point some distance from the centre of the loading plate is measured.

The test can be carried out in an open trench, inside a tunnel or a gallery or in a specially excavated rock cavity. An underground roadway is often preferred because it is easier to support the test rig on its roof or the opposing walls. The loading test can be horizontal or vertical and plates transmitting loads to the rock can be elastic or rigid. The stress distribution under the plate depends upon its shape and elastic constants, E and υ. Rock settlement is measured under the plate or on the axis some distance away (Figure 6.4).

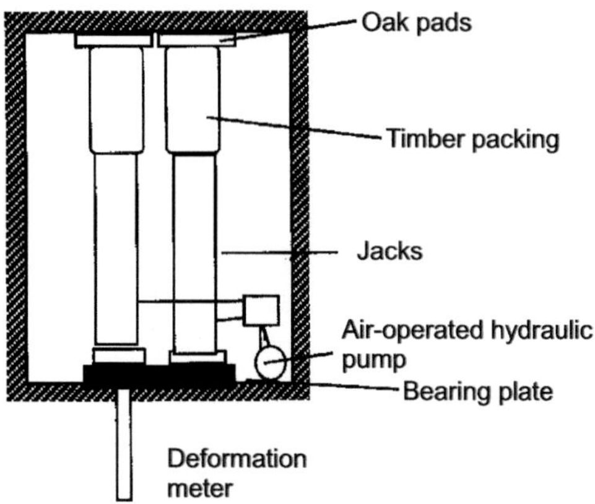

Figure 6.4 Plate bearing test from an underground gallery

The applied force may vary from 50 tonnes to 1200 tonnes (4×300 tonnes) applied by one to four jacks or alternatively through a flat jack cast in concrete in the form of circular hydraulic cushion. The rock affected by plate bearing tests may vary from 0.60 m^2 to 1.2 m^2 using load from 50 to 720 tonnes. The modulus of deformation is calculated as per the following equations (Rocha, 1955; Kujundzic, 1965):

(a) Rigid Circular Plate Measuring Surface Deformation

$$E = \frac{0.54 \times k \times Pr(1 + \upsilon)}{\Delta_r} \tag{6.4}$$

where

Δ_r = average displacement of pad (m)
P = load
υ = Poisson's ratio
r = radius of the plate (m)

(b) Square or Rectangular Plate

Surface displacement can be expressed in the following form:

$$\Delta_{av} = \overline{m}\frac{P(1 - \upsilon^2)}{E\sqrt{A}} \tag{6.5a}$$

Δ_{av} = average displacement of the loaded surface

$\overline{m}$ = coefficient depending on the shape of the loaded surface and stiffness of the plate = 0.96 for circle and 0.8 for square

(c) Deformation under Hydraulic Cushion at Depth (z)

The deformation under the loaded plate or hydraulic cushion at depth (z) is calculated by Habib's equation (Coates and Gyenge, 1966):

$$E = P\frac{(1+\upsilon)}{2\pi r\,\Delta z}\left\{\frac{rz}{(r^2+z^2)} + 2(1-\upsilon)\tan^{-1}\frac{r}{z}\right\} \tag{6.5b}$$

where

P = load

r = radius of the plate or cushion (m)

z = depth (m)

υ = Poisson's ratio

E = modulus of deformation being measured

Δz = deformation measured at depth z (m)

The main disadvantages of the plate bearing test are that the stresses do not penetrate into competent rock mass, but strains of deeper penetration do occur when rock is fractured or fissured.

6.5 Cable Method of *in situ* Testing

This test is devised to load a large volume of rock by a steel cable anchored at depth in a small borehole (Figure 6.5). The minimum anchoring depth should be 8–10 times the plate diameter so that the anchor reaction may not affect the displacement of the rock surface (Zienkiewicz and Staff, 1967).

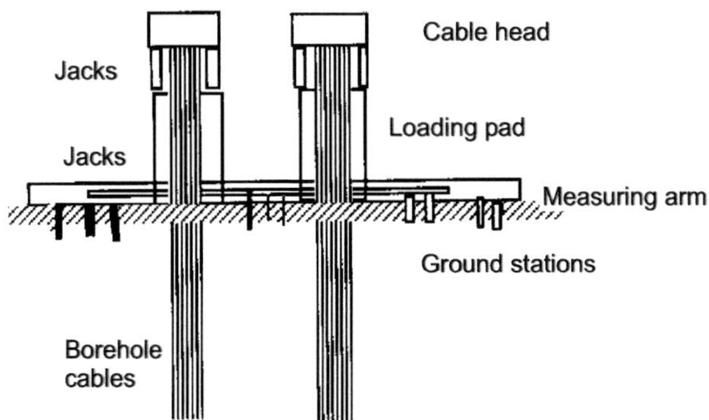

Figure 6.5 Cable method of testing rock

A loading area can be made as large as practicable and loads up to 1200 tonnes can be applied through a single cable to allow for a large volume of rock to be influenced. Several cables could be used to apply even greater loads.

The displacements are measured in loading directions and also in tangential direction. The displacements Δ_V and Δ_R for a square plate with side 'a' are given by

$$\Delta_V = 2.97\, A_1 \frac{P}{a} \tag{6.6a}$$

where

$$A_1 = \frac{1}{2\pi E}\left(1 + \frac{1}{n} + \frac{2}{\sqrt{n}}\right)$$

n = ratio of Young's modulus in horizontal to vertical direction

Similarly,

$$\Delta_R = 2.97\, B_1 \frac{Q}{a} \tag{6.6b}$$

Δ_R = average displacement of the pad in the direction tangential to the loading surface

$$B_1 = \frac{1}{2\pi E}\left(\frac{\sqrt{2(n+1)}}{n}\right) \tag{6.6c}$$

and P and Q are the normal and tangential loads.

6.5.2 Combined Cable and Cylindrical Jack Method

This method is an extension of Zienkiewicz method in which the test cable is passed through the centre of the cylindrical jack applying horizontal loads on the borehole walls (Figure 6.6).

Deformation in both axial and tangential directions is measured and results calculated in terms of Equations 6.6a to 6.6c.

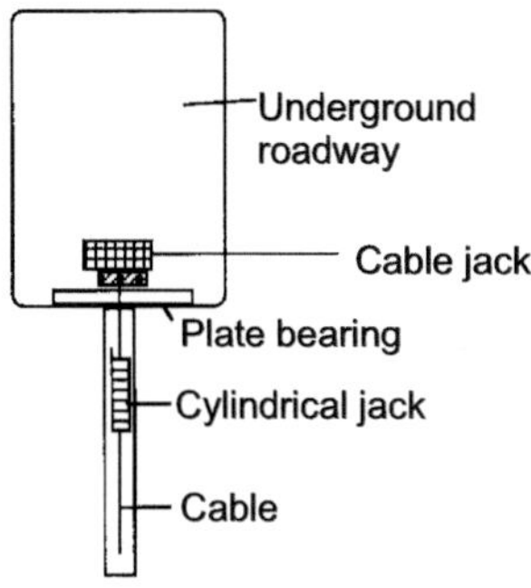

Figure 6.6 Combined cable and cylindrical jack method

The cylindrical jack causes tensile stresses to develop in the circumferential directions and the test yields valuable information on tensile strength of rock *in situ* under triaxial condition and on Young's modulus under similar *in situ* conditions.

6.6 Flat Jack Test

Flat jack used for measuring residual stress (Mayer, 1951; Tincelin, 1952) can also be used for measuring strains, deformations and shear stress *in situ*. The modulus of deformation of rock mass can also be derived from the stress/strain curves.

6.6.1 Stress Measurement

A slot is cut into the rock mass by drilling a series of overlapping drill holes or by using a circular saw. This will relieve the *in situ* stress, originally acting along the slot, on the rock surface and also between two pins A and B, originally grouted 2y distance apart. The displacement between two pins as a consequence of stress relief is given by 2Δv. A flat jack is grouted into the slot (Figure 6.7) so that each face is in uniform contact with the rock. The jack is pressurised until the distance between the pins A and B is exactly 2y, when the fluid pressure is recorded.

The original normal stress σ_n across the slot at cancellation pressure P_c is given by the following equation:

$$\sigma_n = \frac{P_c\,(C_j - d)}{C} \tag{6.7}$$

σ_n = original normal stress to the slot
P_c = cancellation pressure of the jack
C_j = half-width of flat jack
C = half-width of slot
d = inoperative region of flat jack near its edges

6.6.2 Modulus of Elasticity Measurements

The test arrangement is exactly the same as shown in Figure 6.7(a) and (b). A cyclic pressure is applied by a flat jack and displacements are measured between two or more pairs of pins to an accuracy of 0.01 mm in two separate flat jacks set at right angles to each other. The convergence caused by slot cutting 2 ΔV between the measuring pins is given by:

$$2\Delta V = \frac{2C\sigma_n}{E}\left[\left\{(1-\upsilon)\left(1+\frac{y^2}{c^2}\right)^{1/2} - \frac{y}{c}\right\} + (1+\upsilon)\left(1+\frac{y^2}{c^2}\right)^{-1/2}\right] \tag{6.7a}$$

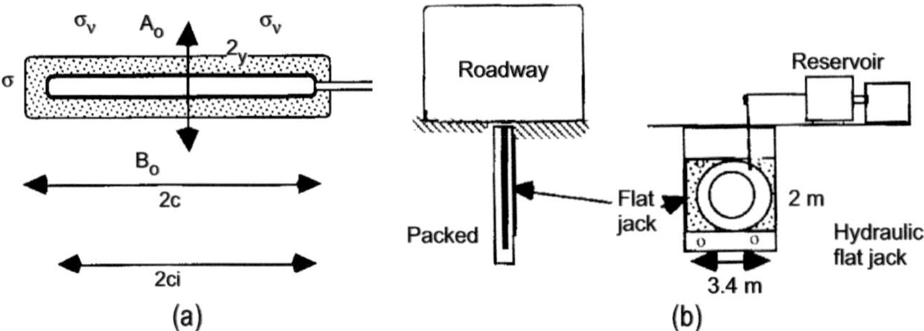

Figure 6.7 (a) Flat jack grouted in rock mass; (b) Flat jack test

By making measurements in two or more different points (y), both Young's modulus E and Poisson's ratio υ can be calculated.

6.6.3 Shear Stress

The displacements parallel to the slot (caused by slot cutting) can be used to estimate shear stress in the plane of the slot.

$$\frac{4G(u+iv)}{Pc} = (x+3)\left(1+\frac{y^2}{c^2}\right)^2 - 2\left(1+\frac{y^2}{c^2}\right)^{1\frac{1}{2}} \tag{6.7b}$$

$$x = (3\text{–}4\upsilon) \text{ for plain strain}$$

$$x = \left(\frac{3-\upsilon}{1+\upsilon}\right) \text{ for plain stress}$$

where

y = original distance between the pin
c = width of elliptical slot
u = displacement parallel to slot
v = displacement right angle to slot
G = modulus of rigidity of slot

6.7 *In situ* Triaxial Test

Gilg and Dietlicher (1965) developed a method of testing rock under a triaxial stress field similar to that conducted in a laboratory. A block of rock is cut out of the rock mass *in situ* and restrained laterally by a steel frame and lateral flat jacks (Figure 6.8). A flat jack restrained by a steel structure applies vertical load. As only one sample is available, it will be necessary to perform a multiple failure state triaxial test to obtain complete peak and residual strengths.

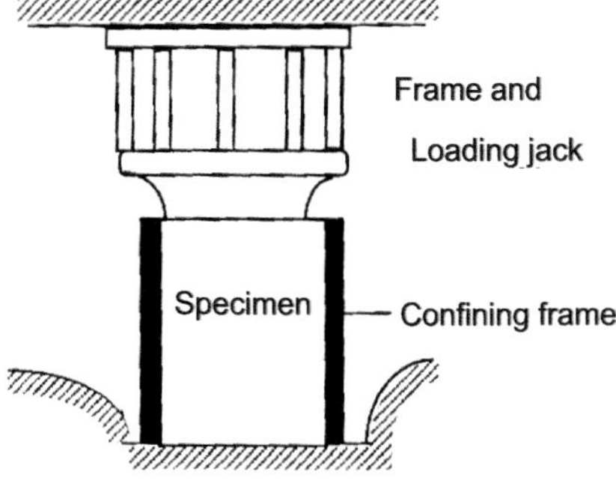

Figure 6.8 *In situ* triaxial test

6.8 *In situ* Shear Test

Purpose of the test

- To measure directly peak and residual shear strength of rock mass as a function of normal load.
- Results used in limiting equilibrium analysis of slope stability problem or stability analysis of dam foundations.

These tests are carried out either on rock *in situ* or on concrete blocks adhered to the rock surface. There are two variations of the test.

(1) Portuguese Test

Field shear tests on rock blocks 0.7 × 0.7 × 0.3 m can be conducted in a manner shown in Figure 6.9(a). A rigid metallic frame surrounds the specimen. The blocks and foundations are saturated with water during the

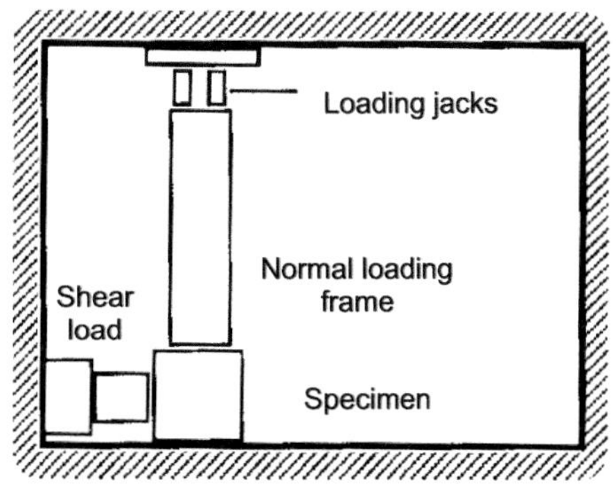

Figure 6.9(a) *In situ* shear test

test. Tangential displacements and normal deformations are measured up to failure.

(2) *In situ* Shear Test with Inclined Jack

A normal stress σ_n is applied till normal deformation is established, then inclined shear force is applied (Fig. 6.9(b)). These are gradually increased until stabilisation of displacement occurred at each step.

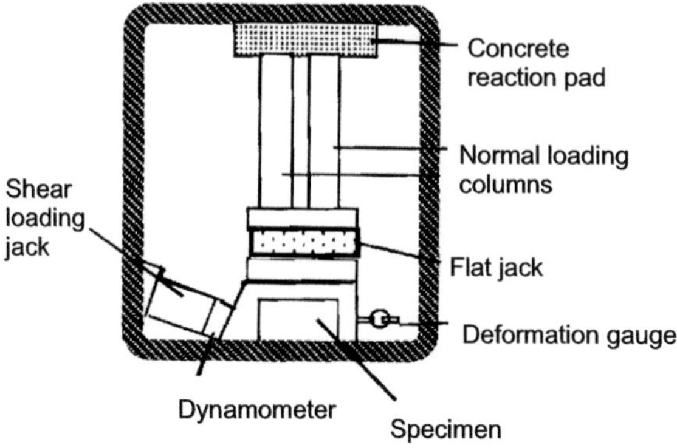

Figure 6.9(b) *In situ* shear test set-up (inclined jack)

The results are interpreted by tracing Coulomb's line on σ, τ diagram. Two different criteria of stress can be employed.

(1) Maximum shear stress τ_{max}
(2) Reversal of the direction of vertical displacements (Figure 6.9c)

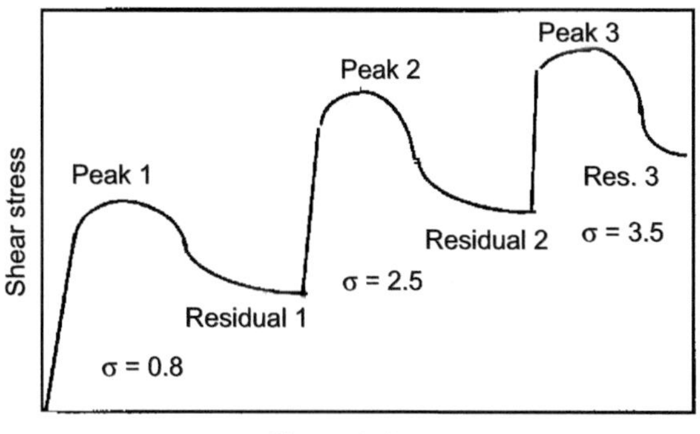

Figure 6.9(c) *In situ* shear test results

In the Portuguese test, the shear stress at inversion was near but less than the maximum shear reached. In one test, the inversion was not so pronounced being one half or less than the maximum stress.

- Specimen size 700 mm × 700 mm × 350 mm *in situ* specimen is casted in spray concrete to smooth out the undulations and loading frame installed.
- Measure shear deformations at eight locations and take the mean value.
- Shear strength test should be carried out on a plane of weakness.
- Shear strength determination should comprise five tests on the same test horizon at different constant stress.

Immediately after the block is sheared, a series of sliding friction tests are conducted for varying normal stress in forward as well as reverse direction on dry or moist rock surfaces to evaluate residual angle of friction ϕ_r.

The test results are interpreted in the same manner as the laboratory shear tests. The main purpose is to determine cohesive strength (c) and the internal angle of friction ϕ_{peak} and ϕ_r (residual (Fig. 6.9(d)).

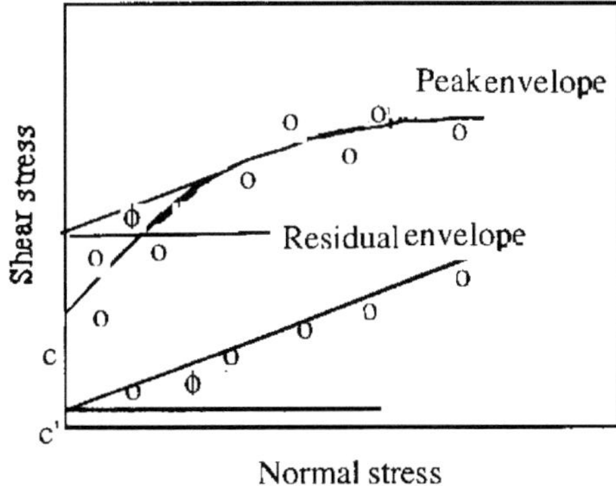

Figure 6.9(d) *In situ* shear strength peak and residual curves

6.9 *In situ* Compression Testing of Coal Pillars

Large-scale *in situ* compressive strength tests on coal pillars are advantageous because of the following factors.

(1) Coal deteriorates considerably upon removal from the mine.
(2) Effects of moisture, humidity and temperature are eliminated and results can be directly used for practical applications.
(3) Underground testing improves the possibility of carrying out tests on the coal specimen *in situ* which could have been lost during the

recovery of specimen, if laboratory testing was considered as an alternative.

(4) *In situ* coal pillars contain separation planes, cleat and bedding planes.

Coal pillars have been tested *in situ* by jacks mounted on top of each specimen (1.5 m × 1.5 m × 0.7 m to 3.0 m height) which are loaded against roof and floor of the seam in a slot. In total, 26 jacks were used during the *in situ* test, the jacks being connected in a special joining system to ensure uniform loading. A maximum pressure of 300 kN/m^2 was exerted on the loading area (Bieniawski, 1969).

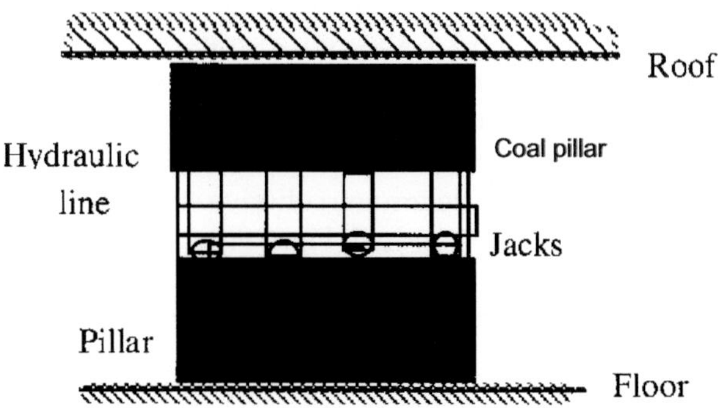

Figure 6.10 *In situ* compressive strength test on coal pillar (Bieniawski, 1967)

The specimens are prepared by trimming the pillar sides by 600 mm to expose fresh coal and a vertical cut is made to separate it from the sides. A horizontal cut is made in the top of the pillar or at mid-height for installing jacks. The specimen is then capped with 75-mm thick concrete pads. Jacks are installed on the top of rigid steel channel arrangement fitted on top of the concrete cap. The instruments are then installed with extensometers on the specimen face. The shape effect is taken into account as follows:

$$\sigma_1 = K\left(\frac{h^n}{D^m}\right) \tag{6.8}$$

6.10 Interpretation of Results from Large-scale Tests

Stress-strain curve derived from a large-scale *in situ* test in compression is indicated in Figure 6.11. The diagram indicates the deformational behaviour of rock in two-and-a-half cycles of loading. The curve OA$_1$ shows loading from 0 stress to stress level σ_1 displaying non-linear stress/deformation characteristic. When unloading the rock mass is carried out to zero stress

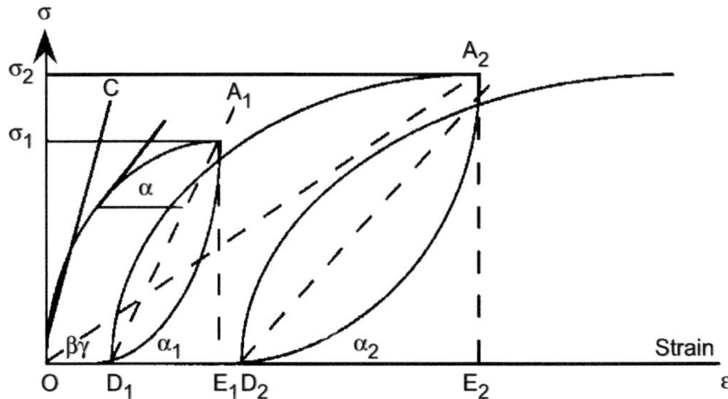

Figure 6.11 Load deformation curve of large-scale test (a) tan α = modulus of elasticity, (b) modulus of total deformation E_{total} = tan γ (modulus of deformation = tan α_1 = tan α_2 (d) k = OD_1/ OE_1 (Jaeger, 1972)

value the strain does not reach zero but retains OD_1 as a residual strain. Reloading the rock mass to higher stress value σ_2 is indicated by the curve D_1A_2 and unloading part as $A_2 D_2$. The permanent deformations OD_1 and OD_2 are caused by closing of small fissures and fractures and considered to be non-reversible.

In Figure 6.11, lines A_1D_1 and $A_2 D_2$ would correspond to average modulus of deformation of fissured rock mass. Line A_1D_1 and A_2D_2 are seldom parallel and it may take several cycles of loading and unloading to settle down to a uniform deformation modulus of tanα_1. It may be noted that the elastic rock deformation may not always be reached. The slope of line OA_2 is the modulus of total deformation denoted by tan γ. While tan α_1, tan α_2 and tan γ are primary characteristics of rock mass, European engineers use the ratio $k = \frac{OD_1}{OE_1}$ as an intrinsic property of rock mass. The load deformation curves from *in situ* test give the basic information regarding load/deformation characteristics under uniform loading. Deformation can be measured in a variety of conditions:

- at the rock surface at the point of application of load
- at a depth in the rock mass
- some lateral distance away from the point of application of loads

Load deformation characteristics depend upon the rate of loading as well as direction of the application of load. Some large-scale loading tests take several weeks of loading.

Figure 6.12 shows the effect of rate of loading on the stress/strain curve in large-scale testing of rock. The lower curve $OA_1\ D_1\ A_2$ indicates the stress-strain curve at low rate of loading while $OB_1\ B_2$ at high rate of loading. The

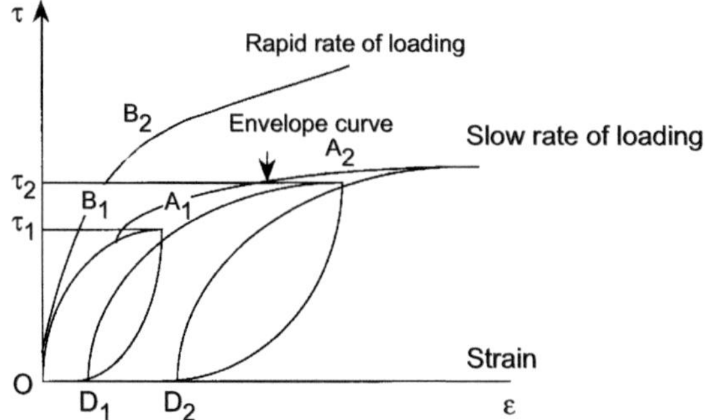

Figure 6.12 Effect of rate of loading on stress/strain curve.

area between curves OB_1B_2 and OA_1 A_2 is the additional work required by the loading system to deform the rock mass at higher rate of loading.

The shape of stress/strain curve envelopes yields important information regarding the structural and deformational behaviour of rock masses. Figure 6.13 shows envelope curves for a range of rock masses showing disparate behaviours. A linear envelope in (a) shows a high strength rock mass with high modulus of deformation. These rock masses do not deteriorate under cyclic loading. Curve (b) shows a discontinuous envelope with point A as a possible point of internal rupture or shear of rock mass when modulus of deformation considerably reduces. Curve (c) shows deviation from linearity, suggesting plastic deformation in the rock mass. In some metamorphic rock types like phyllite-quartzite, envelope curve clearly indicates two points of flexure at points A and B in Figure 6.13(d). These points of flexures represent elastic limits of the different phases of complex metamorphic rocks (Jaeger, 1972).

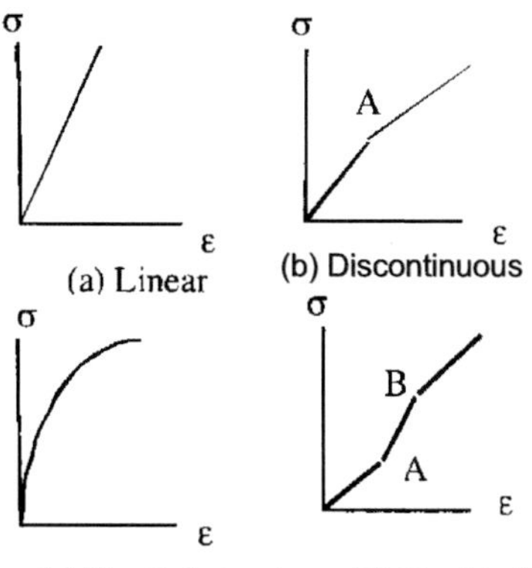

Figure 6.13 Types of load/deformation curves (Jaeger, 1972)

Figure 6.14 shows stress-strain curves for different types of rock masses in cyclic loading. Type 1 rock is a good-quality rock showing high modulus of deformation, low hysteresis and residual deformations on unloading. During unloading cycle, when the load decreases sharply, without corresponding reduction in deformations, there is a possibility of the presence of a discontinuity or indications of possible internal failures. It may also be noted that the loading and unloading modulus of rock mass are of the same order of magnitude.

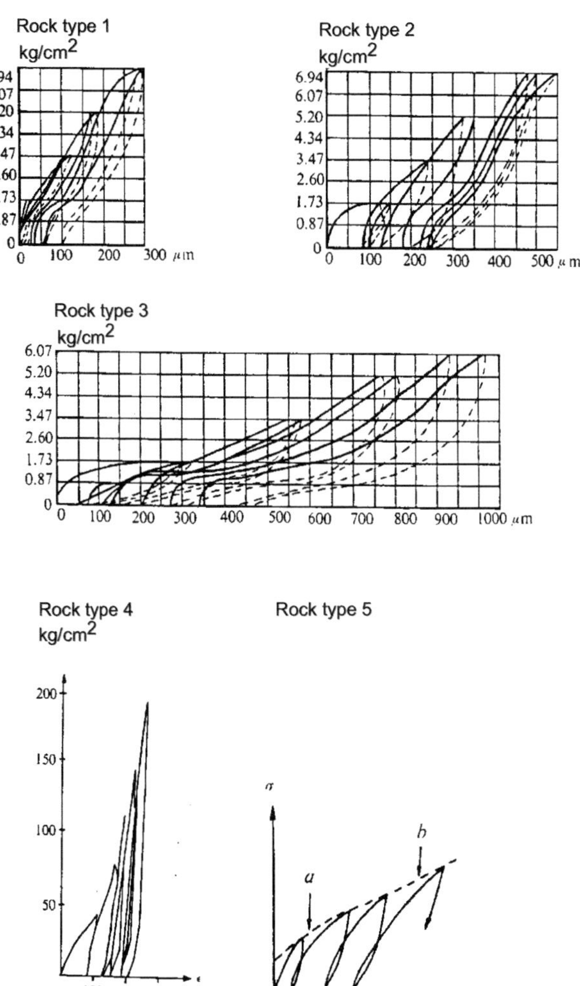

Figure 6.14 Stress/strain curves for different rock types (Oberti, 1960)

Type 2 rock is a medium strength rock mass showing a convex-concave type of envelope curve indicating an improvement in rock mass quality after the initial cycle of loading. The rock mass shows medium modulus of deformation. It may also be noted that for this type of rock the modulus of deformation during loading is higher than the modulus of deformation during unloading indicating that the rock mass has loose internal structure.

Type 3 rock mass is a poor-quality rock showing very low modulus of deformation. It displays a shallow S-shaped envelope curve indicating compaction of rock during initial stage of loading. This is followed by linear characteristics at intermediate load range and ultimately showing a flat stress-strain curve indicating the onset of failure. It may be noted that there is pronounced hysteresis and residual deformation. The unloading curves also show sluggish recovery of elastic deformation which is an indication of possible internal failure or presence of discontinuous rock mass. This may also indicate that some internal areas of rupture no longer behave elastically after failure.

Type 4 rock depicts concave stress-strain envelope curves indicating improvement in the deformation behaviour of rock mass after initial loading.

Type 5 rock is characterised by the stress-strain curve which is at first convex and then straight as shown by points 'a' and 'b' in Figure 7.14(e).

It is usual to load the rock mass at least in two different directions in order to find the planes of weakness. In stratified rock mass, the moduli of deformations in different directions are represented as indicated in Figure 6.15. The variation in the magnitude of the modulus of deformation can be shown by an ellipse with E_1 and E_2 as minor and major axes.

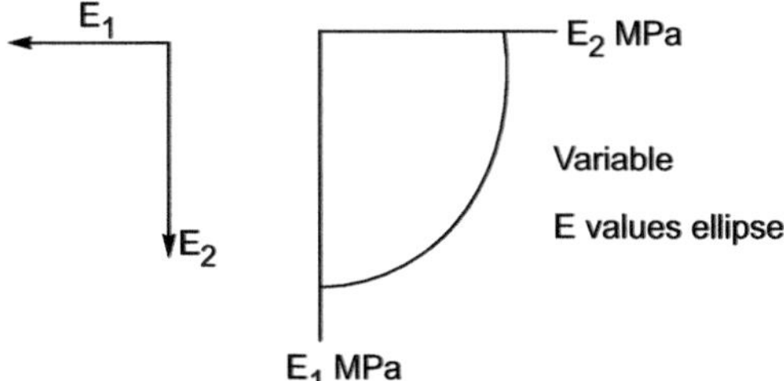

Figure 6.15 Spatial variation of modulus of deformation in stratified rock (E_1 and E_2 are parallel and perpendicular to the bedding plane) (Jaecklin, 1965; Jaecklin, 1966)

A polar diagram shown in Figure 6.16 represents the anisotropy of the rock mass showing the broken contour lines as elastic deformation lines and solid lines as total deformation lines in different directions. Thus representation of the elastic modulus in different directions can express spatial anisotropy of rock mass at a test site.

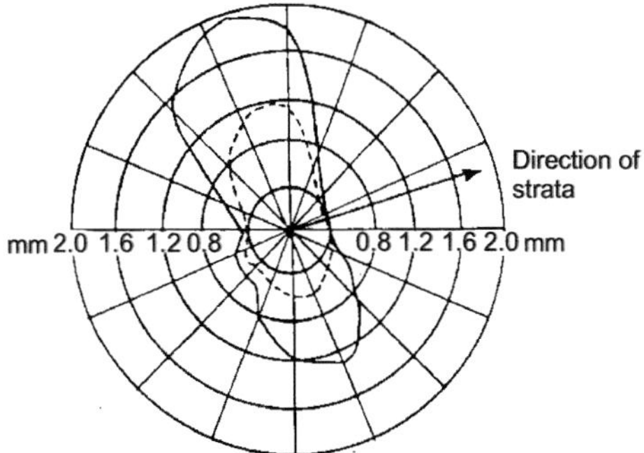

Figure 6.16 Representation of anisotropic properties of rocks.

Figure 6.17 shows the effect of saturation by water on gneiss. It can be seen that the effect of water saturation decreases both the uniaxial compressive strength and to a lower extent the modulus of deformation of rock. It can also be seen in the variation of loading and unloading modulus for gneiss.

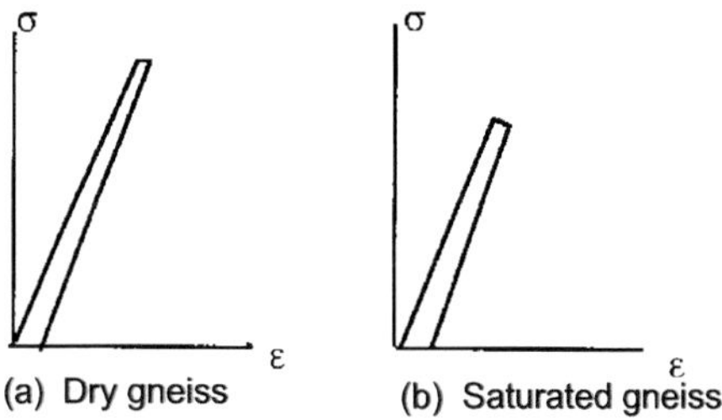

Figure 6.17 Stress-strain curves for dry and saturated rocks (French National Committee on Large Dams, 1964)

6.11 Conclusions

Laboratory tests described in Chapters 1 to 4 are elaborate and require considerable time between sampling process and getting the final report. During the process of obtaining samples, rock is disturbed, weaker material is lost and *in situ* moisture content in the specimen is altered. Large-scale tests involve much larger volume of rock during the test, retain *in situ* moisture conditions and involve the test in weak as well as in intact rock materials, thus giving representative values of strength and deformational properties of rock mass. It is possible to conduct tests in various orthogonal directions, thus taking anisotropy of rock mass into consideration. However, time and expense involved in carrying out these tests practically limit the number of tests to be carried out to obtain meaningful results. Often, conclusions regarding rock mass behaviour have to be derived from the limited number of tests carried out in a spatially variable rock.

References

Bock, H. (ed.) (1978). *An introduction to Rock Mechanics*, Department of Civil and Systems Engineering, University of Northern Queensland, Townsville.

Bleifuss, D.J. (1955). *Proceedings ASCE*, 81,

Bieniawski, Z.T. (1969). '*In situ* large scale testing of coal', *In situ* investigations in soils and rocks, *Proceedings of the Conference by the British Geotechnical Society, London*, 13–15 May, pp. 67–74.

Brewer, G.A. (1952). 'Dilation measurements of steel sphere and rock deformations at Kemano, B.C.', *S.E.S.A Proceedings*, Vol. XIII, No. 2, pp. 63–78.

Coates D.F. and M. Gyenge (1966). 'Plate load testing for rock deformation and strength properties testing'. *Testing Techniques for Rock Mechanics*, Am. Society for Testing Materials, No. 40, Philadelphia.

Dodds, D.J. (1974). 'Interpretation of plate bearing test results'. *Field Testing and Instruments of Rocks*, Special technical publication 554, American Society for Testing Materials, Philadelphia, PA, pp. 20–34.

Dodds, D.J. and Schroeder, W.I. (1974). 'Factors bearing on the interpretation of *in situ* testing results'. American Institute of Mining, Metallurgy and Petroleum Engineers, New York, pp. 394–414.

Gilg, B. and E. Dietlicher (1965). 'Felsmechanische Untersuchungen an der Sperrstelle Punt dal Gall', Schweiz, Bauztg, Vol. 83, No. 43.

Habib, P. (1950). 'Determination of elastic modulus of rock *in situ*'. *Annales Institute Techn. du Batimant*, Paris.

Hibino, S. and M. Motojima (1993). 'Rock mass behaviour during large scale cavern excavation'. *Comprehensive Rock Engineering–Principles, Practice and Projects*, Vol. 4, Editor, J.A. Hudson, Pergamon Press, London, pp. 631–652.

Hucka, V. (1965). 'A rapid method for determining the strength of rocks *in situ*'. *International Journal of Rock Mechanics and Mining Sciences*, Vol. 3.

Jaecklin, F.P. (1965). 'Felsmechanik im Tunnelbau', *Schweiz, Bauztg*, Vol. 83, No. 27.

Jaecklin, F.P. (1966). 'Rock Tunnelling and Optimum Shape', *I^st^ Congress of International Society of Rock Mechanics*, Lisbon, Vol. 2, pp. 397–403.

Jaeger, C. (1972). *Rock Mechanics and Engineering*, ISBN 0 521 07720 6, Cambridge University Press, 1972, 404 pp.

Kujundzic, B. (1965). 'Experimental research into mechanical properties of rock masses in Yugoslavia'. *International Journal of Rock Mechanics and Mining Sciences*, Vol. 2, pp. 75–91.

Krusmanovic, D., M. Tufo and Z. Longof (1966) 'Shear strength of rock masses and possibility of its reproduction on models'. *I^st^ Congress of International Society of Rock Mechanics*, Vol. 2, pp. 537–542.

Mayer, A. (1951). 'Les properietes mecaniques de roches'. *Geotechnique*, 3, p. 329.

Mayer, A. (1963). 'Recent work in rock mechanics', Third Rankine Lecture, Geotechnique, Vol. 5, June, pp. 75-91.

Mazenot, P. (1965). 'Interpretation de nombreuses mesures de deformations executees sur massifs rocheux par E. d.F.', Annales Institute Techn. *du Batimant*, Paris, Vol. 18, No. 206.

Mitchell, J.H. (1900). 'The stress in an Aelotropic Elastic Solid with an Infinite Plane Boundary'. *Proceedings of London Mathematical Society*, 247 pp.

Oberti, G. (1960). 'Experimentelle Untersuchungen uber die charakteristik der Verformbarkeit des Felsen'. *Geologie and Bauwesen*, Vol. 25, pp. 95–113.

Pacher, F. (1969). 'Measurement of deformations in a test gallery as a means of investigating the behaviour of rock mass'. *Felsmechnik und Ingeniurgeologie,* Suppl. 1, pp. 148–161.

Panet, M. (1969). 'Rock mechanics investigations in the Mont Blanc tunnel', *In situ* Investigations in Soils and Rocks. *Proceedings of the Conference by the British Geological Society*, London, 13–15 May, pp. 102–108.

Rocha, M. (1964). 'Mechanical behaviour of rock foundations in concrete dams'. *8th Congress of Large Dams*, Edinburgh, paper R-44, Q-28, pp. 785–832.

Rocha, M., A.F. Da Silveira, N. Grossman and E. Oliveira (1966). 'Determination of deformability of rock mass along boreholes', *Ist Congress of International Society of Rock Mechanics*, Lisbon, Vol. 1, pp. 697–704.

Rocha, M., J.L. Serafim and A.F. Da Silveira (1955), 'Deformability of foundation rocks'. *5th Congress of Large Dams*, Paris, Vol. III, pp. 531–561.

Ruiz, M.D. and F. Camargo (1966). 'A large scale field test on rocks'. *Proc. Ist Congress of ISRM*, Lisbon, Vol. 1, p. 257.

Seeber, G. (1961). 'Auswertung von statischen Felsdehnungsmessungen', *Geologie und Bauwesen*, Vol. 26, pp. 152–176.

Serafim, J.L. (1963). 'Rock mechanics considerations in the design of concrete dams'. *Conference on State of Stress on Earth Crust*, Santa Monica, p. 611.

Serafim, J.L. and J.J.B. Lopez (1961). '*In situ* shear tests and tri-axial tests on foundation rocks of concrete dams'. *Proceedings of 5th International Conference on Soil Mechanics and Foundation Engineering*, Part 1, pp. 533–539.

Stucky, A. (1953). 'Le centre de recherches pour l'etude des barrages', Centenaire de l'Ecole Polytechniques de Lausanne, Switzerland, 106 pp.

Talobre, J. (1962). *Tests, General Report*. Development and Resources Corporation, New York.

Tinecelin, M.E. (1952). 'Mesures des pressions de terrains dans les mines de fer de l'est', Annales de l'Institut du Bâtiment et des Traveaux Publics, No. 58, pp. 972–990.

Zienkiewicz, O.C. and K.G. Staff (1966). 'Cable method of *in situ* rock testing'. *International Journal of Rock Mechanics and Mining Sciences*, Vol. 4, pp. 273–300.

Zienkiewicz, O.C. and K.G. Staff (1967). 'The method of *in situ* testing'. *Ist International Congress of Rock Mechanics*, Lisbon, 1, pp. 667–672.

Assignments

Question 1 In a 5-m diameter unlined circular roadway, the hydraulic pressure was raised to 8 MPa and the corresponding radial deformation measured by the extensometer at a radial distance of 3 m was 0.005 m. Calculate the modulus of deformation if the Poisson's ratio is taken as 0.19.

Question 2 In a plate bearing test, the load applied to the rock mass through a circular rigid plate of 2.0 m in diameter was 100 tonnes. The corresponding average deformation measured below the plate at the plate and the rock-mass interface was 0.0006 m and the deformation measured in a borehole below the centre of the plate at a depth of 5 m was 0.00027 m. Calculate the modulus of deformation of rock at the rock surface and at the anchor point in the observation borehole. Make your own assumptions and comment on the validity of results.

Question 3 (a) Write short notes on the following.

- (i) Pressure chamber tests
- (ii) Plate bearing test
- (iii) Interpretation of creep properties of rocks from plate bearing tests
- (iv) *In situ* shear strength test

Question 4 (a) Give Jaeger's equation as applied to the cylindrical pressure chamber test and explain the terms in Jaeger's equation together with their units.

(b) In a cylindrical pressure chamber test, with the tunnel radius of 5.4 m the final pressure was raised to 10 MPa. If the measured deformation at 5MPa pressure at a radial distance of 4.0 m was 0.0072 m, calculate the modulus of deformation of the rock, assuming that the Poisson's ratio of rock mass is 0.2.

Question 5 (a) Discuss why it is necessary to carry out large scale *in situ* tests in rock mechanics investigation? Outline the importance and limitations of these tests.

(b) Describe with the help of diagrams the shapes of load deformation characteristics of rock masses obtained in pressure chamber tests in cyclic loading for the following rock mass types.

- (i) Massive igneous rock
- (ii) Jointed rock with major joints parallel to the line of loading
- (iii) Medium to soft porous rock liable to micro-fracturing
- (iv) Unconsolidated rock mass prone to settlement

Question 6 Write short notes on:

- (a) Borehole dilatometer
- (b) Factors affecting structural behaviour of tunnel
- (c) Borehole extensometer

Question 7 (a) In a cylindrical pressure chamber test in an unlined roadway the following test results were obtained.

Water pressure (MPa)	*Deformation (m)*	
	Loading	*Unloading*
0.0	0.0	–
0.86	0.001	
1.72	0.002	–
3.45	0.0065	0.0065
1.72	–	0.0049
0.86	–	0.0030
0.00	0.002	0.0020
0.86		
1.72	0.0032	
3.45	0.006	
5.17	0.011	0.0011
3.45		0.0100
3.02		0.0086
1.72		0.0075
0.0		0.0037
0.0	0.0037	
0.86	0.0040	
1.72	0.0051	
3.02	0.007	
3.45	0.0082	
5.17	0.012	
6.44	0.015	
6.89	0.0175	
6.89	0.018	

The diameter of the pressure chamber was 9.0 m, the radial distance of the observation point was 5.6 m and the Poisson's ratio of rock mass was 0.27. Estimate the modulus of deformation of rock.

Indicate how the results of the modulus of deformation test could be used in the design of structures in rock.

Evaluation of Rock Mass Parameters by Borehole Testing

7 Chapter

7.1 Introduction

This chapter examines the measurement of elastic modulus of rock by borehole methods and monitoring of rock deformations in surface and underground rock structures. Borehole methods provide easy access to the rock mass and provide means to investigate rock mass in various orthogonal directions. Measurement of rock displacement in and around mining excavations provides insight into the cause of instability and signposts possible remedial actions.

7.2 Measurement of Elastic Moduli of Rocks by Borehole Methods

7.2.1 Limitations of Large-scale Tests

The large-scale *in situ* tests described in the previous chapter are elaborate and require prolonged preparations in terms of obtaining access to rock masses by driving trenches, shafts and underground galleries. Preparation of test sites or test specimen, installation of monitoring equipment and the test itself has to be planned and executed with extra care and precision. Therefore, these tests are very expensive and it is not possible to carry out the large numbers of tests necessary to incorporate anisotropy, regional variations of rock properties and changes in lithology within the same seam or rock mass.

In the past, many borehole instrumentation systems have been developed and used for measuring *in situ* the elastic modulus or modulus of deformation of rock. Borehole systems offer an advantage in allowing easy access to the rock mass and permitting a large number of tests be carried out in radial directions as well as along the length of the borehole. In addition, tests could be carried in a number of boreholes drilled in orthogonal directions. Thus, the average modulus of deformation can be expediently assessed for reliable design calculation.

7.2.2 Main Features of Modulus of Deformation Measuring Systems in Boreholes

Borehole instrumentation for measuring the *in situ* elastic modulus of rocks differs considerably from stress measuring equipment described in Chapter 8 or borehole deformation measuring systems described elsewhere in this chapter. The main features of the *in situ* modulus of elasticity measurement systems are as follows.

- Exert pressure on the wall of the borehole
- Measure corresponding deformation of the borehole wall
- Use theory of elasticity to calculate the modulus of deformation (Timoshenko and Goodier, 1951; Love, 1944)

7.2.3 Type of Borehole Modulus of Deformation Measuring Instrument Systems

Three distinct types of instruments have been developed which have their own advantages and applications.

(1) Borehole dilatometers or pressure meters
(2) Borehole jacks
(3) Borehole penetrometers

Table 7.1 summarises the three types of borehole instrumentation for measuring the elastic modulus of rock *in situ,* generalised equations for calculation of modulus of deformation and types of instruments developed by various researchers.

7.3 Dilatometers and Pressure Meters

The dilatometer or pressure meter is a cylindrical device which when inserted into a borehole applies uniform internal pressure on the borehole wall. The corresponding borehole deformation is measured either directly by using various types of transducers or indirectly by measuring the change of volume of fluid pumped into the instrumentation system during the test. Figure 7.1 indicates diagrammatically a borehole dilatometer or a pressure meter. In some dilatometers the uniform pressure on the borehole wall is exerted through curved jacks covering the entire periphery of the borehole.

The modulus of deformation of a rock mass can be calculated using the following formula:

$$E = \frac{2(1+\upsilon)}{\delta d} aP \tag{7.1}$$

Table 7.1 Modulus of deformation measuring techniques

Instrument	*Borehole dilatometers*	*Borehole jacks*	*Borehole penetrometers*
Equations	*Competent rock* $E = \frac{P(1+\upsilon)}{\delta d/d}$ *Fissured rock mass* $E = \frac{P}{\delta d}(1+\upsilon + \log\frac{b}{a})$ E = modulus of deformation δd = borehole deformation d = borehole diameter b = radius of broken rock a = radius of borehole	$E = \frac{PK(\upsilon,\beta)}{\delta d/d}$ P = Pressure K (υ, β) = slope of line relating E and $\frac{P}{\delta d/d}$	*Spherical indentor* $\delta d = \frac{9\pi^2 p^2}{16}$ $\frac{(1-\upsilon 1^2)}{E_1} + \frac{(1-\upsilon 2^2)}{E_2}$ *Cylindrical indentor* $\delta d = \frac{(1-\upsilon 2)P}{AE}$ A = area of contact of pin and borehole wall
Instruments	1. Cylindrical pressure cell (Panek et al, 1964) 2. Pressuremeter (Menard, 1963) 3. Geoprobe (Menard, 1967) 4. LNEC dilatometer 5. Kujundzic dilatometer (1965) 6. Menard dilatometer (Menard, 1966)	1. Centex cell, Dawance (1964) 2. NX plate jack, (Goodman et al., 1968) 3. Geo-extensometer, Absi and Senguin (1967) 4. German stress/strain meter (Martini et al., 1964) 5. Photelstatic pre-stressed deformation meter (Singh, 1967) 6. Talobre (borehole jack) 7. Panek borehole jack	1. U.S. Bureau of Mines Borehole Penetrometer (Hult, 1963) 2. Stiff stress meter (Steers 1965) 3. Photo-elastic pre-stressed deformation meter (Singh, 1967)

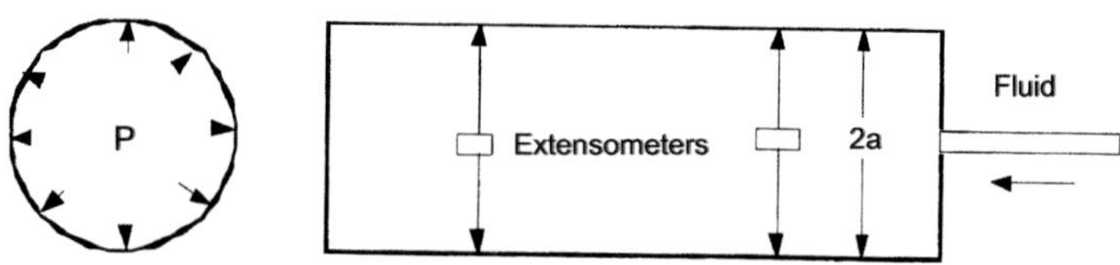

Figure 7.1 Schematic diagram of a dilatometer

where

E = modulus of deformation
υ = Poisson's ratio
P = test pressure in dilatometer (MPa)
δd = borehole diametric deformation (m)
a = radius of the borehole, $d/2$ (m)

Modulus of deformation in fissured rock is given by the following relationship:

$$E = \frac{P}{\delta d}\left(1 + \upsilon + \log\frac{b}{a}\right) \quad (7.2)$$

b = radius of broken rock surrounding the borehole (m)
a = radius of borehole
υ = Poisson's ratio
P = test pressure in dilatometer (MPa)

Dilatometers offer the advantages of carrying out tests in mechanically undisturbed rock masses and at *in situ* moisture conditions. The cost and time required to carry out continuous tests along the borehole length is relatively low, allowing estimation of mean properties which is impossible to achieve with large-scale *in situ* methods. However, the dilatometer tests involve only a small volume of rock in each test as most instrument systems use 75 mm diameter boreholes. Many borehole dilatometers have application in soft rocks where deformations are high and contact pressures at the rock and dilatometer interface are relatively low.

Figure 7.2 shows a classification of dilatometers or pressure meters in use. In general, the dilatometer can be classified as a direct measuring system where change in diameter is measured directly by Linear Variable Differential Transformers (LVDTs) or induction transducers. The LVDTs wider application in the construction of dilatometers than any other transducer can be attributed to ease in application and reliability. Alternatively, the change of diameter in the borehole is indirectly determined by measuring the change of the volume of fluid injected into the dilatometer.

Indirect measuring systems are rather complicated requiring corrections to be applied when measuring indirect deformation as follows:

(i) Error due to compressibility of water, rubber sleeve or air (Figure 7.3a)
(ii) Correction due to change of thickness of the sleeve

The general relationship for calculating the compressibility of a fluid or rubber sleeve is given by the following relationship:

$$v_i = KV\,dp \quad (7.3)$$

where

v_i = reduction in volume
K = compressibility of fluid

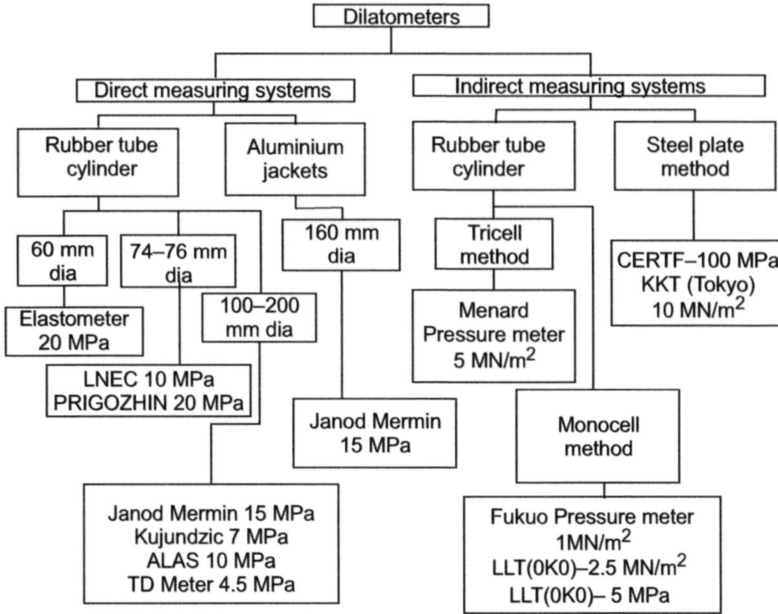

Figure 7.2 Types of dilatometers (Goodman et al, 1968)

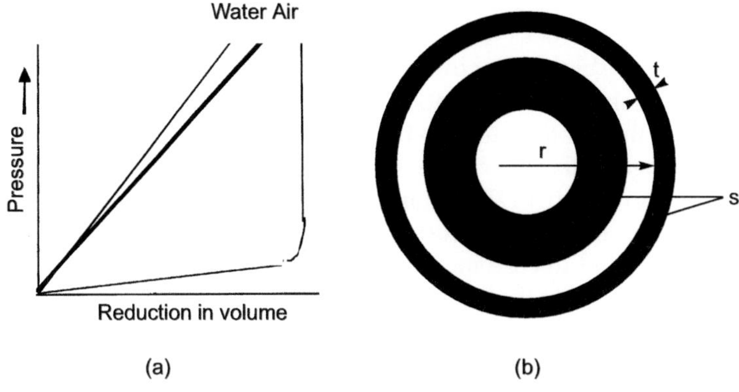

Figure 7.3 Volume correction due to compressibility of fluid and rubber sleeve in the instrumentation

V = original volume
dp = change in pressure

(ii) Correction due to change of thickness of sleeve

As the pressure in the dilatometer increases the rubber sleeve expands in diameter to compress the borehole wall and at the same time the thickness of the sleeve reduces as shown in Figure 7.3b. The correction due to the change in thickness of the sleeve can be calculated by using Equation 7.4.

$$t = r - \left(r^2 - \frac{s}{\pi}\right)^{\frac{1}{2}} \tag{7.4}$$

where

t = thickness of rubber sleeve
r = outside radius of rubber sleeve
s = cross-sectional area of rubber sleeve

However, the combined effect of change of pressure in the dilatometer on the compressibility of the fluid and the rubber sleeve is taken into account by using the sleeve manufacturer's calibration graph.

Figure 7.2 also indicates that the contact pressure developed by various dilatometers changes depending upon the particular application of the deformation instrumentation. A range of instrumentation with contact pressures from 1 MPa to 52 MPa have been used. Maximum contact pressure < 15 MPa is suitable for soft or weathered rock or poor quality rock mass. Higher contact pressures are necessary for determination of the modulus of deformation of intact or massive rocks.

Table 7.2 shows a summary of various modulus of deformation measuring instrumentation. The table describes the method of generating pressure in the instrument, method of measuring pressure and deformation, stress range generated on the borehole wall, diameter and length in each instrumentation system and the loading geometry. The diameter of the dilatometer varies according to chosen design of between 60 to 200 mm. Large diameter instrumentation systems have the advantage of influencing larger volume of rock mass, thus giving more representative modulus of deformation values. Lengths of the dilatometers may vary between 200 mm to 1600 mm with the length-to-diameter ratio being greater than 4 in order to eliminate the end effects produced by the dilatometer.

A brief description of some selected dilatometers or pressure meters is given below.

(1) Cylindrical pressure cell

The cylindrical pressure cell was developed by Panek et al (1964) of the US Bureau of Mines for measuring elastic modulus of rock *in situ*. The cell is designed to be used in 38 mm diameter boreholes and has an overall length of 208 mm. The pressure is developed in the cell by pumping liquid glycerine into it and the inflation of the cell is indirectly measured by monitoring the quantity of liquid pumped into the cell. The pressure developed within the pressure cell is directly measured by a pressure gauge. The range of pressure exerted on the borehole wall is less than 10 MPa. Therefore, the instrument is applicable to soft rocks or very poor quality rock mass.

Table 7.2 Design parameters of various borehole modulus of deformation measuring meters

Instruments	*Method of measuring deformation*	*Method of pressure generation*	*Method of pressure measurement*	*Pressure range on borehole wall* $(MN/m)^2$	*Diameter and length*	*Loading geometry*
Cylindrical pressure cell, (Panek, et al 1964)	Volume of liquid pumped	Glycerine	Pressure gauge	10.5	38 mm × 208 mm	Full circle
Menard pressuremeter (1966)	Volume of liquid pumped	Water	Pressure gauge	51.7		Full circle
Geoprobe	Volume of water	Water	Pressure gauge			Full circle
LNEC dilatometer	Directly 4 LVDT	Liquid	Pressure gauge	14.7	75 mm × 500 mm	Full circle
Kujundzic dilatometer	Directly 2 LVDT	Water	Pressure gauge	6.86		Full circle
Menard dilatometer	Directly 3 LVDT	Water	Pressure gauge	14.7		Full circle
US Bureau of Mines Penetrometer	Movement of wedge	Wedge moving rigid pin	Pressure gauge			Special pins
Stiff stress meter	Movement of wedge	Wedge moving rigid pin	Pressure gauge		54 mm × 230 mm	Special pin
Photo-elastic pre-stressed meter	Movement of wedge	Wedge moving rigid pin	Photoelastic stress meter	133–637	160 mm × 230 mm	Cylindrical and spherical pins

(2) Menard pressure meter

The Menard pressure meters (Menard, 1957; Menard, 1966) comprise two units: a measuring cell and a locking cell. In this borehole instrument, pressure is generated by a combined compressible and incompressible fluid pressure system using water and bottled pressurised gas. Deformations are deduced from the pressure and volume changes at the borehole mouth. Some specifications of the instrumentation scheme are given below.

- Method of measuring deformation – volume of water pumped
- Measurement of pressure by a pressure gauge
- Range of borehole wall stress = 10.34 MN/m^2
- Diameter and length ϕ = 76 mm, length = 515 mm

In general, this type of dilatometer has to be acclimatised at the borehole site in order to take into consideration the ambient temperature variations.

The range of pressure exerted on the borehole wall is less than 10 MPa. Therefore, the instrument is applicable to soft rocks or very poor quality rock masses. The main difficulties encountered with the use of these devices are adverse winter conditions and complicated internal construction not being amenable to field maintenance.

(3) Geoprobe

This instrumentation is a marked improvement on the Menard dilatometer and is designed for use in rugged field conditions. The instrumentation comprises two portable units; the probe and a combined control and recording unit. When testing, these two units are connected together by a single plastic tube that supplies gas to the probe and carries the wire connections from the probe to the signal conditioning and recording unit (Thorley *et al.*, 1969).

Basically, the probe is a single cylindrical load cell that develops pressure within the borehole using a compressible fluid. The corresponding diametral changes in the borehole due to dilation of the cell are monitored electronically by a high resolution LVDT. The variation of resistance of the potentiometers is recorded in terms of displacement readings in the signal control unit. The longitudinal expansion of the cell is restricted by the use two machined cone end restrainers.

The instrumentation has the following specifications:

- Method of measuring deformation – volume change and integrated effect of all diameters
- Liquid-water – gas pressure against water
- Measurement of pressure – pressure gauge
- Range of borehole wall stress = 10.2 MPa

- Diameter and length $\phi = 76$ mm, length = 515 mm

The advantages with this instrumentation system are that the deformations are measured directly at the probe and no incompressible fluid is required in the pressure circuit.

(4) LNEC dilatometer

This dilatometer was developed by Rocha *et al.* (1966) in Portugal to determine modulus of deformation of rock *in situ* along a 76 mm diameter borehole (Figure 7.4). The dilatometer comprises a stainless steel cylinder 66m in diameter and 540 mm in length with a wall thickness of 10 mm. A 4 mm thick neoprene jacket is wrapped round the assembly and a liquid comprising water and oil is pumped into the inter-space between the cylinder and the neoprene jacket to exert pressure on the internal surface of the borehole. The method of applying pressures comprises liquid/gas pressure against water in the cell. The remote end cap of the dilatometer contains a relief valve for releasing fluid pressure by compressed air. The front-end cap of the dilatometer contains an inlet fluid pipe and electrical connections from the LVDTs which are connected to the signal conditioning units. The instrument is installed in the borehole by means of telescopic rods screwed to the front-end cap which are also used to determine the depth and the orientation of the dilatometer in the borehole. The fluid pressure in the dilatometer is measured by a pressure gauge and the pressure is released after each test in order to move the instrument along the borehole.

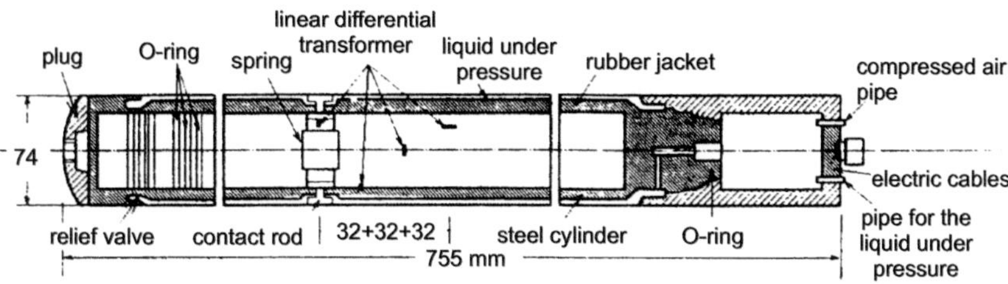

Figure 7.4 LNEC dilatometer (after Rocha *et al.*, 1966)

The deformation of the dilatometer is measured at four cross-sections at a longitudinal interval of 32 mm; thus the deformation is being measured along a length of 96 mm in the borehole. The metallic coil of the LVDT is installed in the dilatometer and the spring-loaded plungers move within the coil assembly and come in contact with the borehole walls by means of compressed air. The range of the LVDT is 5 mm with an accuracy of 1 μm and a sensitivity of 0.1 μm.

The contact stress developed between the dilatometer and the borehole wall is 14.7 MN/m^2. Thus, this instrument is applicable for the measurement of the deformation modulus in weak, weathered rocks and poor quality rock masses.

(5) Kujundzic dilatometer (1964)

Kujundzic (1964) used a rubber jacket in his dilatometer where the borehole deformations were measured by two transducers placed in two orthogonal directions in the same plane. The design parameters of the instrument are given below:

- Method of measuring deformation–2 LVDTs
- Liquid–pumped oil to expand cell
- Measurement of pressure by using a pressure gauge
- Range of borehole wall stress = 6.86 MN/m^2
- Diameter and length ϕ = 200 and 300 mm; length = 1000 and 1200 mm

(6) Menard dilatometer

Menard (1965) modified his earlier model by measuring the deformation by using three LVDTs. Both instruments had limitation of applying lower pressures on the borehole wall thus being suitable for softer rock or weathered rock masses. Other design parameters are as follows:

- Method of measuring deformation–3 LVDTs
- Liquid– water
- Measurement of pressure–pressure gauge
- Range of borehole wall stress–10.6 MN/m^2
- Diameter and length ϕ = 76, 60 mm, length = 515, 502 mm

(7) Comes cell (1965)

This instrumentation scheme is designed for measuring the deformation of borehole by three LVDTs by exerting higher pressures on the borehole walls on a large diameter borehole. The main design parameters of the dilatometer are as follows:

- Method of measuring deformation–3 LVDTs
- Liquid–gas pressure against water
- Measurement of pressure–pressure gauge
- Range of borehole wall stress–16.0 MN/m^2
- Diameter and length ϕ = 160 mm, length = 1600 mm

(8) Yachito tube deformeter

This modulus of deformation measuring system was designed and developed by Yachito Engineering Co. in Japan. The deformeter has a finished diameter of 297 mm to be used in borehole of 300 mm in diameter. The instrument (Figure 7.5) comprises two bakelite plugs assembled at each end of the deformeter with the help of two high tensile steel shafts giving the deformeter an overall length of 1300 mm. A neoprene sleeve covers the entire deformeter and oil pressure is used to expand the neoprene sleeve to load the borehole walls. The diametric expansion of the deformeter is measured by three groups of LVDTs mounted radially to measure deformations in four different directions incorporating eight differential LVDTs in each measuring section.

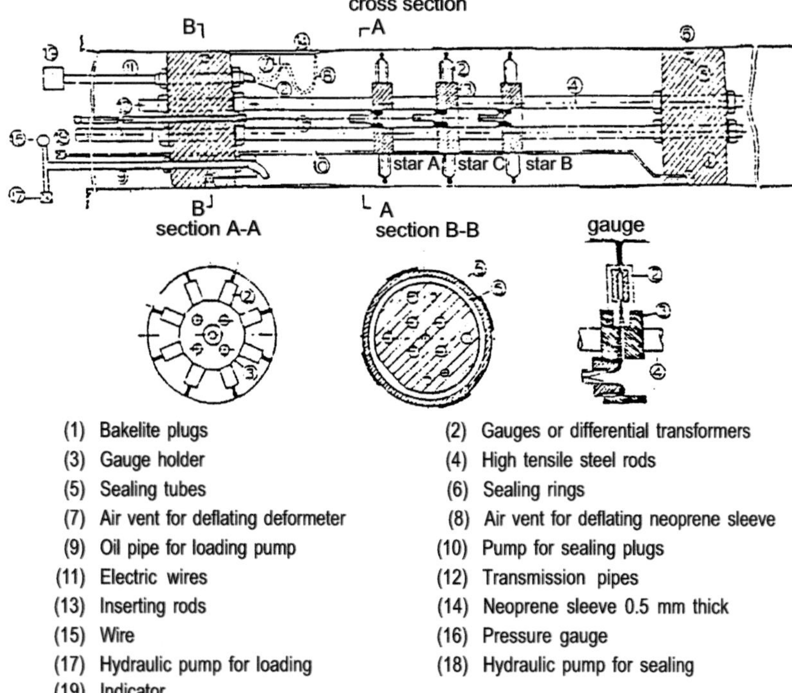

Figure 7.5 Yachito tube deformeter shown without neoprene sleeve

(9) OYO elastometer 200

This dilatometer is 60 mm in external diameter and has an effective length of 520 mm. It is designed to develop contact pressures at the instrument and rock interface up to 20 MPa. Hence, this dilatometer is suitable for testing medium to hard rocks. The instrument has a displacement measurement range of 15 mm, i.e. 25% of its diameter. It can, therefore, also be used in poor quality rock masses. The instrument uses an indirect method of measurement of borehole diameter incorporating a contact balancer system.

(10) CSIRO pressiometer

The main features of this instrumentation system are that it applies uniform stress on the borehole wall and the change in the diameter of the borehole is measured directly by LVDTs. The load on the borehole wall is applied through two opposite segments to be able to split the rock in the walls of the borehole as a possible means of assessing the rock strength and stress utilising a technique similar to hydro-fracturing. The pressiometer itself consists of four quadrant curved jacks mounted on a collapsible but stiff steel core of slightly smaller diameter to facilitate insertion and withdrawal of the pressiometer. Hydraulic connections to the jacks are made by means of small diameter steel tubing to a common hand operated pump and a short bleed tube is attached to facilitate the removal of air bubbles from the hydraulic circuit. A wedge mechanism with a wedge angle of 1:10 ensures that the shoes expand by approximately 6 mm and this also allows the shoes to fit snugly within the borehole wall during the test. This also prevents the inflated jack from deforming excessively and can be used for several cycles of loading.

The borehole deformations are measured in two radial directions at mid-section of the jacks using two DCDT transducers, with an operating range of ±2.5 mm. The transducers are excited from a common 6 volt DC power supply unit. The operating contact stress developed between the dilatometer and the borehole wall is 25–35 MPa. Figure 7.6 shows the CSIRO pressiometer.

7.4 Borehole Jacks

This is a cylindrical device which applies pressure to the wall of a borehole along a limited arc of contact and the corresponding deformation is measured either at the

- centre of the arc of contact, or
- outside the contact area.

The modulus of deformation is obtained by specific formulae developed for each configuration of borehole instrumentation. Figure 7.7 shows a diagrammatic representation of a borehole jack.

A number of borehole jacks have been developed for evaluation of the modulus of deformation of rocks. The variation in design of the specific instrumentation depends upon the method of applying stress on the

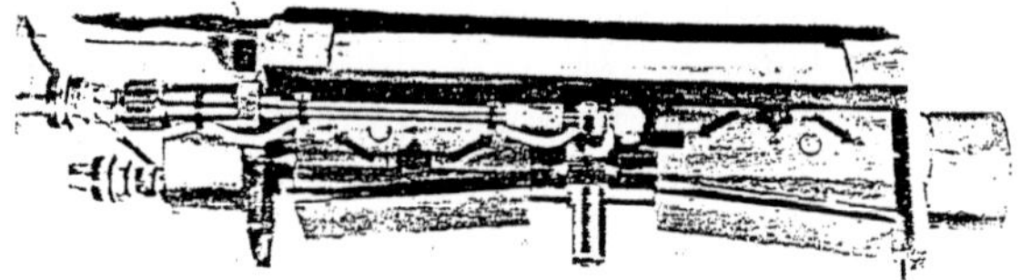

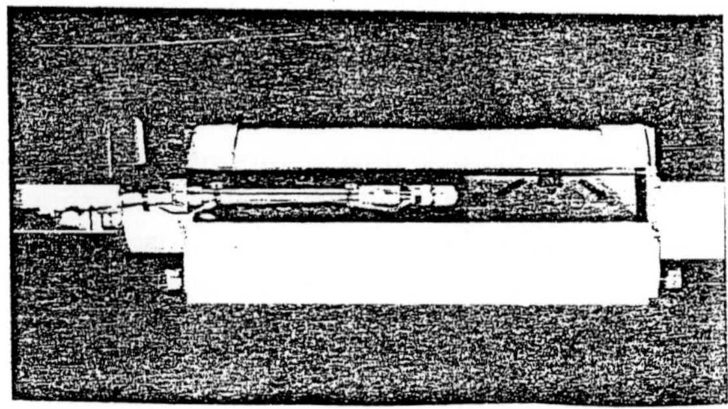

Figure 7.6 CSIRO pressiometer (Worotnicki *et al.*, 1975)

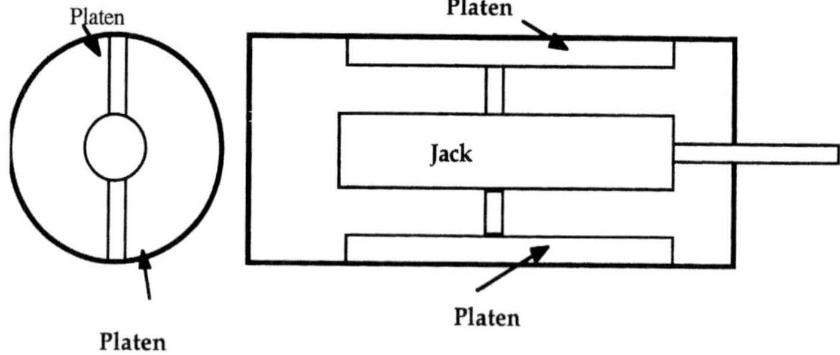

Figure 7.7 Schematic diagram of a borehole jack

borehole wall and the types of loading plates used, either rigid or flexible. Flexible platens enable a uniform distribution of stress on the borehole wall. In the past, the following three methods of applying loads on the platens have been used.

1. Hydraulic piston
2. Wedge
3. Curved or flat jacks

These instrumentation systems enable the tests to be performed in any desired direction in order to evaluate anisotropy or assess the effect of any geological feature on the stability of a rock structure. The modulus of deformation can be calculated by the use of the following generalised formula:

$$E = \frac{\Delta p K(\upsilon, \beta)}{\Delta ud/d} \tag{7.5}$$

where

E = modulus of deformation

Δp = pressure increment (MPa)

Δud = diametrical displacement increment

d = borehole diameter (m)

$K(\upsilon, \beta)$ = constant depending upon υ and E and load applied through angle β

β = angle of loaded arc of the jack

A large number of borehole jacks have been developed and some salient features of selected borehole jacks are presented here.

(1) Centex cell

This borehole jack was developed by Dawance in 1964 for Electricite de France. The Centex cell comprises a split cylindrical sleeve placed over an oil-filled inflatable rubber cylinder. When this is fully inflated, the cylindrical curved jacks establish a contact with the borehole wall through a contact angle of 143°. The method of measuring deformation is through an induction transducer with the instrument sensitivity of 0.01 mm. Pressure is developed in the pressurised rubber of the cell with oil and this pressure is measured with a pressure gauge. The cell has an overall diameter of 76 mm and overall length of 306 mm permitting the instrument to be used in a standard diameter borehole. Thus, no special drilling equipment is needed for the installation and measurement of modulus of deformation with this borehole jack.

The contact pressure developed on the interface of the cell platen and the borehole wall is in the region of 105 MN/m^2. Thus, the instrumentation has application from soft to hard and massive rock types.

(2) NX plate jack

The NX plate jack, developed by Goodman et al., (1972) subtends a contact angle on the borehole wall '2β' equal to 90°. It comprises a cylindrical cell incorporating 10 miniature hydraulic rams pushing two steel platens against the wall of the borehole being tested. The instrument has an external diameter of 76 mm and an overall length of 204 mm. The deformation of the borehole is measured by two pairs of LVDT transducers located at both ends of the platens which are 200 mm in length. The maximum range of the trans-

Table 7.3 Design parameters of various borehole jacks for measuring modulus of deformation of rock *in situ*

Instruments	*Method of measuring deformation*	*Method of pressure generation*	*Method of pressure measurement*	*Pressure range on borehole wall* $(MN/m)^2$	*Diameter and length*	*Loading geometry*
Centex cell (Dawance)	Induction transducers (0.01 mm)	Pressurised rubber with oil	Pressure gauge	105.00	76 mm × 306 mm	
NX plate jack	2 LVDT	12 pistons	Pressure gauge	64 MN/m^2	76 mm × 204 mm	$2\beta = 90°$
Geo-extensometer	2 LVDT	Circular hydraulic pump	Pressure gauge	22	76 mm × 204 mm	$2\beta = 143°$
German stress/strain meter (Martini *et al.*, 1964)	Induction transducer	Wedge mechanism	Calibration graph	—	54 mm × 230 mm	$2\beta = 180°$
Talobre jack	LVDT	Hydraulic piston	Load cell	22	54 mm × 230 mm	$2\beta = 90°$
Panek jack	LVDT	Flat jack	Pressure gauge	14.7	54 mm × 230 mm	$2\beta = 90°$
Photo-elastic pre-stressed meter	Movement of wedge	Wedge moving rigid pin	Photoelastic stress meter	133–637	160 mm × 230 mm	$2\beta = 18°$

ducers is 5.08 mm. The pressure in the hydraulic jacks is measured by a pressure gauge. The range of contact pressure on the borehole wall is up to 64 MN/m^2, thus the instrument is suitable for testing medium strength rocks.

(3) Geo-extensometer

The geo-extensometer is an improved version of the Centex cell developed for softer rock types utilising a cylindrical shaped hydraulic pump. The instrumentation has the following specifications:

- Method of measuring deformation–2LVDTs
- Liquid-circular hydraulic pump
- Measurement of pressure–pressure gauge
- Range of borehole wall stress–22 MN/m^2
- Diameter and length ϕ =76 mm, length = 306 mm
- $2\beta = 143^{\circ}$
- Made by Absi and Seguin (1967)

(4) German stress/strain meter

Developed by Martini *et al.* (1964), the German stress/strain meter utilises a wedge mechanism, driving apart instrument platens by an advancing screw mechanism. The main specifications of the instrument are as follows:

- Method of measuring deformation–induction transducer with the sensitivity of 0.01 mm
- Pressure developed by a wedge mechanism
- Measurement of pressure–calibration graph
- Range of borehole wall stress–very high
- Diameter and length: ϕ = 50 mm, length = 530 mm
- $2\beta = 180^{\circ}$

(5) Photo-elastic pre-stress deformation meter

This instrumentation scheme was developed by Singh (1967) for application to the collar of a 160 mm diameter borehole in an underground excavation. The instrument utilises a photo-elastic stress plug to measure the contact stress at the interface of the borehole wall and the platen. The platens are pushed apart in the instrument by a wedge operated by an advancing screw mechanism. The salient points of the instrumentation scheme are given below:

- Method of measuring deformation–wedge mechanism using calibration graph
- Method of generating pressure–wedge mechanism

- Measurement of pressure–photo-elastic stress gauge utilising fringe order technique
- Range of borehole wall stress–637 MN/m^2
- Diameter and length $\phi = 160$ mm, length = 383 mm
- $2\beta = 19.5^o$

(6) Talobre Jack

Talobre (1964) developed this jack utilising hydraulic pistons to push cylindrical platens on to the borehole wall. The platens have a contact angle 2β of 90°. The specification of the instrument is given below.

- Method of measuring deformation–LVDT
- Method of generating pressure–hydraulic piston
- Measurement of pressure–load cell
- Range of borehole wall stress–22 MN/m^2
- Diameter and length $\phi = 56$ mm , length = 230 mm
- $2\beta = 90^o$

(7) Panek jack

This instrumentation scheme was developed by Panek et al (1964) for measurement of the modulus of deformation in hard rock. The main attributes of this borehole jack are as follows:

- Method of measuring deformation–LVDT
- Method of generating pressure–flat jack
- Measurement of pressure–pressure gauge
- Range of borehole wall stress up to 69 MPa
- Diameter and length $\phi = 54$ mm, length = 230 mm
- $2\beta = 90^o$

7.5 Borehole Penetrometer

7.5.1 Principle of Penetrometer

These types of instruments force a pair of spherical or cylindrical indentors against a borehole wall (Figure 7.8) and measure the corresponding borehole deformation.

The calculation of modulus of deformation is based on the theory of elasticity simulating a sphere indenting a borehole wall. Two cases are given below.

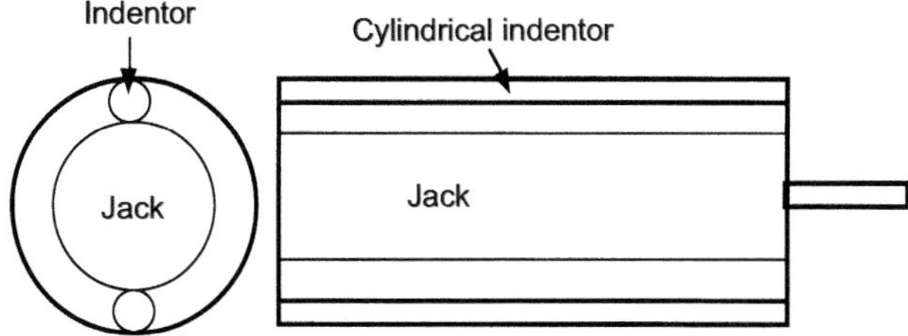

Figure 7.8 Borehole penetrometer

(1) Spherical pin

$$\delta u = \frac{9\pi^2 P^2}{16} + \left(\frac{(1-\upsilon_1^2)}{E_1} + \frac{(1-\upsilon_2^2)}{E_2}\right)\frac{R_1 - R_2}{R R_2}\Bigg]^{1/3} \quad (7.6)$$

R_1 = radius of borehole
R_2 = radius of indenter
υ_2 = Poisson's ratio of indenter
υ_1 = Poisson's ratio of rock
E_1 = elastic modulus of rock
E_2 = elastic modulus of indenter
P = load
δu = measured displacement

(2) Cylindrical pin

$$\delta u = \frac{(1-\upsilon^2)P}{AE} \quad (7.7)$$

A = area of contact
E = modulus of deformation
P = load on the pin
E = Young's modulus
υ = Poisson's ratio
δu = measured displacement

A number of borehole penetrometers has been developed. Their application is mainly concerned with testing of a rock's response to loads exerted by rock bolts. The contact pressure developed in the interface of the rock and the indenters of penetrometers is very high in the order of magnitude of 150 to 675 MPa, while the loaded volume does not exceed 0.0001 m^3. Therefore these instrumentations can only be used to test the relative deformability of rock and do not have much relevance in calculating modulus of deformation of rock masses. Several instrumentation schemes have been developed for measuring the modulus of deformation of rocks utilising penetrometer principles. A description of some of the borehole penetrometers is given in the following sections.

7.5.2 U.S. Bureau of Mines Penetrometer

The U.S. Bureau of Mines penetrometer was made by was made by Hult (1963) and was extensively used by Dryselius (1965) utilising a spherical indentor to measure the deformation of rock under high contact pressures. Thus it was envisaged that comparatively large rock masses surrounding the borehole will be affected and it will therefore represent the material characteristics of a larger area of influence. The instrumentation itself was 54 mm in diameter and 230 mm in length. The method of generating pressure on the borehole wall was by extracting a movable wedge which in turn moved a spherical pin against the wall of the borehole. The load applied to the borehole wall was directly measured by a pressure gauge. The method of measuring deformation was by recording the movement of wedge inside the borehole probe. The range of borehole wall stresses was 290 MPa.

7.5.3 Stiff Stress Meter

The stiff stress meter was developed by Steers (1965) for measuring very hard rock application. The main design attributes of this instrument are as follows.

- Method of measuring deformation–movement of wedge
- Method of generating pressure–rigid pin moving wedge
- Measurement of pressure–pressure gauge
- Range of borehole wall stress–very high
- Diameter and length $\phi = 32$ mm, length = 95 mm
- Pin–spherical pin

7.5.4 Photo-elastic Pre-stressed Deformation Meter

The photo-elastic pre-stress meter, developed by Singh (1967), utilises a stiff spherical indenting pin or a cylindrical pin with line contact on the borehole wall acting as a penetrometer. The design configurations and specifications are given below.

- Method of measuring deformation–screw movement
- Method of generating pressure–screw driven wedge
- Measurement of pressure–photo-elastic stress cell
- Range of borehole wall stress–133–637 MPa
- Diameter and length $\phi = 89$ mm, length = 380 mm
- Cylindrical pin or spherical indenting pin

7.6 Testing Procedure in Using Borehole Deformation Instruments

7.6.1 Testing Procedure

Because the range of deformation measurements in most dilatometers is limited, one of the main requirements of the test procedure is that the borehole deformation measurements are carried out in boreholes 2–3 mm larger than the instrumentation system itself. Therefore special care has to be taken in drilling the boreholes using diamond coring bits. Cores obtained at the site can be utilised for determining Poisson's ratio and other rock properties by laboratory tests. In softer rock, the borehole should be cased immediately after drilling and tests should be carried out from the blind end of the borehole outwards after withdrawing the casing.

During the borehole deformation testing, stress/strain diagrams are obtained in various loading and unloading cycles as a record of the deformation of rock mass caused by various levels of applied stress σ.

The borehole deformation tests can also be used to obtain stress/strain information on the behaviour of fissured rock under stress. Results depend upon (a) method used in obtaining deformation and (b) point of measuring deformation viz. the surface of rock, at depth in the rock mass or some distance laterally from the load. Test results vary depending upon the following test parameters.

- Rate of loading
- Direction of loading

Borehole tests take relatively less time i.e. several hours as opposed to several weeks in the case of large-scale tests. Figure 7.9 shows alternative schemes of recording load deformation characteristics of rock masses in order to derive maximum information from the test results, as follows.

In **scheme A**, loading and unloading is carried out up to several levels of stress recording the stress/time behaviour of the rock. The main purpose of

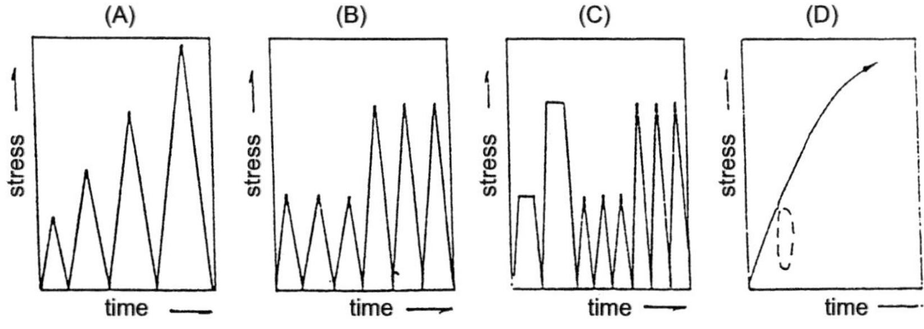

Figure 7.9 Method of loading used in the borehole deformation test (after ASTM)

the test is to observe the response of the rock to various levels of stress, and the effect of consolidation and hysteresis due to repeated loading and unloading.

In **scheme B**, the loading pattern is repeated several times at the same level of stress where more emphasis is given to the observation of permanent settlement effects due to repeated loading cycles. The results are used to calculate modulus of deformation of the rock.

In **scheme C**, the rock is loaded at various levels of stress and the time-dependent deformation at two stress levels is recorded. The results enable calculation of secondary creep rates at various stress levels and the modulus of deformation for subsequent cycles.

In **scheme D**, cyclic loading is taken up to the maximum stress at which fracturing of the borehole takes place; more emphasis is being given to the yield stress of the rock and measurement of the stress in the rock using a modified hydro-fracturing technique.

7.6.2 Interpretation of Results Using the Borehole Penetrometer

The general approach in the interpretation of the penetrometer data is based on the laboratory calibrations. Interpretation of deformation of the borehole by indenters of various shapes is based on the theory of elasticity (Love, 1944; Timoshenko and Goodier, 1951). Application of the above theory to borehole penetrometers indicates that contact pressures on the borehole wall are very high and well beyond the elastic limits of most rock types (Singh, 1967). Attempts have been made to interpret stress deformation curves based on laboratory calibration on nine different rock types.

Figure 7.10 shows the stress/strain curves under line contact loading of rock slabs by the photoelastic pre-stressed meter in a testing frame. It is noted that the slope of the stress strain curves of 10 rock types do not follow the sequence of the elastic moduli of the rocks. Therefore, it is not logical to use penetrometer results for interpreting the modulus of deformation of rocks.

7.7 Methods of Monitoring Rock Movements

The stability of rock structures depends upon a multitude of interactive factors identified in Figure 7.11 (Kovari and Amsted, 1993). The structural behaviour of the rock excavation depends upon the initial state of stress and the rock conditions specified by the strength and deformational properties of the rock. In high stress fields and weak rock, an excavation is likely to undergo excessive deformation. Additional supports may be required to stabilise the rock structure. Methods of excavation such as the use of tunnel

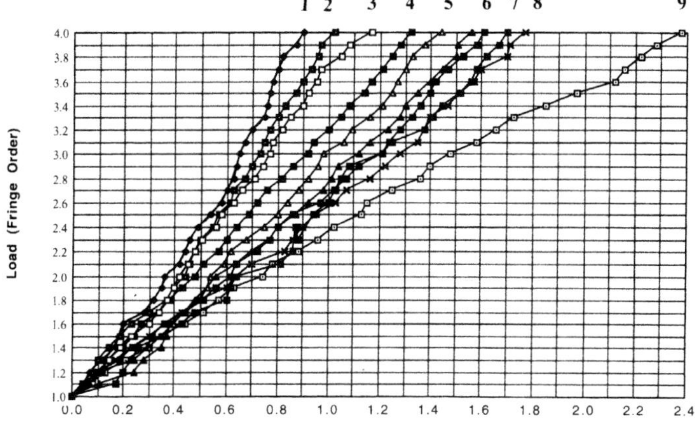

(1) Steel, (2) marble, (3) granite, (4) Ancaster limestone, (5) Aston limestone (6) Portland limestone, (7) white sandstone (8) York sandstone and (9) Roche Abbe

Figure 7.10 Load deformation curves of rock slabs by the photo-elastic pre-stressed meter using cylindrical platens (Singh, 1967)

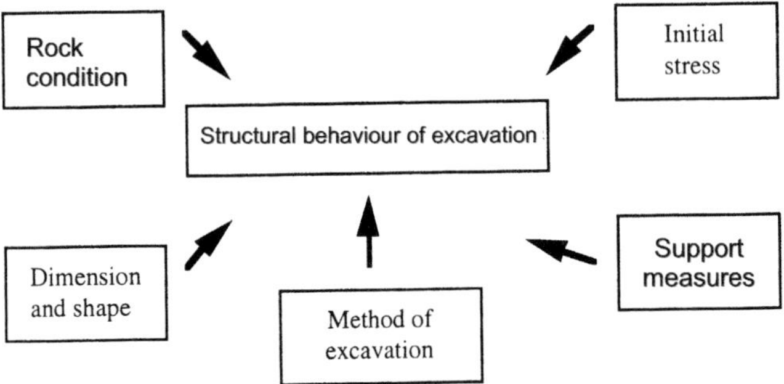

Figure 7.11 Factors affecting stability of desired structures (Kovari and Amsted, 1993)

boring machines or drilling and blasting also affect the redistribution of stress and the development of fracture patterns around the excavation. Thus, structures produced by drilling and blasting display less stability than those excavated by full face boring or cutting machines.

Thus factors influencing the stability of an engineered rock structure can be summarised as follows:

(1) Rock condition characterised by
- intact rock strength and
- strength of rock joints

(2) Stress regime characterised by
- initial state of stress
- stress redistribution caused by dimension and shape of the tunnel and the method of excavation, utilising drilling and blasting or cutting
- machine, which determines uniform loading or point loading of the support system

It has been observed that the point-loaded support undergoes considerably more deformations than in the case of uniformly loaded support.

7.6.1 Possible Adoption of Roadway Shapes to Increasingly High Rock Pressure Regimes

Figure 7.12 indicates that the shape of an underground excavation has significant bearing on the stability of the underground excavations at various depths. Square or rectangular roadway shapes are commonly used in shallow coal mines as main or face access roadways. At greater depths, however, this shape causes stress concentrations in the corners which may result in roof or floor failure. Dimensions and shapes of tunnels are also important from stability considerations as increasingly round profiles promote an arching action on the rock and lining system.

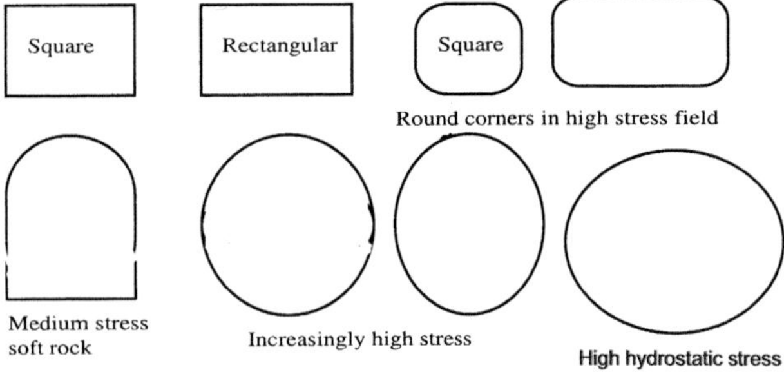

Figure 7.12 Shape of tunnel in increasing high-stress regimes

Figure 7.12 indicates the progressively changing profile of tunnels in increasingly high-stress regimes. Use of invert arches and circular supports are increasingly employed where floor rock is susceptible to heave due to excessive floor pressures.

There are two ways in which rock can respond in the vicinity of an underground structure by de-stressing as a consequence of failure or creep processes:

(1) closure of the opening or
(2) if hindered, by the development of rock pressures.

It has therefore become customary to measure strata displacement for stability evaluation of underground structures during the construction or working life of the project.

7.6.2 Reasons for Measuring Rock Movements

In practice, the measurement of rock displacement of a rock mass or continuous monitoring of rock movement in the vicinity of an underground excavation is necessary for the following reasons.

(i) Displacement measurement and analysis techniques are simple and do not require complex computations.

(ii) Displacement measurement is more relevant to conditions where stability criteria relate to closure.

(iii) Methods can be used to find the origin and cause of instability and to suggest possible remedial action.

7.6.3 Techniques of Measuring Rock Displacements

In the past, various methods have been used for monitoring or measuring strata displacements around an underground excavation.

(i) Surface closure or extension measurement

(ii) Borehole instrumentation, which can be

  (a) Extensometer—to measure axial movements along the axis of the borehole

  (b) Inclinometers—to measure lateral movements

  (c) Three-dimensional movements—three-dimensional probe within the borehole

(iii) Tilt measurement—surface subsidence

The above-mentioned strata displacement measuring or rock mass monitoring techniques can be used to determine the stability status of surface and underground excavations, besides assessing foundation loading and movements in rock slopes.

7.6.4 Criteria for Decision-making Using Displacement Criteria

An engineer has to make decisions with regard to location, alignment, shape and size of the underground structure, method of excavation, support measures, temporary and permanent supports and ground improvement methods. The technical criteria for making correct decisions are often based on displacement measurements in and around the excavation in the rock. These decision criteria may be based on the safety considerations during the construction and the operational life of the structure. From the safety perspective the criteria for displacement may be falling of isolated roof blocks in underground mine or localised roof fall or large-scale collapse or caving during ore extraction. Alternatively, the amount of displacement may be used as a criterion for the design of original dimensions of underground gate roadways (Whittaker and Singh, 1979). Measurement of parameters of subsidence or settlement in a foundation of a high rise building in the vicinity of a shallow tunnel may be used as a criterion of stability (Figure 7.13).

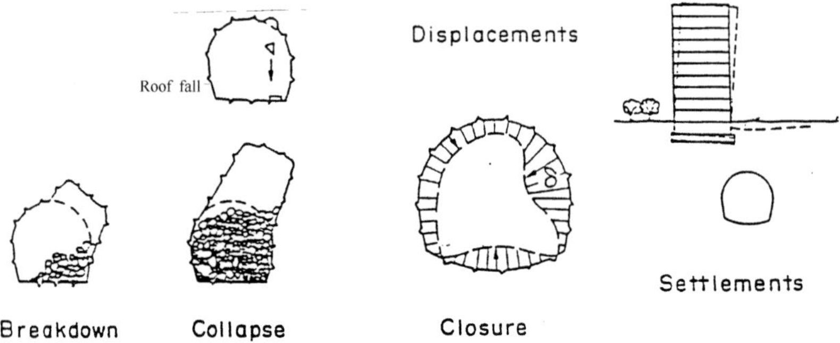

Figure 7.13 Criteria for decision-making in underground construction (Kovari and Amsted, 1993)

7.7 Monitoring of Designed Rock Structures during the Construction and Operational Stages

Extensometers are extensively used in the field of rock engineering for measuring directional displacements, bed separations, settlements, closure, and floor heave. In tunnels, borehole extensometers are used to monitor roof and side wall displacements and to locate the extent of tension zones above the excavation in order to design the length of roof anchorage. Extensometers are also used as monitoring devices for the safety evaluation of mining access roadways, underground excavations, shafts and tunnels, particularly on potentially unstable rock or loose wall slopes and in monitoring the performance of rock support systems. Continuous monitoring of underground excavations can provide warning of potential problems in order to implement remedial measures in time (Hoek and Brown, 1980). Extensometers are also used for measuring precise rock movements to an accuracy of 0.02 mm in large-scale tests and *in situ* plate bearing tests. Several methods of measuring rock mass displacements are available as follows.

7.7.1 Convergence Measurements

Convergence measurements are normally carried out by means of a tape or a tape extensometer placed between targets attached to the opposite walls or roof and floor of an excavation as shown in Figure 7.15. A number of convergence or closure measuring instruments are available.

- Tape extensometers
- Bar extensometers
- Rod extensometers
- Joint extensometers
- Wire extensometers

Deformation surveys of underground access roadways or underground tunnels are routinely carried out for the following reasons.

- To measure the closure of a tunnel for assessing clearance and serviceability of the roadway or tunnel
- To assess the rate of closure of tunnel to establish stability
- To ascertain the cause of instability

The most convenient method of measuring tunnel closure is to establish a station by inserting rock bolts on the opposite side walls and roof and floor of the excavation and to carry out periodic measurements with time. By subtracting the original readings from the subsequent readings, the tunnel's vertical and side closures can be obtained.

Figure 7.14 shows different arrangements for measuring tunnel closure. The actual closure measurements could be carried with the help of a precise measuring tape, a surface rod extensometer with expansion shell anchors or rod extensometers with fully grouted steel stations (Figure 7.15). The last two instrumentation systems are used to measure closure by a dial gauge to the accuracy of a fraction of a mm.

Figure 7.16 shows the tunnel closure in millimetres with time. It can be seen that chord 3 has very little movement while length 1 shows

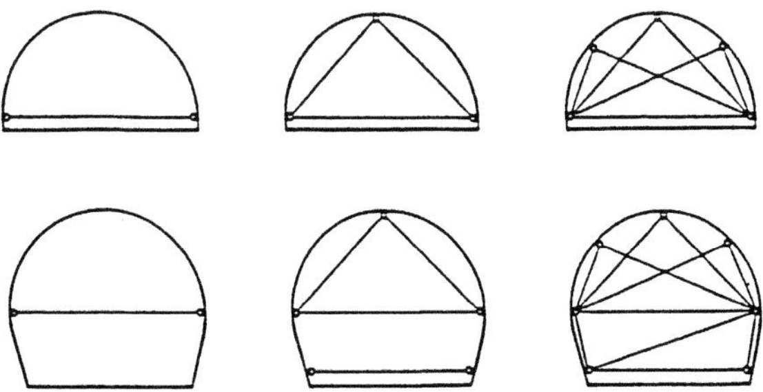

Figure 7.14 Different arrangements of measuring tunnel closure

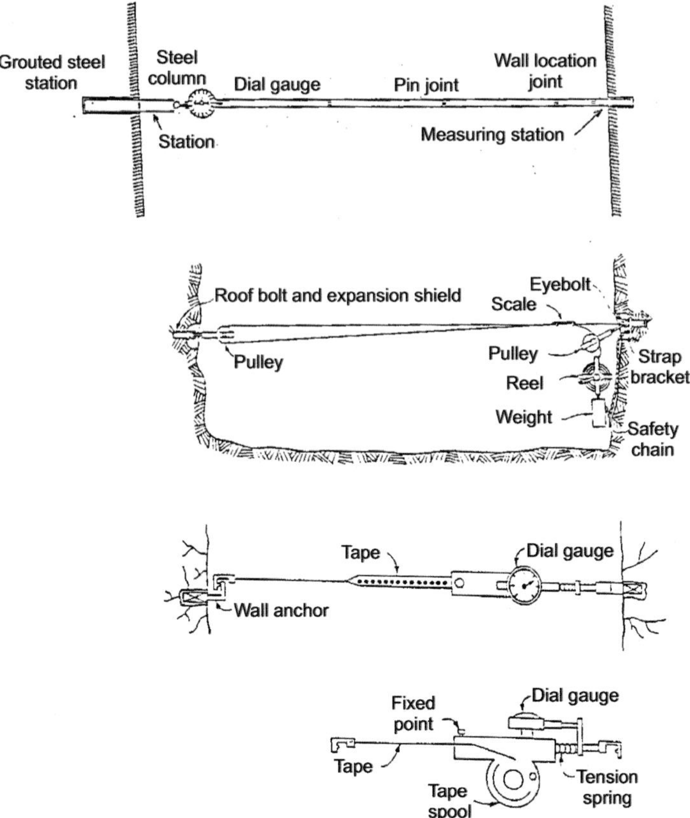

Figure 7.15 Three extensometers for measuring tunnel closure precisely

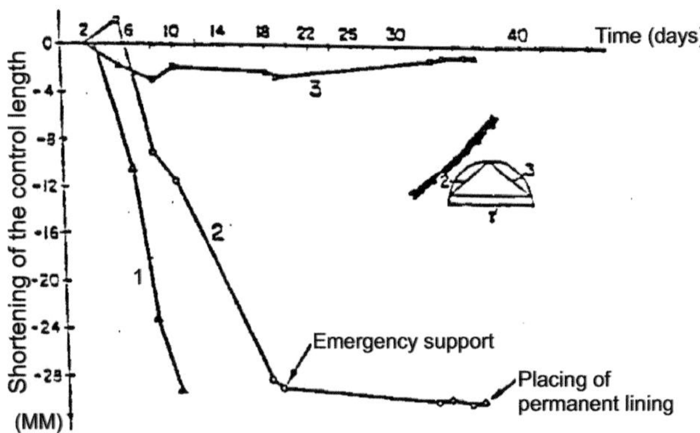

Figure 7.16 Closure-time graph of tunnel to assess the source of instability

uncontrolled movements with time. It can also be seen that chord 2 shows the arrest of uncontrolled closure because of the installation of an emergency support. The excessive and uncontrolled closure of the tunnel was attributed to the presence of a fault in the vicinity of the tunnel.

7.7.2 Borehole Extensometers

Borehole extensometers are designed to monitor directional or axial deformations in the rock mass surrounding a surface or underground excavation. Basically points within the borehole whose movement is to be monitored are installed within the borehole in the form of anchors. Attached to these anchors are individual wires or rods whose relative axial movement is recorded with respect to the bottom and/or collar of the borehole. If cracks are developing in the rock mass surrounding an excavation, then the wire or rod of the extensometer will appear to move inside the borehole by the same amount as the movement of the individual anchor. If, however, the ground around an excavation is compacting then the extensometer wire or rod will appear to move out or extrude out of the borehole. Thus, the precise movement of the borehole anchors with time can be recorded. Types of extensometers used in any investigation will depend upon the field conditions and the scope of the investigation. In simple situations, a single position extensometer may be all that is required while in a more complex situation a sophisticated multiposition extensometer may be necessary to monitor bed separation and crack formation or ground compaction around an excavation. Figure 7.17 shows three types of borehole extensometers used extensively in rock mechanics investigations.

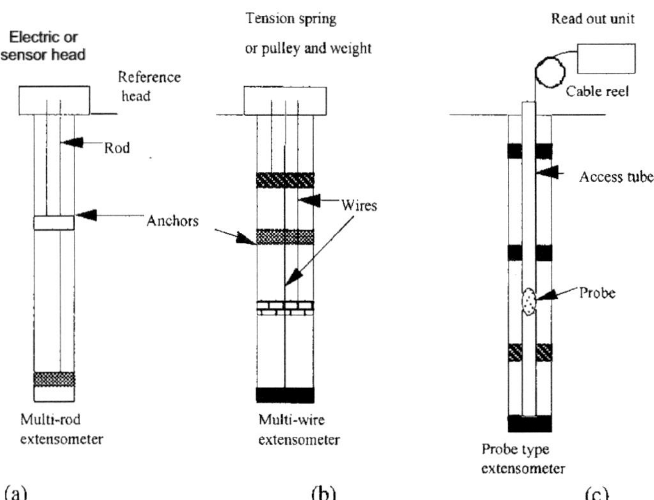

Figure 7.17 Borehole extensometers

- ***Wire extensometer***: A wire type of extensometer utilises a thin stainless steel wire to connect the reference anchor points to be placed in a single borehole. The wires are tensioned by springs or weights extended over a roller sheave and connected to hanging weights.
- ***Multiposition rod extensometer***: The multiposition rod extensometer comprises disc-shaped anchors and 10 mm rods screwed to the individual anchors to form the measuring elements. The rods are housed in the flexible tubes and are free to move. Outer anchors have a suitable number of holes to allow them to pass without causing friction.

A pre-fabricated assembly is normally used to minimise the field work. A standpipe is grouted at the mouth of the borehole before the extensometer assembly is inserted into the borehole and grouted with a weak grout. The initial readings of the rods are taken with a depth micrometer with reference to the frame bolted to the standpipe. Subsequent readings of the rods are used to obtain differential movements of the anchors.

7.7.3 Reed Switch Extensometer

Figure 7.17(c) shows a reed switch extensometer made by Burland *et al.* (1972) and Marsland (1973) comprising a number of permanent magnets installed in the borehole wall as anchors. A reed switch sensor head lowered down the borehole by a graduated cable determines the precise locations of these anchors within the borehole. The procedure for inserting the reference anchors in the borehole wall is to mount a number of ring magnets on spring loaded PVC magnet holders at predetermined positions on the access tube and hold them together by a wire. This instrumentation is used in a 100 mm diameter and up to 30 m long borehole and using up to 12 ring magnet reference points. The whole assembly is then inserted in the borehole and the wire is withdrawn to release the ring magnets onto the borehole wall. The interspaces between the borehole wall and the access tube were backfilled with weak Celtite Selfix grout to prevent collapse of the borehole in soft ground. Two methods of measuring axial movements of magnetic anchors can be used.

1. Lowering the reed switch sensor head into the borehole by a graduated tape measure and locating the precise centre of each ring magnet sensor
2. Permanently placing the reed switch sensor at the centre position of a ring magnet attached to a steel rod (Each switch is connected to a separate electric circuit. Adjusting the rod to the central position of the ring magnet and measuring the movement of the rod by a depth micrometer help determine precise axial movement of each magnetic anchor.)

Figure 7.18 shows a method of presenting the data obtained from an extensometer installed on the sidewall of an underground roadway.

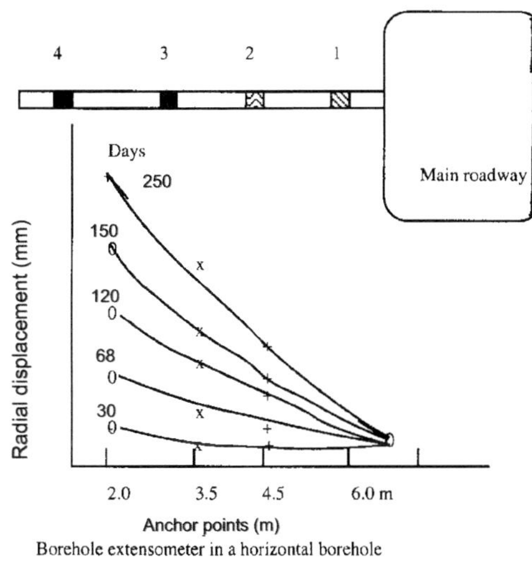

Figure 7.18 Method of plotting extensometer data

7.7.4 Application of Extensometers at Pit Bottom and Shaft Inset for Assessment of Shaft Instability

Surface extensometers shown in Figure 7.15 and borehole extensometers shown in Figure 7.17 have been used to monitor strata instability in shafts insets in coal mines. An example is given of a shaft which was completed in 1914. It had a major fault in close proximity to the shaft pillar (Figure 7.19). In order to improve the productivity of shaft 2, it was decided in 1980 to drive a

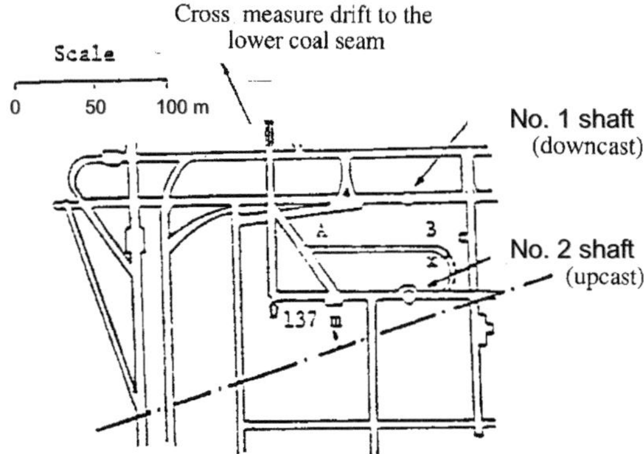

Figure 7.19 Shaft bottom layout at the site of investigation

curved roadway using drill and blast technique in the vicinity of the shaft to complete the pit bottom mine car loop. This construction work initiated ground movement in the vicinity of the shaft pillar in the form of gradual closure of the shaft bottom roadways, the skip pocket and the shaft inset. Side movements were significant and appreciable cracking of masonry and a concrete wall was increasing and roof girders were buckling.

In order to monitor the instability, the following instrumentation scheme was adopted.

- In the shaft inset, five measuring stations were established on which the closure measurements were carried out. Dial gauge extensometer as shown in Figure 7.15(a) and telescopometers were used to measure the closure to an accuracy of 0.1 mm.
- Multi-position rod extensometers were used in 6–8 m deep boreholes. Also two 6 m deep single position rod extensometers were used to monitor the strata movement.

The borehole extensometer design was based on placing four rods in a 43 mm diameter borehole, each rod being positioned in a plastic sleeve allowing it to move freely, and at the end of it was installed a fixed reference anchor. Figure 7.20 shows the instrumentation scheme used in the project and Figure 7.21 shows the closure results over a period of 550 days. After some 140 days of monitoring, it was decided to reinforce the shaft as well as

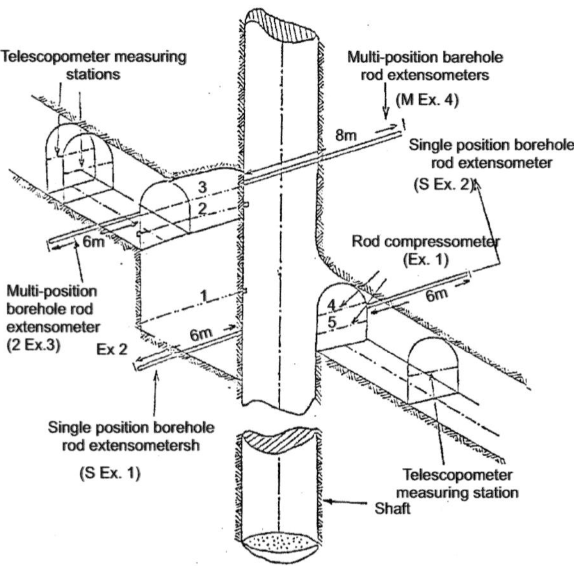

Figure 7.20 Example of use of extensometers to assess instability of shaft insets and pit bottom roadways

the inset by installing 3 m long and 29 mm diameter, fully grouted resin rock bolts placed in a 1.1 m square grid pattern to secure the rock mass.

Figure 7.21 also shows the influence of rock bolts in reducing the rate of closure at a number of stations after installation of the reinforcement system. The results also indicate that it took about 100 days before the full beneficial effect of installing the rock bolt system was realised.

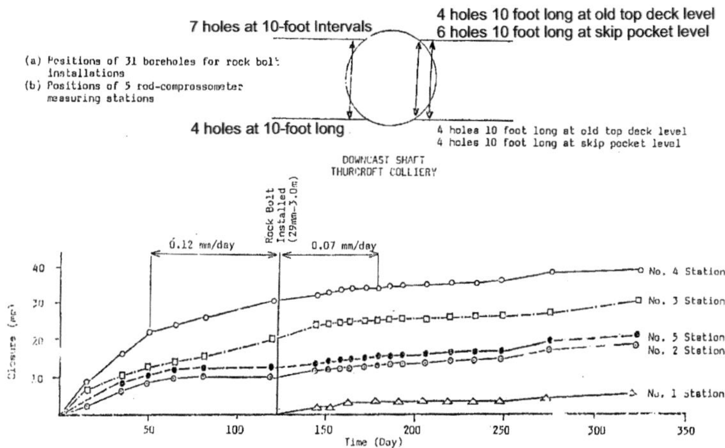

Figure 7.21 Roadway closure versus time at the site of investigation

7.8 Borehole Inclinometers

A borehole inclinometer is a device which when lowered down a borehole measures the inclination of a discrete point within a borehole with respect to gravity. An inclinometer probe comprises a weight actuated cantilever clamped at the top of the probe. An electrical strain gauge rosette installed on the cantilever will give an electronic signal as the cantilever bends with respect to gravity in an inclined borehole. The out-of-balance current is proportional to the inclination of the borehole and a calibration graph can be used to relate an electric output to the inclination readings (Fig. 7.22).

The entire inclinometer system comprises a borehole drilled in the rock surrounding an excavation and lined with a grooved guide tube for the inclinometer probe. The interspace between the guide tube and the borehole wall is backfilled with a low strength grout. The top and the bottom of the inclinometer probe is fitted with guide wheels which permit it to slide in the grooves of the access tube (Figure 7.23). This feature of the instrumentation permits the measurement of inclination by the tilt sensor to be carried out along a vertical plane. The access tube has grooves in two orthogonal directions which permits the borehole survey to be carried out in two

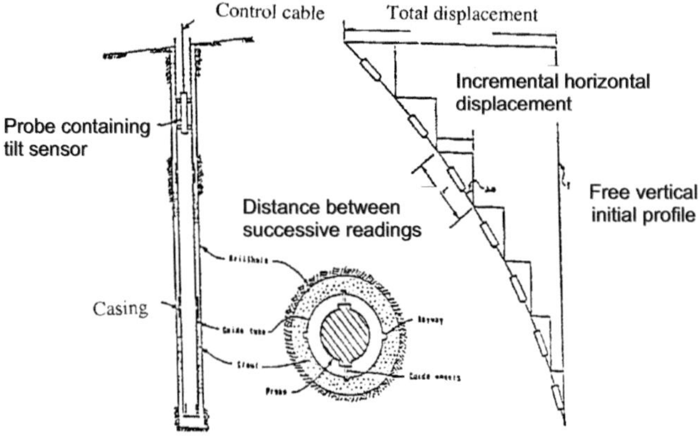

Figure 7.22 Principle of measuring horizontal displacement by using inclinometer

mutually perpendicular directions, thus, enabling the true dip and true dip direction of the borehole to be evaluated.

The inclinometer is lowered in the borehole by equal incremental depths by a graduated cable and the inclination readings are taken. The results are plotted for each depth and spot angular readings are plotted. The horizontal displacement at each incremental depth is taken as L sin $\delta\theta$. Integrated displacements versus depth graph will give the horizontal displacements of discrete points within the borehole (Figure 7.23).

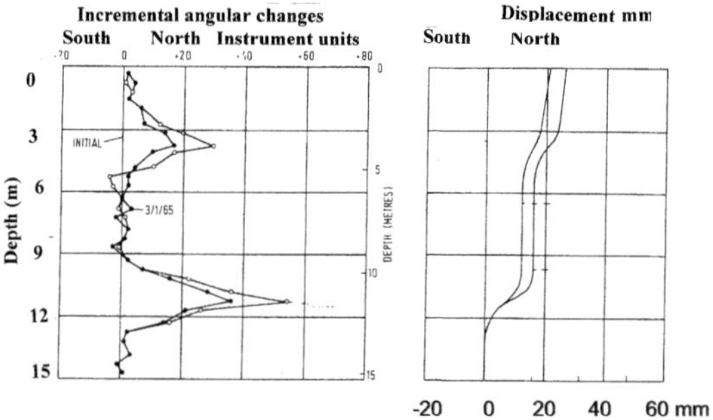

Figure 7.23 Inclinometer's angular readings and integrated horizontal displacements versus depth (ISRM, 1981)

7.9 Measurement of Rock Displacements in Three Dimensions

An instrumentation scheme to monitor three-dimensional strata movements in boreholes was especially developed for coal mining applications. The instrumentation scheme together with the details of the probe design are illustrated in Figure 7.24 (Singh, 1976). The three-dimensional borehole measurement system comprises a borehole drilled in the host rock surrounding an excavation in which a number of discrete magnetic reference points, mounted on suitable holders, are installed along the periphery. The borehole is lined with a grooved access tube which permits a three-dimensional probe to be lowered in the borehole. A Hall probe extensometer transducer locates the magnetic reference points to an accuracy of 0.006 mm and the depth of the probe in the borehole with reference to the borehole mouth is measured by a probe depth monitor (Smart and Singh, 1976). Two tilt transducers, incorporated within the probe at right angle to each other, simultaneously measure the inclination of the borehole at the reference point in two orthogonal directions. This enables the calculation of three-dimensional coordinates of the reference points to be evaluated with reference to the borehole mouth.

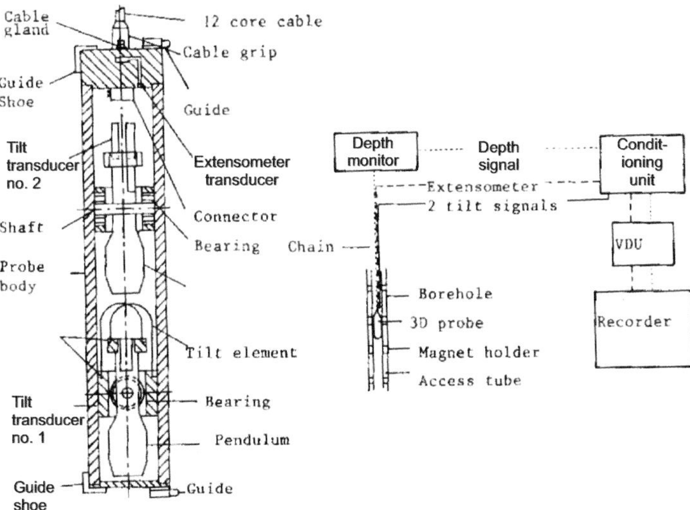

Figure 7.24 Three-dimensional rock displacement probe

The directional control of the probe within the borehole is effected by use of spring loaded slides in the probe engaging into the grooved access tube. This prevents twisting of the chain and the probe in the borehole. The power packs, signal conditioning units and recorders associated with the probe and depth monitoring unit were especially designed for use in the gassy environment in coal mines.

The site of the investigation was situated in a gate roadway supported by the Moll system of arches in the Z-system of longwall mining. The depth

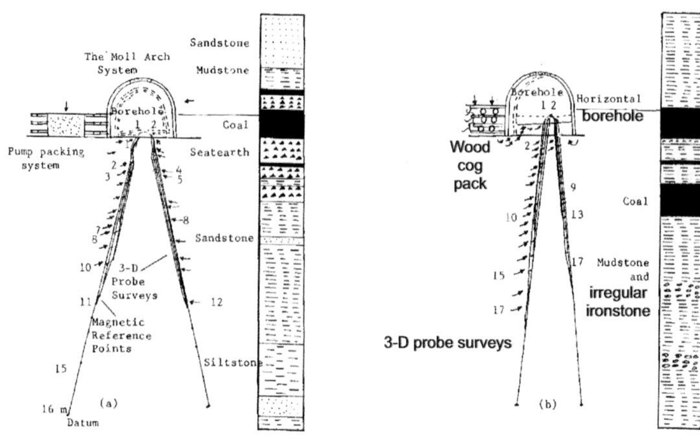

Figure 7.25 Results of three-dimensional rock displacement probe investigation

below the surface was 500 m and the site of the investigation was 1700 m from the pit bottom at the south-east corner of the shaft. The reused roadway was 4.27 m × 3.06 m, supported by arches set at 0.86 m intervals and reinforced by hardwood chocks. A section of this roadway was instrumented some 100 m in advance of the second face using a 3D probe in four borehole some 16 m deep. The results of the 3D probe survey as shown in Figure 7.26 indicate the following.

- A gradual increase in the floor lift as the face approaches the instrumented site. The rate of floor heave was higher near the waste and slightly lower at the rib side.
- Most of the floor heave occurred within the first 3 m of the floor in weak strata like seat earth and mudstone. No appreciable movement of the floor occurred below 7 m of the floor.
- Considerable floor heave towards the waste side indicated a poorly constructed waste side pack causing excessive lowering of the roof and appreciable floor lift.

This field investigation assisted in improving the face design in the subsequent extraction panels.

7.10 Conclusions

The determination of the modulus of deformation by borehole methods can be carried out in relatively undisturbed rock masses in terms of *in situ* state of stress, *in situ* moisture content, the inherent geological weaknesses and structures. Therefore, these tests permit the determination of the realistic rock mass deformation modulus. In contrast, the laboratory testing of the elastic modulus of rocks involves rock sampling processes that disturb the *in situ* state of stress and *in situ* moisture content and cause weaker rock material to disintegrate during the sampling and specimen preparation processes. Laboratory results, therefore, overestimate the rock's elastic modulus.

Dilatometers are full circle devices permitting larger areas of contact within the borehole but generate relatively low contact stress. Thus, such instruments are applicable to a low-strength rock and weathered geological material at shallow depths. In contrast, borehole jacks have contact with the borehole wall at a lower angle of arc and, therefore, generate comparatively higher contact pressures. These instruments enable the properties of rock to be tested in different radial directions. Thus, anisotropy of the rock mass can be quantitatively assessed.

Both of the above instrumentation types permit a large number of tests to be conducted along the entire length of a borehole. Moreover, boreholes can be drilled in orthogonal directions to obtain average rock properties and also to quantify anisotropy.

In general, both, dilatometers and borehole jacks affect comparatively small volumes of rock mass during an individual test as compared to the large-scale tests or laboratory tests. In contrast, large-scale tests affect large volumes of rock mass but only permit a limited number of tests due to the complexity of the test arrangements.

The penetrometers affect very small volumes of rock during the test and therefore develop very high contact stresses well beyond the elastic limits of most rock types. Therefore penetrometers can be used for conducting tests in high-strength host rocks but they have very little use in evaluating the modulus of deformation of rock masses.

Inclinometers have been used for measuring horizontal displacements within boreholes. They are applicable to the investigation of instability in the vicinity of deep foundations, measurements of surface subsidence, investigations of rock slope stability and landslides, and measurements of floor heave in underground roadways.

References

Absi, E. and M. Senguin (1967). *Le Noveau Geoextensometre*, Supplement to Annales de l'Institut Technique du Batimant et das Travaux Public, No. 235–236, July–August, pp. 1151–1158.

Burland, J.B. and J.F.A. Moor (1973). 'The measurements of ground displacements around deep excavations'. *Field Measurements in Geotechnical Engineering*, Butterworth, London, pp. 70–84.

Burland, J.B., J.F.A. Moor and P.D.K. Smith (1972). 'A simple and precise borehole extensometer'. *Geotechnique*, pp. 174–177.

Bromwell, G.L., R.C. Ryan and E.W. Toth (1971). 'Recording inclinometers to measurre soil movements'. *4th Pan American Conference on Soil Mechanics*, Puerto Rico, Vol. 2, pp. 333–343.

Benson R.P., D.K. Murphy and D.R. McCreath (1970). 'Modulus testing of rock at the Churchill Falls Underground Power House'. Determination of *in situ* Modulus of Deformation of Rock, ASTM STP 477, American Society for Testing of Materials, pp. 89–166.

Comes, G. (1965). *Contribution a La Determination Des Characteristiques Mechaniques d'une Foundation Rocheusse*, Travaux.

Cornforth, D.H. (1973). 'Performance characteristics of slope indicators series 200 B Inclinometers'. *Field Measurements in Geotechnical Engineering*, Butterworth, London, pp. 126–135.

Carvalho, O.S. and K. Kovari (1977). 'Displacement measurements as a means for safe and economical tunnel design'. *International Symposium on Field Measurements in Rock Mechanics*, Zurich, Balkema, Rotterdam, pp. 709–721.

Dawance in Group De Travail du Comite National Francis (1964). *Measure des Modulus de Deformation des Massif Rochaeux Dans les Sondages, Proceedings of Eighth Congress on Large Dams*, Vol. 16, No. 28, May, pp. 3–7–320.

Davydova, J.A. (1968). 'Interpretation of pressiometric test results with consideration of loaded section final length'. *International Symposium on Rock Mechanics*, Madrid.

De La Cruz, R.V. (1982). 'External displacement method for determining the *in situ* deformability of rock masses'. *23rd Symposium in Rock Mechanics*, Berkeley, California, Chapter 79, pp. 779–787.

Dunnicliffe, C.L. (1971). 'Equipment for field measurements'. *4th Pan American Conference on Soil Mechanics*, Puerto Rico, Vol. 1, pp. 313–366.

Geoprobe Instruments (1967). Literature by Test Lab. Corps, 216, N. Clinton St., Chicago.

Goodman, R.E., T.K. Van and F. Heuze (1968). 'The measurement of rock deformability in boreholes'. *Proceedings of 10th Symposium in Rock Mechanics*, Vol. 1, Chapter 19, pp. 523–555.

Gould, J.P. and C.J. Dunnicliffe (1971). 'Accuracy of field deformation measurements'. *4th Pan American Conference on Soil Mechanics*, Puerto Rico, Vol. 1, pp. 313–318.

Green, G. (1973). 'Principle and performance of two inclinometers for measuring horizontal ground movements'. *Symposium on Field Measurements*, British Geotechnical Society, Butterworth, London, pp. 166–179.

Hedley, D.C.F. (1969). 'Design criteria for multi-wire extensometer system' *Proceedings of First Canadian Symposium on Mine Surveying and Rock Deformation Measurements*, Fredricton, Oct 24–28, pp. 349–347.

Hoek, E and E.T. Brown, (1980). Underground Excavation in Rock, Inst. Mining and Metallurgy, London.

Hult, S. (1963). 'On the Measurement of Stresses in Solids'. *Transactions Chalmer University of Technology*, Catenburg, Sweden, No. 280.

Jaeger, J.C. and N.G.W. Cook (1964). *Theory and Applications of Curved Jacks, State of Stress in Earth Crust*, (W.R. Judd, Ed.), Elsevier, New York, pp. 381–395.

Kujundzic, B. and M. Stojakovic (1964). 'A contribution of the Experimental Investigation of Change of Mechanical Characteristics of Rock Massive as a Function of Depth'. *Transaction of Eighth Congress on Large Dams*, Edinburgh, pp. 1051–1067.

Kovari, K. and C. Amsted (1993). 'Decision-making in tunnelling based on field measurements'. *Comprehensive Rock Engineering*, Vol. 4, Hudson, J.A. (Ed.), Pergamon Press, Chapter 20, pp. 571–605.

Love, A.E.H. (1944), *A Treatise on the Mathematical Theory of Elasticity*, New York, 643 pp.

Marsland, A. (1973). 'Instrumentation for Flood Defence Bank Along the River Thames'. *Proceedings of the British Geotechnical Society, Symposium of Field Instrumentation*, London.

Martini, M.J., H. Duerbaum, W. Greset, E. Habetha, H. Kleinsonge and M. Langer (1964). 'Method to determine the physical properties of rocks'. Report 46, Theme 28, *Eighth Congress on Large Dams*, Edinburgh, pp. 859–869.

Ménard, L. (1966). 'Use of pressuremeter to study rock masses'. *Rock Mechanics and Engineering Geology*, Vol. 4, pp. 160–171.

Ménard, L. (1966). *Rules for Calculation and Design of Foundation Elements on the Basis of Pressuremeter Investigations of the Ground*. Translated by B.E. Hartman, Adjusted by J.B. Francq, Distributed by Terrametrics.

Ménard, L. (1957). 'Measures des proprietes physiques des sols *in situ*', *Annales des ponts et chaussées*, No. 3, May–June.

Mikkelsen, E.P. (1996). Field instrumentation in landslides investigation and mitigation. Special Report 247, Transport Research Board, National Research Council, USA, Keith A. Turner and R.l. Schuster (eds.), pp. 278–316.

Neff, T.L. (1970). 'An evaluation of several types of extensometers'. *Proceedings of Sixth Canadian Rock Mechanics Symposium*, May.

Panek, L.A., L.A. Hornsey and R.L. Lappi (1964). 'Determination of Modulus of the Rigidity of Rock by Expanding a Cylindrical Pressure Cell in a Drill Hole'. *6th Symposium in Rock Mechanics*, Rolla, University of Rolla, pp. 427–449.

Pearson, G.M. and K.H. Singh (1970). 'A New Device for Rock Displacement Measurements'. *Colliery Guardian*, December, pp. 633–636.

Rocha, M., A.F. Da Silveira, N. Grossman and E. Oliveira (1966). 'Determination of deformability of rock mass along boreholes'. *1st Congress of International Society of Rock Mechanics*, Lisbon, Vol. 1, pp. 697–704.

Rocha, M., A.F. Da Silveira, N. Grossman and E. Oliveira (1970). 'Characterisation of deformability of rock masses by dilatometer tests'. *Proceedings of Second Congress of International Society of Rock Mechanics*, Belgrade, Vol. 1, Paper 2.32.

Rouse, G.C. and G.B. Wallace (1966). 'Rock stability measurements for underground openings'. *Proceedings of First Congress of International Society of Rock Mechanics*, Lisbon, Vol. 2, Paper, 7–20, pp. 335–340.

Singh, R.N. (1967). 'Determination of Elastic Modulli of Rocks *in situ*'. M. Eng. Thesis, Postgraduate School in Mining, University of Sheffield, September.

Singh K.H.D. (1970). 'EH-extensometer, a Practical Instrument for Rock Mechanics Measurements'. *Canadian Mining Journal*, pp. 45–47.

Singh, R.N. (1976). 'Measurement and Analysis of Strata Deformation Around Mining Excavations'. Ph.D. Thesis, Department of Mineral Exploitation, University College, Cardiff, pp. 35–39.

Singh, R.N. and N.I. Aziz (1983). 'Instrumentation for stability evaluation of coal mine tunnels and excavations'. *Field Measurements in Geomechanics*, Vol. 2, pp. 1191–1204.

Singh, R.N. and A.M. Heidarieh-Zadeh (1982). 'Rock bolt reinforcement system to stabilise shaft intersections and pit bottom roadways during underground reconstruction'. *2nd Symposium in Rock Mechanics,* University of Berkeley, California, 25–27 August, pp. 961–970.

Singh, R.N., S.M. Reed, B. Denby and D.B. Hughes (1985). 'An investigation into groundwater recovery and backfill consolidation in British surface coal mines'. *National Symposium on Surface Mining, Hydrology, Sedimentology and Reclamation,* Lexington, Kentucky, 6 pp.

Smart, B.G.D., R.N. Singh and A.K. Isaac (1976). 'Design and Development of Magnetic Cantilever Extensometer'. *International Journal of Rock Mechanics and Mining Science,* Vol. 15, pp. 269–276.

Smart, B.G.D. and R.N. Singh (1976). 'A Borehole Instrumentation System for Monitoring Strata Displacement in Three Dimensions'. *International Journal of Rock Mechanics and Mining Science,* Vol. 15, pp. 77–86.

Smart, B.G.D., A.K. Isaac and N.I. Aziz (1977). 'The Design and Trial Applications of a Multi-Reference Point Borehole Extensometer'. *Proceedings of Conference on Rock Engineering,* Newcastle-Upon-Tyne, U.K., 4–7 April, pp. 211–243.

Steers, J.M. (1965). 'Evaluation of a electrometer for estimating roof bolt anchorage'. U.S. Bureau of Mines, R I 6646.

Takano, M. and Y. Shidomoto (1966). 'Deformation test on mudstone enclosed in a foundation by means of tube deformometer'. *Proceedings of First International Congress of ISRM,* Lisbon, Vol. 1, pp. 761–764.

Talobre, J.A. (1964). 'La Mesure In-situ des Proprietes Mecaniques des Roches at la Securite des Barrages de Grande Hauteur'. *Proceedings of Eighth Congress of Large dams,* Vol. 2, Theme No. 28, pp. 397–399.

Thorley, A., Y. Broise, M.L. Calhoon, Z.P. Zeman and W.G. Watt (1969). 'Borehole instruments for economical strength and deformation *in situ* testing'. *In situ* investigations in soils and rocks, *Proceedings of the Conference by BGS,* London, 13–15 May, pp. 155–165.

Timoshenko, S. and J.N. Goodier (1951). *Theory of Elasticity.* McGraw Hill, New York, 506 pp.

Wardell G.G. (1964). 'Application of instrumentation in determining rock behaviour during stoping at the Star Mine, Burks, Idaho'. *Proceedings of the Sixth Symposium on Rock Mechanics,* Rolla, Missouri, pp. 23–42.

Worotnicki, G., J.R. Enver and A. Spathis (1975). 'A Pressiometer for determining deformation modulus of rock *in situ.*' Research Paper 261, Division of Applied Geomechanics, CSIRO, Symposium *in situ* testing for design parameters, Australian Gemechanics Society, Victoria Group, November, 11 pp.

Whittaker, B.N., R.N. Singh and A.M.H. Zadeh (1980). 'Underground excavation structural stability with reference to construction work within shaft pillar'. *International Symposium on Safety in Underground Works,* Brussels, May.

Whittaker, B.N. and D.R. Hodgkinson (1970). 'Strata displacement measurements on Multi-wire borehole instrumentation'. *Colliery Guardian,* pp. 445–449.

Whittaker, B.N. and Singh, R.N. (1979). 'Evaluation of the design requirement and performance of gate roadways'. *Mining Engineer,* pp. 535–578.

Measurement of Stress in Rock

8 Chapter

8.1 Introduction

An understanding of the *in situ* state of stress in a rock mass is necessary in the design of rock structures for overcoming stress-controlled instability. This chapter briefly describes the components of stress in the rock mass and their implications in the design of major excavations in rock. The main design components for an underground excavation are the redistribution of stress around the excavation as a consequence of mining, selection of the orientation of the structure with respect to the stress field and design of suitable support measures to ensure stability. The following facets of stress measurement in rock are examined:

- State of stress in the rock mass
- Reasons why *in situ* stress measurement is necessary in any rock mechanics investigation
- Estimation of geo-stresses and the direction of stress in relation to geological and geo-technical features
- Methods of measuring *in situ* stress fields
- Stress distribution around a mine roadway
- Stress distribution around a longwall face

8.2 Stress in Rock

In solid mechanics, stress is an abstract concept which can be calculated by dividing the load by the area of cross-section of the structure on which the stress is acting. However, it is difficult to estimate the stress within the earth's crust due to lack of definition of the load acting on the rock and the corresponding area of cross section. One of the methods used to describe the stress in rock is to define the components of the principal stresses existing in the rock mass due to the following factors (Goodman, 1976):

1. Weight of the overlying strata
2. Horizontal stress induced due to horizontal constraints in the rock
3. Stress history arising from the formation of geological structures such as:
   - Erosion process
   - Formation of fault
   - Presence of other geological structures such as dykes and folds

The estimation of stress in rock in the above-mentioned circumstances is discussed in Section 8.4.

8.3 Reasons for Measuring *in situ* State of Stress

The stress measurement of rock *in situ* is necessary to deal with any of the following purposes (Denkhaus, 1968; Goodman, 1976).

(i) To decide on the orientation of large excavations or tunnels In a high horizontal stress field, the large dimensions of an underground structure should not be oriented perpendicular to the direction of the major principal stress as excessive movement across the longer span will cause premature failure of the excavation. Figure 8.1 shows the layout of a large underground power station, where the main machine hall is 180 m long, 25 m wide and some 50 m high at a depth of 300 m below the surface. In order

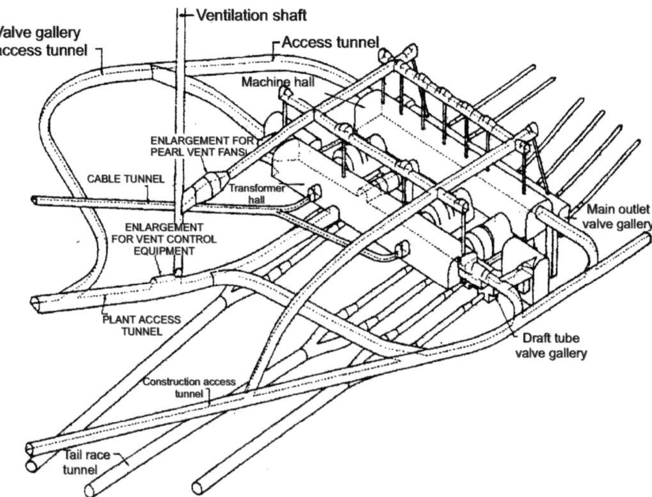

Figure 8.1 Layout of large underground chamber and direction of horizontal stress (Hoek and Brown, 1982)

to ensure stability of this opening, it was necessary to keep the axis of the hall parallel to the highest horizontal stress.

However, in a jointed rock mass the orientation of the major axis of a tunnel is controlled by the rock structures and care should be taken that the direction of the strike line of intersection of two major joint sets should not be oriented parallel to the tunnel axis. Figure 8.2 shows the effect of the orientation of the underground excavation upon the size of unstable wedges formed by the intersection of major structural discontinuities (Hoek and Brown, 1982).

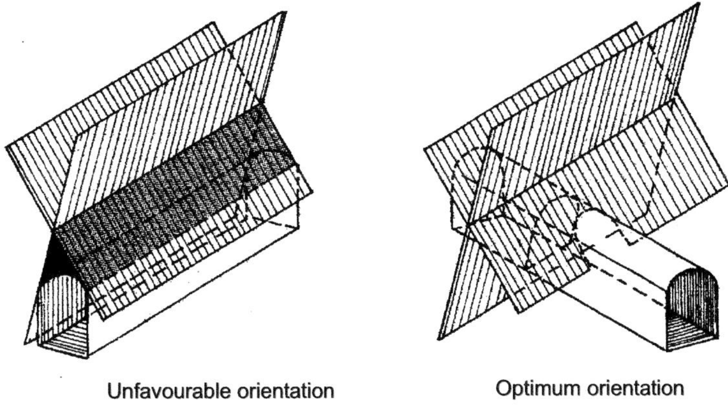

Figure 8.2 Direction of a tunnel controlled by rock structures (Hoek and Brown, 1982)

(ii) Shape of tunnel in high stress field In a high stress field, the shape of an underground excavation plays an important role in the stress concentration and initiation of rock failure around the opening. It is, therefore, pertinent that in the presence of a high stress field to keep the corners of the excavation rounded so as not initiate stress-controlled fractures in the rock due to stress concentration. Knowledge of the stress distribution in the rock mass is therefore necessary to decide on the shape of the excavations (see Figure 7.12) and their orientation with respect to the stress field.

(iii) In situ stress relative to the strength of the rock In a mountainous region, the stress in the rock changes with increase in depth below the surface and the presence of geological structures inducing high horizontal stress fields. The following patterns of stress induced instability may take place.

- If the stress in the rock increases to more than 15% of the ultimate uniaxial compressive strength of the rock (i.e. $\sigma_v > 0.15\ \sigma_c$) slabbing type of failure may occur.

Thus, the condition of slabbing failure is:

$$\sigma_v > 0.15\,\sigma_c$$

where σ_v = vertical stress

σ_c = uniaxial compressive strength

- If the stress in the rock is greater than 25% of its uniaxial compressive strength then tensile cracks or over-breaks will appear on the crown of an underground excavation.

  The condition for tensile failure is as follows:

$$\sigma_v > 0.25\,\sigma_c$$

- When stress in rock exceeds the uniaxial compressive strength in the case of massive and hard rock ($\sigma_v = \sigma_c$) isolated violent detachment and violent rock burst may take place.

Figure 8.3 shows a cross-section along the Mont Blanc tunnel in the French Alps where the depth of cover over the tunnel exceeds 2,540 m at the Aiguille du Midi and 2,100 m under the Aiguille de Toule. At these points the rock within the tunnel was highly stressed due to the weight of overlying rocks and possibly residual tectonic stresses resulting in violent rock bursts at the former site and the slabbing type of failures at the latter site (Panet, 1969).

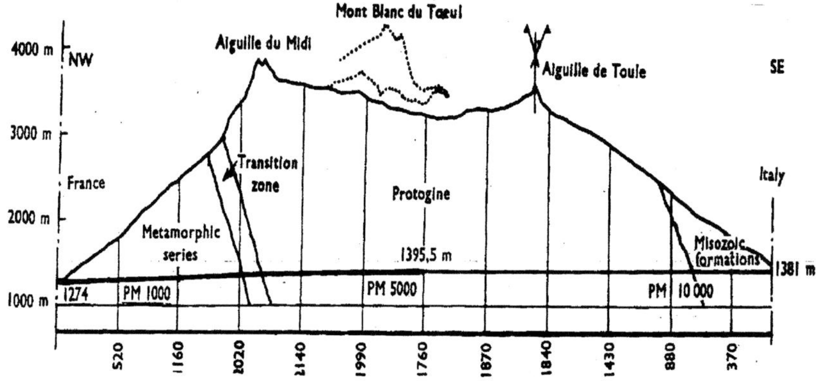

Figure 8.3 A cross-section through the Mont Blanc tunnel showing site of stress-controlled instabilities (Panet, 1969)

(iv) If the virgin strata stress is greater than the hydraulic stress, no tunnel lining is required.

(v) In surface mining, the direction of cut should be parallel to the direction of the highest horizontal stress field.

In surface mining, the orientation of the cut should be parallel to the direction of the high horizontal stress in order to assist the blasting action of the explosives in the blast holes. An orientation of the horizontal stress at an angle to the cut will result in creating over-breaks of the blast holes (Figure 8.4), thus reducing the blasting efficiency.

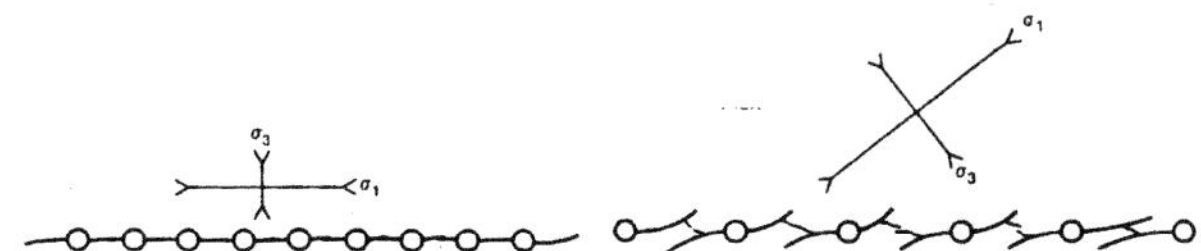

Figure 8.4 Orientation of open cut in relation to the direction of high horizontal stress

(vi) In order to interpret the results of displacement measurements in underground excavations, a knowledge of the state of stress in rock is often necessary.

8.4 Estimation of Geo-stress

8.4.1 Vertical Stress

The vertical stress in the rock mass occurs due to the weight of the overlying rock. Consider a cube-shaped rock with a side 'a' metre at a depth 'z' metres below the surface as shown in Figure 8.5.

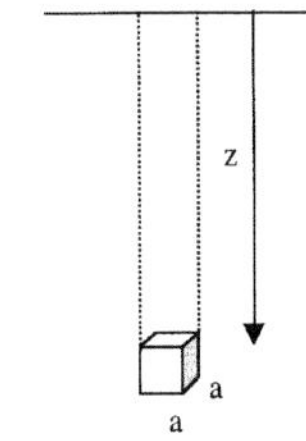

Figure 8.5 Vertical stress in rock

The load acting on the top of the cube is given by

$$\text{Load} = \gamma \cdot z\, a^2 \qquad 8.1(a)$$

$$\sigma_v = \text{vertical stress} = \text{load/area} = \gamma z\, a^2/a^2 = \gamma z \qquad 8.1(b)$$

where

σ_v = vertical stress (MPa)
γ = average unit density of rock
z = depth below surface (m)

For general rock mass the vertical stress (Figure 8.6) is given by:

$$\sigma_v = \gamma z = 0.027\, z\ (\text{MPa}) \qquad (8.2c)$$

8.4.2 Horizontal Stress

The horizontal stress on the earth can be calculated by the general formula as follows:

$$\sigma_h = k\sigma_v \qquad (8.3)$$

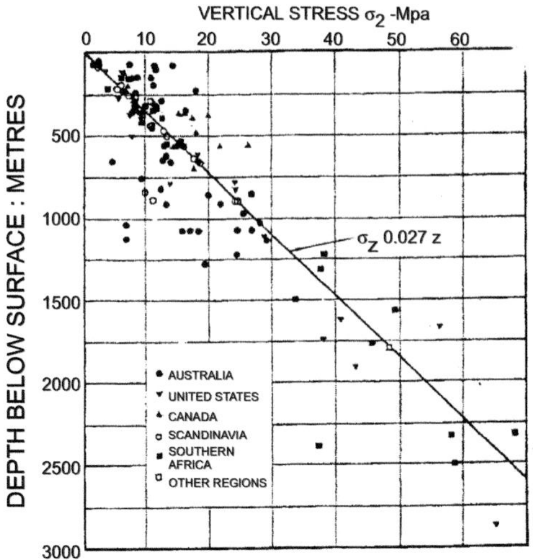

Figure 8.6 Distribution of vertical stress with depth (Hoek and Brown, 1982)

where

σ_h = horizontal stress

σ_v = vertical stress

k = ratio of the average horizontal stress to vertical stress

It may be noted that the value of 'k' varies from place to place and its value is 1 for soft rocks at great depth.

Figure 8.7 shows the variation of the ratio of horizontal to vertical stress ratio in rock with depth below surface. The lower bound and the upper bound of the above data have been calculated by carrying out a regression analysis and the following equations have been derived to estimate k_{max} and k_{min} as follows (Hoek and Brown, 1982):

$$k_{min} = \frac{100}{Z} + 0.3 \qquad k_{max} = \frac{1500}{Z} + 0.5 \tag{8.4}$$

where

k_{min} = the lowest value of the ratio of average horizontal stress to vertical stress in rock

k_{max} = the highest value of the ratio of horizontal to vertical stress in rock

8.4.3 Horizontal Stress at the Site of Erosion

Figure 8.8 shows the horizontal stress distribution at the site of erosion. It can be appreciated that vertical stress due to the weight of the rock before erosion is given by Equation 8.5(a)

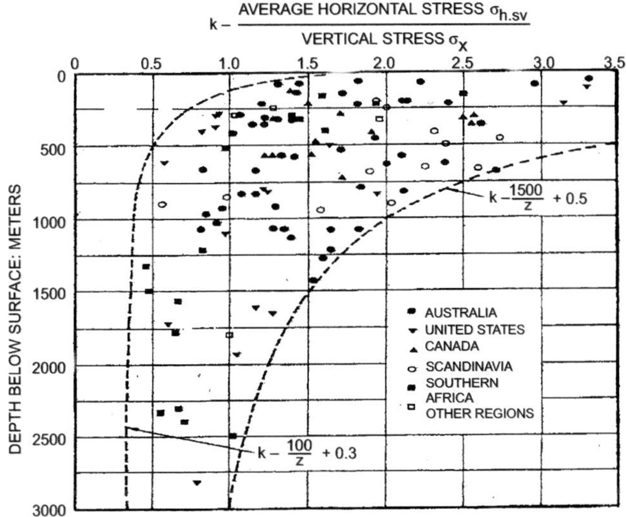

Figure 8.7 Variation of 'k' with depth (Hoek and Brown, 1982)

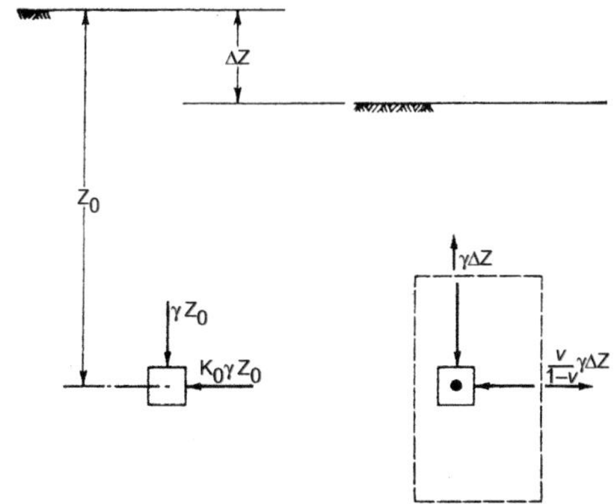

Figure 8.8 Effect of erosion on stress at depth

$$\sigma_v = \gamma Z_0 \tag{8.5(a)}$$

Z_0 = depth at the point of interest before erosion

γ = average density of rock (kN/m^3)

The horizontal stress before the erosion is given by:

$$\sigma_h = k_0 \sigma_v \tag{8.5(b)}$$

where

k_0 = ratio of the average horizontal to vertical stress before erosion

It follows that the ratio of average horizontal stress to vertical stress at the site of an erosion of Δz metres is given by (Goodman, 1980):

$$k_{(z)} = k_0 + \left[\frac{\Delta_z}{z}\left\{k_0 - \left(\frac{\upsilon}{1-\upsilon}\right)\right\}\right] \tag{8.3c}$$

$k_{(z)}$ = ratio of horizontal stress to vertical stress at the site of erosion
k_0 = ratio of average horizontal stress to vertical stress before erosion
υ = Poisson's ratio
z = depth below surface (m)
Δz = depth of erosion (m)

8.4.4 Estimation of Horizontal Stress at the Site of Fault

(i) Normal fault

Faulting in the earth takes place when the shear stress due to any tectonic load at the site exceeds the shear strength of the rock masses. At this point, the magnitude of vertical stress is much higher than the horizontal stress and the movement of rock occurs along the fault plane due to gravity. The average horizontal stress to vertical stress ratio at the site of normal fault k_a is given by the following equation:

$$k_a = \cot^2\left(\frac{\pi}{4} + \frac{\phi}{2}\right) - \frac{\sigma_c}{\gamma}\cot^2\left(\frac{\pi}{4} + \frac{\phi}{2}\right)\frac{1}{z} \tag{8.6}$$

(ii) Reverse fault

At the site of a reverse fault the magnitude of horizontal stress is much higher than the vertical stress and it exceeds the shear strength of the rock mass. The rock in this case moves against gravity up the fault plane causing

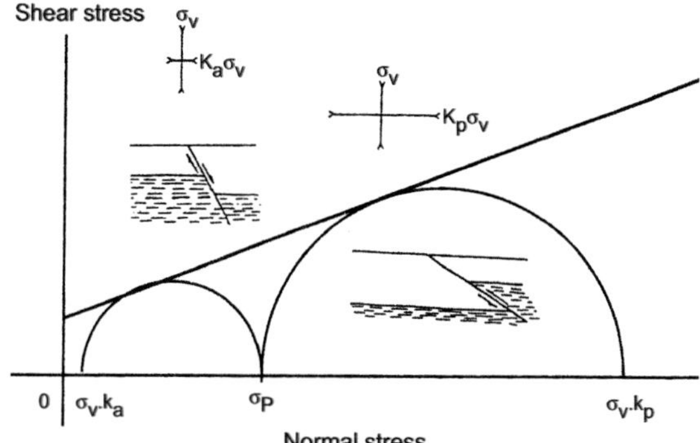

Figure 8.9 (a) Mohr circle of stress distribution on the earth before the onset of fault; (b) stress vectors at normal and reverse fault (Goodman, 1980)

some overlapping of strata. The ratio of the horizontal to vertical stress at the site of reverse fault k_p is given by the following equation:

$$k_p = \tan^2\left[\frac{\pi}{4}+\frac{\phi}{2}\right]+\frac{\sigma_c}{\gamma}\cdot\frac{1}{z} \tag{8.7}$$

8.4.5 Direction of Stress in Relation to Geological Features

(i) Stress directions in relation to surface topography

For the sake of simplicity, the local earth surface is taken as horizontal where the vertical stress in rock is caused by the weight of the overlying rock and is small in magnitude. The corresponding horizontal stress is significantly greater than the vertical stress due to horizontal constraints. For depths exceeding 1000 m, the average vertical stress tends to be equal to the horizontal stress due to creep deformations in the rock. However, in a mountainous area the horizontal stress is considered parallel to the ground and the vertical stress normal to the topography (Figure 8.10).

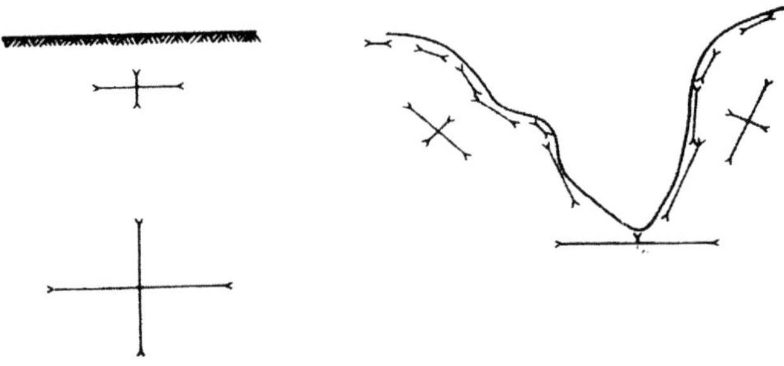

(a) Stress in level ground (b) Stress in mountainous region

Figure 8.10 Effect of topography on stress direction (Goodman, 1980)

(ii) Stress directions in relation to geological features

Figure 8.11 shows the direction of the principal stress in relation to geological structures.

In the case of a normal fault, the horizontal stress σ_3 is normal to the fault plane and the vertical stress is less than the horizontal stress in magnitude. In the case of a reverse fault, the principal horizontal stress is significantly higher than the vertical stress and acts in a direction normal to the fault plane while in the case of a strike fault the principal stresses act at an angle to the fault plane.

Similarly, in the case of a dyke a high principal stress will act in the strike direction of the intrusion inducing cracks through which lava flow may take

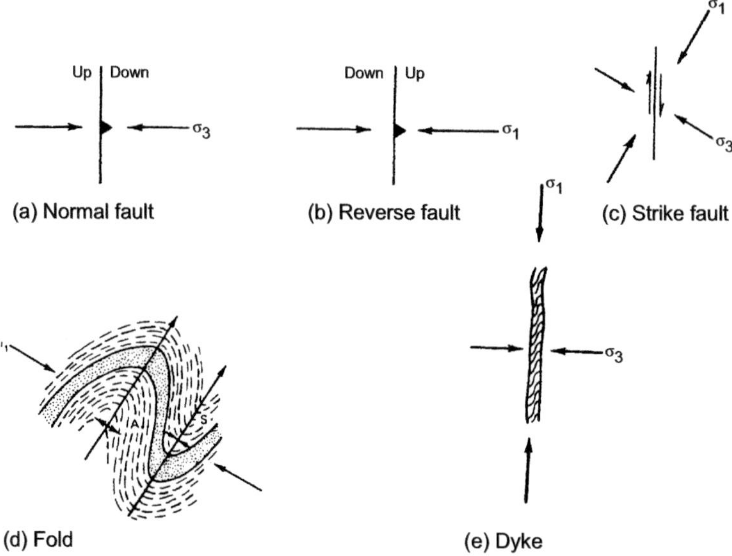

Figure 8.11 Directions of stress inferred from geological structures (Goodman, 1980)

place to form a dyke. Therefore, the principal horizontal stresses are parallel and at right angles to a dyke. In the case of a fold, the principal stress acts normal to the direction of the axes of the folds.

8.5 *In situ* Stress Measurement Techniques

As mentioned earlier, stress is an abstract concept and cannot be directly measured in the rock mass. However, change in stress in a rock can induce some changes in other properties of the rock which can be directly measured and can be correlated to changes in the stress components in the rock. With knowledge of the theory of elasticity, the principal stress components in the rock can thus be calculated.

In the past 45 years, a number of techniques have been developed to measure *in situ* stress components in rock. Some of the most popular techniques of *in situ* stress measurements are as follows.

1. Stress restoration techniques
   (i) Flat jack technique
   (ii) Cylindrical jack technique
   - Soft inclusion technique
   - Rigid inclusion technique

2. Strain relief methods
   (i) Curved jack fracturing
   (ii) Hydraulic fracturing
3. Stress relief method
   (i) Over-coring methods
   (ii) Borehole deformation gauge method
   (iii) Borehole strain gauge method
   (iv) Borehole end strain measurements
4. Miscellaneous stress relief methods
   (i) Method using elastic recovery of strain
   (ii) Borehole deepening method
5. Indirect methods
   (i) Methods based on detection of yield point
   (ii) Estimation of *in situ* stress by index tests
   (iii) Core discing method

Some of the more commonly used methods of *in situ* stress measurement in rock are described below.

8.5.1 Flat Jack Method of Stress Measurement

The method comprises the use of a hydraulic flat jack capable of being inserted in a slot in the rock and developing pressures up to 35 MPa. Access to the rock mass is obtained by an underground roadway or tunnels and a flat surface of the rock is prepared for the test. The first step is to install one or more sets of measuring pins designated as the reference pins on the face of the rock 2y distance apart shown in Figure 8.12 (typically 150 mm apart) or inside a pair of shallow boreholes. A deep slot is cut perpendicular to the rock face mid-distance between the reference pins by drilling overlapping sets of boreholes or by using a diamond saw. As a consequence of cutting the slot, the rock in between the pins is de-stressed and the separation distance between the two pins is reduced by an amount 2Δv. A flat jack is inserted in the slot, cemented into position and pressurised. When the pins return to the original position the pressure in the jack 'P_c' is measured (Fig. 8.13) which can be related to the initial normal stress to the plane of the flat jack as follows:

$$\sigma_n = \frac{P_c(C_j - d)}{C} \tag{8.8}$$

where

C = width of the slot
C_j = width of the jack
d = curved thickness of the jack

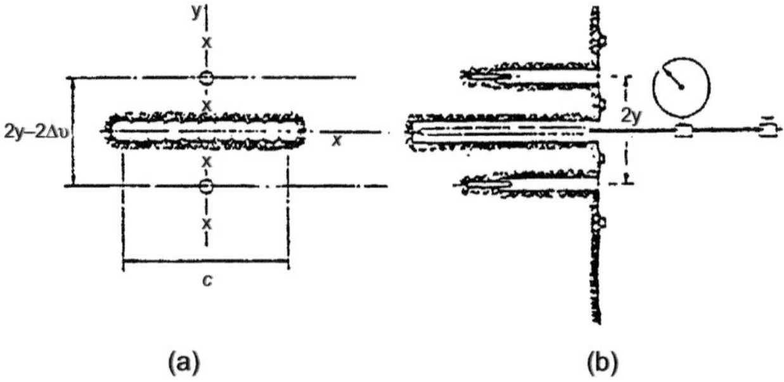

Figure 8.12 Flat jack method: (a) test arrangement; (b) normal pressure at cancellation

P_c = cancellation pressure
σ_n = stress normal to the slot

The relationship between the normal stress across the slot and the convergence $2\Delta v$ is given in Equation 8.9:

$$2\Delta v = 2c\frac{\sigma_n}{E}\left\{(1-\upsilon)\left[\sqrt{\left(1+\frac{y^2}{c^2}\right)}-\frac{y}{c}\right]+\frac{(1+\upsilon)}{\sqrt{\left(1+\frac{y^2}{c^2}\right)}}\right\} \tag{8.9}$$

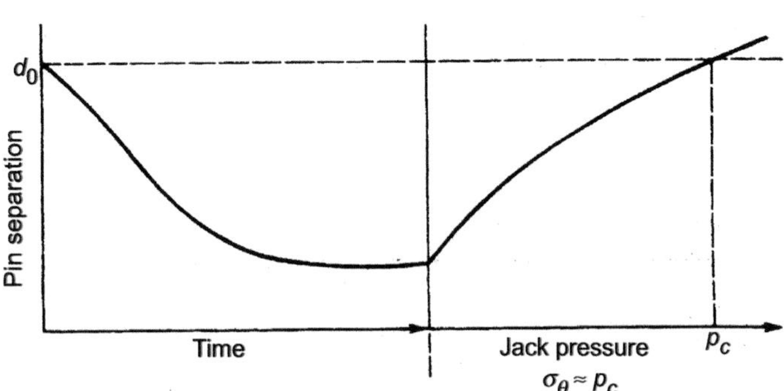

Figure 8.13 Test technique with the flat jack test

If the jack is placed at the roof and sidewall of a circular roadway, where the initial stresses are known to be horizontal and vertical, then the stress in the rock is given by the Equation 8.10.

$$\begin{vmatrix}\sigma_{\theta,W}\\ \sigma_{\theta,R}\end{vmatrix} = \begin{vmatrix}-1 & 3\\ 3 & -1\end{vmatrix}\begin{Bmatrix}\sigma_h\\ \sigma_v\end{Bmatrix} \tag{8.10}$$

$$\sigma_v = \frac{3}{8} \cdot \sigma_{\theta,W} + \frac{1}{8}\sigma_{\theta,R}; \qquad \sigma_h = \frac{1}{8}\sigma_{\theta,W} + \frac{3}{8}\sigma_{\theta,R} \tag{8.11}$$

In order for the above equation to be valid, the following assumptions are made.

1. The underground opening is perfectly circular in an homogeneous and isotropic rock. Initial stresses are known to be vertical and horizontal.
2. The radius of the tunnel is much larger than the width of the flat jack.
3. The rock is perfectly elastic and the value of the Young's modulus of rock for loading and unloading cycle is the same.
4. The flat jack and the slot have the same dimensions.

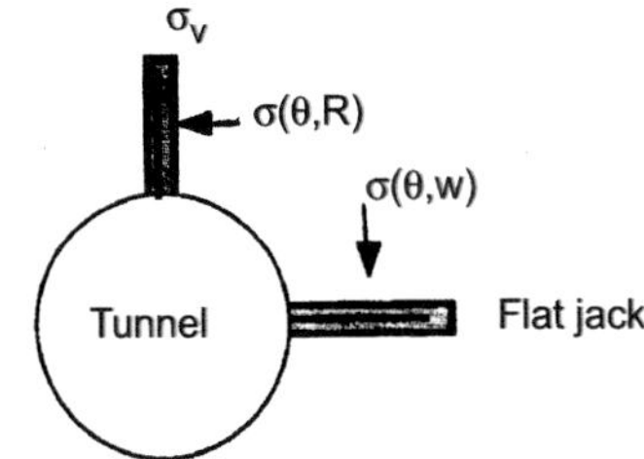

Figure 8.14 Flat jack test in a circular roadway

Figure 8.15 shows two alternative arrangements for the flat jack tests.

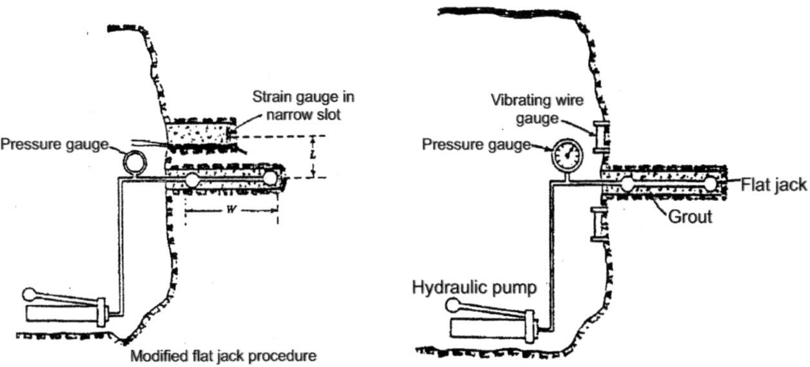

Figure 8.15 Alternative flat jack test arrangements

Worked Example

Two flat jacks, 305 mm^2 in dimensions are placed in the wall and roof of a circular roadway which is 2.45 m in diameter. The cancellation pressure at each of the flat jacks is as follows:

$$\text{Flat jack 1 (horizontal)} = 17.5 \text{ MPa}$$
$$\text{Flat jack (vertical)} = 6.21 \text{ MPa}$$

Estimate the initial vertical and horizontal stresses.

Solution

$$\sigma_{\theta W} = 17.5 \text{ MPa} \qquad \sigma_{\theta R} = 6.21 \text{ MPa}$$

$$\begin{vmatrix} \sigma_{\theta,W} \\ \sigma_{\theta,R} \end{vmatrix} = \begin{bmatrix} -1 & 3 \\ 3 & -1 \end{bmatrix} \begin{Bmatrix} \sigma_h \\ \sigma_v \end{Bmatrix}$$

$$\sigma_v = \frac{3}{8} \times \sigma_{\theta,W} + \frac{1}{8}\sigma_{\theta,R}; \qquad \sigma_h = \frac{1}{8}\sigma_{\theta,W} + \frac{3}{8}\sigma_{\theta,R}$$

$$\sigma_h = \frac{1}{8}\sigma_{\theta,W} + \frac{3}{8}\sigma_{\theta,R}$$

$$\sigma_h = \frac{1}{8}\sigma_{\theta,W} + \frac{3}{8}\sigma_{\theta,R}; \qquad \sigma_h = \frac{1}{8}17.5 + \frac{3}{8}6.21 = 4.51 \text{ MPa}$$

$$\sigma_v = \frac{3}{8} \times \sigma_{\theta,W} + \frac{1}{8}\sigma_{\theta,R}; \qquad \sigma_v = \frac{3}{8}17.5 + \frac{1}{8}6.21 = 7.33 \text{ MPa}$$

8.5.2 Stress Relief Technique

A borehole under stress is deformed when the surrounding rock is relieved of the pre-existing stress by drilling a larger concentric borehole. The deformation of the inner borehole which is a function of the *in situ* stress field, is measured to calculate the original stress components acting on the rock using the theory of elasticity.

(a) Over-coring technique

The over-coring technique comprises drilling a 150 mm diameter borehole on the wall of an underground roadway. At the bottom of this borehole, an EX diameter borehole is drilled to install a stress measurement device. In practice, three different ranges of instrumentations can be used to measure the borehole deformation.

(1) Diametric deformation of borehole
(2) Strain measurement at the bottom of the borehole
(3) Measurement of strain in an elastic inclusion

A diamond core drill is used at the bottom of the borehole for over-coring the instrument in order to relieve the stress on the borehole, thereby, deforming it (Fig. 8.16).

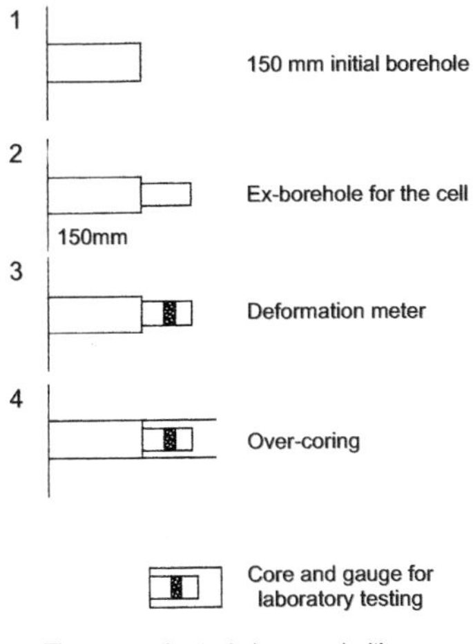

Figure 8.16 Measurement of stress in a tunnel by over-coring technique

In a homogeneous and isotropic rock, the deformation of a borehole can be related to the principal components of stresses using the theory of elasticity as follows:

$$u = \frac{d}{E}[(\sigma_1 - \sigma_2) - \upsilon\sigma_3 + 2(\sigma_1 - \sigma_3)(1 - \upsilon^2)\cos 2\theta] \tag{8.12}$$

where

u = diametric deformation of the borehole

d = borehole diameter

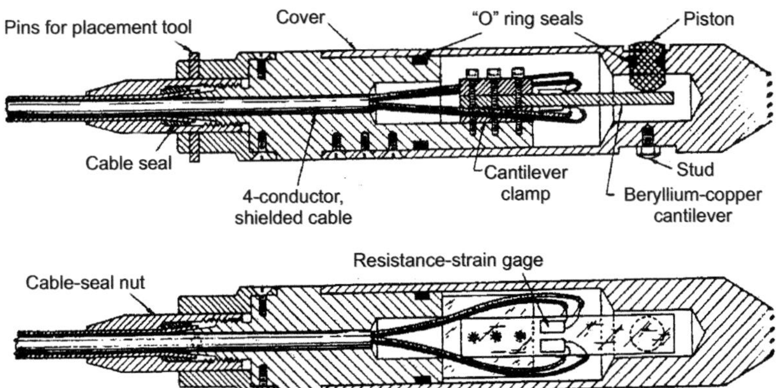

Figure 8.17 USBM borehole deformation meter

$\sigma_1, \sigma_2, \sigma_3$ = principal stresses acting on the rock
E = Young's modulus
υ = Poisson's ratio
θ = angle clockwise from the vertical stress σ_1 to the direction in which deformation is measured

Thus, if changes in the deformation are measured across the three orthogonal diameters of the borehole (Fig. 8.17) and E, υ and θ are known, then the magnitude and directions of principal stresses can be calculated. In the past, many borehole deformation meters have been fabricated and used as presented in Table 8.1.

Table 8.1 Types of borehole deformation stress meters

Single component equipment	*Biaxial borehole deformation meters*	*Triaxial deformation/stress meters*
Maihak vibrating wire cells	CSIR cell	U.S. Bureau of Mines cell (Fig. 8.17)
		Cantilever and ring transducers, Griswold
		USMB 4-component cell
U.S. Bureau of Mines cell	White pine cell Seebek cell	University of Lehigh cell

8.5.3 Measurement of the Diametric Borehole Deformation in Three Directions Simultaneously

The USBM deformation gauge comprises three units of the borehole deformation meters coupled together for measuring the deformation of the borehole in three different directions at 120° to each other. The instrument's sensor uses strain-gauged cantilever to give a voltage output proportional to the deformation of the borehole. There are three opposite pairs of carbide buttons each pressing against a cantilever arm fixed to the base plate and tightened against the borehole wall by springs. The instrumentation is installed at the bottom of the borehole. When the borehole is over-cored, the borehole deformation in three directions is measured simultaneously. The results of borehole deformation are plotted against the drilled distance shown in Figure 8.18.

The relationship between the borehole deformation and the principal stress components is given as follows:

$$\Delta_{d(\theta)} = f_1 \cdot \sigma_x + f_2 \cdot \sigma_y + f_3 \cdot \sigma_z + f_4 \tau_{xz} \tag{8.13a}$$

$$f_1 = d(1 + 2\cos 2\theta)\frac{(1-\upsilon^2)}{E} + \frac{d \cdot \upsilon^2}{E} \tag{8.13b}$$

$$f_3 = d(1 - 2\cos 2\theta)\frac{(1-\upsilon^2)}{E} + \frac{d \cdot \upsilon^2}{E} \tag{8.13c}$$

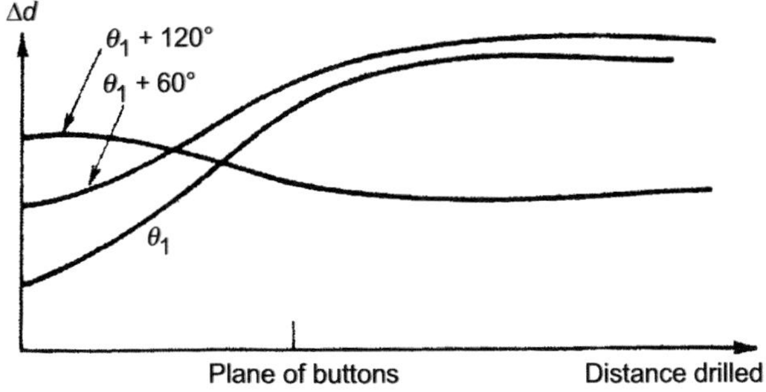

Figure 8.18 Stress measurement by three component borehole deformation technique

$$f_2 = -\frac{d \times \upsilon}{E} \qquad (8.13d)$$

$$f_4 = d(4\sin 2\theta)\,\frac{(1-\upsilon^2)}{E} \qquad (8.13e)$$

where σ_z is vertical and 2θ is measured in counter-clockwise direction to one pair of buttons.

Disadvantages

- Linear dependence of the stresses upon the elastic constants
- Use of large diameter core drills and limitation of drilling long straight boreholes

Desirable characteristics of the instrumentation

- Stability of the instrument
- Simple installation procedure
- Obtain accurate stress/strain data
- Cost-effective method
- Application to coal measure strata
- Application in the vicinity of a fault zone

8.5.4 Borehole Strain Gauge Method

This method is based on the determination of strains on the wall of boreholes induced by over-coring a strain cell comprising strain gauge rosette installed in a concentric borehole. If a sufficient number of strain tensors is measured during the stress relief operation, the six components of the stress tensors can be calculated from the field results using the solution procedure developed by Leeman and Hayes (1966) based on the theory of elasticity. The hollow soft inclusion cell developed by Worotonicki and Walton (1976) is a very popular device used in Australia since it permits determination of all components of the field stress tensors to be made in a single borehole over coring operation. The CSIRO triaxial strain cell (Fig. 8.19) comprises at least three strain gauge rosettes mounted on a thin-walled shell. The cell itself is bonded on the borehole wall using a suitable epoxy or polyester resin. The main modes of stress measurement operations are as follows.

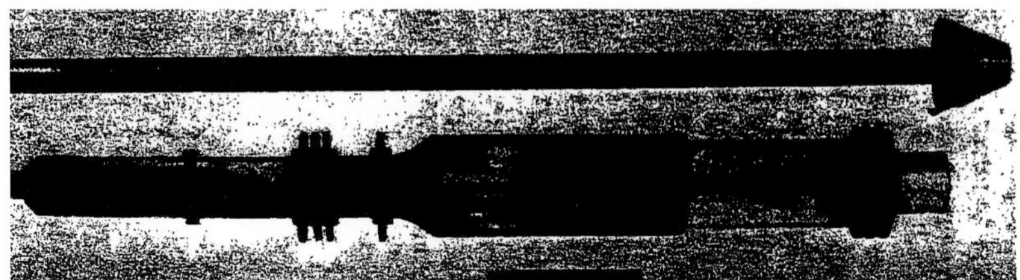

Figure 8.19(a) CSIRO triaxial stain gauge cell

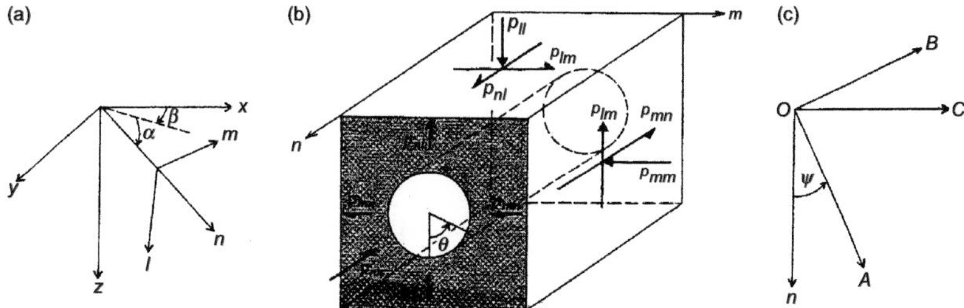

Figure 8.19(b) Relationship between stress and strain components

- Stress relief in the vicinity of the strain cell induces strains in the gauges of the rosettes equal in magnitude to the component of the original stress existing in the borehole wall.
- During the over-coring operation, the strain gauge readings are taken at intervals of every 50 mm of drilling. Once the over-coring operation is completed, the drill water is run in the over-coring borehole until strain readings in the strain monitors are stabilised.
- These borehole strain readings are used to deduce the local state of stress in rock prior to drilling the boreholes using the theory of elasticity and measured value E and υ of the rock surrounding the borehole.
- The expression for the stress concentration around the borehole is given by the following equations.

$$E \times \varepsilon_A = P_{ll}\, a_1 + a_2\, p_{mm} + a_3\, p_{nn} + a_4\, p_{ln} + a_5\, p_{lm} + a_6\, p_{mn} \tag{8.14}$$

where

$a_1 = 1/2\,\{[(1-\upsilon)-(1+\upsilon)\cos 2\psi] - (1-\upsilon^2)(1-\cos 2\psi)\cdot\cos\theta\}$

θ = orientation level of the local stress to strain rosette

ψ = position of local stress in relation to principal stress

υ = Poisson's ratio of the rock

Similarly, a_2 to a_6 are remaining stress constants for the rock.

Thus, six independent observations of strain are made and six simultaneous equations are established. Solution of these simultaneous equations is obtained by using numerical techniques (Larson, 1992; Mindata, 1991). This enables the calculations of stress tensors around the borehole to be obtained. A further elastic solution is used to relate these stress tensors for calculating the principal stress components using the theory of elasticity.

Advantages

- The construction of the strain cell ensures that it is extremely rugged and fully waterproof; this ensures that there will be no malfunctions of the strain gauges in wet conditions.
- The presence of drilling debris in the borehole does not affect the installation of the cell ensuring full contact between the strain gauge rosettes and borehole walls.
- The installation procedure of the strain cell also ensures full contact between the strain gauge rosette and the borehole wall in the presence of
  - drilling debris
  - joint, fracture or imperfect rock material in the borehole wall.

Disadvantages

This instrument is known to creep over a long period and, therefore, the monitoring period should be restricted to one week.

8.5.5 Measurement of Strains at the End of Borehole

The instrument consists of a four-element 45° rosette strain gauge affixed to the end of a borehole that is ground flat and made smooth using an impregnated diamond bit. The borehole is then over-cored over a distance of two to three times its diameter in order to relieve the stresses from the end of the borehole containing the strain gauge rosettes.

A correction factor must be applied to the measured strain readings due to the stress concentration on the base of the borehole. Elastic constants of rock mass E and υ are required to calculate the stress on the plane normal to the axis of the borehole (Figure 8.20). Three orthogonal boreholes are required to determine the complete tensor of a three-dimensional stress field. This

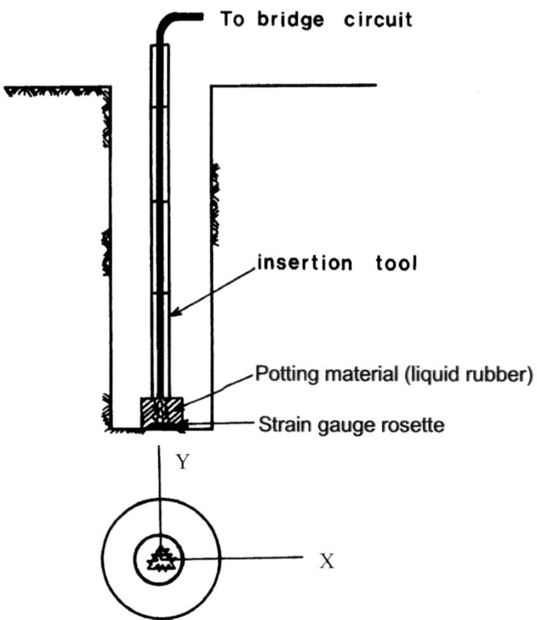

Figure 8.20 Strain gauge rosette at the end of borehole

method can be used in deep underground mine in hot and humid environment. Care should be taken to ensure that there is a complete bond between the strain gauge rosettes and the borehole end surface in hot and humid mining conditions. In wet conditions, there is difficulty in applying this method in a down borehole application where there is a possibility of accumulation of water at the bottom of the borehole. Due to the use of rosette strain gauges together with a fast-setting epoxy resin, this method offers a relatively inexpensive way of measuring the field stress in rock *in situ*.

8.5.6 Hydraulic Fracturing Technique

This method utilises a vertical borehole to get access to the rock mass without resorting to the use of underground roadways or tunnels. An inflatable packer is lowered down the borehole by means of a hollow pipeline to isolate a section of the borehole. A fluid is injected in the borehole section under test and the fluid pressure is increased until the rock is fractured. Figure 8.21 shows a schematic diagram of the hydraulic fracturing technique of stress measurement together with the plot of the vertical and tangential stress induced in the rock due to the hydraulic fracturing operation.

Fig. 8.22 shows the borehole pressure-time plot in hydraulic fracturing, with the diagnostic pressures at P_{c1}, the breakdown pressure, the shut-in

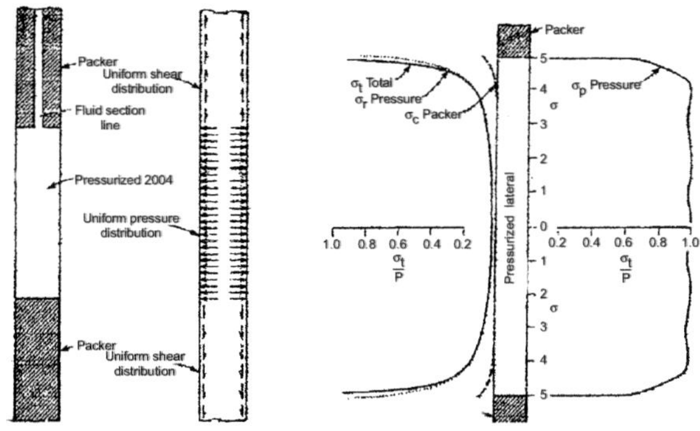

Figure 8.21 Hydraulic fracturing technique of measuring stress in rock

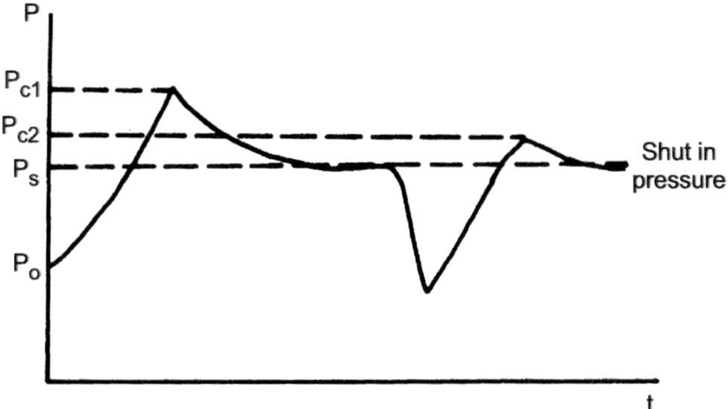

Figure 8.22 Pressure history during hydraulic fracture tests

pressure at P_s and the rock reopening pressure at P_{c2}. Using Kirsch's solution, the tangential stresses on the wall of the borehole at the sides will be

$$\sigma_\theta = 3\sigma_{h,min} - \sigma_{h,max}$$

At the time of the hydraulic fracture

$$3\sigma_{h,min} - \sigma_{h,max} - P_{c1} = -\sigma_t$$

A crack once formed will continue as long as the pressure applied is greater than σ_θ.

Stress normal to fracture plane

$$\sigma_{h,min} = P_s$$

where, P_s = shut in pressure

New peak pressure on cyclic loading P_{c2}

$$\sigma_t = P_{c1} - P_{c2} = \text{tensile strength}$$

Assuming vertical stress = γz

A condition is that the crack orientation should be vertical which can be checked either by television camera or an impression packer.

Example A vertical fracture was initiated in a borehole at a depth of 1,000 m. Water pressure above the hydraulic pressure was 5.89 MPa and the shut in pressure was 0.89 MPa above the hydrostatic pressure. After a day, the pressure was raised to 0.81 MPa above the shut in pressure. Calculate the vertical stress, the horizontal stress and the tensile strength of rock.

Solution

$$\sigma_v = \gamma H = 0.027 \times H = 27 \text{ MPa}$$

$$p_o = \rho g h = 9.81 \times 1000 = 9.81 \text{ MPa}$$

$$p_{c1} = 9.81 + 5.89 = 15.70 \text{ MPa}$$

$$p_s = 9.81 + 0.89 = 10.70 \text{ MPa}$$

$$p_{c2} = 10.70 + 0.81 = 11.51 \text{ MPa}$$

$$\sigma_{h,min} = p_s = 10.70 \text{ MPa}$$

$$\sigma_t = p_{c1} - p_{c2} = 15.70 - 11.51 = 4.19 \text{ MPa}$$

$$3\,\sigma_{h,min} - \sigma_{h,max} - p_{c1} = -\sigma_t$$

$$3 \times 10.70 - \sigma_{h,max} - 15.70 = -4.19$$

$$\sigma_{h,max} = 32.10 - 15.70 + 4.19 = 20.59 \text{ MPa}$$

8.6 Application of Stress Measurements

The main application of stress measurement is to confirm the results of theoretical stress analysis. Two cases of stress distribution are presented here as follows.

(i) Stress around a rectangular access roadway in rock at a depth
(ii) Stress around a single longwall face

8.6.1 Vertical Stress Distribution Around a Rectangular Roadway

Figure 8.23 shows the total load on the strata 'od' due to pre-mining vertical stress around a rectangular roadway, given by load $A_1 = \sigma_1 \times od$. As a consequence of driving the roadway, the strata above the roadway in Figure 8.23 is de-stressed and the load is transferred as an abutment load equally on both sides of the excavation. It can be seen in Figure 8.23 that the peak stress develops at the edge of the excavation and diminishes in a negative exponential manner over the solid strata until it reaches the cover load. It can be shown that:

$$A_2 = \sigma_1 \times od = 2\,A_3$$

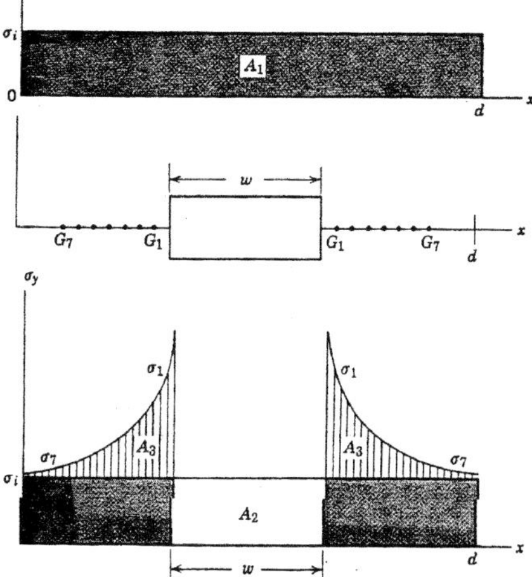

Figure 8.23 Stresses around a rectangular roadway in hard and massive ground

If the stress on the edge of the pillar exceeds the uniaxial compressive strength of the rock, the rock will break and transfer the peak stress to the strata in-bye. The rock on the outer boundary of the pillar will be in a broken state and applying a confining pressure to the rock lying immediately in-bye. Thus, the in-bye rock in the pillar will be in a triaxial state of stress capable of bearing additional vertical loads. It may be observed that the abutment load on the elastic rock is sharp.

8.6.2 Vertical Stress Distribution Around a Longwall Face

Figure 8.24 shows the vertical stress distribution around a 210 m wide longwall face at a depth of 470 m based on field observations in the East Midland Coalfield, UK (Whittaker and Singh, 1979). Stresses are expressed as a ratio of vertical stress to the initial geostatic stress in the pre-mining conditions. As a consequence of extracting coal at the longwall face the initial geo-stress along the face line OZ becomes zero. The pre-existing strata load is transferred as a front abutment stress ahead of the face and as side abutment on the flank zones of the longwall face on the pillars adjoining the main and tail gates.

Stress distribution along the central line of the face

It can be seen that the peak front abutment load ahead of the face reaches four to five times the cover load some three to five metres ahead of the face. This load gradually diminishes in a negatively exponential manner and reaches the cover load some 60 m in advance of the face. In the front abutment zone,

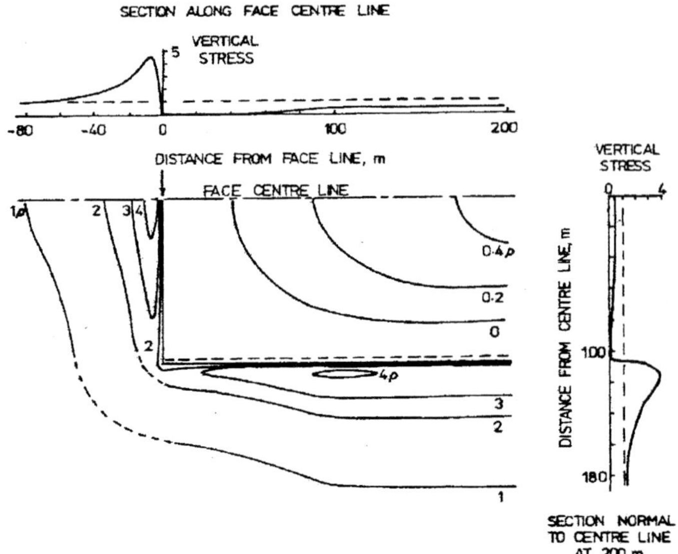

Figure 8.24 Stress distribution around a longwall face (Whittaker and Singh, 1979)

the vertical pressure at the central line of the face is at a maximum and reduces to twice the cover load at the corners of the face. The strata overlying the goaf area immediately behind the longwall face are de-stressed and are caving downwards. Further behind the face, the caved strata starts compacting and starts bearing some cover load. At a distance 1.4 times the width-to-height ratio, the vertical stress reaches the cover load.

Stress distribution of the side abutment zone

On the flank pillars, the strata pressure at the corner of the face is about twice the cover load which gradually increases to four times the cover load at a distance 90 m behind the longwall face.

Stress distribution along the cross section of the face

The side abutment stress on the flank pillar reaches the peak abutment stress some 8–10 m from the rib because of the crushing of the edge of the rib pillar. The peak abutment pressure is about four times the cover load and diminishes to the cover load some 60–70 m along the flank pillar. In the goaf area, immediately adjoining the gate roadway the roof strata are de-stressed and undergoing caving. The broken goaf starts bearing some cover load and reaches 0.4 times the cover load at the centre line of the longwall panel.

8.7 Conclusions

A rock mass *in situ* is in a state of finite stress. The vertical component of the stress can be attributed to the weight of the overlying rocks which permits the horizontal stress to develop due to the constraints of the confining rock. However, this stress pattern is modified in the area of changed surface topography and also near the area of geological structures and a site of erosion.

Knowledge of *in situ* state of stress in rock mechanics is necessary for the design of major structures in rock, especially in selecting the orientation of the major underground excavations in relation to high stresses. It is possible to estimate the magnitude and direction of stresses with a fair degree of accuracy at many construction sites in rock. However, measurement of *in situ* states of stress is necessary where there are major mining or civil engineering developments or to verify the assumptions made during the rock engineering design.

In this chapter, several methods of determining *in situ* state of stress in rock are presented.

The borehole strain gauge method using a soft inclusion stain cell and utilising the borehole over-coring method is a very popular method of stress measurement in Australia (Worotonicki and Walton, 1976). This method utilises six independent observations of strain on the borehole wall and establishes six simultaneous equations expressing the stress and strain relationships. The solution of these simultaneous equations is obtained by using a computerised numerical solution method developed by Mindata (1991).

Measurement of *in situ* state of stress is the primary data necessary for the design of structures in rock.

References

Ageton, R.W. (1967). 'Deep mine stress determination using flat jack and borehole deformation methods'. US Bureau of Mines Report of Investigation, 6887.

Buchbinder, G.C.R., E. Nyland and J.C. Blanchard (1966). 'Measurement of stress in borehole'. *Canadian Geological Survey Paper*-66-13.

Denkhaus, H.G. (1968). 'The significance of stress in rock masses'. *Proceedings of the 5th International Symposium on Rock Mechanics*, Madrid, pp. 263–271.

Fairhurst, C. (1965). 'Measurement of *in situ* Stresses with Particular Reference to Hydraulic Fracturing'. *Felsmechanik*, 11 (3–4)

Gay, N.C. (1975). '*In situ* Stress Measurement in South Africa'. *Tectonophysics*, 29, pp. 447–459.

Goodman, R.E. (1976). *Methods of Geological Engineering*. West Publishing Company, St Paul, Minnesota, 472 pp.

Goodman, R.E., (1980), Introduction to Rock Mchanics, John Wiley & Sons. New York, 478 pp.

Hast, N. (1958). 'The Measurement of Rock Pressures in Mines'. *Sveriges Geol. Undersokn. Arsbok, Serc.*, Vol. 52, No. 3.

Helal, H. and R. Schwartzmann (1983). '*In situ* Stress Measurement with Cerchar Dilatometric Cell'. *International Symposium on Field Measurement in Geo-mechanics*, Zurich, September 5–8, pp. 127–136.

Haimson, B.C. (1978). 'The Hydrofracturing Stress Measuring Technique Method and Recent Field Results'. *International Journal of Rock Mechanics and Mining Sciences*, Vol. 15, pp. 167–178.

Haimson, B.C. (1983). 'The State of Stress at the Nevada Test Site—A Demonstration of Reliability of Hydro-fracture and Over-coring Technique'. *International Symposium on Field Measurement in Geomechanics*, Zurich, September 5–8, pp. 115–126.

Hast, N. and T. Nilsan (1964). 'Recent pressure measurements and their Implications in dam building'. *8th Congress of Large Dams*, Edinburgh, Vol. 1, 60 pp.

Hawkes, I. and V.E. Hooker (1974). 'The Vibrating Wire Stress Meter'. *3rd Congress of International Society of Rock Mechanics*, Denver, Vol. 2A, pp. 439–444.

Herget, G. (1973). 'Variation of rock stress at Canadian Iron Ore Mine'. *International Journal of Rock Mechanics and Mining Sciences*, Vol. 10, pp. 35–51.

Hoek, E. and E.T. Brown (1982). *Underground excavation in rocks*. The Institution of Mining and Metallurgy, pp. 98–101, 193, 196.

Hooker, V.E., Bickel, and J.B. Aggeson (1972). *In situ Determination of Stress in Mountainous Topography*. US Bureau of Mines Report of Investigation, 7654.

Jaeger, J.C. (1969). *Elasticity, Fracture and Flow*. Methuen & Co., London, 3rd Edition, 268 pp.

Jaeger, J.C. and N.G.W. Cook (1976). *Fundamentals of Rock Mechanics*. Chapman and Hall, London, 585 pp.

Judd, W.R. (ed.) (1964). *State of Stress in Earth Crust*. New York.

Kehle, O.K. (1964). 'The determination of tectonic stresses through analysis of hydraulic well fracturing'. *J. Geophysical Research*, Vol. 69, Number 2.

Larson M. (1992). 'Stress out—A data reduction program for inferring stress state of rock having isotropic material properties'. *A user's manual*, U.S. Bureau of Mines Information Circular 9302, 168 pp.

Leeman, E.R. and D.J. Hayes. (1966). 'A technique for determining the complete state of stress in rock using a single borehole, Proc. Ist. Conf. International Society of Rock Mechanics, Lisbon, vol. 2., pp. 17–24.

Leeman, E.R. (1969). 'The Doorstopper and Triaxial Rock Stress Measuring Instruments'. Developed by CSIR, South African Institute of Mining and Metallurgy, Vol. 67, pp. 305–339.

Leeman, E.R. (1968). 'The determination of complete state of stress in rock in a single borehole—laboratory and underground measurements'. *International Journal of Rock Mechanics and Mining Sciences*, Vol. 5, pp. 31–56.

Morgan, T.A. and L.A. Panek (1963). *A Method of Determining Stress in Rock*. US Bureau of Mines, RI 6312.

Morgan, T.A., W. Fischer and W. Sturgis (1965). 'Stress distribution on West Vaco Mine as Determined by Borehole Stress Relief'. US Bureau of Mines, RI 6675.

Mindata Ltd, (1991), *Stress 91 Users Manual*, 1990. Available from Mindata Ltd, Seaford, Australia.

Nag, D.K., M. Seto and V.S. Vutukuri (1998). '*In situ* Rock Stress Measurements Using Acoustic Emission Behaviour of Rock Core Specimen'. *International Conference on Geomechanics/Ground Control in Mining and Underground Construction*, 14–17 July, pp. 1087–1094.

Olsen, O.J. (1957). 'Measurement of residual stress by the strain relief method'. *2nd Am. Symposium on Rock Mechanics*, Colarado School of Mines, 52, No. 3.

Obert, L. (1961). *Determination of Stress in Rock. A State of Art Report*, ASTM.

Obert, L. (1962). '*In situ* Determination of Stress in Rock'. *Mining Engineer*, Vol. 14, No. 8.

Obert, L. and W.I. Duvall (1967). *Rock Mechanics and Design of Structures in Rock.* John Wiley and Sons, New York, 650 pp.

Panet, M. (1969). 'Rock mechanics investigations in the Mont Blanc Tunnel. *In situ* investigations in soils and rocks'. *Proceedings of the Conference by the British Geotechnical Society*, London, 13–15 May, pp. 101–108.

Rocha, M., Da Silveira, A.F, (1969), A new method for the complete determination of the state of stress in rock masses. *Geotecnic*, Vol. 19, pp. 116-132.

Tesarik, D.R., J.B. Seymour, T.R. Yanske and R.W. McKibbin (1995). 'Stability analysis of a backfilled room and pillar mine'. US Bureau of Mines, Report of investigation/1995, RI, ISSN 1066-5552, 20 pp.

Tesarik, D.R., J.B. Seymour and J.D. Vickery (1989). 'Instrumentation and modelling of Cannon's mine B-North orebody'. Proceedings of Innovation in Backfill Technology, *Proceedings of the 4th International Symposium on Mining with Backfill*, Montreal, 2–5 October, pp. 119–128.

Whittaker, B.N. and R.N. Singh (1978). 'Evaluation of design requirements and performance of gate roadways in longwall mining'. *Mining Engineer*, 135, July, pp. 535–548.

Whittaker, B.N. and R.N. Singh (1979). 'Design and Stability of Pillars in Longwall Mining'. *Mining Engineer*, July, pp. 59–73.

Worotonicki, and R.J. Walton (1976). 'Triaxial hollow inclusion gauges for determination of rock stress *in situ*'. *Proceedings ISRM Symposium on investigation of stress in rock—Advances in stress measurement*, Sydney, Supplement, pp. 1–8.

Worotonicki, and D. Denham (1976). 'The state of stress on the upper part of the Earth's crust in Australia according to Measurement in Mines, Tunnels and from Seismic observations'. *Proceedings of ISRM Symposium on Investigation of Stress in Rock—Advances in Stress Measurement*, Sydney, Australia, August, pp. 72–82.

Assignments

Question 1 A tunnel is being driven in a competent rock mass at a depth of 1,000 m. The average vertical stress in the rock is 0.027 MPa per metre depth and the ratio of horizontal to vertical stress is (σ_h/σ_v) is 0.3. The maximum compressive stress which can be sustained by the tunnel wall, is 2.7 times the maximum principle stress. The cohesive strength of the rock is 14 MPa and the angle of internal friction is 42°.

Draw Mohr's circle of the state of stress before and after the excavation, the equation of the stress envelope and estimate the radial stress applied by the support to the tunnel wall to stabilise the excavation.

Question 2 A natural slope rises at 45° for 1,000 m horizontally and then levels off. The rock has an unconfined uniaxial compressive strength of 50 MPa. An adit has to be driven for an underground pressure tunnel beginning at the portal at the base of the slope and continuing directly into the mountain. At what distance from the entrance would you first expect to encounter rock pressure problems?

Question 3 A zone of active thrust faulting (low angle reverse faulting) occurs in a rock with $\phi = 30°$ and $\sigma_n = 6.9$ MPa and unit weight of 0.027 MN/m^3. Estimate the major and minor principal stresses at the depth of 945 m assuming conditions of faulting. Compare your estimates with that given by the Hoek and Brown formulae.

Question 4 A rock mass at a depth of 4,500 m had a value of 'k' equal to 0.8. If the Poisson's ratio of the rock is 0.25, what should the horizontal stress at the site of erosion of 2,000 m of rock be?

9 Chapter
Design of Structures in Rock

9.1 Introduction

The construction of any excavation in rock results in the redistribution of stress in the host rock. The response of the rock mass to the change in stress distribution depends upon the strength and deformation properties of the rock mass as discussed in the previous chapters. The first step in any stability analysis of any simple mining excavation is therefore to determine the pattern of stress distribution around the excavation of specific shape and aspect ratio with respect to the *in situ* stress.

This chapter deals with the qualitative assessment of stability of simple excavation shapes using closed-form mathematical solutions.

9.2 Scope of Design of Mining Extraction Systems

The design of a structure in rock mass involves the prediction of the response of rock mass to mining activities. Of particular concern are the following.

- Location and design of the main service and access openings
- Extraction procedure and sequence
- Stability performance of the opening throughout the life of the mine
- General stability after mining

The design of a structure in rock may involve the following.

(1) Design of permanent opening and access roadways
(2) Design of pillar dimensions
(3) Design of stope dimensions
(4) Pillar layout
(5) Stope mining sequence
(6) Pillar extraction sequence

(7) Overall direction of mining, advance or retreat
(8) Type and timing of placing backfill

The design process for permanent openings and mine service openings should include the following procedure.

(1) Develop the excavation design with respect to required duty and size requirements.
(2) Calculate the boundary stresses.
(3) Compare the tangential stress $\sigma_{\theta\theta}$ with uniaxial compressive strength and tensile strength.
(4) Modify design to limit failure of boundary rock.
(5) Determine stresses at interior points.
(6) Determine the extent of the potential failure zone surrounding the excavation and assess its implications for mining.
(7) If the zone of failure is acceptable, then design the support system.
(8) If the zone of failure surrounding the opening is unacceptable, then modify the design to reduce the zone of failure and adopt an acceptable support system.

Fig. 9.1 shows the algorithm for design.

9.3 Predictive Methods

There are two major approaches to carry out the assessment of structural stability of mining excavations.

(1) Study of physical model
(2) Study of mathematical models involving closed loop solutions

Physical Models

The main aim of the physical model is to identify conditions contributing to extensive failure of a mining structure. Usually a model, scaled down both in dimensions as well as strength and deformation properties, is used. Although scaling down of the model in dimensions is relatively easy, the main difficulty arises in maintaining the following attributes in the model:

- Similitude in material properties
- Load ratio in the model and prototype.

These problems can be overcome by the use of *centrifugal models* but these facilities are expensive to construct and operate. The centrifugal model is essentially a research tool rather than an appliance for routine design applications.

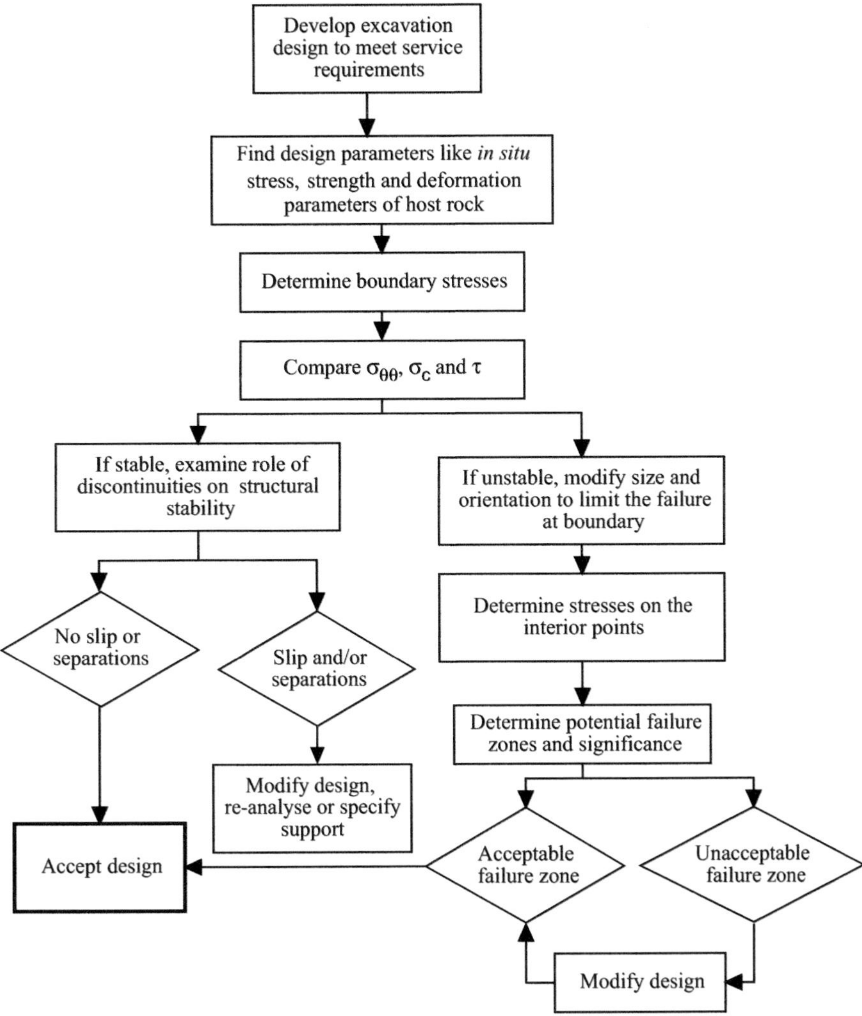

Figure 9.1 Geomechanics design of mining systems (after Brady and Brown, 1985)

The main disadvantages of the physical models are as follows.

(1) Expensive and time consuming
(2) Inherently limited in their application to the surface of the model (They provide virtually no information on state of stress and displacements in the interior of the model medium).
(3) The use of the model can be justified in a single, confirmatory study or to verify mine structure design.

Closed Loop Solutions for Design of an Opening in an Infinite Medium

An analysis of stress and displacement field in and around a body under applied load can be examined in cartesian co-ordinats or polar co-ordinates.

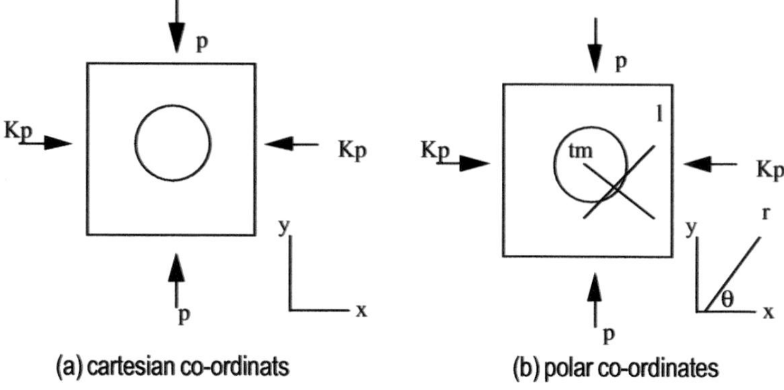

Figure 9.2 Long horizontal opening in an elastic medium

Practical Approach

(i) Select an existing closed-form solution.
(ii) Verify the accuracy.
(iii) Apply suitable acceptability tests as follows.
  (a) Boundary conditions for the problems
    - impose state of tension
    - displacement at excavation surface
    - far field stress
  (b) Differential equations of equilibrium
  (c) Constitutive equations for material
  (d) Strain compatibility equations

9.4 Closed-form Solution of Circular Shape

Far field biaxial stress The stress remote from the excavation boundary is given as follows:

$$P_{yy} = p, \qquad P_{xx} = K \cdot p$$

Stress distribution around circular opening (Kirsch equation) The radial stress σ_{rr}, the tangential stress $\sigma_{\theta\theta}$ and the shear stress $\sigma_{r\theta}$ around a circular opening is given by following equations (Kirsch, 1898):

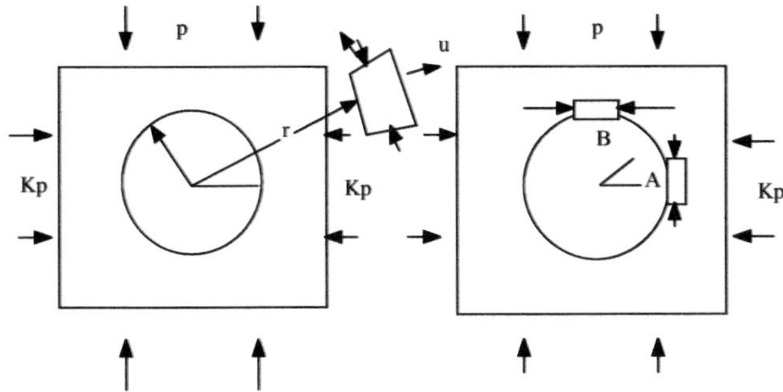

Figure 9.3 Stress at the boundary of a circular opening

$$\sigma_{rr} = \frac{p}{2}\left[(1+K)\left(1-\frac{a^2}{r^2}\right) - (1-K)\left(1-4\frac{a^2}{r^2}+\frac{3a^4}{r^4}\right)\cos 2\theta\right]$$

$$\sigma_{\theta\theta} = \frac{p}{2}\left[(1+K)\left(1-\frac{a^2}{r^2}\right) + (1-K)\left(1+\frac{3a^4}{r^4}\right)\cos 2\theta\right]$$

$$\sigma_{r\theta} = \frac{p}{2}\left[(1-K)\left(1+2\frac{a^2}{r^2}-\frac{3a^4}{r^4}\right)\sin 2\theta\right]$$

Displacement The radial displacement U_r and the tangential displacement U_θ around a circular opening are given by the following equations:

$$U_r = -\frac{pa^2}{4G\cdot r}\left[(1-K)-(1-K)\left\{2\left(1-2\upsilon\frac{a^2}{r^2}\right)\cos 2\theta\right.\right]$$

$$U_\theta = \frac{pa^2}{4G\cdot r}\left[(1-K)\left\{2(1-2\upsilon)+\frac{a^2}{r^2}\right\}\cos 2\theta\right]$$

where

G = modulus of rigidity
U_r, U_θ = induced displacements
$\sigma_{rr}, \sigma_{\theta\theta}, \sigma_{r\theta}$ = total stress generated by opening
υ = Poisson's ratio
a = radius of the tunnel
r = radial distance at which stress and displacement are being computed

Stress at the excavation boundary Putting $r = a$ in the Kirsch equation, the stresses at the excavation boundaries are

$$\sigma_{\theta\theta} = p[(1+K) + 2(1-K)\cos 2\theta]$$
$$\sigma_{rr} = 0, \text{ and } \sigma_{r\theta} = 0$$

This establishes that the excavation boundary is traction-free.

Far field stresses

Putting $\theta = 0$ and r large

$$\sigma_{rr} = Kp, \qquad \sigma_{\theta\theta} = p, \sigma_{r\theta} = 0$$

Boundary stresses around a circular opening (Fig. 9.4)

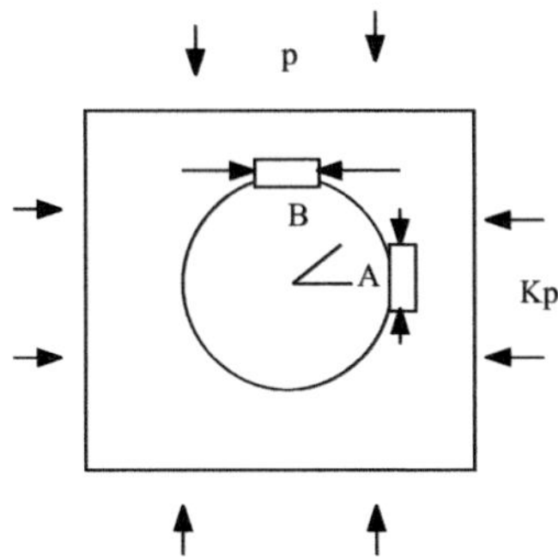

Figure 9.4 Stress at the boundary of circular tunnel

The Kirsch equations (1898) define the state of stress on the boundary of a circular opening in terms of coordinate angle θ. The surface is traction-free and only the non-zero stress component is $\sigma_{\theta\theta}$. For $K < 1.0$, the maximum and minimum boundary stresses occur at the side of the wall ($\theta = 0$) and crown ($\theta = \pi/2$) of the excavation, these stresses are defined by the following equations.

At point A $\quad \theta = 0, \quad (\sigma_\theta)_A = p\,(3 - K)$

at point B $\quad \theta = \pi/2, \quad (\sigma_{\theta\theta})_B = p\,(3K - 1)$

These expressions show that in a *uniaxial stress field with* K = 0, the maximum and minimum boundary stresses are

$$\sigma_A = 3p\text{ , and } \sigma_B = -p$$

These values represent upper and lower limits for stress concentration at the boundary. If $K > 0$, the side wall stress is less than 3 p and the crown stress is greater than – p. The existence of tensile boundary stresses in a compressive stress field is worth noting.

Hydrostatic Stress Field (K = 1)

$$\sigma_{\theta\theta} = 2\,p$$

The boundary stress takes the value 2 p, independent of the coordinate angle θ. This represents the optimum distribution of the boundary stress, since the boundary is uniformly compressed over the complete excavation periphery.

The Kirsch equations are considerably simplified for hydrostatic stress conditions.

$$\sigma_{rr} = p\left[1 - \frac{a^2}{r^2}\right]$$

$$\sigma_{\theta\theta} = p\left[1 = \frac{a^2}{r^2}\right]$$

$$\sigma_{r\theta} = 0$$

The independence of the stress distribution of the coordinate angle θ and the fact that $\sigma_{r\theta}$ is zero, indicate that the stress distribution is axisymmetric.

9.5 Stress Around an Opening of Elliptical Shape

Bray (1977) produced a set of simplified formula for calculating the state of stress at points into the medium surrounding an elliptical opening taking into considerations the effect of local boundary curvature on the boundary stress (Fig. 9.5).

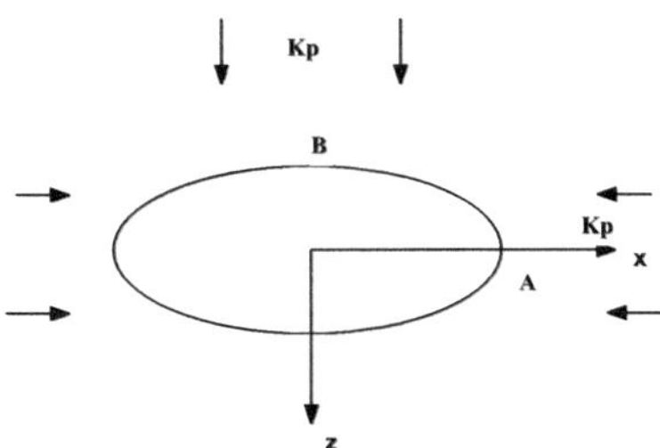

Figure 9.5 Stress around a circular opening

Ellipse major axis = 2a and minor axis 2b.

Radius of curvature found by simple analytical geometry

$$\rho_A = \frac{b^2}{a} \qquad \rho_B = \frac{a^2}{b}$$

$$q = \frac{W}{H} = \frac{a}{b}, \qquad q = \sqrt{\frac{W}{2\rho_A}}, \qquad \frac{1}{q} = \sqrt{\frac{H}{\rho_B}}$$

Sidewall stress in the ellipse boundary

$$\sigma_A = p(1 - K + 2\sqrt{q}) = p\left\{1 - K + 2\sqrt{\frac{W}{2\rho_A}}\right\}$$

Crown stress

$$\sigma_B = \left[K - 1 + \frac{2K}{q}\right] p \left[K - 1 + 2K\sqrt{\frac{2\rho_B}{H}}\right]$$

Formulae for the ideal excavation shapes such as circle and ellipse can be used to establish useful working ideas of the state of stress around regular excavation shapes.

9.6 Excavation Design in Massive Rock

Different design considerations are given to the two types of mining excavations as follows:

(i) Service openings whose working life approaches the life of the ore body. These openings are:
- mine accesses
- ore haulage drives
- airways
- crusher chambers
- underground workshop space

(ii) Low cost and high operation life of the openings

These openings are exemplified as follows:

(i) Production openings such as:
- stopes
- drill headings
- stope access
- ore extraction and service ways

(ii) Openings with temporary role

The design methodology involves the following approaches.

- Design of single excavations.
- Opening will not be mined in the zone of influence of any existing opening.
- Rock mass is massive and intercepted by only one or two persistent structural features.
- Rock mass strength is defined by a compressive failure criterion.

Design of mining structures assumes following factors.

(i) The existence of an extensive zone of failed rock around the periphery of an opening is common in mining practice.

(ii) Basic mining problem is not to prevent rock mass failure, but to ensure that large, uncontrolled displacement of peripheral rock does not occur.

This can be achieved by paying attention to

- excavation shape,
- excavation practice and
- application of one or several rock support or reinforcement procedures.

General problems of mine excavation design can be devolved into the following areas.

(i) Excavation location and geometry
(ii) Development of excavation sequence and support specifications

Problems of location and geometry are considered here under.

9.7 Interaction from Other Mine Workings

Stress distribution around a circular hole in a hydrostatic stress field of a magnitude 'p', is given below

$$\sigma_{rr} = p\left[1 - \frac{a^2}{r^2}\right]$$

$$\sigma_{\theta\theta} = p\left[1 + \frac{a^2}{r^2}\right]$$

$$\sigma_{r\theta} = 0$$

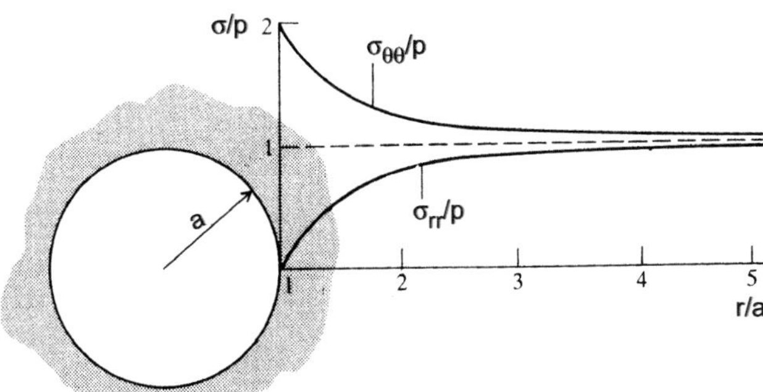

Figure 9.6 Stress around a circular opening.

This shows that the stress distribution is axisymmetric. It is readily calculated that for r = 5a,

$$\sigma_{\theta\theta} = 1.04\ p$$
$$\sigma_{rr} = 0.96\ p$$

on the surface defined by 5a the stress state is not significantly different from the field stresses.

Interaction of two excavations

If two excavations of the same diameter are 6a distance apart, the pre-mining stress field will be equal to the virgin stress field.

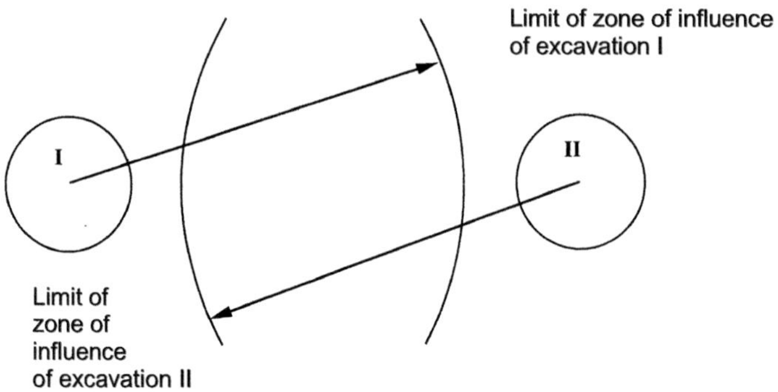

Figure 9.7 Interaction of two excavations

Thus for circular openings of same radius, the distances I and II are > 6a (Fig. 9.7).

In an overlap region the state of stress is produced by pre-mining stresses and the stress increment induced by each of excavations I and II.

In each of the other sections of each zone of influence, the state of stress is due to the particular excavation.

Factors affecting the zone of influence are as follows.

1. Excavation shape
2. Pre-mining stresses
3. Number of interacting excavations
4. Interaction between different sized excavations

9.8 Interaction due to Different Size Openings

Figure 9.8 shows a large diameter opening I with a small diameter opening II in its zone of influence.

Excavation I: It is outside the area of influence of excavation II. The stress distribution around excavation I can be derived from the stress distribution around a single opening.

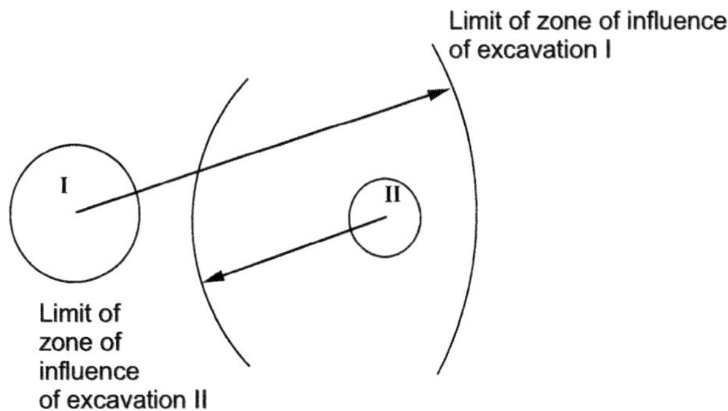

Figure 9.8 Interaction of two excavations (Brady and Brown, 1985)

Excavation II: The field stresses are those due to the presence of excavation I. Calculate the stress at the centre of excavation II due to the presence of excavation I. This introduces the far-field stress in the Kirsch equations to calculate the required boundary stresses for the smaller excavation.

Example Location of haulages, access and service openings in the zone of influence of the major production openings (Brady and Brown, 1985).

Figure 9.9 shows access openings on the footwall side of the inclined ore body. The zone of influence could be defined for an ellipse inscribed in the stope cross-section and the stope is outside the zone of influence of each access drive.

Thus, stress distribution around an elliptical excavation in a biaxial stress field may be a useful design tool.

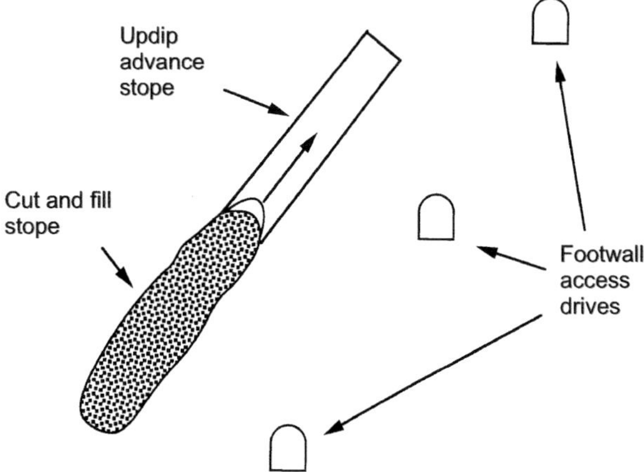

Figure 9.9 Footwall access to a stope

Thus, it is the domain within which the stress > 0.05 p_{max}

or $$|\sigma_3 - p_{min}| > 0.05\, p_{max}$$

where

p_{max} = larger of the field stresses p and Kp
p_{min} = smaller of the field stresses p and Kp

The zone is contained within an ellipse of overall width W_1 and height H_1 (Fig. 9.10).

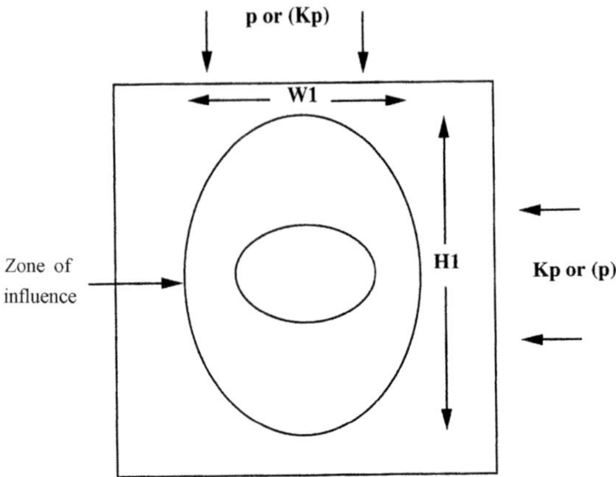

Figure 9.10 Zone of influence of an elliptical opening

The dimensions W_1 and H_1 are given by the following expressions derived by Bray (1977), noting that, of the two values of W_1 and H_1, the larger value of each are selected.

$$W_1 = H\sqrt{10\alpha\left|q(q+2) - K(3+2q)\right|}$$

$$W_1 = H\sqrt{[\alpha\,\{10(K+q^2) + Kq^2\}]}$$

$$H_1 = H\sqrt{[10\alpha\left|K(1+2q) - q(3q+2)\right|]}$$

or

$$H_1 = H\sqrt{\{\alpha\,[10(K+q^2)+1]\}}$$

where

$$\alpha = 1, \text{ if } K < 1 \text{ and } K = \frac{1}{a} \text{ if } K > 1$$

For extreme values of q and K. For K > 5 and q > 5, W_1 must be increased by 15%. If K< 0.2 and q < 0.2, H_1 must be increased by 15%.

9.9 Excavation Shape and Boundary Stress

Figure 9.11 shows a long opening of elliptical cross-section, with axes parallel to pre-mining stresses.

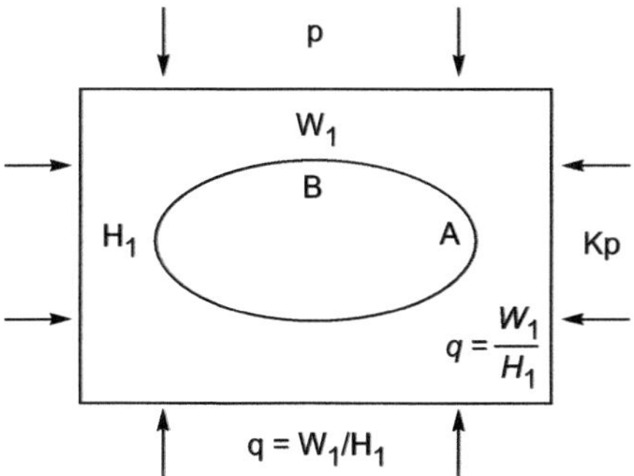

Figure 9.11 Stress around an elliptical opening

Stress at points A and B around an elliptical opening is given as follows (Lamb, 1956):

$$\sigma_A = p(1 - K + 2\sqrt{q}\,) = p\left[1 + K + \sqrt{\frac{2W}{\rho_A}}\right]$$

$$\sigma_B = p(K - 1 + \frac{2K}{q}) = p\left[K - 1 + 2K\sqrt{\frac{2H}{\rho_B}}\right]$$

Where

σ_A = boundary circumferential stress at the side-wall of the excavation
σ_B = boundary circumferential stress at the crown
ρ_A = radius of curvature at point A
ρ_B = radius of curvature at point B

If the radii of curvatures at A and B are small then the stresses at the excavation boundary are high.

9.10 Applications: Example 1

(a) Ovaloid opening in a bi-axial stress field (Brady and Brown, 1985)

An ovaloidal opening with its major axis perpendicular to pre-mining principal stress is considered. If W/H ratio = 3, and radius of curvature of sidewall $\rho_A = H/2$, and K= 0.5, the sidewall boundary stress is given by the following relationship:

$$\sigma_A = p(1 - K + 2\sqrt{q}) = p\left[1 - K + \sqrt{\frac{2W}{\rho_A}}\right]$$
$$= p[1 - 0.5 + \{2 \times 3H/(H/2)\}^{0.5}]$$
$$= 3.96\ p$$
$$\sigma_B = p(K - 1 + 2K/q) = p\ (0.5 - 1 + 1/3)$$
$$= -0.167\ p$$

This shows that the excavation aspect ratio (say W/H), as a boundary curvature, can be used to develop a reasonably accurate picture of the state of stress around an opening.

(b) Square hole with rounded corner

Example 2 Given a square opening with rounded corners, each with radius of curvature $r = 0.2$ B in a hydrostatic stress field (Figure 9.12). Assume that the excavation can be approximated as an ovaloidal excavation and estimate boundary stress at the sidewall of the excavation.

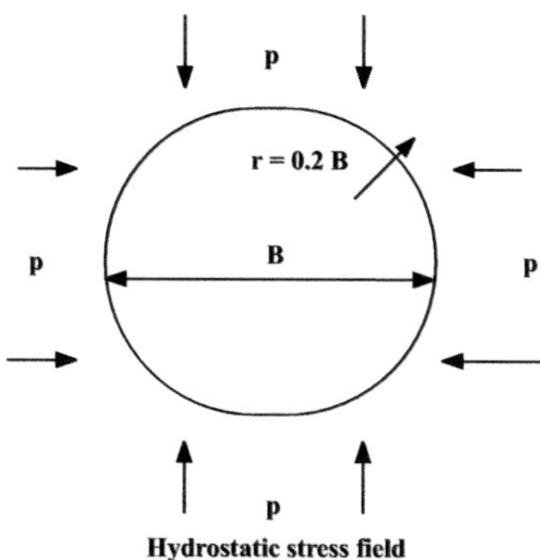

Figure 9.12 Ovaloidal opening in a hydrostatic field (Brady and Brown, 1985)

Simple geometry will show that the width of the inscribed ovaloid is given by the following relationship:

$$W = 2B\left[2^{1/2} - 0.4\left(2^{\frac{1}{2}} - 1\right)\right]$$
$$= 2B\ [\ 1.41 - 0.4\ (1.41 - 0.164)$$
$$= 2B \times 1.246 = 2.492\ B$$

$$\sigma_A = p(1 - K + 2\sqrt{q})$$

$$= p\left[1 - K + \sqrt{\frac{2W}{\rho_A}}\right]$$

$$= p\left[1 - 1 + \sqrt{\frac{2W}{0.2B}}\right] = p\left[1 - 1 + \sqrt{\frac{2.49B}{0.2B}}\right] = 3.53\ p$$

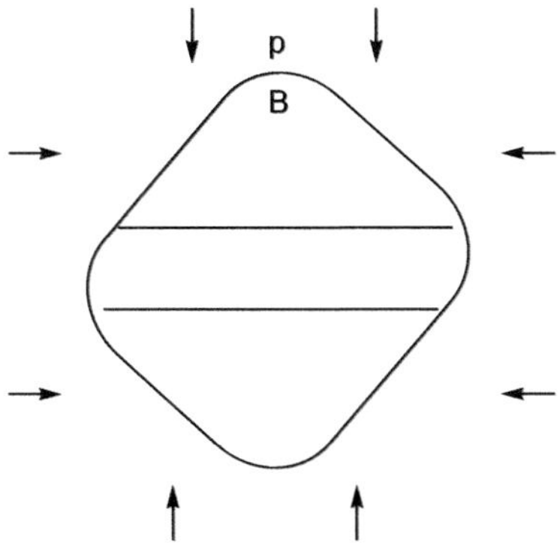

Figure 9.13 Stress around an ovaloidal opening

Example 3 A haulage drive has an arched opening with a width of 4.0 m and a height of 4.5 m Figure 9.14. The field stress ratio K is 0.3. Calculate the sidewall stress (Brady and Brown, 1985).

Solution

$$\sigma_A = p(1 - K + 2\sqrt{q}) = p\left[1 - K + \sqrt{\frac{2W}{\rho_A}}\right]$$

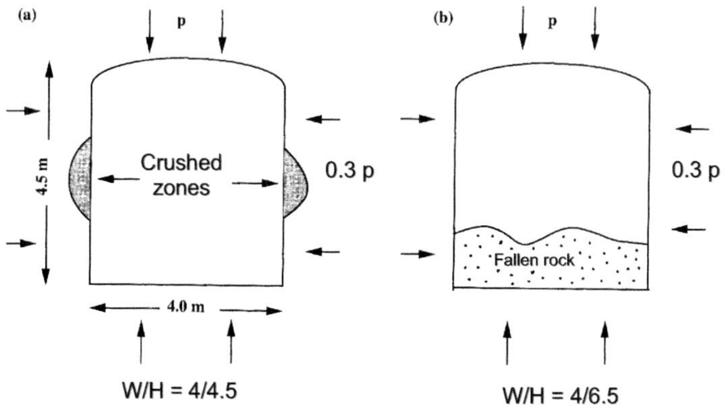

Figure 9.14 Stress around an arched roadway

$$\sigma_B = p(K - 1 + 2K/q) = p\left[(K-1) + K\sqrt{\frac{2H}{\rho_B}}\right]$$

$$\sigma_A = p(1 - K + 2q) = p\,(1 - 0.3 + 2 \times 4.0/4.5) = p\,(0.77 + 1.77) = 2.47\,p$$

By inspection if W/H ratio is reduced to 0.5, then σ_A will be 1.7 p. Therefore, opening dimensions should be increased in the direction of major principal stress. Mining an opening with a low width-to-height ratio and leaving a bed of broken rock on the base of the excavation is a feasible design method.

Example 4 In underground metalliferous mining, such excavations can be exemplified as:

- underground workshop,
- crusher station, and
- battery charging station.

An example given by Brady and Brown (1985) calculates stress around underground workshop for the following conditions as follows (Figure 9.15)

Width-to-height ratio (W/H) = 2/3
Pre-mining stress ratio (K) = 0.5

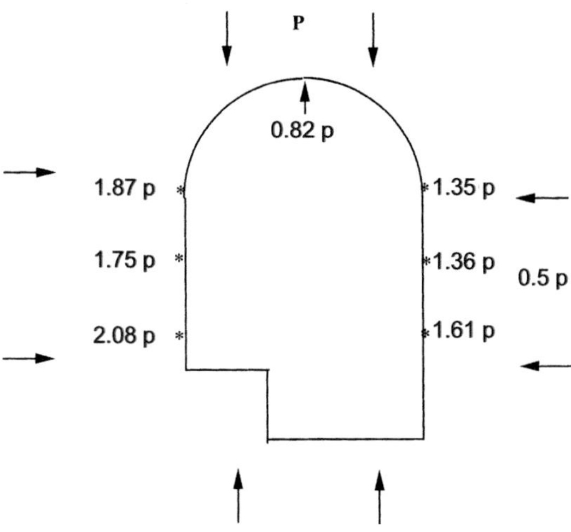

Figure 9.15 Stresses around underground workshop

Stress concentration at the crown σ_B and side of excavation σ_A are given as follows:

$$\sigma_A = p(1 - K + 2\sqrt{q}) = p\left[1 - K + \sqrt{\frac{2W}{\rho_A}}\right]$$

$$\sigma_B = p(K-1+2K/q) = p\left[(K-1) + K\sqrt{\frac{2H}{\rho_B}}\right]$$

$$\sigma_A = p(1-K+2q) = p\{1-0.5+(2\times 2/3) = 1.83\,p$$

$$\sigma_B = p(K-1+2K/q)$$

$$= p\left\{0.5-1+\frac{(2\times 0.5\times 3)}{2}\right\} = p$$

9.11 Rock Failure Zone Around Circular Boundary

Example 5 Failure criterion of rock mass

$$\sigma_1 = F(\sigma_3)$$

Circular excavation in biaxial field

$$\sigma_{\theta\theta} = P[1+K+2(1-K)\cos 2\theta]$$

If a rock has a uniaxial compressive strength of 16 MPa, the radial stress is given by:

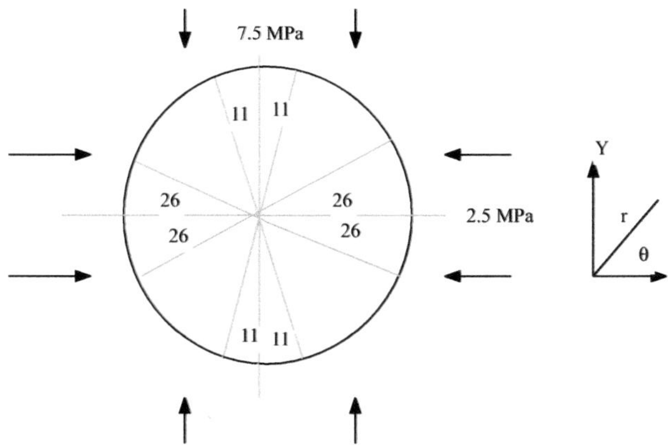

Figure 9.16 Zone of fracture around circular roadway

$$\sigma_{\theta\theta} = 7.5[1+0.3+2(1-0.3)\cos 2\theta]$$

$$= 7.5(1.3+1.4\cos 2\theta) \geq 16$$

or

$$\cos 2\theta \geq 0.595$$

θ is given by

$$-26° \leq \theta \leq 26° \quad \text{or} \quad 154° \leq \theta \leq 206°$$

At the crown, the failure will be in tensile mode:

$$7.5(1.3+1.4\cos 2\theta) \leq 0$$

$$79° = \theta = 101° \qquad \text{or} \qquad 259° \,\delta\, \theta \,\delta 281°$$
$$\sigma_A = p(1 - K + 2q^{0.5}) = p\,\{1 - K + (2W/\rho_A)^{0.5}\}$$
$$\sigma_B = p(K - 1 + 2k/q) = p\,\{K - 1 + K(2H/\rho_B)^{0.5}\}$$

This shows that the circular opening will fail. The proper design of the tunnel will be to increase the height of the opening.

9.12 Support and Reinforcement in Massive Rock

Example 6 The sidewall stress and roof stress around the elliptical opening in Fig. 9.17 are given by the following equations:

$$\sigma_A = p(1 - K + 2q^{0.5}) = 20\,(1 - 0.4 + 2 \times 2) = 92 \text{ MPa}$$
$$\sigma_B = p\,(K - 1 + 2K/q) = 20\,(0.4 - 1 + 2 \times 0.4/4)$$
$$= 20\,(-0.6 + 0.2) = -8 \text{ MPa}$$

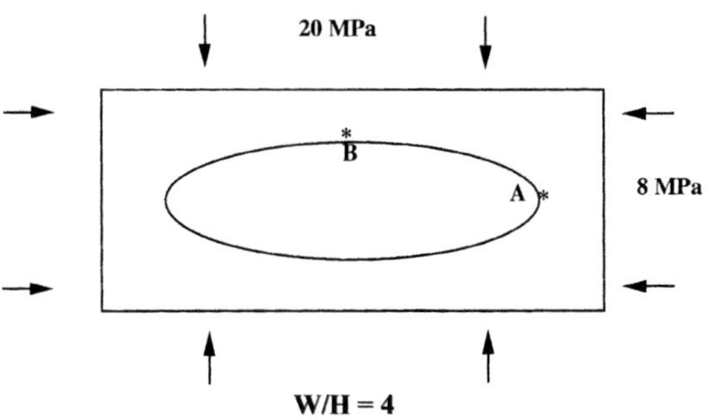

Figure 9.17 Support in massive rock

If a set of vertical support is installed sufficient to generate a vertical load of 1 MN/m^2 uniformly distributed over the excavation surface, the boundary

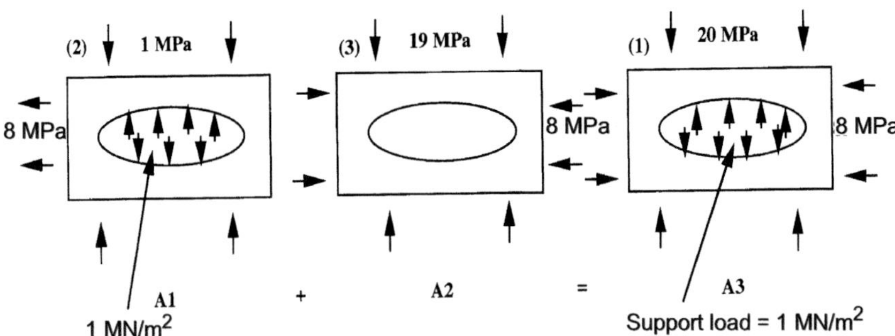

Figure 9.18 Support in massive rock

stresses around the excavation can be determined by superimposition of stresses.

$$\sigma_{A1} = \sigma_{A2} + \sigma_{A3}$$

$$= 1 + 19\,(1 - \frac{8}{19} + 2 \times 4) = 1 + 163 = 164 \text{ MPa}$$

$$\sigma_{B1} = \sigma_{B2} + \sigma_{B3} = 0 + P\left[\frac{8}{19} - 1 + \left\{\frac{2 \times 8}{19} \times \frac{1}{4}\right\}\right]$$

$$= 0 + 19\,(0.42 - 1 + 0.21) = -7.03 \text{ MPa}$$

(a) This example indicates that the support does not modify the elastic stress distribution around an underground opening.
(b) If the failure of a rock mass is possible, installation of support is unlikely to modify the stress distribution sufficiently to preclude the development of failure.
(c) It is therefore necessary to consider other mechanisms to explain the mode of action of support and reinforcement system.

9.13 Stress Around an Elliptical Opening and a Long Stope

It has been shown in Section 9.5 that the stress concentration around an elliptical opening depends upon the aspect ratio of the structure (W/H ratio), initial virgin stress of rock and the ratio of horizontal to vertical stress (σ_h/σ_v). Table 9.1 shows the stress concentration at the sidewall (σ_A) and on the crown of the tunnel (σ_B) for various ratio of horizontal to vertical stress K ranging from 0.25 to 5. It may also be noted that W/H ratio of 1 represents stress concentration around a circular tunnel.

Table 9.1 Stress concentration and aspect ratio of wide elliptical tunnels

Q = W/H	1		2		3		4		5	
K	σ_A	σ_B	σ_A	σ_B	σ_A	σ_B	σ_A	σ_B	σ_A	σ_B
0.25	2.75 p	–0.25 p	3.6 p	–0.5 p	4.21 p	–0.58 p	4.75 p	–0.63 p	5.22 p	–0.65 p
0.30	2.7 p	–0.1 p	3.55 p	–0.4 p	4.16 p	–0.50 p	4.7 p	–0.55 p	5.17 p	–0.58 p
0.50	2.5 p	0.5 p	3.32 p	0 p	3.96 p	–0.17 p	4.5 p	–0.25 p	4.96 p	–0.3 p
0.75	2.25 p	1.25 p	3.10 p	0.5 p	3.71 p	0.25 p	4.25 p	0.13 p	4.72	005 p
1	2 p	2 p	2.8 p	p	3.46 p	0.66 p	4 p	0.5 p	4.48	0.4 p
2	p	5 p	1.84 p	3 p	2.46 p	2.33 p	3 p	2 p	3.82 p	1.8 p
3	0 p	8 p	0.82 p	5 p	1.4 p	4 p	2 p	3.6 p	2.82 p	3.2 p
4	–p	11 p	–0.20 p	6 p	–0.4 p	5.6 p	1 p	5 p	1.48 p	4.6 p
5	–2 p	14 p	–1.76 p	9 p	–0.6 p	7.33 p	0.42 p	6 p	0.48 p	6 p

It can be seen in Figure 9.19 that for the K (σ_h/σ_v) value of 0.5 the sidewall stress is considerably higher than the crown stress. As the span increases, the sidewall stress increases and reaches a value of five times the cover load at a W/H ratio of 5. The crown stress decreases as the W/H ratio of the tunnel increases and its value changes from compressive to tensile stress at a W/H ratio of 2. It may be noted that the crown of a tunnel fails under a tensile stress.

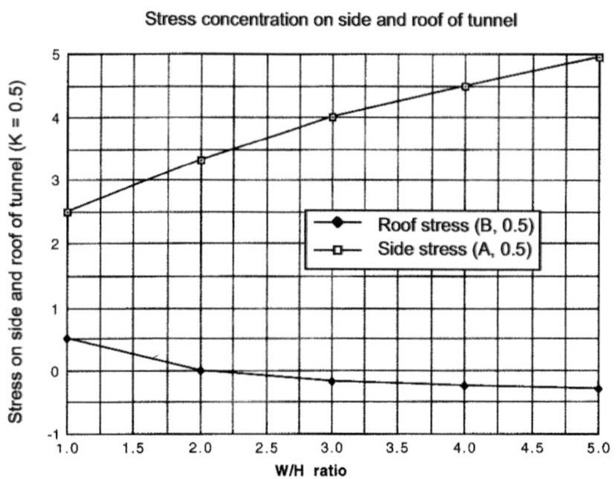

Figure 9.19 Stress around an elliptical tunnel for K = 0.5

Figure 9.20 shows the stress concentration around a circular tunnel (W/H ratio 1) and an elliptical tunnel (W/H > 1) for K = 3 representing *in situ* stress conditions prevailing in some collieries in NSW, Australia. It can be seen that the crown stress in the tunnel is much higher than the corresponding side stress and it decreases as the span of the tunnel increases

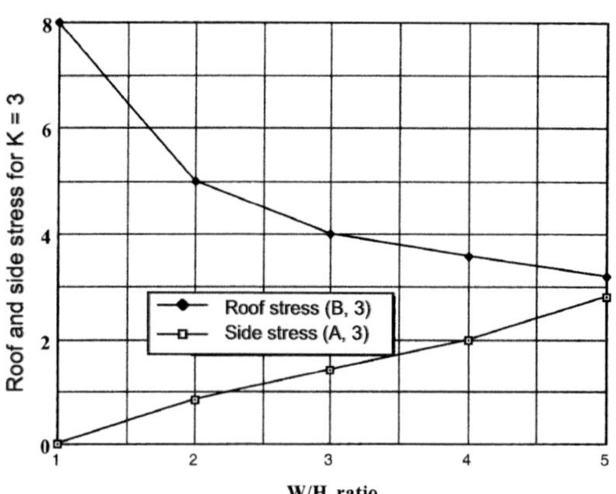

Figure 9.20 Stress concentration on roof and side of tunnel for K = 3

while the side stress increases with the span. It is noteworthy that all roof stresses and the side stress in the tunnel are compressive.

At the geostatic pressure (K = 1), the crown stress is equal to the side wall stress for a circular roadway. As the span of the tunnel increases the sidewall stress increases while the stress at the crown of the roadway decreases. Figure 9.21 indicates that as the horizontal stress increases, then stress at the crown of the tunnel becomes high compressive stress while the sidewall stress reduces from compressive to tensile stresses.

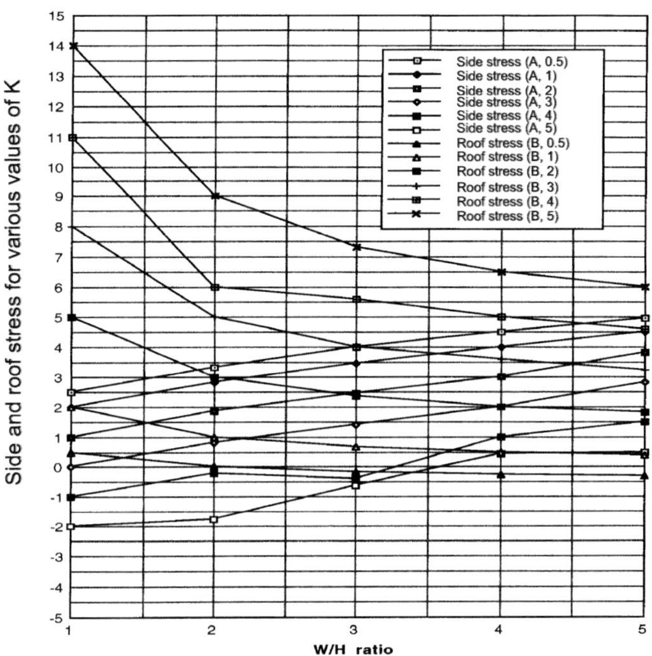

Figure 9.21 Stress concentration on roof and side of tunnel for various values of K and aspect ratio of tunnels

In very high horizontal stress regimes, increase in the tunnel span results in decrease in sidewall stress and at a higher aspect ratio an increase in compressive stress results. There is considerable reduction in the corresponding crown stress with an increase in the aspect ratio.

Table 9.2 shows the stress concentration around circular and elliptical openings for an aspect ratio of 0.05 to 1 for a variety of values of $K(\sigma_h/\sigma_v)$. This shape of opening may be akin to some stopes.

It can also be seen in Table 9.2 that if a circular opening is made in ground having K = 0.25, the initial crown stress at the roof of the opening will be

Table 9.2 Effect of aspect ratio q and horizontal to vertical stress ratio (K) on stress concentration around elliptical and circular openings

q = W/H	*1*		*1/2 = 0.5*		*1/4 = 0.25*		*1/5 = 0.2*		*1/10 = 0.1*		*1/20 = 0.05*	
	σ_A	σ_B	σ_A	σ_B	σ_A	σ_B	σ_A	σ_B	σ_A	σ_B	σ_A	σ_B
K = 0.25	2.75 p	–0.25 p	0.66 p	0 p	1.75 p	1.25 p	1.6 p	1.1 p	1.38 p	2.5 p	1.19 p	9.25 p
0.3	2.7 p	–0.1 p	0.7 p	0.2	1.7 p	1.7 p	1.5 p	23 p	1.33 p	5.3 p	1.14 p	11.3 p
0.5	2.5 p	0.5 p	0.9 p	1.0 p	1.5 p	3.5 p	1.4 p	4.5 p	1.13 p	9.5 p	0.94 p	19.5 p
0.75	2.25 p	1.25 p	1.16 p	2 p	1.25 p	5.75 p	1.1 p	7.25 p	0.88 p	4.7 p	0.69 p	29.7 p
K=1	2 p	2 p	1.4 p	4 p	1 p	8 p	0.9 p	10 p	0.63 p	20 p	0.44 p	40 p
K=2	p	5 p	2.4 p	9 p	0	17 p	–0.1	21 p	–0.37	41 p	–0.56 p	81 p
K=3	0.8 p	8 p	3.4 p	14 p	–p	26 p	–1.1	32 p	–1.37	62 p	–1.56 p	122 p
K=4	–p	11 p	4.4 p	19 p	–2 p	35 p	–2.1	41	–2.37	83 p	–2.56 p	163 p
K=5	–2	14 p	5.5 p	24 p	–3 p	44 p	–3.1	54 p	–3.4	104 p	–3.56 p	204 p

tensile (–0.25 p) and at the side of the opening will be 2.75 p compressive. If the roof of this stope is changed until the aspect ratio becomes 0.05, the crown stress will become 9.25 p and the side wall stress of the stope will be 1.19 p. The effect of inducing a high horizontal stress on a circular tunnel is to reduce the side wall stress from 2.75 p compressive to –2 p tensile and to increase the crown stress from tensile stress –0.25 p to 14 p compressive. It may also be noted that the crown pillar stress in a high stress field (K = 3) with width to height aspect ratio of 0.05 will be 122 p and side stress –1.56 p tensile.

9.14 Conclusions

Closed-loop solutions for stress in rock around openings of idealised shape can give an important insight to stress induced instability problems in the rock. Kirsch's (1898) solution of stress around a circular hole in an infinite medium and various solutions of stress distribution around elliptical openings by Greenspan (1944), Poulos and Davis (1974) and Jaeger and Cook (1979) can be used to solve many stress-related problems in rock mechanics. This chapter presents some worked-out examples of selected problems in mining.

It is customary to use square or rectangular roadways in tabular ore deposits and also coal mining. Therefore, many authors have presented specific solutions to stress distribution problems around square hole by Brock (1958) and around rectangular openings by Heller *et al.* (1958). Worked-out examples presented in this chapter are largely due to Brady and Brown (1985).

References

Bray J.W. (1977) quoted in Brady, B.H.G. and E.T. Brown (1985). *Rock Mechanics for Underground Mining*. George Allen and Unwin, London, 527 pp.

Brock, J.S. (1958), *Analytical determination of the stresses around square holes with rounded corners* David Taylor Model Basin, Report 1149.

Greenspan, M. (1944). 'Effect of a small hole on the stresses in a uniformly loaded plate'. *Quarterly Applied Mathematics*, 2, pp. 60–71.

Heller, S.R. Jr., J.S. Brock and R. Bart (1958). 'The stresses around a rectangular opening with rounded corner in a uniformly loaded plate'. *Transaction of the Third U.S. Congress on Applied Mechanics*, AIME.

Jaeger, J.C. and N.G.W. Cook (1979). *Fundamentals of rock mechanics*, 3rd Edition, London, Chapman and Hall.

Kirsch, G. (1898). 'Die Theorie der Elastizitat und die Bedürfnisse der Festigkeitlehre,' V.Duet. Ing., Vol. 42 (28), pp. 797–807.

Ladanyi, B. (1974). 'Use of long-term strength concept in the determination of ground pressure on tunnelling', *Advances in Rock Mechanics, Proceeding of the Third Congress in Rock Mechanics,* Denver, Published by Natl. Acad. Sci. 2B, pp. 1150–56.

Lamb, H. (1956). *Infinitesimal calculus*, Cambridge University Press, Cambridge.

Love A.V.H. (1944). *The Mathematical Theory of Elasticity*, Dover Publications, New York.

Mindlin, R.J. (1939). Stress distribution around a tunnel', American Society of Civil Engineers, April, pp. 619–642.

Muskhelishvili, N.I. (1953). 'Some basic problems in the mathematical theory of elasticity', Translated by J.R.M. Radok, Noordhof, Groningen.

Prescott, J. (1946). *Applied Elasticity*, Dover Publications, New York.

Poulos, H.G. and E.H. Davis (1974). *Elastic solutions for soil and rock mechanics*, Wiley, New York.

Savin, G.N. (1961). *Stress concentration around holes*, Pergamon Press, New York.

Terzaghi, K.Van and F.E. Richart (1952). 'Stresses in rocks about cavities' *Geotechnique*, Vol. 3, pp. 57–90.

Timoshenko, S. and J.N. Goodier (1951). *Theory of Elasticity*, Chapter 4, McGraw Hill Book Co., New York.

Stability of Underground Openings by Mathematical Modelling

10
Chapter

10.1 Introduction

The stress analysis techniques incorporating closed loop solutions discussed in the previous chapter are important design tools for assessing stress induced instabilities around underground excavations under idealised conditions. However, in order to assess practical ground control problems in the vicinity of underground mining excavations more complexities need to be built into the mathematical models in order to design support systems and to calculate the amount of closure encountered in the excavations. In this chapter, various approaches for the design of excavations in rock taking into account various shapes of openings, different rock properties and criteria of failure are discussed. These methods permit the calculation of stress and displacement around the opening at any point and allow estimation of support loading. An empirical method of support design in jointed rock mass is also discussed together with the application of a ground-support reaction curve for roof support design.

10.2 Methods of Approach

In the past, various methods have been used for the stability analysis of underground mine openings as follows.

(1) Stress concentration theory

This approach enables the calculation of stress concentration around circular and elliptical holes in an infinite plate of idealised elastic and isotropic material. These idealised solutions enable calculations of the following parameters:

(a) Stress distribution

(b) Elastic displacement

Such solutions provide qualitative analysis of the stress distribution around mining excavations and assist in formulating the principles for design of excavations in rock.

(2) Arch theories for excavations in shallow depths
   (a) Load on support
   (b) Estimation of support type by rock classification
   Example – Terzaghi classification

(3) Theory of buried pipe – for shallow to medium depths

(4) Yield zone theory
   (a) Ladanyi – stress distribution in tunnels in yielding rocks
   (b) Bray – elasto-plastic theory
   (c) Wilson – elastic theory

   Mathematical solutions by Ladanyi (1974), Bray (1967), and Wilson (1980) utilising the yield zone theory seek to calculate the following parameters around a circular tunnel:
   (a) stress
   (b) tunnel closure

(5) Design of support using ground/support reaction curve

(6) Empirical methods of support design in jointed rock

10.3 Design of Structures in Rock and Development of Support Design Criteria

The principal types of mining excavations are designed on the basis of the life of the structure, use, safety requirements and cost of the support. The following types of mining excavations have different life spans and method of support as discussed below.

(1) Shafts, access roadways for men and materials, ventilation roadways and main ore-passes which require stability throughout the life span of the mine.

   The nature of the support used here is permanent concrete or brick lining especially in ventilation roadways where very low resistance to ventilation current to minimise power cost of the ventilation fan is important.

(2) Roadways in solid ground

   These are permanent roadways whose life is in excess of five years up to the complete life of the mine. The types of support required will be steel arches with laggings or rock bolts with wire mesh to provide safety against roof falls.

(3) Roadways adjacent to working faces or stopes
These roadways are subjected to high stress fields and particularly to the dynamic abutment stresses caused by adjacent mine workings. These roadways are frequented by men and machines, and the life of such roadways is limited to one to two years requiring cost effective support systems. Examples of such roadways are as follows:
  - Gate roadways in longwall mining
  - Haulage developments in block caving

(4) Working faces and stopes
Safety of men and machines and ground control against catastrophic failure are the main criteria of stability during the limited life of these workings.

10.3.1 Criteria of Stability

The foregoing discussion leads to the conclusion that the criteria of stability in underground openings are as follows:

- Failure of opening
- Possibility of excessive closure

The degree of acceptable closure may change according to usage and expected life of the opening e.g. winding shafts will need to take into consideration clearance for cages or skips and may have to take into consideration the amount of shaft convergence as a design parameter. Ventilation shafts have to weigh the possible loss of ventilation efficiency against the cost of lining. In the case of ore passes, excessive failure is not acceptable but relatively high closure is tolerable. Therefore, an underground structure must be designed so that the amount of closure experienced comes within the acceptable limits specified. For example, travelling roadways in coal mining are not permitted to be less than 1.6 m in height, so the amount of maximum acceptable closure in an initially 3.06 m high gate roadway for longwall advance mining is 1.46 m.

10.3.2 Design Principles

Past experience has shown that

- closure of openings can be related to stress, and
- the shape and orientation should be designed to give the best stress distribution in the surrounding rocks.

From the previous discussions, the principles for the design of excavations in biaxial stress fields and elastic rock can be expressed as follows.

1. Avoid sharp corners.

2. In high stress regimes in hydrostatic stress fields (K = 1), the best shape of the roadway is circular.
3. In a stress field with a ratio of horizontal to vertical stress of K > 1, the stress is lowest on the boundary of an opening having the largest radius of curvature and the stress is highest normal to the lowest span of the opening.
4. When K ≠ 1 the best shape is ovaloidal e.g. if the opening has to be excavated with the height to width ratio of 1 : 2 with K = 0.5, the best shape is elliptical with the higher stress parallel to the longer axis of the opening Fig. 10.1.
5. When K < 0.33, tensile stresses occur on the boundaries of all excavation shapes.

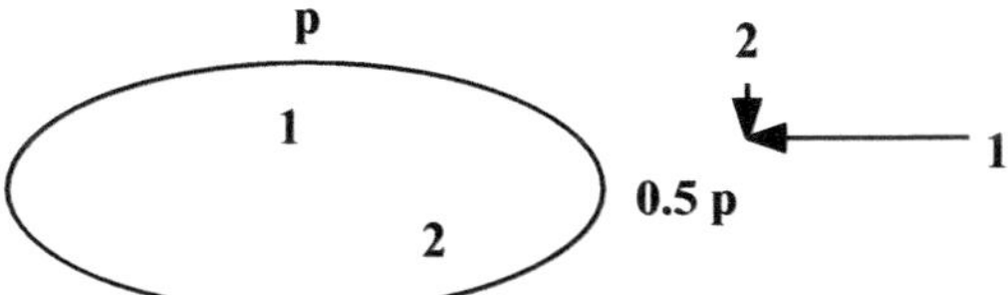

Figure 10.1 Effect of W/H ratio and K on orientation of excavation

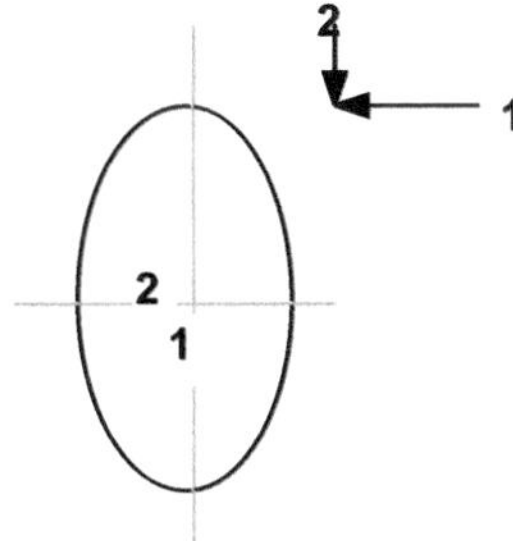

Figure 10.2 Orientation of elliptical opening

6. Boundary stresses in an ovaloidal opening can be reduced to a minimum if the axis ratio of the opening can be matched to that between the *in situ* stresses.
7. In jointed rock, the orientation and size of an opening is important in reducing the size of unstable wedges formed above the excavation and in restricting the volume of unstable rock formed above the opening.

10.4 Terzaghi's Ground Arch Theory

This theory is based on observations made by earlier workers that the load on the standing support inside a mining excavation or a shallow tunnel is only a fraction of the load due to the weight of the overlying rock. As a consequence of forming a roadway or tunnel the majority weight of the overlying rock is transferred to the solid rock surrounding the tunnel as an abutment load. The load on the standing support is only due the pressure exerted by loosening rock surrounding the tunnel on the support.

According to Terzaghi (1946) (Fig. 10.3) the rock within the area 'abcd' becomes loosened during excavation and moves towards the opening. It

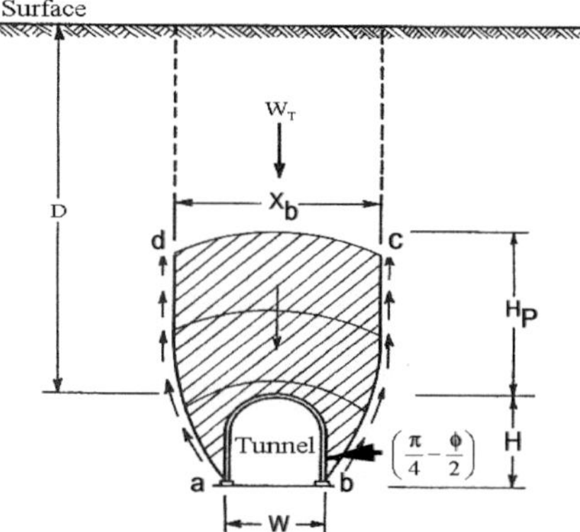

Figure 10.3 Terzaghi's arch theory for shallow tunnel

was assumed that the loosened rock behaves as granular material and detaches itself from the side of the tunnel by an angle $\left(\frac{\pi}{4}-\frac{\phi}{2}\right)$. It is further assumed that the movement is resisted by the frictional force which transfers the major proportion of the overburden pressure W_T to the surrounding rock. The support system is required to bear the load of a block with height equal to H_p and width X_b.

The span of the broken rock arch depends upon the characteristics of the surrounding rock and the tunnel's dimensions ($H \times W$).

Thus the span of the arch is given by the following equation:

$$X_b = W + 2H \tan\left(\frac{\pi}{4}-\frac{\phi}{2}\right)$$

The vertical support pressure is given by:

$$\sigma_V = \frac{\gamma X_b}{2K \tan \phi}\left[\left(1-e^{-\left(\frac{K \tan\phi \cdot 2D}{X_b}\right)}\right)\right]$$

The horizontal pressure is given by the following equation:

$$\sigma_H = 0.3\left(\frac{1}{2}H + \sigma_V\right)$$

where

H = height of tunnel (m)
W = width of tunnel (m)
D = depth below the surface (m)
γ = rock mass density
ϕ = internal angle of friction

$$K = \text{virgin stress factor} = \frac{\sigma_H}{\sigma_V}$$

X_b = span of the pressure arch

H_p = height of the pressure arch

Based upon the above theory Terzaghi (1946) suggested a subjective method of estimating the load on the standing support based on the following assumptions:

$$\phi = 37°; \qquad D = 63\text{ m}; \qquad \sigma_V = 0.67\,(W + H);$$

$$\frac{\sigma_V}{\gamma} = \text{Constant}\,(W + H)$$

Hard and intact rock $\quad \frac{\sigma_V}{\gamma} = 0 \quad$ No support required

Blocky and seamy rock $\quad \frac{\sigma_V}{\gamma} = 0.1 \text{ to } 1.1\,(W + H)$

Squeezing rock $\quad \frac{\sigma_V}{\gamma} = 1.4 \text{ to } 4.5\,(W + H)$

This approach provided the first rock mass classification system for design of roof supports in shallow tunnels.

Limitations

The limitations of this approach for the design of tunnel support are as follows.

(i) if H = W = 7 m no further support pressure is required below 63 m depth, even in squeezing rock.

(ii) In soft rock, roadway closure does take place.

(iii) Yielding support takes less load than rigid support.

Thus, an alternative method of designing the support for the main drivages in rock is required.

Example 1 A horizontal tunnel at a depth of 100 m, with an excavated dimension 6.5 m wide and 4.5 m high, is proposed to be driven in a volcanic rock. It is given that the density γ is equal to 2500 kg/m^3, internal angle of friction of rock ϕ is 29° and the ratio of horizontal to vertical stress K is equal to 0.5. Determine (i) vertical and horizontal support loading and (ii) suggest types of temporary and permanent support required to stabilise the tunnel.

Solution $X_b = 6.5 + 2 \times 4.5 \tan (45 - 29/2)$

$= 6.5 + 2 \times 4.5 \times 0.58 = 11.8$ m

$$\sigma_V = \frac{\gamma X_b}{2K \tan \phi}\left[\left(1 - e^{-\left(\frac{K x \tan\phi \cdot 2xD}{Xb}\right)}\right)\right]$$

$$\sigma_V = \frac{2500 \times 11.8}{2 \times 0.5 \tan 29}\left[1 - e^{-\frac{0.5 \times 0.554 \times 2 \times 100}{11.8}}\right]$$

$$= 53219.4 \times 0.99 = 52692.7 \text{ N/m}^2 = 0.5269 \text{ MPa}$$

$$\sigma_H = 0.3\left(\frac{1}{2}H_\gamma + \sigma_V\right)$$

$$= 0.3\left(\frac{1}{2} \times 4.5 \times 2500 + 52692.7\right)$$

$$= 0.3\ (5625 + 52692.7) = 0.175 \text{ MPa}$$

Temporary support – rock bolts and mesh
Permanent support – steel arches with lagging

***Example* 2** A coal mine roadway has excavated dimensions of 5.4 m × 3.12 m at a depth of 150 m below the surface. The rock density is 2300 kg/m^3, internal angle of friction ϕ is 35° and the ratio of horizontal to vertical stress (K) is 3. Calculate the vertical and horizontal loading of the support.

Solution $X_b = 5.4 + 2 \times 3.12 \tan\left(45 - \frac{35}{2}\right) = 5.4 + 6.24 \times 0.52 = 8.65 \text{ m}$

$$\sigma_V = \frac{2300 \times 8.65}{2 \times 3 \times \tan 35°}\left(1 - e^{-\frac{3 \times \tan 35 \times 2 \times 150}{8.65}}\right)$$

$$= 4735 \times (1 - e^{-72.8}) = 4739 \text{ N/m}^2$$

$$= 0.0474 \text{ MPa}$$

$$\sigma_H = 0.3\left[\frac{1}{2}3.12 \times 2300 + 4740\right] = 0.3\ (3588 + 4740) \text{ N/m}^2$$

$$= 0.025 \text{ MPa}$$

10.5 Theory of Buried Pipe due to Muir Wood

Muir Wood (1979) applied an elastic analysis to shallow tunnels, assuming that a lined tunnel will behave like a buried pipe surrounded by a zone of failed rock (yield zone) around the opening. Other assumptions made in the above analysis were the following.

(i) The roadway is unlined.

(ii) The roadway is circular.
(iii) The ground is isotropic and homogeneous.
(iv) K is a function of Poisson's ratio v.

The formula predicts the radial closure of the tunnel but the limitations of this analysis are that it is sensitive to the elastic modulus of rock and it tends to give much lower values of tunnel closures than observed in mining practice. This method has, therefore, very limited application in mining.

10.6 Circular Opening in a Yielding Rock Subjected to Hydrostatic Stress

Closed loop solutions of stress distribution around a circular opening give elastic displacements that are of much lower value than those observed in the field. In practice it is experienced that a zone of broken rock known as the yield zone surrounds an opening and also the amount of tunnel closure are much higher than predicted by the elastic theory.

Therefore, attempts have been made by various workers to incorporate complexities to the mathematical models for calculating realistic roadway closure.

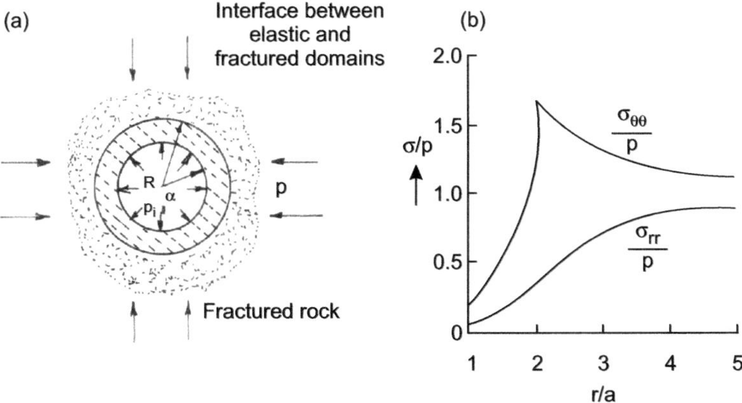

Figure 10.4 Stress around a circular opening in yielding rock (Ladanyi, 1974)

Ladanyi (1974) assumed that the field stress is such that an annulus of broken rock or a yield zone is generated in the excavation periphery (Fig. 10.4). Other assumptions made in his analysis were as follows:

p_i = applied support pressure (MPa)
R = radius of yield boundary (m)

The strength of the rock mass is described by the Mohr-Coulomb criterion as follows:

$$\sigma_1 = \sigma_3 \frac{1+\sin\phi}{1-\sin\phi} + 2c\frac{\cos\phi}{1-\sin\phi}$$

or

$$\sigma_1 = K \times \sigma_3 + \sigma_c$$

The strength of the fractured rock is purely frictional and is defined as follows:

$$\sigma_1 = \sigma_3 \frac{1+\sin\phi_f}{1-\sin\phi_f}$$

$$\sigma_1 = K_1\,\sigma_3$$

Stress distribution in the yield zone is given by the following expression:

$$\sigma_{rr} = p_i\left[\frac{r}{a}\right]^{K_1-1}$$

$$\sigma_{\theta\theta} = K_1\,p_i\left[\frac{r}{a}\right]^{K_1-1}$$

The radial stress at the elastic boundary is given by the following equation:

$$p_1 = p_i\left[\frac{R}{a}\right]^{K_1-1} \quad \text{or} \quad R = a\left[\frac{p_1}{p_i}\right]^{\frac{1}{K_1-1}}$$

where p_1 is the equilibrium radial stress at the annulus of the outer boundary.

The stress distribution at the elastic zone at radial distance r is given as follows:

$$\sigma_{\theta\theta} = p\left(1+\frac{R_e^2}{r^2}\right) - p_1\frac{R^2}{r^2}$$

$$\sigma_{rr} = p\left(1-\frac{R^2}{r^2}\right) + p_1\left(\frac{R^2}{r^2}\right)$$

At the inner boundary of the elastic zone the state of stress is defined as

$$\sigma_{\theta\theta} = 2p - p_1; \qquad \sigma_{rr} = p_1$$

and

$$p_1 = -\frac{2p-\sigma_c}{1+K}$$

$$R = a\left[\frac{2p-\sigma_c}{(1+K)p_i}\right]^{\frac{1}{K_1-1}}; K = \frac{1+\sin\phi}{1-\sin\phi} = \text{triaxial stress factor for intact rock}$$

Example 3 Given that the internal angle of friction of intact rock as well as that of fractured rock is 35° and the uniaxial compressive strength is 0.5 of the initial *in situ* hydrostatic stress. If the radius of the yield zone is equal to 1.99 times the radius of the circular excavation, calculate the stress field around the opening in the yielding rock.

Solution The stress field around the opening is shown in the Figure 10.4(b).

$$\sigma_1 = \frac{1+\sin\phi}{1-\sin\phi}\cdot\sigma_3 + 2C\frac{1-\sin\phi}{1-\sin\phi}$$

$$\sigma_1 = 3.68\,\sigma_3 + 1.9\times 2\,c = 3.68\,\sigma_3 + 3.84\,c$$

Strength of fractured rock

$$\sigma_1 = 3.68\,\sigma_3$$

Equilibrium at the boundary of the yield zone

$$p_1 = 0.05\,p\left[\frac{R}{a}\right]^{2.68} = 0.32\,p$$

$$R = a\left[\frac{p_i}{0.05p}\right]^{\frac{1}{2.68}} = 1.99\,a$$

In elastic zone the circumferential and radial stresses can be calculated as follows:

$$\sigma_{\theta\theta} = p\left[1+\frac{(1.99a)^2}{r^2}\right] - 0.32p\,\frac{(1.99a)^2}{r^2}$$

$$= p\left[1+3.96\,\frac{a^2}{r^2}\right] - 0.32\,p\times 4\times\frac{a^2}{r^2} = p\left[1+2.72\,\frac{a^2}{r^2}\right]$$

$$\sigma_{rr} = p\left(1-\left\{\frac{R}{r}\right\}^2\right) + p_1\left\{\frac{R_e}{r}\right\}^2 = p\left\{1-0.68\,\frac{R^2}{r^2}\right\}$$

The above equations permit the radial stresses and the normal stresses to be calculated around a circular tunnel as shown in Table 10.1.

Table 10.1 Stress distribution around a circular tunnel

r	a	$1.99a$	$2a$	$3a$	$4a$	$5a$	$6a$
σ_{rr}/p	0.05	0.32	0.33	0.7	0.85	0.89	0.93
$\sigma_{\theta\theta}/p$	0.18	1.14	2	1.68	1.17	1.1	1.07

The main features of the stress distributions are as follows.

1. The high and increasing gradient in the radial variation of $\sigma_{\theta\theta}$ both in the absolute sense and also compared with σ_{rr}.

2. A significant step increase in $\sigma_{\theta\theta}$ at the interface between the fractured and intact rock.
3. The purpose of support in massive rock is follows.
   - Provides radial continuity of contact between rock fragments in the yield zone.
   - Serves to generate a radial confining stress at the excavation boundary.
   - The mode of action of the support is to generate and maintain high triaxial stress in the fracture domain, by mobilising the frictional action between the surfaces of the rock fragments.

10.7 Bray's Elasto-plastic Solution for Tunnel Stability

In order to predict realistic tunnel closure, Bray (1967) assumed that the zone of yielded rock around a roadway at depth is plastic. This yield zone is intercepted by fractures which are spiral in shape and they subtend an angle 'δ' to the wall of the tunnel (Fig. 10.5).

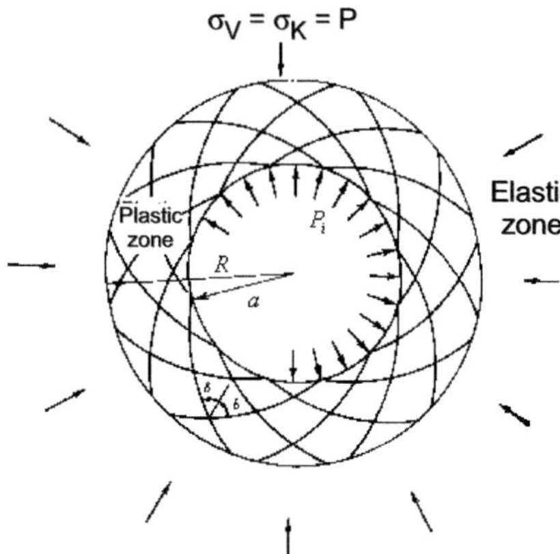

Figure 10.5 Bray's elasto-plastic solution for tunnel stability (Goodman, 1980)

It was also assumed that the joints have shear strength characteristics as given by the following equation:

$$\tau_p = C_j + \sigma \tan \phi_j$$

Stress in elastic zone at a point at radial distance 'r' is given by the following equations:

$$\sigma_r = p - \frac{b}{r^2}, \qquad \sigma_\theta = p + \frac{b}{r^2}$$

where

$$b = \left[\frac{\left\{\tan^2\left(\frac{\pi}{4}+\frac{\phi}{2}\right)-1\right\}p+\sigma_c}{\left\{\tan^2\left(\frac{\pi}{4}+\frac{\phi}{2}\right)+1\right\}}\right]R^2$$

$$U_r = \frac{1-\upsilon}{E}\left\{p_i\frac{r^{q+1}}{a^q}-pr-\frac{t}{r}\right\}+\frac{t}{r}$$

$$Q = \frac{\tan\delta}{\tan(\delta-\phi_j)}-1, \qquad \text{and}$$

$$t = \frac{1-\upsilon}{E}R_e^2[(p+c_j\cot\phi_j)-(p_i+c_j\cot\phi_j)\left\{\frac{R}{a}\right\}^Q+\left\{\frac{1-\upsilon}{E}\right\}b$$

Stress in the plastic zone at a point at distance 'r' is given by the following equation:

$$\sigma_r = (p_i+c_j\cot\phi_j)\left(\frac{r}{a}\right)^Q - c_j\cot\phi_j$$

and

$$\sigma_\theta = (p_i+c_j\cot\phi_j)\frac{\tan\delta}{\tan(\delta-\phi_j)}\left[\frac{r}{a}\right]^Q-(c_j\cot\phi_j)$$

where

δ = angle of break to the tunnel wall
Q = a variable depending upon the angle of the fracture to the tunnel wall and the internal angle of friction of the fractures
U_r = radial closure of tunnel at radius 'r'
σ_c = uniaxial compressive strength (MPa)
$p = \sigma_V = \sigma_h$ = Initial stress of rock (MPa)
p_i = support pressure (MPa)
r = radius at which the stress is being considered (m)
R = outer radius of the plastic zone (m)
a = Radius of tunnel (m)
c_j = cohesive strength of joints
ϕ_j = internal angle of friction of the joints
b = a variable

Assumptions

The following assumptions are made in deriving the above equations.

- Hydrostatic stress field

- Mohr-Coulomb criterion of failure
- Shape of the fracture is spiral
- Rock is homogeneous and isotropic

Example 4 A given rock mass has properties $E = 1200$ MPa, $\upsilon = 0.3, \phi = 40^{\circ}$, and $\sigma_c = 10$ MPa. The tunnel radius is 2.5 m and the yield zone radius is 11 m. The *in situ* state of hydrostatic stress results are p = 30 MPa and support pressure $p_i = 0.25$ MPa. Fractures have the following properties:

$$\phi = 30^{\circ}, \quad \delta = 48^{\circ} \quad \text{and} \quad C_j = 0.$$

Calculate the inward radial displacement of the tunnel U_r.

Solution At the tunnel wall of radius a = 2.5 m

$$Q = \frac{\tan \delta}{\tan (\delta - \phi_j)} - 1, \qquad = \frac{\tan 48}{\tan (48 - 30)} - 1 = 2.42$$

$$b = \left[\frac{\left\{ \tan^2\left(\frac{\pi}{4} + \frac{\phi}{2}\right) - 1 \right\} p + \sigma_c}{\left\{ \tan^2\left(\frac{\pi}{4} + \frac{\phi}{2}\right) + 1 \right\}} \right] R^2$$

$$= \left(\frac{\{\tan^2 (45 + 40/2) - 1\}\, 30 + 10}{\tan 2\, (45 + 40/2) + 1} \right) 11^2 = 2550$$

$$t = \frac{1 - \upsilon}{E} R^2 [(p + c_j \cot \phi_j) - (p_i + c_j \cot \phi_j) \left\{ \frac{R}{a} \right\}^Q + \left\{ \frac{1 - \upsilon}{E} \right\} b$$

$$t = \frac{0.7}{1200} \times 11^2 \left[30 - 0.25 \frac{11}{2.5} \right] Q + \frac{1.3 \times 2550}{1200} = 4.24$$

$$U_r = \frac{1 - \upsilon}{E} \left(p_i \times \frac{r^{Q+1}}{a^Q} - p \times r \right) + \frac{t}{r}$$

$$= \frac{1 - 0.3}{1200} \left(0.25 \times \frac{25^{2.42+1}}{2.5^{2.42}} - 30 \times 2.5 \right) + \frac{4.25}{2.5}$$

$$= -0.043 + 1.696 = 1.653 \text{ m}$$

Example 5 A 5 m diameter tunnel is driven in shale. The modulus of deformation of rock is 1000 MPa and Poisson's ratio 0.25. The initial state of stress was 27.9 MPa (hydrostatic) and estimated support pressure 0.3 MPa. The fractures intersect the wall of the tunnel at 45° and the internal angle of friction of the fractures is 30°. Given that t = 3.45, calculate the closure of the tunnel.

$$U_r = \frac{1-\upsilon}{E}\left(p_i \times \frac{r^{Q+1}}{a^Q} - p \times r\right) + \frac{t}{r}$$

Polar coordinates of a point where stress is required is r, $\tilde{\theta}$. Hence

$$\frac{r^{Q+1}}{a^Q} = \frac{a^{Q+1}}{a^Q} = a, \text{ pi} = 0.3, \text{ p} = 27.9 \text{ MPa}$$

$$U_r = \frac{1-\upsilon}{E}(\text{pi} - \text{p})\, a + \frac{t}{r}$$

$$= \frac{1-0.25}{1000}(0.3-27.9)\times 2.5 + \frac{3.45}{2.5} = -0.05175 + 1.38 = 1.328 \text{ m}$$

10.8 Drift Support Design

10.8.1 Support of Roadway in Caving Ground

Factors affecting the design of a drift in caving ground are as follows:

- Influence of rock joints and fractures
- Rock load on the support
- Abutment loading
- Ground movement, expected repair and desirable flexibility of support

In hard rock mining, an essential element of support design is bolt, mesh and/or shotcrete.

1. *Estimation of the strength and deformational properties of rock mass*

In order to make an estimation of the cohesive strength and internal angle of friction of a rock mass, an estimate is made by visual observations of the proportion of intact rock and the proportion of rock joints taking part in the deformation. Thus,

Rock mass strength = strength of fracture surface + fraction of intact rock strength

$$C = C_f + R\,(C_r - C_f)$$

$$\tan\phi = \tan\phi_f + R\,(\tan\phi_r - \tan\phi_f)$$

C, ϕ = assigned cohesive strength and internal angle of friction of rock mass

C_f, ϕ_f = cohesion and friction angle of fractures

C_r, ϕ_r = cohesion and internal angle of friction of intact rock

R = percentage contribution to strength by joints

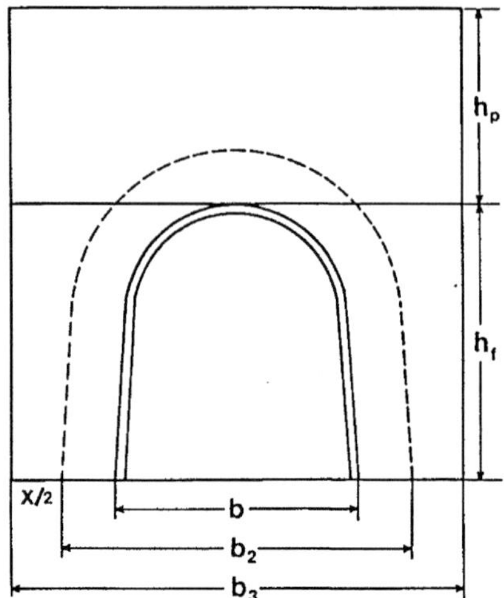

Figure 10.6 Estimation of tunnel support loading

2. Rock mass strength from Mohr-coulomb criteria

$$\sigma_1 = 2C \tan\left(\frac{\pi}{4} + \frac{\phi}{2}\right) + \sigma_3 \tan^2\left(\frac{\pi}{4} + \frac{\phi}{2}\right)$$

$$\sigma_m = \frac{\sigma_1}{SF}$$

σ_1 = maximum principal stress
σ_3 = minimum principal stress
ϕ = internal angle of friction of rock mass
σ_m = strength assigned to rock mass
SF = scale factor, laboratory to field strength

Estimation of load on the support is based on the concept developed by Terzaghi (1946) which assumes that the vertical stress developed on the support is caused by the gravitational load of the block acting on the support (Fig. 10.6). The height of the block resting on the support depends upon the width and height of the tunnel and the condition of the ground described by k_1 as follows:

$$H_p = k_1 (b + H_f)$$

k_1 = 1 for solid ground
= 2 or 3 for caving ground

The width of the loosened block is given by the width of the tunnel plus twice the length of rock bolts, together with the double thickness of loosened ground. The loosened rock mass is assumed to extend X/2 distance beyond the depth of the reinforced arch and is equal to the thickness of anchor of a point anchored bolt. Thus, the area of the block to be the supported by the tunnel arch is given below:

$$A = b_3 \times 1\ m^2$$
$$A = (b + 2r + x)$$

In equilibrium, the load of the loosened block is supported by reaction in the steel arch together with the thrust provided by the reinforced rock arch. The equilibrium thrust provided by the arch is given by the following equation:

$$\gamma \times H_p \times A = \sigma_s A_s + \sigma_m A_m$$

10.8.2 Drift Support in the Area of Active Caving (Fig. 10.7)

$$\sigma_V = \sigma_2 + \Delta\sigma_2 \approx 2\sigma_2$$

σ_V = vertical stress

σ_2 = virgin strata stress

$= 0.027 \times \text{depth (m)}$

$\Delta\sigma_2$ = increased abutment stress

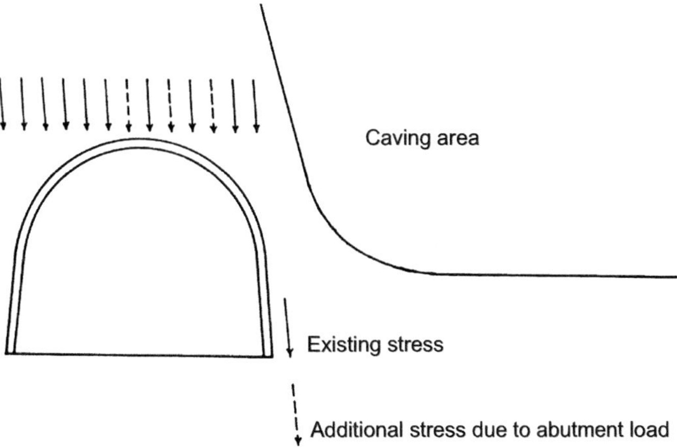

Figure 10.7 Tunnel support loading in the area of abutment load

Modified equation

$$\gamma (H_p \times A) \frac{\lambda}{100} = \sigma_s (A_s + A_a) + \sigma_m A_m$$

λ = percentage increase in the existing stress due to abutment

A_a = additional area of artificial support required

Example 5 Design a support system for a mining tunnel in fractured and altered rock using the following data:

$$\text{Rock mass density } (\gamma) = 2700\ \text{kg/m}^3$$
$$\text{Uniaxial compressive strength } (\sigma_c) = 69\ \text{MPa}$$
$$\text{Cohesion of intact rock (c)} = 10.4\ \text{MPa}$$
$$\text{Cohesive strength of fractures } (c_f) = 0.069\ \text{MPa}$$
$$\phi_r = 45°$$
$$\phi_f = 12°$$
$$\text{Drift size} = 3.66\ \text{m} \times 3.66\ \text{m}$$

Rock reinforcement is used to pin big loose wedges so that no account is taken of the contribution of the rock arch. Calculate the required thickness of the poured concrete lining. The strength of the concrete lining is 34.5 MPa, the factor of safety for concrete is 2.5, and assume that R = 0.16.

Solution

$$H_p = 1.0\{b + H_f\} = 7.32\ \text{m}$$
$$\gamma H_p A = \sigma_m A_m + \sigma_s A_s$$

Neglecting $\sigma_m A_m$, we have

$$\gamma H_p A = \sigma_s A_s$$

Estimation of rock mass cohesion and friction

$$C = C_f + R(C_r - C_f)$$
$$= 0.69 + 0.16\,(10.4 - 0.069) = 1.72\ \text{MPa}$$
$$\tan\phi = \tan\phi_f + R(\tan\phi_r - \tan\phi_f)$$
$$= 0.21 + 0.16\,(1.0 - 0.21) = 0.3364; \quad \phi = 18.59°$$

The maximum principal stress the rock mass can support is:

$$\sigma_1 = 2C\tan\left(\frac{\pi}{4} + \frac{\phi}{2}\right) + \sigma_3\tan^2\left(\frac{\pi}{4} + \frac{\phi}{2}\right)$$
$$= 2 \times 1.72\tan 54.3° = 4.78$$

Strength assigned to rock mass $\sigma_m = \sigma_1/3 = 1.59$ MPa

Equilibrium equation:

$$\gamma H_p A = \sigma_m A_m + \sigma_s A_s$$

Neglecting $\sigma_m A_m$, we have

$$\gamma H_p A = \sigma_s A_s$$

$$\sigma_s A_s = 2.7 \times \frac{9.81}{1000} \times 7.32 \times 3.66 \times 1 = 0.709\ \text{MPa per metre length}$$

$$A_s = \frac{0.709}{\sigma^3} = \frac{0.709}{34.5/2.5} = 0.0513\ \text{m}^2$$

The thickness of concrete required is 26 mm each side.

For abutment loading

$$\gamma(H_p \times A)\frac{1}{100}\lambda = \sigma_s(A_s + A_a) + \sigma_m A_m$$

$$\sigma_s(A_s + A_a) = 2\left[\frac{2.7 \times 9.81}{1000} \times 7.332 \times 3.66 \times 1\right] = 1.418\ m^2$$

$$(A_s + A_a) = \frac{1.418}{34.5/2.5} = 0.1027\ m^2$$

51 mm of concrete per side is needed.

Example 6 Design a standing support for a mining roadway whose dimensions are 3.66 m × 3.66 m in dimension driven in a rock with uniaxial compressive strength 175 MPa, cohesive strength 30 MPa, and internal angle of friction 40°. The rock mass unit weight is 2650 kg/m^3, the cohesive strength of discontinuities is 1.25 MPa and the internal angle of friction for fractures is 20°. Field investigation has indicated that 15% of failure surface is made up of the intact rock. The existing mine supports are in a 1.2 m × 1.2 m pattern of full column resin grouted rock bolts, 22 mm in diameter and 1.5 m long.

Calculate the size of the additional standing supports which may be required to stabilize the roadway.

Solution

(a) Estimation of rock cohesion and friction:

$$\begin{aligned} C &= C_f + R\,(C_r - C_f) \\ &= 1.25 + 0.15\,(30 - 1.25) = 5.56\ \text{MPa} \\ \tan\phi &= \tan\phi_f + R\,(\tan\phi_r - \tan\phi_f) \\ &= \tan 20° + 0.15\,(\tan 40° - \tan 20°) = 0.436 \\ \phi &= 23.55° \end{aligned}$$

(b) Maximum stress the rock mass can support:

$$\sigma_1 = 2\,C\tan(45° + 11.75°) = 14.5\ \text{MPa}$$

(c) Strength assigned to rock mass

$$\frac{\sigma_1}{SF} = \frac{14.5}{3} = 4.83\ \text{MPa}$$

(d) Rock load to be supported:

$$H_p = 1.0\,(W + H) = 7.32\ \text{m}$$

(e) $\gamma H_p A = \sigma_m A_m + A_s \sigma_s$

$$\frac{1}{1000} \times 2.650 \times 9.81 \times 7.32 \times 3.66 \times 1 = 4.83 \times 1.5 \times 2 \times 1 + A_s\sigma_s$$

$$A_s\,\sigma_s = -\,12.9 + 0.696 = -12.204\ \text{MPa}$$

No additional support is required.

10.9 Yield Zone Around a Circular Roadway in Coal Measure Rocks

This concept was developed by Wilson (1977) for prediction of closure of coal mining roadways in solid ground in the UK. The method is based on the premise that a yield zone is formed around a roadway at depth and the purpose of the roadway support is to provide confining pressure to the yielding rock surrounding the roadway. The frictional forces and adhesive forces generated by the broken rock in the yield zone supports the strata surrounding the roadway (Fig. 10.8).

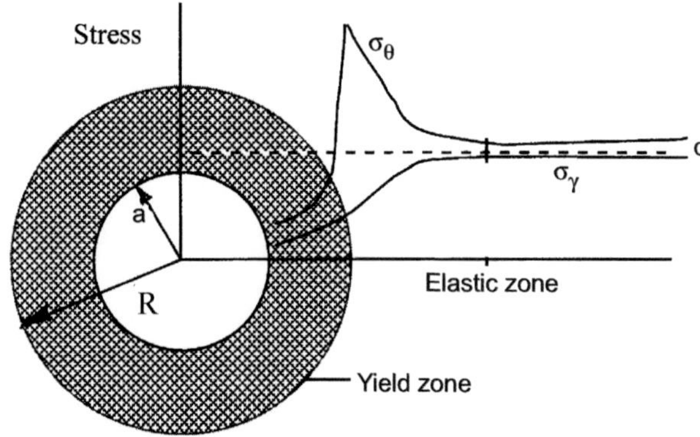

Figure 10.8 Stress and displacement around a circular roadway in a yielding ground (Wilson, 1977)

The basic assumptions made in this analysis are:

(a) Circular roadway of radius 'a'
(b) Homogeneous rock of uniaxial compressive strength σ_c before failure
(c) The failed rock within the yield zone has equivalent cohesion (p′) after failure
(d) Hydrostatic stress field of magnitude (p)
(e) Plain strain condition exists

Calculations showed that the tangential stress at the boundary of the roadway is 2 p. If, this stress exceeds the uniaxial compressive strength of coal measures rock at depth σ_c, a yield zone surrounding the excavation will form.

Stress distribution at a radial distance 'r'

Stress conditions at the yield zone are:

$$\sigma_1' = \sigma_c' + K\sigma_3$$

Failure criteria is given by:

$$\sigma_1 = \sigma_c + K\sigma_3$$

In the yield zone, the stresses are:

$$\sigma_r = (p_i + p')\left[\frac{r}{a}\right]^{K-1} - p'$$

$$\sigma_{\theta\theta} = K(p_i + p')\left[\frac{r}{a}\right] K - 1 - p'$$

where

$$p_i = \text{support pressure}$$

$$p' = \frac{\sigma'}{K-1}$$

In the elastic zone, the radial stress and normal stresses are given as follows:

$$\sigma_{rr} = p - A\left[\frac{a}{r}\right]^2$$

$$\sigma_{\theta\theta} = p + A\left[\frac{a}{r}\right]^2$$

where

$$A = \left[\frac{(K-1) + \sigma_c}{(K+1)}\right]\left[\frac{2p - \sigma_c + p'(K+1)}{(p_i + p')(K+1)}\right]$$

From the above analysis, the following parameters are deduced.

(i) The depth below which the **yield zone** will form is given by the following equation:

$$H' = \frac{1}{2\gamma}\,[p_i(K+1) + \sigma_c]$$

(ii) The radius of the yield zone is given by:

$$R = a\left[\frac{2p - \sigma_c + p'(K+1)}{(p_i + p')(K+1)}\right]^{\frac{1}{K-1}}$$

Displacement of a point at a radial distance 'r' at the yield boundary is given by:

$$U_r = Aa\left[\frac{a}{r}\right]\left[\frac{1+\upsilon}{E}\right]$$

υ = Poisson's ratio

E = modulus of elasticity

U_r = displacement at a point of radial distance (r)

10.9.2 Wilson Formula

As the strata yield, the stress decreases, and rock will expand. If expansion ε of the strata in the direction of opening is proportional to lateral distortion, then

$$c = d\,\frac{1+\upsilon}{E}\left[\frac{(K+1)p+\sigma_c/f}{K+1}\right]\left[\frac{2p-\sigma_c/f}{(pi+p')(K+1)}\right]^{\frac{2+\varepsilon}{K-1}}$$

where

$c =$ diametral closure of tunnel (m)
$d =$ driven diameter of tunnel $= 2a$ (m)
$\upsilon =$ Poisson's ratio
$E =$ modulus of elasticity (MPa)
$K =$ triaxial stress factor
$p =$ cover load (MPa)
$\sigma_c =$ unconfined compressive strength in laboratory
$f =$ scale factor
$p_i =$ resistance offered by lining, (MPa)
$p' =$ apparent adhesion of broken rock, MPa (0.1 MPa)
$\varepsilon =$ expansion factor
$0 < \varepsilon < 0.5$ (probably 0.2)

if υ and E are not known

$$\upsilon = 0.3, \quad E = 3.5\times10^3\ \text{MPa}, \quad \frac{1+\upsilon}{E} = 0.37\times10^{-3}$$

Other strata

$$\upsilon = 0.25, \quad E = \sigma\times0.31\times10^3\ \text{MPa}$$

$$\frac{1+\upsilon}{E} = \frac{4}{\sigma_c}\times10^{-3}\ \text{MPa}$$

10.8.1 Examples of Application (after Wilson, 1981)

(1) Circular drivage, diameter = 4 m (given)
Siltstone compressive strength $\sigma = 70$ MPa (measured)
Free of joints $f = 3$ (estimated)
$K = 3.5$ (measured)
$H = 480$ m (given)
$p = 0.025\times480 = 12$ MPa (estimated)
Circular ring support
$p_i = 0.2$ MPa, $\upsilon = 0.25$ (estimated)

$p' = 0.1; \quad E = 21.7 \times 10^3$ MPa (Given)

$\varepsilon = 0.2$

$$c = d\,\frac{1+\upsilon}{E}\left[\frac{(K+1)p+\sigma_c/f}{K+1}\right]\left[\frac{2p-\sigma_c/f}{(p_i+p')(K+1)}\right]^{\frac{2+\varepsilon}{k-1}}$$

$$= 4\cdot\frac{1+0.25}{21.7\times10^3}\left[\frac{(3.5-1)12+70/3}{(3.5+1)}\left[\frac{2\times12-70/3}{(0.2+0.1)(3.5+1)}\right]^{\frac{2+\varepsilon}{K-1}}\right]$$

$$= \frac{0.23}{10^3}\left[\frac{53.33}{4.5}\right]\left[\frac{0.66}{1.35}\right]^{0.88} = \left(\frac{0.23\times11.85\times0.53}{10^3}\right) = \left(\frac{1.45}{10^3}\right)$$

$$= 1.45\text{ mm}$$

Visual inspection shows no distortion of rings.

Case 2 Major inset - Abernant Colliery (after Wilson, 1983)

$d = 9.$ m

Very weak strata $\sigma_c = 10$ MPa (estimated)

Major fault zone $f = 7$ (estimated)

$K = 3$ (tested)

$H = 730$ m (known)

$p = 18$ MPa (estimated)

Temporary arch $p = 0.005$ MPa (estimated)

$p' = 0.1 \quad \upsilon = 0.25$

$\varepsilon = 0.2, \; E = 3.1 \times 10^3$ MPa (estimated)

$$c = d\,\frac{1+\upsilon}{E}\left[\frac{(K+1)p+\sigma_c/f}{K+1}\right]\left[\frac{2p-\sigma_c/f}{(p_i+p')(K+1)}\right]^{\frac{2+\varepsilon}{k-1}}$$

$$= \left(\frac{3.62}{10^3}\right)\left(\frac{37.42}{4}\right)\left(\frac{34.57}{0.6}\right)$$

$$= \frac{3.62}{10^3}\times9.355\times86.41 = 2.93\text{ m}$$

Observed closure was in excess of 3 m.

10.8.2 Closure of Main Roadways in the Selby Coalfield, U.K. (Wilson, 1980)

Estimation of the effect of depth and strength of lining in main drivages in soft rock.

$d = 6$ m, $\quad \sigma_c = 40$ MPa

$$f = 5, \quad K = 3, \qquad p = \frac{\text{depth}}{40}\ \text{MPa}$$

Case 1

$$p_i = 0.1, p' = 0.1 \qquad \text{Depth} = 1200\ \text{m} \qquad E = 12.4 \times 10^3\ \text{MPa}$$
$$p = 30\ \text{MPa}$$

$$c = 6 \times \left(\frac{1.25}{12.4 \times 10^3}\right)\left(\frac{(3-1)\,30 + 40/5}{(3+1)}\right)\left(\frac{60-8}{4 \times 0.2}\right)^{\frac{2.2}{2}}$$

$$= \frac{0.6 \times 17 \times 98.67}{10^3} = 1.0\ \text{m}$$

Case 2

$$P = 1.0 \qquad \text{Depth} = 1000\ \text{m} \qquad p = 25$$
$$E = 12.4 \times 10^3$$

$$c = d\,\frac{1+\upsilon}{E}\left[\frac{(K+1)p + \sigma_c/f}{K+1}\right]\left[\frac{2p - \sigma_c/f}{(p_i + p')(K+1)}\right]^{\frac{2+\varepsilon}{K-1}}$$

$$= \frac{6 \times 1.25}{12.4 \times 10^3}\left(\frac{58}{4}\right)\left(\frac{42}{4.4}\right)^{1.1}$$

$$= 104.5\ \text{mm}$$

10.8.3 Application of the Wilson Method to Non-circular Roadways

Closure equation gives the total convergence from full development of the yield zone.

- It takes time to develop full closure
- Close to face has end effects
- If support is installed some distance away from the face then some closure has already taken place

1. Law relating closure and distance to road end

$$\frac{c}{B} = 1 - e^{\frac{-d}{K_1}}$$

c = closure (m)
d = tunnel diameter (m)
K_1 = constant

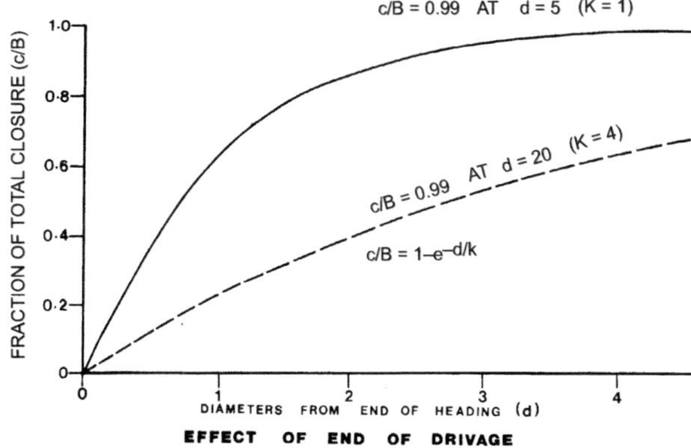

Figure 10.9 Estimation of closure behind the face line

Without the time effect, closure will be 99% at a distance 5 d from the face (Fig. 10.9).

Curve I $\frac{c}{B} = 1 - e^{-1} = (1 - 0.36) = 0.63$

Rock flowage with time may significantly increase the distance over which closure occurs.

Curve II $\frac{c}{B} = 1 - e^{-20/4} = 0.99$

$$\frac{c}{B} = 1 - e^{-d/4} = 1 - e^{-1/4} = 0.22$$

2. *Shape of the roadway*

If the width of the roadway is approximately equal to height, the boundary of the yield zone is circular. Experience has shown that by inserting equivalent diameter D_e in the Wilson's equation, a good estimation of closure can be obtained.

$$D_e = 2 \times \sqrt{\frac{\text{Actual cross-section area}}{p}}$$

3. *Unsupported floor*

Use modified support resistance Pe $= 0.8\,(p_i + p')$
Estimated vertical closure $= 1.2\,C_e$
Estimated side closure $= 0.85\,C_e$
C_e is the equivalent closure of a circular roadway of diameter D_e.
Other effects of the unsupported floor are:

- Floor material breaks and bulks – 30% more closure = $1.51\,C_e$
- Rigid plate breaks in middle = $2.4 - 3.6\,C_e$

4. *Stratification*

Weighted mean of strata properties using perimeter length (Fig. 10.10)

$$\sigma = \left(\frac{l_1 s_1 + l_2 s_2 .... + 2l_7 s_7 + l_8 s_8}{2pr}\right)$$

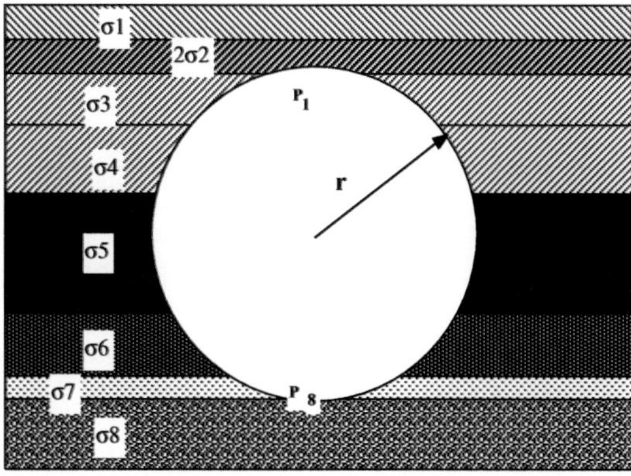

Figure 10.10 Estimation of strength of strata in a circular roadway in stratified rocks

5. *Overbreaks in the roof*

Overbreaks may occur in the roof before the support is set.

- Also in the case of poured concrete it is difficult to get the concrete to the top of the roof space.
- In loose ground if the overbreaks are not adequately grouted then failure of the lining may occur due to over-compression. Examples of such occurrence has been experienced at Abernant, Cadley Hill and Wolstanton collieries in the UK.

10.9 Ground-support Reaction Curve for the Design of Support Systems

10.9.1 Concept of Ground-support Reaction Curve

A unique method for designing support systems for an underground mining roadway is the use of the ground-support reaction curve. The essence of this approach is that due to formation of the roadway the rock surrounding the excavation destresses and starts expanding towards the opening. The standing support provides a reaction and arrests this ground movement. As a consequence, resistance to movement of strata increases against the ground movement thereby stabilising the excavation. The purpose of the support is to provide early resistance to rock movement with adequate control of the stiffness of the support.

Figure 10.11 shows the cycle of operations in formation and supporting an arch-shaped roadway utilising the drill and blast method and using standing supports. The cycle of operations are as follows (Hoek and Brown, 1985):

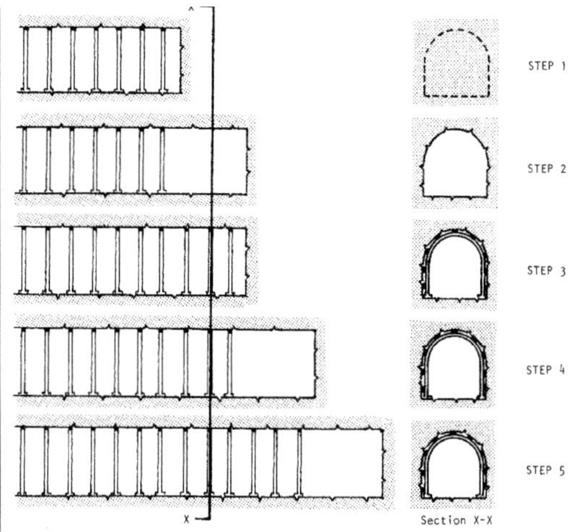

Figure 10.11 Cycle of operation in driving and erecting arch supports in a roadway (Hoek and Brown, 1985)

Step 1: Initial state of stress in undisturbed rock at point XX can be estimated as $p = 0.025 \times$ depth (m) MPa as shown as point A in Figure 10.12.

Step 2: As the excavation is formed by drilling, blasting and mucking operations, the surrounding rock is destressed but the load on the support is zero. However, the tunnel roof would not collapse due to limited movement of the roof provided by the proximity of the tunnel face.

Step 3: Support is set after the tunnel has undergone rock deformation as indicated by points A and B in Figure 10.12.

Step 4: As the tunnel advances about one-and-a-half times the tunnel diameter beyond XX, further radial deformation in the tunnel roof is indicated by the lines BF and corresponding side movement by the line CE. Standing support resistance increases with the radial deformation of the tunnel as indicated by line BS.

Step 5: If no support is installed, then the tunnel wall will deform up to EG and the tunnel roof from F to H.

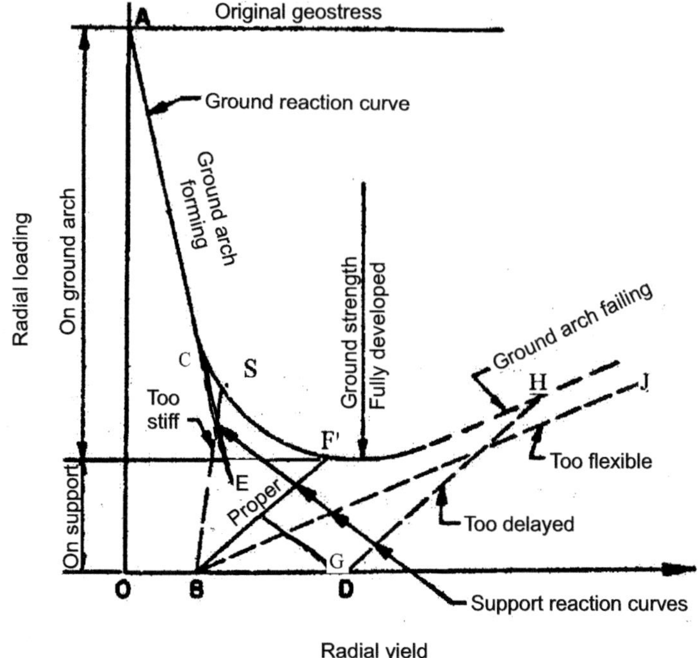

Figure 10.12 Ground reaction curve for an arched shaped roadway (Hoek and Brown, 1985)

Figure 10.12 also shows deformation of stiff support BS which will result in destruction of the support. Line BF indicated the deformation of a properly installed support whereas DH shows a support which is installed too late and BJ as support providing flexible resistance to ground movement.

10.9.2 Theoretical Development of the Ground-reaction Curve for a Rock Mass

Stress around a mine opening and the resulting deformations are the consequence of interaction between the rock mass and the support. Some discrete points on the ground reaction curves can be calculated for idealised conditions for a given criterion of failure. For the idealised solution, the following assumptions are made.

- Mohr-Coulomb criterion of failure
- Elasto-plastic behaviour of rock
- Linear post-failure volume expansion factor
- Idealised isotropic and homogeneous material

The material behaviour of rock used in the mathematical model encompassing the above conditions is presented in Figure 10.13. The notation used in the relevant equations is given below:

a = radius of the tunnel

R = radius of the elasto-plastic boundary

K = triaxial stress factor for intact rock = $\tan^2\left(\frac{\pi}{4}+\frac{\phi}{2}\right)$

K_1 = triaxial stress factor for the rock mass = $\tan^2\left(\frac{\pi}{4}+\frac{\phi_r}{2}\right)$

p′ = residual cohesion of rock

ε = volume expansion factor and its value depends upon dilation angle ψ

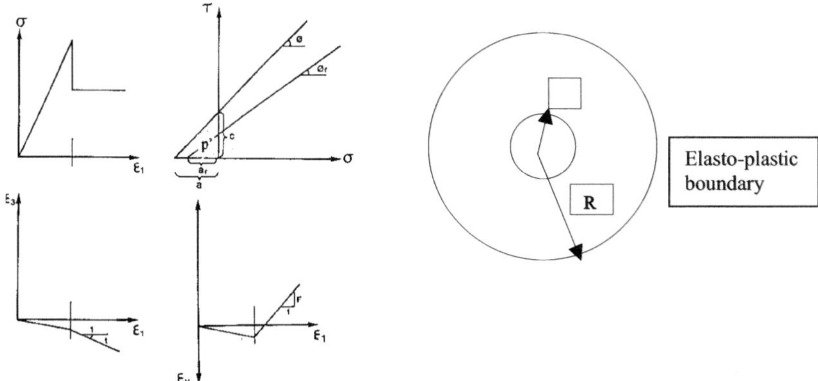

Figure 10.13 Rock properties used for creating a theoretical ground-support reaction curve

The radius of the boundary of the yield zone can calculated as follows:

$$R = a\left[\frac{\frac{2}{1-K}\{(p+p')-c+p'\}}{(p_1+p')}\right]^{\frac{1}{K_1-1}}$$

In order to calculate discrete points in the ground- support reaction curve using elasto-plastic theory, the following steps are taken.

1. Initial state of stress p in MPa is calculated as 0.0025×depth (metres) MPa
2. Stress and displacement at the end of a purely elastic deformation shows a point on the curve when the ground arch just starts forming.

The stress and the elastic deformation at this point are given as follows:

$$p_1 = \frac{2}{K+1}(p-p')-p'$$

The corresponding elastic deformation is given by the following equation:

$$u_e = a\left(\frac{1+\upsilon}{E}\right)(p-p_1)$$

3. The radius of the elasto-plastic boundary at the zone is given by the following equation:

$$R = a\left[\frac{\frac{2}{1-K}\{(p+p')-c+p'\}}{(p_1+p')}\right]^{\frac{1}{K_1-1}}$$

The displacement at the elasto-plastic boundary is given as follows:

$$u_1 = \frac{aA}{1+K}\left[2\left(\frac{R}{a}\right)+(K-1)\right]$$

$$A = \left(\frac{1+\upsilon}{E}\right)(p-p_1)$$

and

$$p_1 = \frac{2}{1+k}[(p+p')]-p'$$

4. The gravity expansion of the roof in an arch-shaped roadway is given by:

$$u_g = \rho g\,(R-a)$$

5. After establishing the salient points on the ground deformation curve for the circular opening, a complete stress deformation curve of the support system is superimposed on the tunnel deformation curve to form a ground-support deformation curve for support design.

10.10 Conclusions

(1) Closed-loop solutions for stress concentration around a circular and elliptical hole in an infinite plate can solve many stress-induced instability problems in mining. In particular, problems related to shape, orientation and stress interactions by various underground mine workings with respect to initial *in situ* state of stress can be satisfactorily resolved.

(2) Pressure arch theory can be used to estimate stress induced in the support in shallow tunnels in granular or bedded deposits with the ratio of horizontal to vertical stress K being < 1. This approach also helps in estimating support loading in a variety of ground conditions that forms a basis for designing support systems for tunnels in intact as well as caving grounds.

(3) In the quest to calculate the closure of a tunnel many complexities have to be built into the mathematical model in relation to rock properties and the shape of the excavation in order to obtain realistic closure results. The mathematical models developed by Bray (1967) for elasto-plastic rock, Ladanyi (1974) for yielding rocks and Wilson (1980) for soft coal measures rock are noteworthy in that they give realistic results in a variety of mining conditions.

(4) A novel approach for the design of a support system for a circular tunnel in ideal conditions is the use of the ground-support reaction curve. This approach permits the superimposition of the support characteristics on the tunnel deformation curves to aid design of the support system.

References

Bray, J.W. (1967). 'A study of jointed and fractured rock', *Rock Mechanics and Engineering Geology*, 5, pp. 197–216.

Goodman, R.E. (1980) Introduction to Rock Mechanics, John Wiley, 478 pp.

Kastner, L. (1949). 'Uber den echten Gebirgsdruck beim Bau...', Tunnel, *Osterreich Bauzeitschrift*. Vol. 10, No. 11.

Labasse, H. (1948, 1949). *Les pressions de terrains*, Revue Universelle des Mines, Series 9, Vol. 5 and 6.

Ladanyi, B. (1974). 'Use of the long-term strength concept in determination of ground pressure on tunnel lining', *Advances in Rock Mechanics, Proceedings of Third Congress of International Society of Rock Mechanics*, Denver, Washington, DC, National Academy of Science, pp. 1150–56.

Muir Wood, A.M. (1979) 'Ground behaviour and support for mining and tunnelling', 14th Sir Julius Wernher Memorial Lecture, Tunnelling '79, Inst. of Mining & Metallurgy, London, XI – XXII.

Talobre, J. (1957). *La Mecanique des Roches*, Paris: Dunod.

Terzaghi, K. (1946). 'Rock defects and loads on tunnel support', in R.V. Proctor and T.L. White, *Rock Tunnelling and Steel Supports*, Commercial Shearing and Tamping Co., Youngstown, Ohio.

Ward, W.H. (1978). 'Ground support for tunnels in weak rocks', *Geotechnique*, 28, pp. 133–71.

Wilson A.H. (1977). 'The effect of yield zones on the control of ground', *Proceedings of 6th International Strata Control Conference*, Banff, Paper 3.

Wilson, A.H. (1980). 'The stability of underground workings in soft rocks of the coal measures., Ph.D. thesis, University of Nottingham (unpublished).

Wilson, A.H. (1981). 'Stress and stability of coal rib sides and pillars', *Proceedings of Ist Annual Conference on Ground Control in Mining*, West Virginia University, S.S. Peng (ed.), pp. 1–12.

Wilson, A.H. (1983). 'The stability of underground workings in soft rocks of the coal measures', *International Journal of Mining Engineering*, 1, pp. 91–181.

Assignments

Question 1 A 100 m deep tunnel is being driven in weak mudstone which has excavated dimensions of 6 m in width by 4.5 m in height. The average density of rocks between the tunnel and the surface is 2,150 kg/m^3 and the angle of internal friction (ϕ) is 22°. The ratio of horizontal to vertical stress is 0.7. Determine the anticipated vertical and horizontal support pressures. Illustrate the likely mode of loading transmitted to the support system. Also describe a suitable form of support for the excavation taking into account short- and long-term requirements.

Question 2 Describe with the help of a diagram the vertical stress distribution around a rectangular roadway in an underground coal mine at medium depth. Also show the effect of stress distribution on strata failure.

Question 3 Draw the vertical stress distribution around a longwall face at a depth of 500 m below the surface.

Question 4 Design a thickness of concrete lining for a mining tunnel in a fractured and altered rock mass with the following properties:

Unit weight of rock= 2,700 kg/m^3

Unconfined uniaxial compressive strength = 69 MPa

Cohesive strength of intact rock C_r = 10.4 MPa

Cohesive strength of fracture and altered rock = 0.069 MPa

Angle of friction of intact rock ϕ_r = 45°

Angle of friction of altered rock ϕ_f = 12°

The size of tunnel is 3.33 m × 3.66 m, strength of concrete is 34.5 MPa and its factor of safety is 2.5. The general fracture persistence is 84%. Calculate the thickness of poured concrete lining of the drift neglecting the contribution made by rock bolts in supporting the tunnel.

Question 5 (a) Bray's elasto-plastic solution gives the radial displacement as follows:

$$U_r = \frac{1-\upsilon}{E}\left[\left\{p_i \frac{r^{q+1}}{a^q} - pr\right\} - \frac{t}{r}\right] + \frac{t}{r}$$

$$Q = \frac{\tan \delta}{\tan(\delta - \phi_j)} - 1, \text{ and}$$

$$t = \frac{1-\upsilon}{E} R_e^2 [(p + c_j \cot \phi_j) - (p_i + c_j \cot \phi_j)]$$

$$\left\{\frac{R}{a}\right\}^Q + \left\{\frac{1-\upsilon}{E}\right\} b$$

$$b = \left[\frac{\left\{\tan^2\left(\frac{\pi}{4} + \frac{\phi}{2}\right) - 1\right\} p + \sigma_c}{\left\{\tan^2\left(\frac{\pi}{4} + \frac{\phi}{2}\right) + 1\right\}} \right] R^2$$

Describe the notations in the above equations and list the conditions under which the equation is valid and indicate the shape of the fractures.

(b) A 5 m diameter tunnel is driven in shale with rock properties: $E = 1000$ MPa, $\upsilon = 0.25$. The initial stress is 27.9 MPa and the support pressure is 0.3 MPa. The fractures have properties $\phi_j = 30°$, $\phi = 45°$. Given that the shear strength is 3.45 MPa, calculate the closure of the tunnel wall ($r = a$).

Question 6 Draw the general ground-support reaction curve for a circular tunnel showing (1) line of stable support, (2) that for the delayed support, (3) that for a too flexible support and (4) the critical displacement.

Question 7 The excavated dimensions of a 180 m deep horizontal tunnel, which is being driven in weak seat-earth, are 4.5 m high and 6 m wide. The average unit weight of the overlying rock mass is 2,350 kg/m^3, internal angle of friction is 27° and the ratio of horizontal to vertical stress is 0.5. Determine the anticipated vertical and horizontal support loading and type of permanent support suitable for the tunnel.

Question 8 A 9 m diameter tunnel at a depth of 750 m is to be driven in a very weak strata intersected by a major fault. The estimated uniaxial compressive strength of rock is 10 MPa and the scale factor is 7. The triaxial stress factor as determined in the laboratory test is 3. The estimated lithostatic stress is 18 MPa, resistance offered by the support is 0.005 MPa, p′ is 0.1 MPa, the volume expansion factor of rock is 0.2. The modulus of deformation of rock mass is 3100 MPa and Poisson's ratio is 0.25. Calculate the tunnel closure.

11 Chapter Design and Stability of Pillars and Associated Structures

11.1 Introduction

The use of pillars as natural structural support elements dates back to prehistory. The need for the design of such pillars for safety has been prompted by the use of room and pillar mining for coal, limestone and trona around the world, occasionally accompanied by major and catastrophic failures, including subsidence of working and abandoned mines. The disaster that struck Coalbrook North Colliery in South Africa in 1960 with the collapse of an area of approximately 320 ha and a death toll of 432 people prompted research aimed at developing a reliable method of pillar design (Salamon and Munro, 1967). Despite major advances in the understanding of pillar mechanics, the mining industry continues to be beleaguered with such collapses, as evidenced by cascading pillar failure over 13 panels in a room and pillar trona mine in Wyoming, USA in 1995 (Zipf and Swanson, 1999)

It is necessary to reiterate that pillar design cannot be treated in isolation as it is only one entity in the interactive system comprising stopes/rooms, the pillars themselves and the peripheral rock over the stopes, all embedded in a rock mass environment; the stability of the components hinges upon their interdependence (Figure 11.1). Design of stable pillars necessarily calls for appropriate design of a safe limit for room span on which the performance of the peripheral rock mass critically depends. The stability of the rock mass surrounding the room and pillar array is governed, therefore, by the following.

- Design for the stability of rooms/stopes
- Design of the pillars in between the openings based on their load bearing capacity

The effective performance of pillars depends upon the following factors.

(1) Dimensions of individual pillars
(2) Geometric disposition of the pillars within the ore body

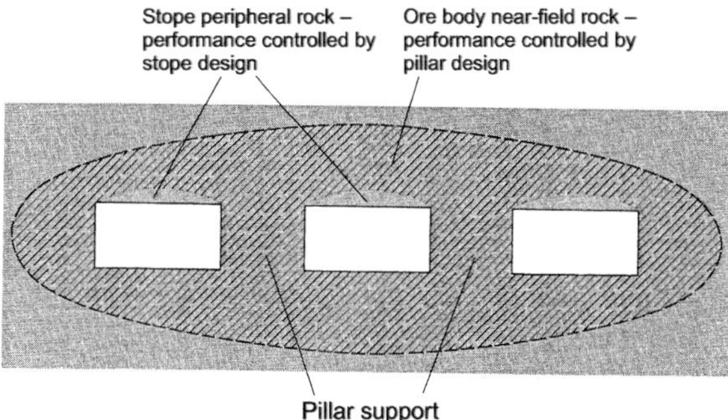

Figure 11.1 Global stability of a room and pillar mining system (Brady and Brown, 1985)

(3) Pillar dimensions such as to support the overlying strata and minimise the sterilisation of reserves
(4) Provision of global support to the mine structure without causing catastrophic subsidence or pillar burst.

11.2 Types of Mine Pillars

11.2.1 Load-bearing Pillars

The type of mine pillars used depends upon the interaction of a multitude of factors. Some of the most important factors are given below.

(1) Mining system
(2) Method of pillar formation
(3) Method of loading and confinement

The nature of the ore body and its geometrical disposition are the major factors in selecting an appropriate mining technique. For example, the room and pillar mining method is applicable to bedded deposits where the rooms at all of the four sides surround the pillars and the pillar loading is essentially vertical as shown in Figure 11.2(a). In the open stoping method or in a long hole stoping method in massive or highly inclined deposits the pillars formed by the stopes are biaxially loaded by the surrounding country rock. For example in Figure 11.2(b), pillar A is a horizontal transverse pillar while pillar B is a horizontal longitudinal pillar. Pillar B is also called the sill pillar for stope 1 and the crown pillar for the stope 2.

11.2.2 Isolation Pillars in the Panel System of Mining

The sustained load on the pillars may cause the failure of pillars which may, in turn, cause caving of the overlying strata. If the volume of the extraction is

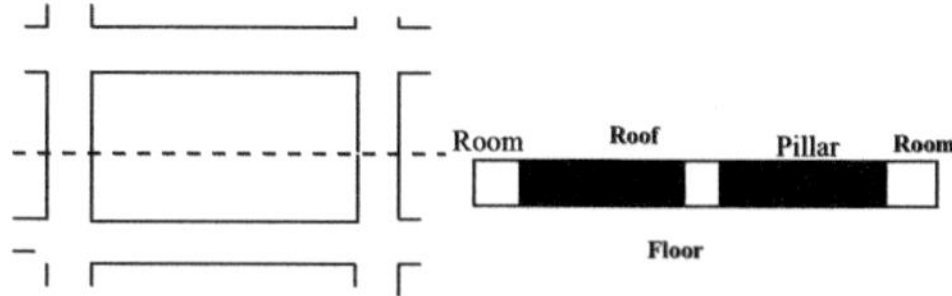

(a) Pillar loaded in uniaxial stress field

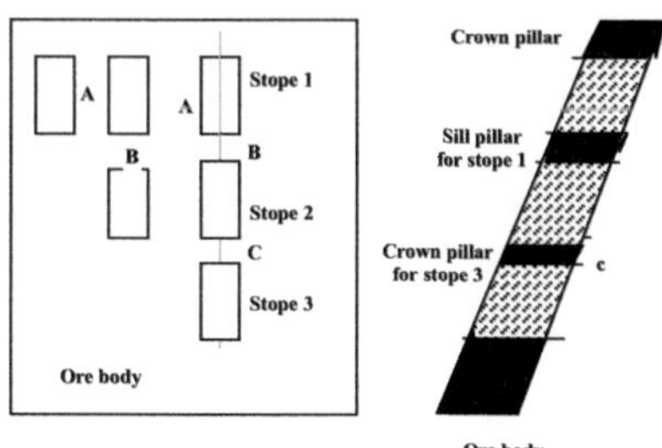

(b) Pillars in biaxial stress field

Figure 11.2 Methods of loading of pillars in level and inclined ore bodies

extensive, a collapse, if it occurs, may propagate through the pillar structure. The panel system of mining is designed to contain the unstable strata or roof collapse within the panel by providing the barrier pillars which are virtually indestructible (Figure 11.3).

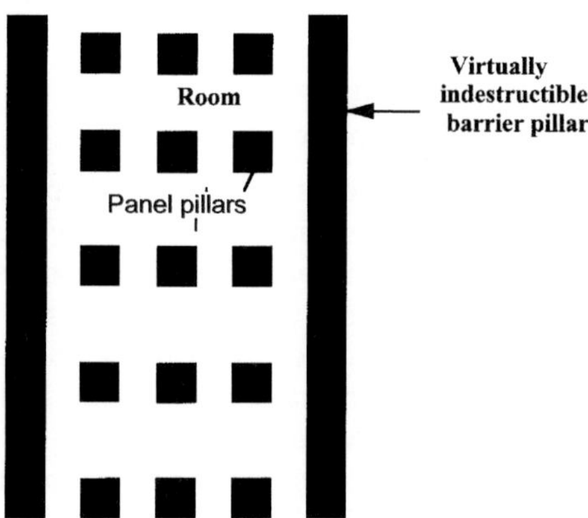

Figure 11.3 Panel layout showing panel pillars and barrier pillars

11.2.3 Mode of Failure of Pillars

The performance of pillars in practice has been observed by various workers both in coal mines by Bunting (1911), Greenwald *et al.* (1939, 1941), Wagner

(1974), Van Heerden (1975); in oil shale mines by Hardy and Agapito, and in metalliferous mines by Brady (1975, 1977) and Wagner (1974). Figure 11.4 presents the various modes of failure of uniaxially loaded pillars. The following distinct modes of failure can be identified.

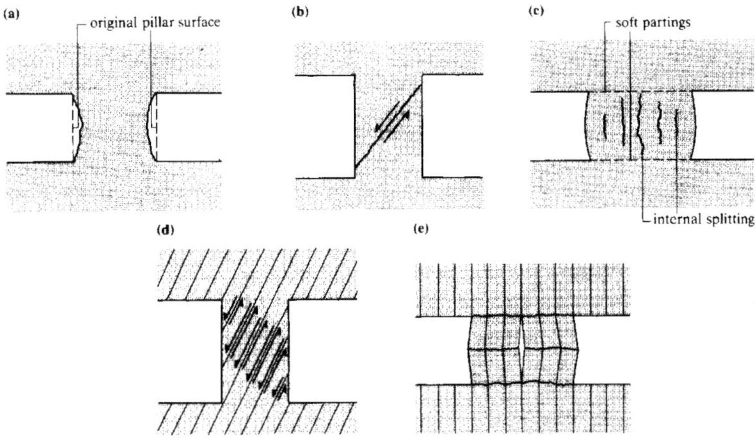

Figure 11.4 Modes of failure of uniaxially loaded pillars (Brady and Brown, 1985)

(1) Fretting or necking of pillars

This type of failure is caused by local overstressing of pillar material, especially in massive rock with low height/width ratio of the pillar. The failure occurs as spalling of the pillar as a consequence of overstressing.

(2) Inclined shear failure

This mode of failure occurs in regularly jointed rock masses and is characterised by the following factors.

- High height/width ratio of the pillars
- Failure along single inclined shear plane

(3) Internal splitting failure

With a highly deformable interface between the pillar and the wall rock, soft rock may yield and this is followed by internal axial splitting of the pillar as shown in Figure 11.4(c).

(4) Structurally controlled failure

Shear failure along the defined structures such as bedding planes, joints etc. is often encountered in the field, as illustrated by Figure 11.4(d).

(5) Structurally controlled buckling failure

Well developed foliation or schistosity parallel to the principal axis of loading may be characterised by buckling failure as shown in Figure 11.4(e).

Similar inferences can be drawn for the failure modes for the biaxially loaded pillars (Wagner, 1974; Brady, 1975, 1977).

11.3 Design of Pillars in Room and Pillar Partial Extraction System

Design of a stable pillar layout consists of estimating the pillar load and pillar strength and checking for stability compliance in terms of the factor of safety. The schematics of the design process is illustrated in Figure 11.5.

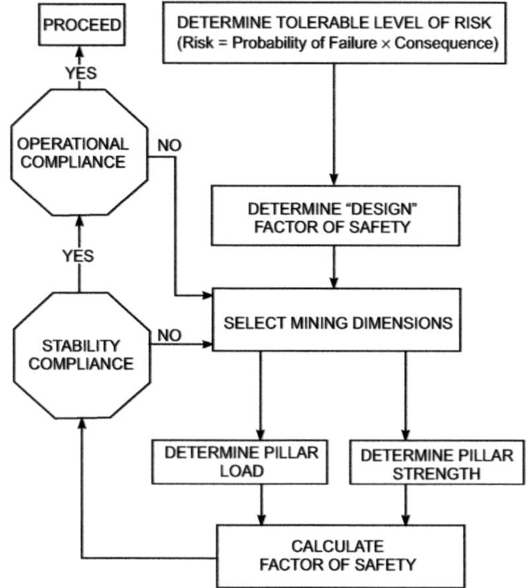

Figure 11.5 Pillar design decision process.

11.3.1 Estimation of Pillar Load

The estimation of pillar load is simplistically based on the tributary area theory which rests on the premise that given a uniform array of pillars in a level seam, the pillars carry the overburden loads besides the load imposed by the tributary area adjacent to the pillar (Figure 11.6).

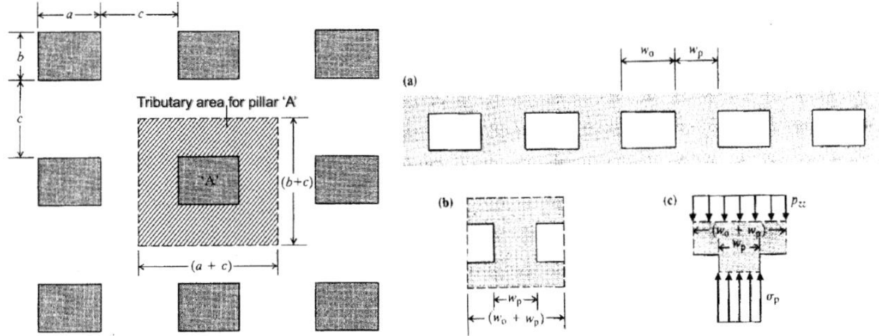

Figure 11.6 Partial extraction layout (Brady and Brown, 1985)

Thus, the magnitude of the stress on the pillars will depend on the ratio of the pillar size to the room size. In the case of an array of square pillars of identical size, the average vertical principal stress in the pillar is defined by:

$$\sigma_p = \sigma_v \left(\frac{W_p + W_o}{W_p}\right)^2 = \gamma H \left(\frac{W_p + W_o}{W_p}\right)^2 \tag{11.1}$$

where,

γ = unit weight of the overlying rock mass
H = depth of the pillar below surface
W_p = width of the pillar
W_o = width of the opening or roadway

The equation may be rewritten as

$$\sigma_p = \frac{\sigma_v}{1-e} = \frac{\gamma H}{1-e} \tag{11.2}$$

where,

e = the extraction ratio which for the square pillars may be defined as

$$e = 1 - \left(\frac{W_p}{W_p + W_o}\right)^2 \tag{11.3}$$

For simple mining geometry with regular arrays of pillars, the estimates provided by the tributary area theory are acceptable provided that the diameter (in plan) of the array of pillars being considered exceeds the depth below surface. The load on the pillars is determined by both the stiffness of the surrounding strata and the stiffness of the pillars. When a mining panel is narrow compared to its depth, the overlying roof rock arches across the panel and the total dead weight of the overburden is not transferred to the pillars. As the panel width to depth ratio increases, the stiffness of the overlying cover reduces until it is incapable of arching across the panel and the panel pillars have to carry the dead load of the overburden. For panels which are narrow compared to depth of cover, and for irregular arrays of pillars, resort to numerical techniques becomes necesarry, using the displacement-discontinuity (DD) variety of the Boundary Element Method.

11.3.2 Estimation of Pillar Strength

The strength of pillars depends, *inter alia*, on:

(1) strength of the pillar material,
(2) the width and height ratio of the pillar,
(3) the pillar volume or size,
(4) end conditions i.e. the roof-pillar and/or floor-pillar interaction,
(5) the influence of water and weathering processes and
(6) method deployed in excavating the pillar, which affects the creation of cracks in the pillar periphery.

Although the strength characteristics of pillars have been studied by many, the general approach to pillar strength estimation has rested on the following.

(1) Laboratory strength measurements on variously sized coal specimens and extrapolation to full-scale pillars (Greenwald, Howarth and Hartman, 1939).
(2) *In situ* testing of large pillars (Cook *et al.*, 1971; Bienawski, 1968)
(3) Back analysis of failed and intact pillars (Salamon and Munro, 1967; Salamon, 1999).

The first and second approaches have largely been supplanted by empirical strength equations derived from back-calculations. The strength of the rock mass apart, shape exerts the most decisive influence on the strength of pillar. Thus a squat pillar will be stronger than a tall slender pillar because of the greater degree of confinement which acts in the centre of the squat pillar. This is exactly analogous to the results obtained in the laboratory testing of rock specimens. The basic concepts related to confinement within coal pillars was first developed by Wilson (1972), and is now widely accepted. The simple empirical formula that have been proposed thus far for pillar strength estimation are either linear or power formulae. The linear equation defines pillar strength as a linear function of the width to height ratio:

$$S = K\left[r + (1 - r)\frac{W}{h}\right] \tag{11.4}$$

where

K = compressive strength of coal
r = dimensionless constant

W = width of the pillar
h = height of the pillar

Using R as the ratio of $\frac{W}{h}$, Equation 11.4 reduces to:

$$S = K[r + (1 - r)R] \tag{11.5}$$

According to this formula, geometrically similar pillars have the same strength regardless of their actual dimensions. Based on the testing of small-scale *in situ* pillars in South Africa, Bienawski produced a linear formula to describe the pillar strength for South African coals:

$$S = 62\left[0.64 + 0.36\frac{W}{h}\right] \text{MPa} \tag{11.6}$$

The second commonly used power law formula, based on the laboratory studies of Holland and Gaddy who evaluated the pillar strength model, where the pillars were square in plan, was as follows:

$$S = Kh^{\alpha}W^{\beta} \tag{11.7}$$

where

K = strength of unit volume of coal
h = the height of the pillar
W = width of the pillar
α and β = dimensionless constants

In a dimensionally correct form, the equation can be expressed as follows:

$$S = K_1\left(\frac{W}{W_o}\right)^{\alpha}\left(\frac{h}{h_o}\right)^{\beta} \tag{11.8}$$

where 'W' and 'h' are the linear dimensions of the pillar, and K_1 is the strength of a reference body of coal of height h_o and a square cross-section with side lengths W_o. The reference body is taken to be a cube of unit volume for the sake of convenience, in which case h_o and W_o are both unity and can be omitted from the formula. In contradistinction to the linear strength equation, the power law equation is volume sensitive. The power law has found wide usage in South Africa and was based on a statistical back analysis of pillar performance using the maximum likelihood method. The field data base comprised 27 instances of pillar collapse and 98 unfailed cases which were singled out for analysis. The starting postulate was that failure occurs when the strength equals the actual load on the pillar. Expressed in terms of the conventional safety factor this reduces to:

$$\text{F.S.} = \frac{\text{Strength}}{\text{Load}} \qquad 0 < \text{F.S.} < \text{infinity} \tag{11.9}$$

Clearly, when F.S. = 1 collapse takes place while no failure results if F.S.>1. Since the field data base included all parameters for calculating the load, equating S to unity, equations could be constructed to compute the unknown values of pillar strength, with the critical factor of safety clustered around unity (Salamon and Munro, 1967). The statistical analysis produced the following formula:

$$S = 7.2 \frac{W^{0.46}}{h^{0.66}} \tag{11.10}$$

This well-known formula has been used almost exclusively to design over 1.5 million pillars in South Africa and validated by field performance. Since pillars are frequently non-square in plan, the concept of effective width was introduced by Wagner (1974) where

$$W_e = 4 \frac{A_p}{C_p} \tag{11.11}$$

where A_p is pillar area and C_p is its circumference.

Limitations

(1) Calculated stress is not same as real stress distribution.
(2) This analysis takes only the pre-mining axial stress component into consideration. Other components of stress acting on the pillar are not tenable.
(3) The effect of location of the pillar within the ore body is ignored.

11.3.3 Area Extraction Ratio

Figure 11.7 shows the relationship between the normalised pillar loading versus area extraction ratio 'r'.

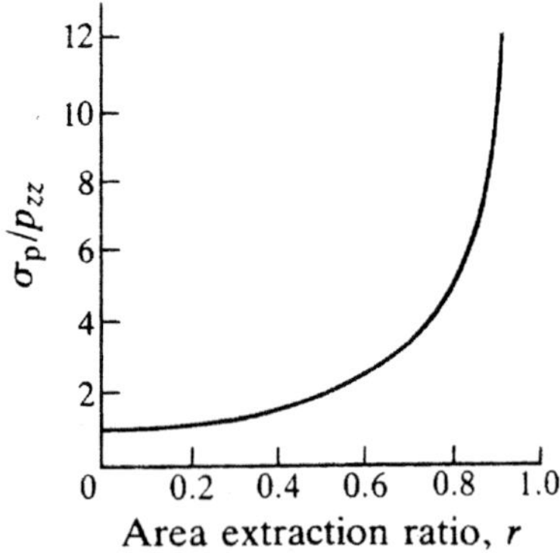

Figure 11.7 Extraction ratio and stress concentration on the pillar (Brady and Brown, 1985)

The following main observations can be made with respect to the 'r' extraction ratio.

(1) Axial stress on the pillar can be calculated in terms of stope and pillar layout.
(2) For a uniform layout, the average axial stress can be determined by the extraction ratio.
(3) Extraction ratios greater than 0.75 are rare.
(4) For a change in 'r' from 0.9 to 0.91, the stress concentration factor increases from 10 to 11.1.

11.3.4 Pillar Strength Formula for Oil Shale

Pillar strength formula for oil shale is given by the following expression:

$$S = S_1\, V^{-0.118}\left(\frac{W_p}{H}\right)^{0.833} \tag{11.12}$$

Using a scaling relationship, this formula can be used as follows:

$$S_p = S_s\left(\frac{V_p}{V_s}\right)^{-0.118}\left(\frac{W_p}{H_p}\bigg/\frac{W_s}{H_s}\right)^{0.833} \tag{11.13}$$

where

S_p = strength of pillar
S_s = strength of specimen
W_p = width of pillar
H_p = height of pillar
W_s = width of specimen
H_s = height of specimen
V_s = volume of specimen
V_p = volume of pillar

11.4 Design of Stope Layout

11.4.1 Types of Mining

This method of pillar design is applicable to mining bedded type ore deposits with such methods as room and pillar using a partial extraction system. This technique could also be applied to the following methods of mining as follows.

- Long hole stoping
- Vertical crater retreat
- Open stoping

11.4.2 Design Requirements

The following design parameters should be selected for the design of pillars.

- Highest possible extraction ratio
- Local stable stope spans
- General control of catastrophic failure of the adjoining rock mass (subsidence control, no catastrophic failure of pillars)
- Others, including the following:
    - o *In situ* vertical stress
    - o Working height (H)
    - o Stope span W_o
    - o Pillar width W_p
    - o Factor of safety against failure (FS)

11.4.3 Representation of Partial Extraction in Terms of Total Area Extraction

In order to optimise the best pillar design strategy involving all mining geometries satisfying the geo-mechanical stability criteria, an attempt should be made to examine the possibility of maximising the percentage extraction.

Once the method of representing partial percentage extraction of a pillar is expressed in terms of extracting the equivalent height of coal over the entire width of the pillar ($W_o + W_p$) as shown in Figure 11.8, it can be shown that:

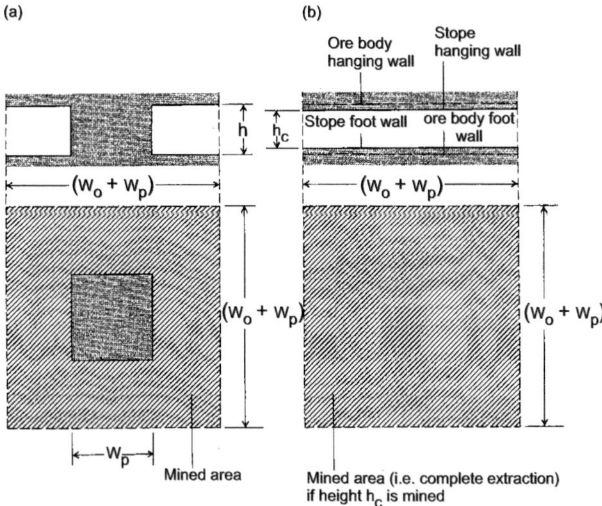

Figure 11.8 Representation of a partial extraction operation in terms of equivalent volume to the total area of extraction (Brady and Brown, 1985)

$$h_e(W_o + W_p)^2 = V_e = h[(W_o + W_p)^2 - W_p^2]$$

It follows that equivalent working height:

$$h_e = h\left[1 - \left(\frac{W_p}{W_o + W_p}\right)^2\right] \quad \text{and} \quad h_e = h[1-(1-e)^2] \tag{11.14}$$

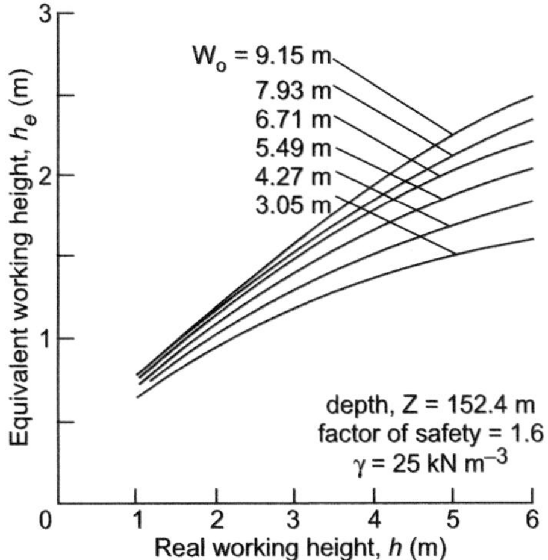

Figure 11.9 Relationship between equivalent working height and real height for an ore body (Brady and Brown, 1985)

11.4.4 Volumetric Extraction Ratio

Volumetric extraction ratio is the ratio of the equivalent working height and ore body thickness.

$$R = \frac{h_e}{M} = \frac{h}{M} = \left[1 - \left(\frac{W_p}{W_o + W_p}\right)^2\right] \tag{11.15}$$

Fig. 11.10 shows the relationship between R and depth, for various extraction thicknesses.

11.4.5 Inferences from the Above Example

(i) For a given depth, the maximum volumetric extraction ratio decreases.
(ii) Low-maximum percentage extraction is possible from a thick seam.

$$h_e = h\left[1 - \left(\frac{W_p}{W_o + W_p}\right)^2\right] \tag{11.16}$$

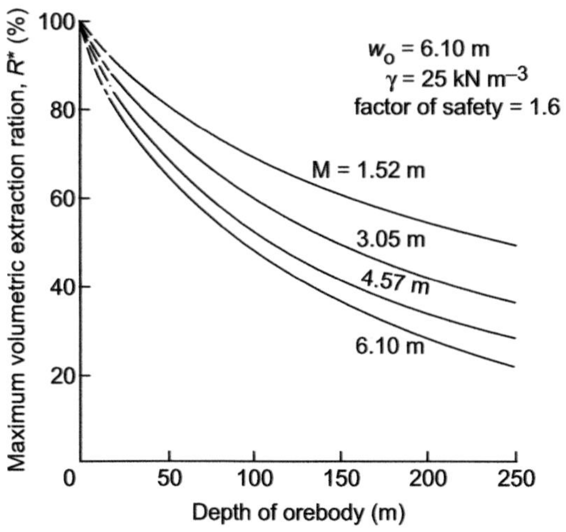

Figure 11.10 Maximum volumetric extraction ratio for various depths of ore body thickness (Brady and Brown, 1985)

The recovery of ore from an ore body is maximised if the following conditions are met.

(1) If the maximum height of the ore body is mined.

(2) Maximum room span W_o consistent with the local stability of the wall rock is mined.

11.5 Design Process by Examples

Example 1 (Brady and Brown, 1985)

Given a horizontal stratified ore body at

$$\text{Depth, d} = 150\text{ m}$$
$$\text{Stope span, } W_o = 6\text{ m}$$
$$\text{Pillar width, } W_p = 7.0\text{ m}$$
$$\text{Pillar height, H} = 3.0\text{ m (extracted)}$$
$$\text{Strength, S} = 10.44\, H^{-0.7}\, W_p^{\,0.5}$$
$$\text{Density} = 22.5\text{ kN/m}^3$$

(1) Vertical stress

$$\sigma_v = \rho\gamma h = 0.0225 \times d = 0.0225 \times 150 = 3.375\text{ MPa}$$

(2) Average stress on the pillar

$$\sigma_p = \sigma_v\left(\frac{W_o + W_p}{W_p}\right)^2 = 3.375\left(\frac{6+7}{7}\right)^2 = 11.69\text{ MPa}$$

(3) Strength

$$S = 10.44 \frac{W_p^{0.5}}{H^{-0.7}} = 10.44 \frac{7^{0.5}}{3^{-0.7}} = 12.80 \text{ MPa}$$

(4) Factor of safety

$$F = \frac{S}{\sigma_p} = \frac{12.80}{11.64} = 1.09$$

$$\text{Area extraction ratio} = \left(\frac{W_o}{W_o + W}\right)^2 = \left(\frac{6}{6+7}\right)^2 = 21\%$$

The low factor of safety demands that redesign is necessary to achieve a factor of safety of 1.6. Options are as follows.

(1) Reduce roof span W_o to reduce induced stresses.
(2) Increase pillar width; reduce induced stress.
(3) Reduce stoping height; increase pillar strength.

$$F = \frac{S}{\sigma_p} = 10.44\, H^{-0.7} \frac{W_p^{0.5}}{\sigma_v \left(\frac{W_o + W_p}{W_p}\right)^2}$$

$$1.6 = \frac{10.44}{3.375} H^{-0.7} \frac{W_p^{2.5}}{(W_o + W_p)^2}$$

$$W_p^{2.5} - W_p^2 - 12W_p - 36.89 = 0 \qquad W_p = 9.2 \text{ m}$$

Example 2

Ore body thickness = 3 m
Depth, d = 75 m
Extracted span, W_o = 5.0 m
Pillar size, W_p = 7.0 m
Extracted height, H = 2.2 m
Factor of safety = 2.0

Assess the feasibility of extracting additional 0.6 m coal from the roof. Given that

$$S = 7.5\, H^{-0.55} W_p^{0.46} \qquad \text{Density} = 25 \text{ kN/m}^3$$

Solution

$$\sigma_v = 0.025 \times 75 = 1.875 \text{ MPa}$$

$$S = 7.5\,H^{-0.55}\,W_p^{0.46} = 7.5 \times 2.2^{-0.55}\,(7)^{0.46} = 10.90 \text{ MPa}$$

$$\sigma_p = \sigma_v \left(\frac{W_o + W_p}{W_p}\right)^2 = 5.51 \text{ MPa}$$

$$F = \frac{S}{\sigma_p} \quad F = 1.98$$

If roof is stripped

$$H = 2.8 \text{ m}$$

$$S = 7.5\,(2.8)^{-0.66} \times (7)^{0.46} = 9.3 \text{ MPa}$$

$$F = \frac{S}{\sigma_p} = \left(\frac{9.3}{5.51}\right) = 1.69$$

Example 3 A horizontal seam is extracted in a single operation having the following conditions:

Depth of the seam roof = 96.6 m

Height of the seam = 1.8 m

Bord width, W_o = 6.0 m

Number of bords in the panel = 9

Find the required pillar size and percentage extraction in the panel if the factor of safety is 1.6. What is the overall extraction if a barrier of 12 m in width is maintained between adjacent panels?

Solution

d = depth below surface, m = 97.5 m

F = factor of safety = 1.6

r = extraction ratio

W_o = width of the bord or opening, m = 6.0

W_p = width of the pillar, m = ?

S = pillar strength, MPa

σ_v = vertical geo-stress, MPa

σ_p = pillar stress, MPa

H = thickness of seam extracted, m = 1.8 m

$$S = 7.18\,W_p^{0.46}\,H^{-0.66}$$

$$= 7.18\,W_p^{0.46}\,1.8^{-0.66}$$

$$\sigma_v = 0.02488 \times 97.5 = 2.4258 \text{ MPa}$$

$$\sigma_p = \sigma_v \left(\frac{W_o + W_p}{W_p}\right)^2$$

$$\sigma_p = 2.4258 \left(\frac{6.0 + W_p}{W_p} \right)^2$$

$$\sigma_p = \sigma_v \left(\frac{1}{1 - r} \right)^2$$

$$F = S/\sigma_p$$

$$1.6 = \frac{(7.18\, W_p^{0.46}\, 1.8^{-0.66})}{2.4258\,[(6.0 + W_p)/W_p]^2}$$

$$1.4\, W_p^{2.46} - W_p^2 - 12\, W_p - 36 = 0$$

Newton–Raphson method of iteration leads to

$$W_p = 7.6 \text{ m}$$

$$\text{Area extraction ratio} = \left[1 - \left(\frac{W_p}{W_o + W_p} \right)^2 \right] = \left[1 - \left(\frac{7.6}{6.0 + 7.6} \right)^2 \right]$$

$$= 0.6877 = 68.8\%$$

Overall extraction after allowing for 12 m barrier (Fig. 11.11)

$$\text{Overall extraction ratio} = r_0 = \frac{(A - A_p)}{A} = \frac{\text{Area xyzt} - A_p}{\text{Area xyzt}}$$

$$= \frac{[1 - \{C \times 1 + (n - 1)\, W_p^2\}]}{[C\,\{1 + (n - 1)\, C + W_o\}]} = 0.637 = 63.7\%$$

Newton–Raphson iterative method

$$1.25\, W_p^{2.46} - W_p^2 - 12\, W_p - 36 = 0$$

Let

$$f(W_p) = 1.25\, W_p^{2.46} - W_p^2 - 12\, W_p - 36 = 0$$

$$f'(W_p) = 3.075\, W_p^{1.46} - 2\, W_p - 12 = 0$$

Now the general iteration formula is:

$$W_{P_{n+1}} = W_{P_n} - \frac{f(W_{P_n})}{f'(W_{P_n})}$$

Let the initial value $W_{P_o} = 9.6$

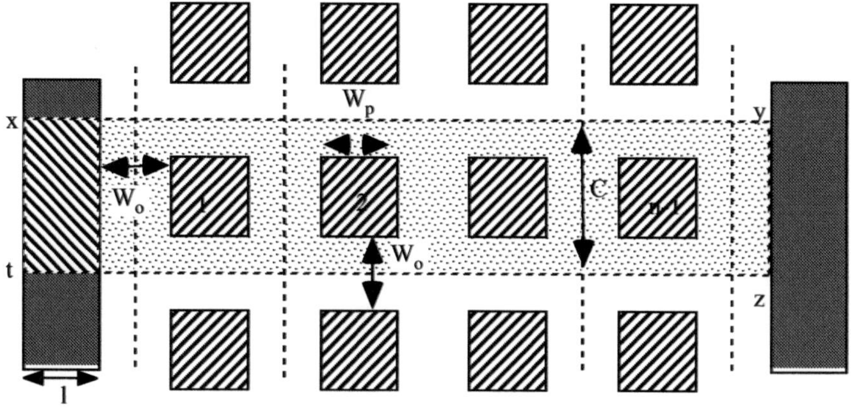

$c = W_p + W_o$ = pillar width centre to centre

Figure 11.11 Percentage extraction in a bord and pillar panel

$$W_{P_1} = 9.6 - \frac{1.25 \times 9.6^{2.46} - 9.6^2 - 12 \times 9.6 - 36}{3.075 \times 9.6^{1.46} - 2 \times 9.6 - 12}$$

$$= 9.6 - \frac{115.2 \times 9.6^{0.46} - 243.76}{29.52 \times 9.6^{0.46} - 31.2} = 9.6 - 1.9 = 7.70 \text{ m}$$

Second iteration

$$W_{P_2} = 7.7 - \frac{1.25 \times 7.7^{2.46} - 7.7^2 - 12 \times 7.7 - 36}{3.075 \times 7.7^{1.46} - 2 \times 7.7 - 12}$$

$$= 7.7 - \frac{9.38}{33.1} = 7.7 - 0.28 = 7.41 \text{ m}$$

$$W_{P_3} = 7.41 - \frac{(1.25 \times 7.41^{2.46} - 7.41^2 - 12 \times 7.41 - 36)}{(3.05 \times 7.7^{1.46} - 2 \times 7.41 - 12)} = 7.41 - \left(\frac{-7.37}{29.91}\right)$$

$$= 7.41 + 0.24 = 7.65 \text{ m}$$

Example 4 Two workable seams lie in close proximity where the parting is too thin to allow workings to be carried out in both seams. Find which seam should be extracted to give maximum yield for the following data.

Depth from the surface to the upper seam	= 51.5 m
Height of upper seam	= 3.5 m
Height of lower seam	= 4.5 m
Number of openings in the panel	= 8

Width of the opening W_o	= 6.5 m
Factor of safety	= 1.6

The lower seam has a fragile roof and therefore 0.6 m of coal is left to support the roof. It is also necessary to leave a sufficient barrier between adjacent panels; the width of barrier is taken to be six times the working height.

Solution

Seam 1 $H_1 = 55\text{ m} \qquad W_o = 6.5\text{ m}$

$h_i = 3.5\text{ m}$

$F = 1.6$

$S = 7.18\, W_p^{0.46}\, h^{-0.66}$

$= 7.18\, W_p^{0.46}\, 3.5^{-0.66} = 3.14\, W_p^{0.46}$

$\sigma_v = 0.002488 \times 55 = 1.3684\text{ MPa}$

$$\sigma_p = \sigma_v \left(\frac{W_o + W_p}{W_p}\right)^2 = 1.3684 \left(\frac{(6.5 + W_p)}{W_p}\right)^2$$

$$\sigma_p = \sigma_v \left(\frac{1}{(1-r)}\right)^2$$

$f = S/\sigma_p$

$1.6 = 3.14\, W_p^{0.46} / 1.3684\, [\,(6.5 + W_p)\, /\, W_p]^2$

$W_p^{2.46} - 0.69\, W_p^2 - 8.97\, W_p - 29.15 = 0$

$W_p = 7.2\text{ m}$

$h_e = h_{max}\, [\,1 - \{\, W_p/(W_p + W_o)\}^2\,]$

$= 3.5\, [\,1 - \{\, 7.2/(7.2 + 6.5)\}^2\,]$

$= 2.53\text{ m}$

Where
S = pillar strength (MPa)
σ_v = vertical geo-stress (MPa)
σ_p = pillar stress (MPa)
H = depth below surface (m)
F = factor of safety
r = extraction ratio
W_o = width of the bord or opening (m)
W_p = width of the pillar (m)
h = thickness of seam extracted (m)
h_e = equivalent height (m)
h_{max} = maximum height of extraction (m)

Seam 2 $H = 63\text{ m},\ W_o = 6.5\text{ m},\ h = 3.9\text{ m}$

$\sigma_v = 0.02488 \times H = 0.02488 \times 63 = 1.5674\text{ MPa}$

$$\sigma_p = \sigma_v [(W_o + W_p)/W_p]^2 = 1.5674 [(6.5 + W_p)/W_p]^2$$
$$S = 7.18\, W_p^{0.46}\, h^{-0.66}$$
$$= 7.18\, W_p^{0.46}\, 3.9^{-0.66}$$
$$= 2.924\, W_p^{0.46}$$
$$F = S/\sigma_p = 2.924\, W_p^{0.46}/1.5674 [(6.5 + W_p)/W_p]^2$$
$$W_p^{2.46} - 0.857\, W_p^{2} - 11.14\, W_p - 36.21 = 0$$
$$W_p = 8.5 \text{ m}$$
$$h_e = h_{max} [1 - \{W_p/(W_p + W_o)\}^2]$$
$$= 3.9 [1 - \{8.5/(8.5 + 6.5)\}^2]$$
$$= 2.64 \text{ m}$$

From the calculations it appears that slightly higher yield can be obtained by mining the lower seam. The equivalent working height of the seam is as follows:

$$\left(\frac{2.64 - 2.53}{2.53}\right) = 4.35\% \text{ higher than the upper seam}$$

$$h_e = h_{max}\left[1 - \left(\frac{W_p}{W_o + W_p}\right)^2\right]$$

Example 5 A coal seam 6.2 m thick occurs at a depth of 82 m. It is intended to extract the total thickness of the seam in two steps; in the primary extraction the mining height is 4 m working on the natural floor of the seam. In the secondary operation the rest of the 2.2 m thick coal is extracted to a true roof. A final factor of safety of 1.4 is decided on at the width of opening of 6 m. Find the initial and final volumetric extraction ratios and equivalent working heights. Also ascertain if the factor of safety after primary extraction is sufficiently high to prevent a possible pillar failure.

$$S = 7.18\left(\frac{W_p^{0.46}}{h^{0.66}}\right)$$

$$\sigma_v = 0.002488 \times H$$

$$\sigma_p = \sigma_v\left(\frac{W_o + W_p}{W_p}\right)^2; \quad \sigma_p = \sigma_v\left(\frac{1}{1-r}\right)^2$$

$$F = S/\sigma_p$$

where S = pillar strength (MPa)

σ_v = vertical geo-stress (MPa)

σ_p = pillar stress (MPa)

H = depth below surface (m)
F = factor of safety
r = extraction ratio
W_o = width of the bord or opening (m)
W_p = width of the pillar (m)
h = thickness of seam extracted (m)
h_e = equivalent height (m)
h_{max} = maximum height of extraction (m)
$h_e = h_{max}\,[1-\{W_p/(W_p+W_o)\}^2]$

Solution Pillar design on the final geometry

$H = 88.2\text{ m} \qquad F = 1.4$
$h = 6.2\text{ m}$
$W_o = 6.0$
$\sigma_v = 0.02488\,H = 0.02488 \times 88.2 = 2.194\text{ MPa}$
$F = S/\sigma_p = 7.18 \times 6.2^{-0.66} \times W_p^{0.46}/2.194\,[(6.0+W_p)/W_p]^2$
$= 2.15\,W_p\,0.46/2.194\,[(6.0+W_p)/W_p]^2$
$0.7\,W_p^{2.46} - W_p^2 - 12\,W_p - 36.0 = 0$
$W_p = 12.2\text{ m}$

After primary mining the following conditions will apply.

$H = 88.2$
$h = 4$
$W_p = 12.2$
$W_o = 6$
$F = ?$
$F = S/\sigma_p = 7.18 \times 4^{-0.66} \times W_p^{0.46}/2.194\,[(6.0+W_p)/W_p]^2$
$= 2.87\,W_p\,0.46/2.194\,[(6.0+12.2)/12.2]^2$
$= 2.87 \times 3.16/4.88 = 1.85$

This factor of safety is considered to be sufficiently safe to preclude the possibility of pillar failure.

Area extraction ratio $r = 1 - \{W_p/(W_p+W_o)\}^2$

$$= 1 - \left[\frac{12.2}{18.2}\right]^2 = 55.07\%$$

$$h_e = 6.2\left[\left(1 - \frac{12.2^2}{18.2^2}\right)\right]^2 = 3.41$$

$$h_e = 4\left[1 - \left\{\frac{12.2}{18.2}\right\}^2\right] = 2.2$$

Example 6 For the following conditions calculate the required pillar sizes for a proposed contiguous seam extraction project.

Depth from the surface to the upper seam	= 122 m
Thickness of upper seam	= 4.3 m
Thickness of inferior coal at the base of the top seam	= 1.8 m
Parting between the seams (shale and sandstone)	= 5.3 m
Thickness of lower seam	= 2.7 m
Number of openings in the panel	= 8
Width of the opening W_o	= 6.5 m
Factor of safety	= 1.7

Solution Because two seams are being mined, a higher factor of safety is warranted.

$$S = 7.18\, W_p^{0.46}\, h^{-0.66}$$

$$\sigma_v = 0.002488 \times H$$

$$\sigma_p = \sigma_v \left(\frac{W_0 + W_p}{W_p} \right)^2; \qquad \sigma_p = \sigma_v \left(\frac{1}{1-r} \right)^2$$

$$F = S/\sigma_p$$

$$1.7 = \left(\frac{7.18\, W_p^{0.46}\, 4.3^{-0.66}}{0.02488 \times 126.3\, [W_p + 6.5]} \right)^2$$

$$W_p = 13.8 \text{ m}$$

where
S = pillar strength (MPa)
σ_v = vertical geo-stress (MPa)
σ_p = pillar stress (MPa)
H = depth below surface (m)
F = factor of safety
r = extraction ratio
W_o = width of the bord or opening (m)
W_p = width of the pillar (m)
h = thickness of seam extracted (m)
h_e = equivalent height (m)
h_{max} = maximum height of extraction (m)
$h_e = h_{max}\, [1 - \{ W_p/(W_p + W_o)\}^2]$

11.5 Floor Heave and Punching Failure of Roof

Figure 11.12 shows the loading of floor strata by a mine pillar causing consequential floor lift if the load exceeds the bearing capacity of the floor. Where the floor strata is plastic in nature in relation to the pillar, the pillar may penetrate the floor causing heave of floor rock adjacent to the pillar edges. Thus, the loading of the floor can be simulated either as line loading or as area loading as discussed in the subsequent sub-sections.

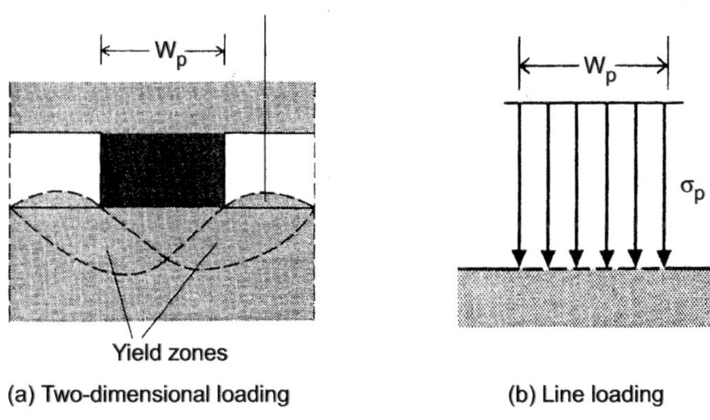

(a) Two-dimensional loading (b) Line loading

Figure 11.12 Yield of country rock under pillar load

11.5.1 Bearing Capacity of Roof and Floor Rock

Strip loading in plastic rock is given by the following expression (Figure 11.13[b]):

$$Q_b = \frac{1}{2}\gamma W_p N_r + CN_c \tag{11.17}$$

where $N_c = (N_q - 1)\cot\phi$

$$N_r = 1.5\,(N_q - 1)\tan\phi$$

$$N_q = e^{\pi\tan\phi}\tan^2\left[\frac{\pi}{4} + \frac{\phi}{2}\right]$$

ϕ = internal angle of friction

Q_b = bearing capacity (MPa)

N_c, N_r = bearing capacity factors

N_q = a variable

C = cohesive strength (MPa)

11.5.2 Bearing Capacity of a Long Chain Pillar

The bearing capacity of the chain pillar, ($W_p \times l_p$) in dimensions can be simulated as the loading of a rectangular area on the floor strata as follows:

$$Q_b = \frac{1}{2}\gamma W_p N_r S_r + C\cot\phi\, N_q S_q - C\cot\phi \tag{11.18}$$

S_q, S_r = shape factor

$$S_q = 1.0 + \sin\phi\left(\frac{W_p}{l_p}\right) \tag{11.18a}$$

$$S_r = 1.0 - 0.4\frac{W_p}{l_p} \tag{11.18b}$$

Example 7 Given that

$$\begin{aligned}\text{Mining depth} &= 150\text{ m}\\ \text{Ore body thickness} &= 3\text{ m}\\ \text{Unit weight} &= 22.5\text{ kN/m}^3\\ C &= 1.2\text{ MPa}\\ \phi &= 28^o\\ W_o &= 6\text{ m}\\ W_p &= 10\text{ m}\end{aligned}$$

Calculate the factor of safety against floor failure.

Solution

$$Q_b = \frac{1}{2}\gamma W_p N_r S_r + C\cot\phi\, N_q S_q - C\cot\phi$$

$$N_q = e^{\pi\tan\phi}\tan^2\left[\frac{\pi}{4}+\frac{\phi}{2}\right] = e^{3.14\tan 28^\circ}\times\tan^2[45+14] = 14.71$$

$$\begin{aligned}N_r &= 1.5\,(N_q - 1)\tan\phi = 1.5\,(14.7-1)\tan 28^\circ = 10.93\\ N_c &= (N_q-1)\cot\phi\\ &= (14.71-1)\cot 28\\ &= 13.71\times 1.88 = 25.77\end{aligned}$$

$$S_q = 1.0 + \sin\phi\left(\frac{W_p}{l_p}\right) = 1 + \sin 28^\circ\times 1 = 1.46$$

$$S_r = 1.0 - 0.4\frac{W_p}{l_p} = 1 - 0.4\times 1 = 0.6$$

$$Q_b = \frac{1}{2}\gamma W_p N_r S_r + C\cot\phi\, N_q S_q - C\cot\phi$$

$$= \frac{1}{2}\, 0.0225 \times 10 \times 10 \times 10.93 \times 0.6 + 1.2 \times \cot 28 \times 14.71 \times 1.46$$
$$- 1.2 \times \cot 28$$
$$= 0.72 + 48.22 - 2.256 = 46.58 \text{ MPa}$$

$$\sigma_p = \sigma_v \left(\frac{W_o + W_p}{W_p} \right)^2 = 0.022 \times 150 \left(\frac{16}{10} \right)^2 = 8.45 \text{ MPa}$$

$$FS = \frac{Q_b}{\sigma_p} = \frac{46.58}{8.5} = 5.5$$

11.6 Stope and Pillar Design in Irregular Ore Bodies

In contrast to the design of coal pillars in a stratified deposit like coal seams as discussed earlier, in metalliferous mining ore bodies of irregular shape in close proximity of each other require careful consideration for the sequence of extraction and pose a much more intractable rock mechanics design problem. As such, recourse to observational methods is often made involving iterative design through interpretation of geomechanical conditions in the ore body and that of the surrounding rock mass and their impact on the location of the pillar in the ore body. The design problem becomes more

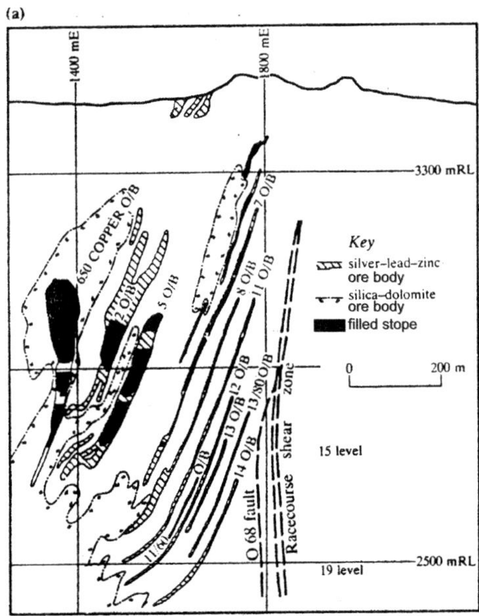

Figure 11.13 Cross-section through Mount Isa lead/zinc ore body

complex in closely spaced, parallel ore bodies where superimposition of pillar axes becomes imperative. The geometric complexity of the mine structure in such ore body configurations is illustrated in Figure 11.13 which presents cross-sections through the northern part of the Mount Isa lead/zinc mine in Australia and the spatial disposition of stopes and pillar structure in the ore body.

Figure 11.14 outlines a general algorithm for opening and pillar design in irregular ore bodies using inputs from the boundary element method for

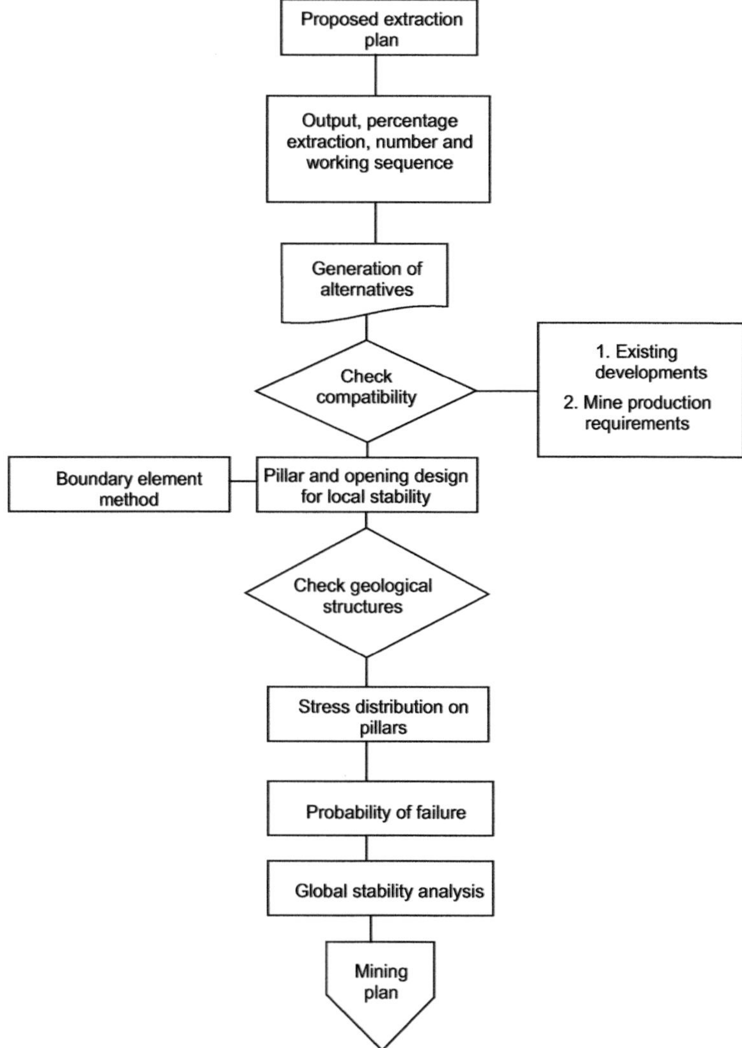

Figure 11.14 Opening and pillar design in irregular ore body

analysis of stress distribution in the proposed ore body structure to map regimes of tensile stress or regimes in which compressive stresses exceed the rock mass strength. One can also consider the concept of local safety factor, based on local strength/stress, which is essentially natural.

Figure 11.15 shows a plot of local factor of safety against compressive failure in an inter-ore body remnant pillar.

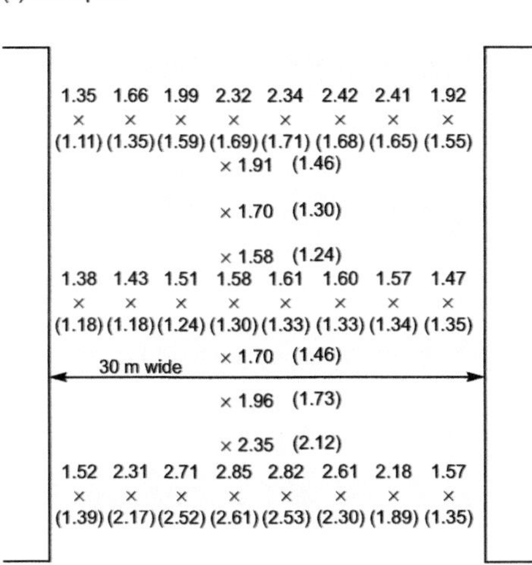

Figure 11.15 Local factor of safety against compressive failure (Brady, 1977)

11.6 Global Stability

In mining, the state of equilibrium changes due to the following conditions.

- Changes in geo-stress with depth
- Transient stresses imposed due to blasting
- Periodic falls or bumps
- Changes in energy stored in rocks due to volume changes e.g. stoping operations

It is necessary to evaluate the impacts of the changes as well as of local failures on the global/more extensive failure in the mine structure and near-field domain.

11.6.1 Global Stability

A body is in equilibrium under a set of applied loads P_i. Let a set of small probing loads ΔP_j be applied at various part of the structure, resulting in a set of displacements ΔU_j. The work done by the small probing forces acting through the incremental displacements is given by:

$$\ddot{W} = \frac{1}{2}\Sigma\,\Delta P_j \Delta U_j \tag{11.19}$$

For global stability

$$\ddot{W} > 0$$

$\ddot{W}$ = second order variable of the total potential energy

$\ddot{W} < 0$ unstable equilibrium

Figure 11.16 shows the interactive elements of different stiffness in a mining domain. The load-convergence characteristics of the system elements are defined as:

$P_r = f(S)$ Pillar

$P_s = K(\gamma - S)$ Loading system (country rock)

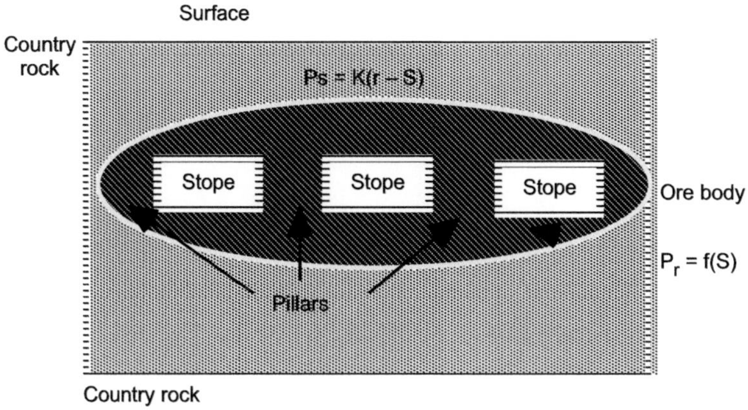

Figure 11.16 Global stability of pillar systems

where γ = displacement of loading system
K = stiffness of country rock
S = displacement of pillar
λ = stiffness of the pillar

If a probing force causing incremental displacement ΔS is given by

$$\Delta P_r = f(S)\cdot \Delta S = \lambda \cdot \Delta S$$

Net probing force causing incremental displacement ΔS is given by

$$\Delta P = \Delta P_r - \Delta P_s = (K + \lambda)\cdot \Delta S$$

$$\ddot{W} = \frac{1}{2}\Delta P \cdot \Delta S = \frac{1}{2}(K + \lambda)\cdot \Delta S^2$$

It follows that in order to satisfy the stable equilibrium (Fig. 11.17) condition the following relationship must be satisfied:

$$K + \lambda > 0 \tag{11.20}$$

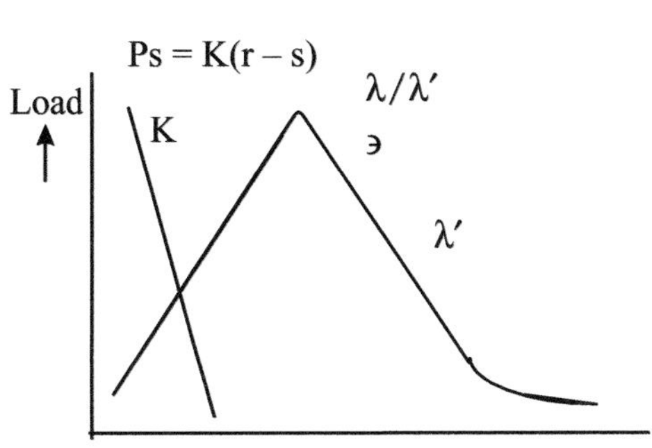

Figure 11.17 Comparative stiffness of ore body and the surrounding rocks

Example 8 (Brady and Brown, 1985)

A pillar with a width to height ratio of 1.2 is to be subjected to stress levels exceeding the peak rock mass strength. In the elastic range, the calculated stiffness of the pillar is 20 GN/m^2. The ratio of post-failure (λ') stiffness to elastic stiffness (λ) varies with width/height (W/H) ratio of the pillar.

For various W/H ratios, *in situ* tests gave the following results:

W/H ratio	1.0	1.33	1.85	2.14
λ'/λ	–0.60	–0.43	–0.29	–0.23

Analysis of convergence at the pillar position for the distributed normal load 'P' of magnitude applied at the pillar position yielded the following results:

Load P (MN)	0.0	125.0	220.0	314.0
Conv. ($m \times 10^{-3}$)	32.0	22.0	14.00	6.0

Assess the global stability of the pillar.

Solution Plotting the ratio of stiffness against the W/H ratio in Figure 11.18 and carrying out a regression analysis, we get:

$$-\frac{\lambda}{\lambda'} = -0.72 + 2.32\frac{W}{H}$$

$$\text{Stiffness ratio of the pillar} = -\lambda/\lambda' = 2.3 \times 1.2 - 0.718 = 2.04$$

$$\lambda' = -\lambda/2.04 = -20/2.04 = -9.8035$$

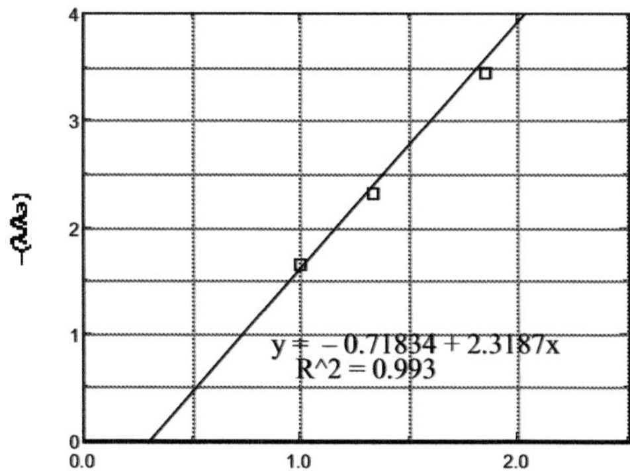

Figure 11.18 Relative stiffness of ore body and width to height ratio

Figure 11.19 is a plot of load P in MN against the convergence in the country rock surrounding the ore body. It can be shown that:

$$K = 12.089 \text{ and}$$

$$[K+\lambda'] = [12.089 - 9.8035] = 2.28$$

Thus,

$$[K+\lambda'] > 0 \text{ which would ensure a stable equilibrium.}$$

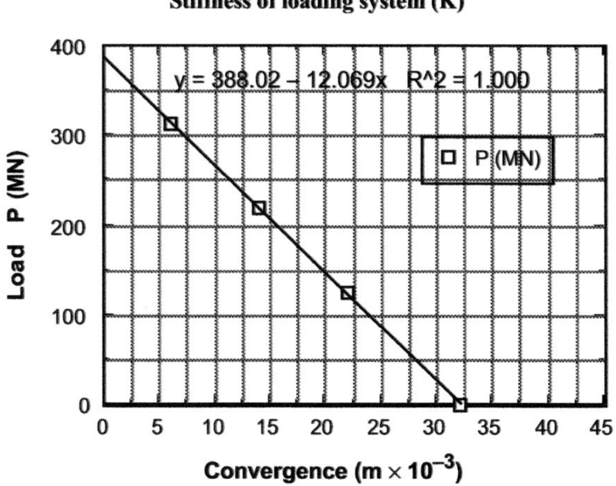

Figure 11.19 Stiffness of surrounding rock

11.7 Rib Pillar Design in Highwall Mining

11.7.1 Highwall Pillars

Highwall mining is an innovative mining system for the recovery of coal in open cut mines by excavating a series of parallel rooms into the exposed coal seam in the highwall leaving rib pillars to support the overburden. The system has found application in the USA and Australia where once the economic limit of open cut mining has been reached, the recovery of highwall coal starts using remotely operated equipment to drive into the highwall coal. In designing the layout for a highwall mining system, it is necessary to specify the safe roof span, the rib pillar width, number of rib pillars in a panel and finally the barrier pillar width separating the panels. There have been many instances, both in Australia and USA, when catastrophic failure of pillars *en masse* has led to abandonment of large coal reserves and in some case damage or loss of equipment used for drivages. The use of barrier pillars helps limit the overburden stresses on the rib pillars and thereby increases their safety factor besides arresting domino-type rib pillar failure.

As in the classical pillar design approach, the strength of rib pillars is estimated using two different formulae based on the pillar's $\frac{W}{H}$ ratio as follows:

Case 1 For $\frac{W}{H}$ ratio < 5: suggested by Mark and Iannachhione (1992)

$$S_p = \sigma_{IN}\left[0.64 + 0.54\frac{W}{H} - 0.18\frac{W^2}{HL}\right] \tag{11.21}$$

where,

W = rib pillar width (m)
H = rib pillar height (m)
S_p = strength of the pillar (MN/m^2)
σ_{IN} = *in situ* strength of the pillar (MPa)
L = rib or barrier pillar length (m)

***Case* 2** For W/H ratio greater than 5

$$S_p = \sigma_{IN}\left[\left(\frac{W_o}{H_o}\right)^b \Big/ V^a\right] \times \left[\frac{b}{e}\left\{\left(\frac{W}{H}\right)\Big/\frac{W_o}{H_o}\right\}^e - 1 + 1\right] \tag{11.22}$$

where,

$\frac{W_o}{H_o}$ ratio = critical width to height ratio = 5

V = rib or barrier pillar volume = W^2H, m^3
$e = 2.5$, $b = 0.5933$ and $a = 0.0667$

$\frac{W}{H}$ = width to height ratio of the rib pillar or barrier pillar

The load on pillar according to tributary area theory is given by the following equation:

$$\sigma_{RP} = \gamma D \frac{(W_{RP} + W_E)}{W_{RP}} \tag{11.23}$$

where,

σ_{RP} = stress on the rib pillar
$\gamma = 0.025\ kN/m^3$
D = depth below surface (m)
W_{RP} = width of the rib pillar (m)
W_E = width of the opening (m)

$$F = \frac{\text{Strength}}{\text{Stress on pillar}}$$

= (Equation 11.21 or eq. 11.22)/Equation 11.23

11.7.2 Barrier Pillar

The barrier pillars besides serving to limit the overburden stresses on the rib pillar also help arrest the domino type of pillar failure. The spacing between the barrier pillars, if they are left after every 'N' pillars, is given by:

$$W_{PN} = N(W_{RP} + W_E) + W_E \tag{11.24}$$

where,

W_{PN} = spacing between panel pillars (m)
N = number of rib pillars in the panel
W_{RP} = width of the rib pillar (m)
W_E = width of the entry (m)

11.7.3 Mechanics of Domino Type Rib Pillar Failure

Pillar load bearing capacity decreases either due to

(1) progressive failure, or
(2) catastrophic failure.

Local mine stiffness criterion developed by Salamon (1970) can be used to define the global stability. Non-violent failure occurs when:

$$K_{LMS} > K_P$$

where,

K_{LMS} = local mine stiffness
K_P = post-failure pillar stiffness

On the other hand, unstable catastrophic failure characterised by violent disruption occurs when conditions are as follows:

$$K_{LMS} < K_P$$

where,

K_{LMS} = local mine stiffness

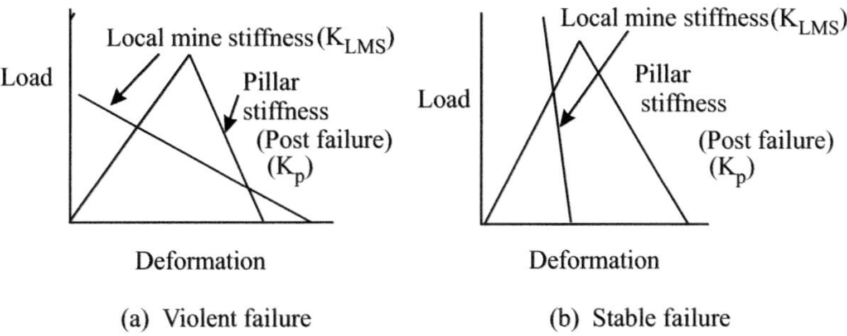

(a) Violent failure (b) Stable failure

Figure 11.20 Criteria of violent and stable failure of a pillar based on the post-failure characteristics of coal (Zipf, 1998)

Figures 11.20 shows the nature of failure occurring according to the global stability criterion. It can be seen that:

(1) Condition for violent failure $K_{LMS} < K_p$
(2) Conditions for stable failure $K_{LMS} > K_P$

where

K_P = post-failure pillar stiffness on load convergence curve of a pillar

11.7.4 Post-failure Stiffness K_p and Width to Height Ratio

Figure 11.21 shows the effect of width to height ratio on the post-failure strength of rock. It can be seen that as the width to height ratio increases the following characteristic trends occur in the rock specimen.

(1) Peak strength increases
(2) Residual strength increases
(3) Pillar stiffness reduces
(4) Failure mode changes from strain softening to strain hardening.

Figure 11.22 pools the data from available measurements on post-failure modulus of full-scale coal pillars.

It may be noted in Figure 11.22 that the post-failure modulus of coal is zero at a width to height ratio of between 3 and 4. At this stress the pillar fails but carries the load with continuous deformation. Based on the data, the post-failure modulus is found to be:

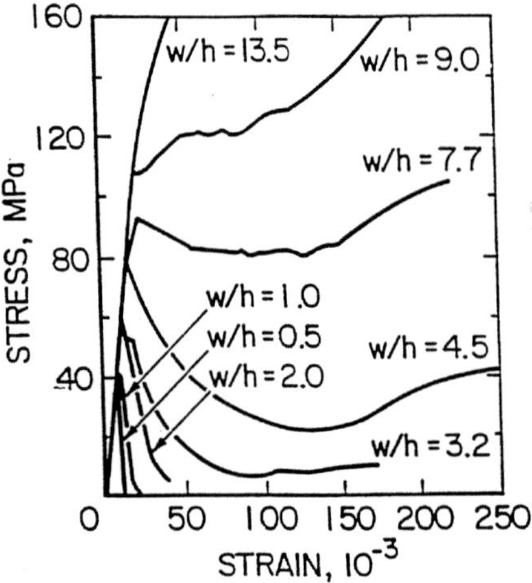

Figure 11.21 Variation of post-failure strength of coal with width to height ratio of pillar

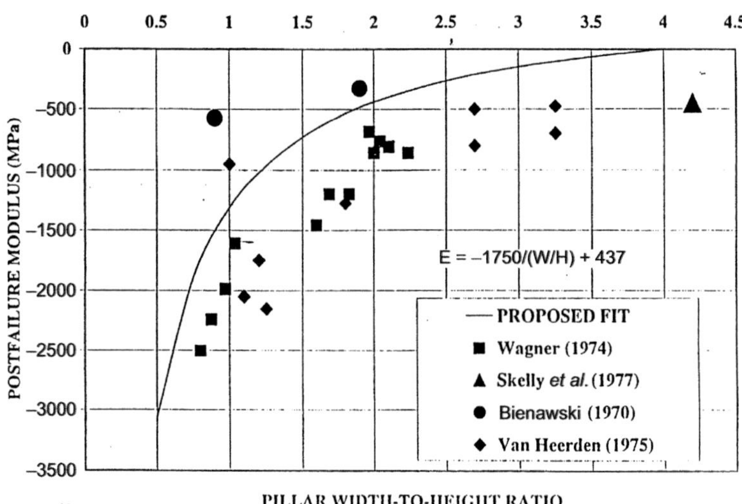

Figure 11.22 Post-failure modulus and the width to height ratio of a pillar (Zipf, 1998)

$$E_p = -1750\left(\frac{W}{H}\right)^{-1} + 437, \text{ MPa} \tag{11.25}$$

Assuming a unit width for the pillar, the expression for stiffness of the pillar is:

$$K_p = E_p\frac{W}{H} = -1750 + 437\left(\frac{W}{H}\right), \text{ MN/m} \tag{11.26}$$

where,

K_p = post-failure stiffness of the pillar (MN/m)
E_p = post-failure deformation modulus of pillar (MPa)
W = width of the rib pillar (m)
H = height of rib pillar (m)

Local mine stiffness K_{LMS} is the other key factor defining the stability criterion. This depends upon two factors.

(1) Modulus of the rock mass, E
(2) Geometry of the mining excavation

Figure 11.23 illustrates the behaviour of local mine stiffness for different mine layouts' with the number of pillars increasing from 3 to 15. For an infinitely wide panel, the local mine stiffness tends to zero which is conducive to unstable violent failure. Figure 11.25 shows local mine stiffness as a function of panel width from which for practical design methodology Zipf (1996) proposed the following approximate expression:

$$\frac{K_{LMS}}{E \times 0.5 \times W_p} = -0.20\left[\frac{3}{N}\right]^{1.5} \tag{11.27}$$

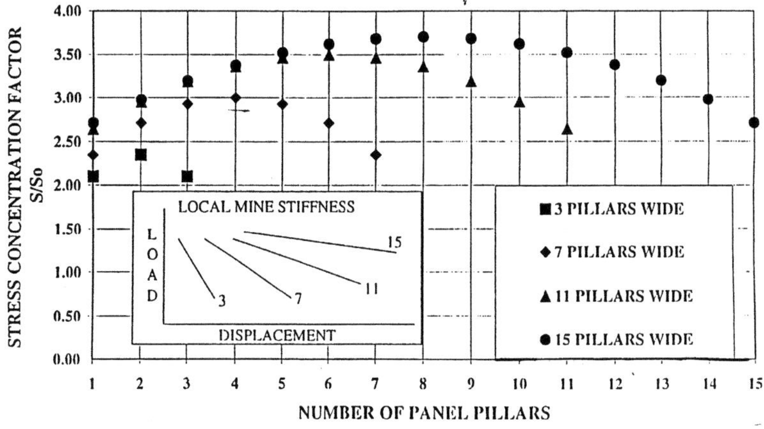

Figure 11.23 Local mine stiffness and number of rib pillars in the panel (Zipf, 1996)

This equation is valid for elastic, homogeneous and isotropic rock, with the panel pillar centre-to-centre half width.

$L_2 = 0.5; W_p$ = rib pillar half width

Local mine stiffness decreases in magnitude monotonically and approaches zero as the panel width increases.

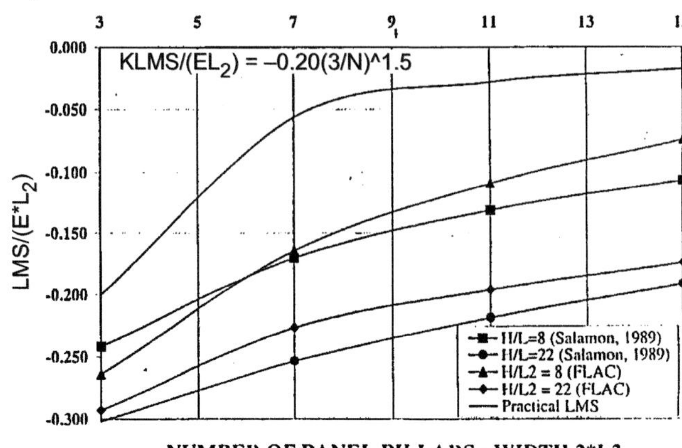

Figure 11.24 Local mine stiffness and number of pillars in the panel (Zipf, 1996)

Worked examples

Example 1 Calculate the factor of safety against rib pillar failure for the following conditions in a highwall mine layout.

$$\text{Width of extraction } W_E = 3.66\text{ m}$$
$$\text{Width of rib pillar } W_{RP} = 1.8\text{ m}$$
$$\text{Height of extraction } H = 3.0\text{ m}$$
$$\text{Depth of workings} = 80\text{ m}$$

Dip of the seam 4 to 8°.

Solution

$$S_p = \sigma_{IN}\left[0.64 + 0.54\frac{W_{RP}}{H} - 0.18\frac{W_{RP}^2}{HL}\right] \tag{11.21}$$

where,

W_{RP} = rib pillar width (m)
H = rib pillar height (m) = 3
S_p = Strength of the pillar (MN/m^2)
σ_{IN} = *in situ* strength of the pillar (MPa) = 6.4
W_E = width of extraction (m) = 3.66
L = length of pillar = W_{RP}

$$\sigma_{RP} = \gamma D\frac{(W_{RP} + W_E)}{W_{RP}} \tag{11.23}$$

$$S_p = \sigma_{IN}\left[0.64 + 0.54\,\frac{W_{RP}}{H} - 0.18\,\frac{W_{RP}^2}{HL}\right]$$

$$= \sigma_{IN}\left[0.64 + 0.54\,\frac{W_{RP}}{H} - 0.18\,\frac{W_{RP}}{H}\right]$$

$$= 6.4\left[0.64 + 0.54\,\frac{1.8}{3.0} - 0.18\,\frac{1.8}{3.0}\right]$$

$$= 6.4\,[0.64 + 0.324 - 0.108] = 5.48 \text{ MPa}$$

$$\sigma_{RP} = \gamma D\,\frac{(W_{RP} + W_E)}{W_{RP}}$$

$$= 0.025 \times 80\,\frac{(1.8 + 3.66)}{1.8}$$

$$= 2 \times 3.03 = 6.06 \text{ MPa}$$

$$\text{Factor of safety} = \frac{\text{Strength}}{\text{Stress on the rib pillar}} = \frac{S_p}{\sigma_{RP}} = \frac{5.48}{6.06} = 0.9$$

Example 2 Calculate the factor of safety against rib pillar failure for the following conditions in highwall mining:

Width of extraction, W_E	= 1.30 m
Width of rib pillar, W_{RP}	= 0.8 m
Height of extraction, H	= 1.3 m
Depth of workings	= 35 m
σ_{IN}	= 3.0 MPa
Dip of the seam	= level

Solution

$$\sigma_{RP} = \gamma D\,\frac{(W_{RP} + W_E)}{W_{RP}}\;\sigma_{RP}$$

$$S_p = \sigma_{IN}\left[0.64 + 0.54\,\frac{W_{RP}}{H} - 0.18\,\frac{W_{RP}^2}{HL}\right]$$

$$= \sigma_{IN}\left[0.64 + 0.54\,\frac{W_{RP}}{H} - 0.18\,\frac{W_{RP}}{H}\right]$$

$$= 3.0\left[0.64 + 0.54\,\frac{0.8}{1.30} - 0.18\,\frac{0.8}{1.30}\right]$$

$$= 3.0\,[0.64 + 0.332 - 0.11]$$

$$= 3.0 \times 0.862 \text{ MPa} = 2.586 \text{ MPa}$$

$$\sigma_{RP} = \gamma D \frac{(W_{RP} + W_E)}{W_{RP}}$$

$$= 0.025 \times 35 \frac{(0.8 + 1.3)}{0.8}$$

$$= 0.875 \times 2.625 = 2.3 \text{ MPa}$$

$$\text{Factor of safety} = \frac{\text{Strength}}{\text{Stress on the rib pillar}} = \frac{2.586}{2.3} = 1.12$$

Example 3 Calculate the factor of safety against rib pillar failure for the following conditions in highwall mining:

Width of extraction, W_E = 3.5 m
Width of rib pillar, W_{RP} = 3.5 m
Height of extraction, H = 2.0 m
Depth of workings = 70
σ_{IN} = 3.0 MPa
Dip of the seam = 8° to 10°

Solution

$$S_p = \sigma_{IN}\left[0.64 + 0.54\frac{W_{RP}}{H} - 0.18\frac{W_{RP}^2}{HL}\right]$$

$$= \sigma_{IN}\left[0.64 + 0.54\frac{W_{RP}}{H} - 0.18\frac{W_{RP}}{H}\right]$$

$$= 3.0\left[0.64 + 0.54\frac{W_{RP}}{H} - 0.18\frac{W_{RP}}{H}\right]$$

$$= 3.0\left[0.64 + 0.54 \times \frac{3.5}{2.0} - 0.18 \times \frac{3.5}{2.0}\right]$$

$$= 3.0 \times [0.64 + 0.945 - 0.315] = 3.0 \times 1.27 = 3.81$$

$$\sigma_{RP} = \gamma D \frac{(W_{RP} + W_E)}{W_{RP}} = 0.025 \times 70 \frac{[3.5 + 3.5]}{3.5} = 3.5 \text{ MPa}$$

$$\text{Factor of safety} = \frac{\text{Strength}}{\text{Stress on the rib pillar}} = \frac{3.81}{3.5} = 1.09$$

Example 4 The local mine stiffness and the post-failure pillar stiffness are given by the following formulae:

$$K_p = -1750 + 437\frac{W_{RP}}{H} \qquad 11.26$$

$$K_{LMS} = -0.2\,EL_2 \left[\frac{3}{N}\right]^{1.5} \qquad 11.27$$

Calculate the local mine stiffness and the post-failure stiffness of the pillars. Table 11.2 shows the local mine stiffness and pillar stiffness calculated for the following three case histories.

Table 11.2 Global stability analyses of pillars

Items	*Case A*	*Case B*	*Case C*
Input data			
Width of opening W_E (m)	3.66	1.30	3.50
Width of rib pillar W_{RP} (m)	1.80	0.80	3.50
Pillar height (m)	3.0	1.30	2.00
Depth below surface (m)	80	35	70
Calculations			
$\frac{W_{RP}}{H}$ ratio	0.60	0.62	1.75
K_P Equation (11.26)	–1489	–1459	–985
$L_2 = \frac{(W_{RP}+W_E)}{2}$	2.73	1.55	3.50
E_P (MPa) from Figure 11.22 or Equation 11.25	–2500 MPa	–2450 MPa	–600 MPa
N-failed pillar numbers	75	45	14
$\left[\frac{3}{N}\right]^{1.5}$	0.04156	0.0172	0.0992
K_{LMS} Equation 11. 27	– 57	–13	–208
K_P Equation 11.26	–1488	–1479	–985
Criteria	$K_{LMS} < K_P$	$K_{LMS} < K_P$	$K_{LMS} < K_P$
Mode of failure	Violent	Violent	Violent

11.8 Conclusions

Pillars are used in room and pillar mining in order to offer immediate local support to the adjoining rooms and to provide the global stability to the mine workings with respect to mine subsidence and catastrophic pillar failure. The procedure for the design of pillars in room and pillar mining is based on the tributary area theory which assumes uniform distribution of the abutment stress on the pillars. Although this method of design is a simplistic approach it provides insight into the stability of pillars in the room and pillar partial extraction system.

Field observations show that the mode of failure in uniaxially loaded pillars is either controlled by the stress or by pillar configurations such as the width to height ratio of the pillar, or by geological structures such as cleat or joints.

Structural stability of the floor strata can be determined satisfactorily by the use bearing capacity formulae used in foundation engineering using formulae for line loading as well as two-dimensional loading.

The design of irregular pillars in metalliferous mining is based on the stress distribution on pillars and the determination of the local factor of safety against failure at various points. This is followed by the determination of global stability of the entire mine structure.

This method of analysis is further extended to the design of rib pillars and barrier pillars in highwall mining. The technique has been successfully applied to the design of rib pillars in the Central Queensland mining province by Zipf (1996).

References

Bienawski, Z.T. (1968). 'The effect of Specimen Size on Strength of Coal Pillars', *Int. Jour. Rock Mechanics and Min. Sci.*, Vol. 5, pp. 325–335.

Bienawski, Z.T. and Van Heerden, W.L., (1975). 'The significance of *in situ* tests on large rock specimen'. *Int. Jour. Rock Mechanics and Min. Sci.*, Vol. 12, No. 4, pp. 101–114.

Brady, B.H.G. (1975). *Rock Mechanics Aspects of 1100 Orebody*, Technical Report, Mount Isa Mines Ltd.

Brady, B.H.G. (1977). 'An analysis of rock behaviour in an experimental stoping block at the Mount Isa Mines, Queensland, Australia', *International Journal of Rock Mechanics and Mining Science*, 14, pp. 59–66.

Brady, B.H.G, and Brown, E.T., (1985), Rock Mechanics for Underground Mining, George Allen & Unwin, London, pp. 316.

Bunting, D., (1911), Chamber pillars in deep anthracite mines, *Trans American Inst. Mining and Metallurgy*, Vol. 130, pp. 314–332.

Brinch Hansen, J. (1970). 'Bearing capacity'. *Danish Geotechnical Institute Bulletin*, No. 28, pp. 5–11.

Coates, D.F. (1981). *Rock Mechanics Principles*, 3rd edition, Mines Branch Monograph 874, Ottawa: Information Canada.

Cook, N.G.W., K. Hodgson and J.P.M. Hojem (1971). 'A 100 MN Jacking System for Testing Coal Pillars Underground', *J. S. African Institute of Mining Metallurgy*, 71, 215–224.

Gaddy, F.L.A. (1956). 'A study of ultimate strength of coal as related to the absolute sizes of cubical specimens tested, *Bulletin of Virginia Polytechnic Institute*, Eng. Exp., Stn. Ser., vol. 112, pp. 1–27.

Gale, W. (1998). 'The Application of Field and Computer Methods for Pillar Design', *International Conference on Geomechanics and Ground Control in Mining and Underground Construction*, 14–17 July, Wollongong, New South Wales, Australia, pp. 243–261.

Greenwald, H.P., H.C. Howarth and I. Hartman (1939). *Experimental Strength of Small Coal Pillars in Pittsburgh Seam,* US Bureau of Mines, Technical Paper No. 605.

Greenwald, H.P., H.C. Howarth and I. Hartman, (1941). *Experimental Strength* of *Small Coal Pillars in Pittsburgh Seam,* US Bureau of Mines, RI 3575.

Hustrulid, W.A. (1976), 'A review of Coal Pillar Strength Formulas', *Rock Mechanics,* Vol. 8. No. 2, pp. 115–145.

Mark, C. (1992), *Analysis of Longwall Pillar Stability,* US Department of Interior, IC 9314, pp. 238–249.

Mark, C., D. Su, and K.A. Heasley (1998). 'Recent Development in Coal Pillar Design in the US', *International Conference on Geomechanics and Ground Control in Mining and Underground Construction,* 14–17 July, Wollongong, New South Wales, Australia, pp. 309–324.

Mark, C. and A. Iannacchione (1992). 'Coal Pillar Mechanics Theoretical Model and Field Measurements Compared', *Proceedings of Workshop in Pillar Mechanics and Design,* US Department of Interior, IC 9315, pp. 78–93.

Merwe, J.N. van der (2003), 'New pillar strength formula for South African coal', J. S. Afrc. Inst. Min. Metall, June, pp. 281–292.

Merwe, J.N. van der (2003), 'Predicting coal pillar life in South Africa', J. S. Afrc. Inst. Min. Metall., June, pp. 293–301.

Pariseau, W.G. 'Limit design of mine pillars under uncertainty', Design Methods in Rock Mechanics, *US Rock Mechanics Symposium,* pp. 287–301.

Salamon, M.D.G. and A.H. Munro (1967). 'A study of strength of coal pillars', *J. South African Institute of Mining and Metallurgy,* Vol. 68.

Salamon, M.D.G. (1967). 'A Method of Designing Board and Pillar Working', J.S. Afrc. Inst. Min. Metall., Vol. 68, pp. 68–78.

Salamon, M.D.G. (1970). 'A Method of Designing Board and Pillar Working', *International Jour. Rock Mechanics and Min. Sci.,* Vol. 7, pp. 613–631.

Salamon, M.D.G. and H. Wagner (1979). 'Role of Stabilising Pillars in the Alleviation of Rock Burst Hazards in Deep Mines', *4th Congress of International Society of Rock Mechanics,* Montreaux, Vol. 2, pp. 561–566.

Salamon, M.D.G. (1999). 'Strength of coal pillars from back calculations', In: *Rock Mechanics for industry,* (Edited Amedai *et al.,* Balkema, pp. 29–36).

Sheorey, P.R. and B. Singh (1974). 'Strength of Rectangular Pillar in Partial Extraction', *International Journal of Rock Mechanics and Mining Science,* Vol. 11, pp. 41–44.

Stears, P.A. (1954). 'Strength and Stability of Pillars in Coal Mines'. *J. Chem. Metall. Min. Soc. South African,* Vol. 54, pp. 309–325.

Wilson, A.H. (1972). 'An Hypothesis Concerning Pillar Stability', *Mining Engineer,* London, June, Vol. 131, No. 141, pp. 409–417.

Van Heerden, W.L. (1975). '*In situ* Complete Stress/Strain Characteristics of Large Coal Specimen', *J. S. Afrc. Inst. Min. Metallurgy,* vol. 75, pp. 207–217.

Wilson, A.H. (1981). 'Stress and Stability in coal rib side and pillars', *Proceedings of Ist Annual Conference in Ground Control in Mining,* West Virginia University, Editor S.S. Peng, pp. 1–12.

Wagner, H. (1974). 'Determination of complete load deformation characteristics of coal pillars', *3rd Congress of International Society of Rock Mechanics,* Denver, Colorado, USA, pp. 1076–1081.

Whittaker, B.N. and R.N. Singh (1979). 'Design and Stability of Pillars in Longwall Mining', *Mining Engineer*, Vol. 139, No. 214, pp. 59–73.

Wilson, A.H. (1972). 'An Hypothesis Concerning Pillar Stability', *Mining Engineer*, London, June, Vol. 131, No. 141, pp. 409–417.

Zipf, A.K. (1992). *Design Method to control violent pillar failure in room and pillar mines*, Bureau of Mines IC 9321, 48 pp.

Zipf, A.K. (1996). 'Analysis and Design Method to Control Cascading Pillar Failure in Room and Pillar Mines', *in: Milestone in Rock Engineering*, (Ed. Bieniawski), Balkema, pp. 225–264.

Zipf, R.K. and P. Swanson (1999). 'Description of large catastrophic failure in a' South Western Wyoming Trona Mines', *in: Rock Mechanics for Industry* (Editor Amedai *et al.*) Balkema, pp. 293–298.

Zipf, A.K. (1998). 'Case histories of high wall web pillar collapse and preventive design methods', *Conference on Highwall Mining*, Unpublished.

Design and Stability of Rib Pillars and Chain Pillars in Longwall Mining

12
Chapter

12.1 Introduction

Longwall mining is the most efficient coal mining system available for the mining of deep-seated coal seams, amenable to mechanisation and automation. Very large investment in modern longwall faces in the order of tens of million of dollars demands effective rock mechanics design of the longwall panel as a prerequisite for this major capital expenditure. This chapter describes the design of rib pillars in single entry longwall mining and also chain pillars in multi-heading longwall mining.

12.2 Importance of Longwall Mining

Longwall mining is the most widespread system of underground coal mining used the world over. This is largely due to effective strata control at the face and high productivity reaching up to 6 to 8 million tonnes per face per year. One of the most important factors affecting the overall success of a longwall mining operation is to ensure continuity of operation in order to realise economical viability of the highly capital-intensive longwall mining operations. A stoppage of a longwall face due to roof fall in gate roadways and intersections may cost revenue of some $420 per minute in terms of lost production. One of the most important elements of a longwall face is the chain pillar or rib pillar providing stability to the gate roadways, which are essential for access to the coal face, provide airflow and supplies to the face and transport of the coal outbye. Adequately sized longwall pillars are necessary to provide effective ground control of gate roadways guarding against roof fall hazards, provide a safe escape route from the longwall face to the surface in case of emergency and keep mining problems confined to within the longwall panel.

12.3 Purpose of Coal Pillars

Figure 12.1 shows a layout of a single entry longwall face used for advance or retreat mining. The main purpose of pillars in this system of mining is as follows.

1. Roadway protection in the vicinity of caving strata
2. Longwall isolation
3. Barrier pillars against main development roadways
4. Control of subsidence at the surface
5. Pillars below remnant pillars
6. Avoidance of major geological fault

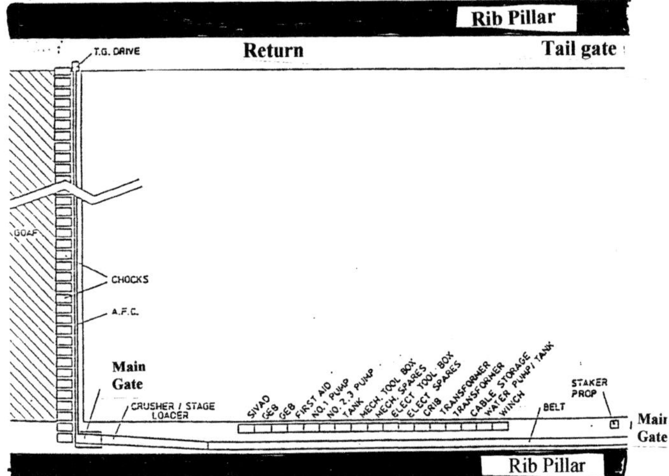

Figure 12.1 Single entry longwall operation supported by rib pillars

Figure 12.2 shows a layout of a double entry longwall retreat face. In many countries the multi-entry longwall face is a statutory requirement to provide alternate escape routes and ease in ventilation of the headings during the longwall development. Although the multi-entry system offers many operational advantages, it has the following drawbacks.

1. Strata control in three-way and four-way intersections is difficult and requires prompt and additional support.
2. The chain pillars in this system of mining undergo additional loading at different phases of mining and need special design considerations.
3. Multiple headings are more expensive per linear metre of advance than the single entry development.

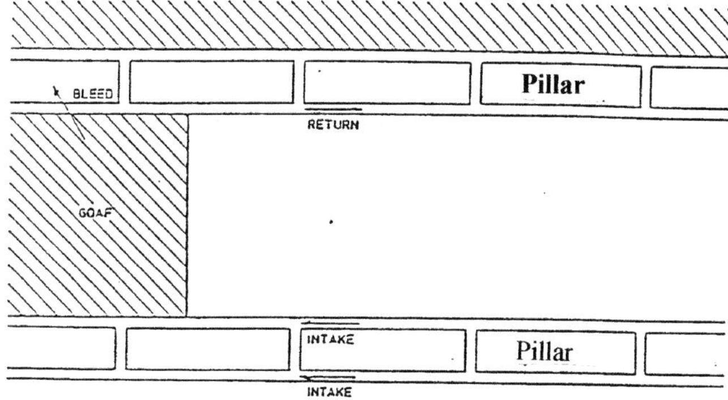

Figure 12.2 Longwall retreat face protected by chain pillars

4. Chain pillars are not amenable to longwall isolation as a number of fire/ventilation stoppings have to be built to seal a face.

12.3.1 Pillar Design Techniques

In the past many methods of sizing mining pillars have been used, including the following.

1. Empirical methods
2. Ultimate strength technique
3. Progressive failure technique
4. Stress balance method or yield zone technique
5. Chain pillar design

12.4 Ultimate Strength Approach

This method of pillar design is based on the subsidence theory requiring the following discrete steps (Singh and Whittaker, 1979):

1. Estimation of the ultimate strength of a coal pillar
   A most commonly used pillar strength formula (Salamon and Munro, 1967) is as follows:

$$\sigma_P = K\frac{W_p^a}{m^b}$$

$K = 7180$

σ_P = Pillar strength (MPa)

m = Extracted seam height (m)

w_p = Pillar width (m)

$a = 0.46$

$b = 0.66$

2. Decision on the factor of safety (F)
3. Calculation of the load on the pillar
4. Calculation of the factor of safety, $F = \dfrac{\text{Pillar strength}}{\text{Average pillar stress}}$ which should be between 1.3 and 1.9.

12.4.1 Pillar Loading for Sub-critical Area of Extraction

Figure 12.3 shows the loading of longwall pillars based on the subsidence theory for sub-critical area of extraction when the subsidence due to longwall extraction does not fully reach the surface.

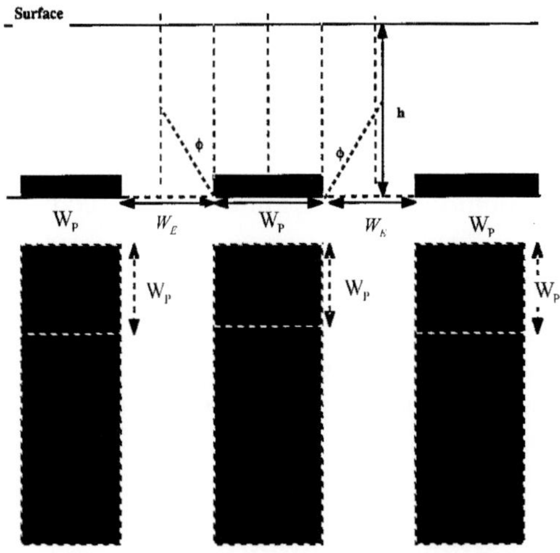

Figure 12.3 Load on longwall pillars (Whittaker and Singh, 1979)

The total volume of overburden strata resting on the longwall pillars can be determined by simple geometry as follows:

$$\text{Total volume resting on pillar} = \left[\left\{(W_P + W_E)h - \frac{W_E^2 \cot\phi}{4}\right\}W_P\right]$$

$$\text{Pillar loading} = \frac{9.81\gamma}{1000}\left[\left\{(W_P + W_E)h - \frac{W_E^2 \cot\phi}{4}\right\}W_P\right]$$

$$\text{Average stress} = \frac{1}{W_P^2}\,\frac{9.81\gamma}{1000}\left[\left\{(W_P + W_E)h - \frac{W_E^2 \cot\phi}{4}\right\}W_P\right]$$

where,

h = depth below surface (m)
W_P = width of pillar (m)
W_E = width of extraction (m)
ϕ = internal angle of friction
γ = density of overburden rock
m = thickness of extraction (m)

12.4.2 Critical and Supercritical Area of Extraction

Figure 12.4 shows the load on the longwall rib pillar for supercritical area of extraction when a finite area at the centre of the extraction panel undergoes full subsidence.

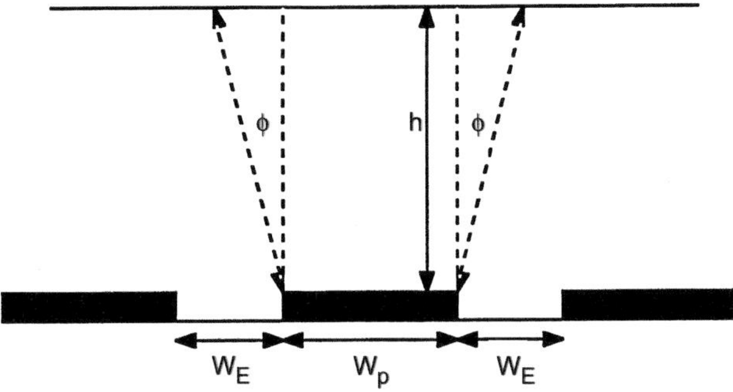

Figure 12.4 Load on pillar for supercritical condition (Whittaker and Singh, 1979)

The volume of the overlying rock loading the longwall rib pillar is given by the following equations:

$$\text{Total volume resting on the pillar} = (W_P h + h^2 \tan\phi)$$

$$\text{Average pillar stress} = \frac{9.81\,\gamma}{1000\,W_P^2}\{W_P(W_P h + h^2 \tan\phi)\}$$

$$\frac{7180\,\sigma_P^{0.46}}{m^{0.66}} = F\frac{9.81\,\gamma}{1000\,W_P^2}\{W_P(W_P h + h^2 \tan\phi)\}$$

$$\frac{7180\,\sigma_P^{0.46}}{m^{0.66}} - F\frac{9.81\,\gamma}{1000\,W_P^2}\{W_P(W_P h + h^2 \tan\phi)\} = 0$$

$$h^2 \tan\phi + W_P h - \frac{1}{W_P}\left[\frac{1000 \times W_P^2 \times 7180 \times W_P^{0.46}}{9.81 \times \gamma F \times m^{0.66}}\right] = 0$$

For $\frac{W}{h} \geq 2 \tan \phi$; $\phi = 31°$ $\gamma = 2350 \text{ kg/m}^3$ $F = 1.3$

Based on the above set of equations a pillar design chart has been constructed for 100 m to 300 m wide longwall faces for a range of conditions as shown in Figure 12.5.

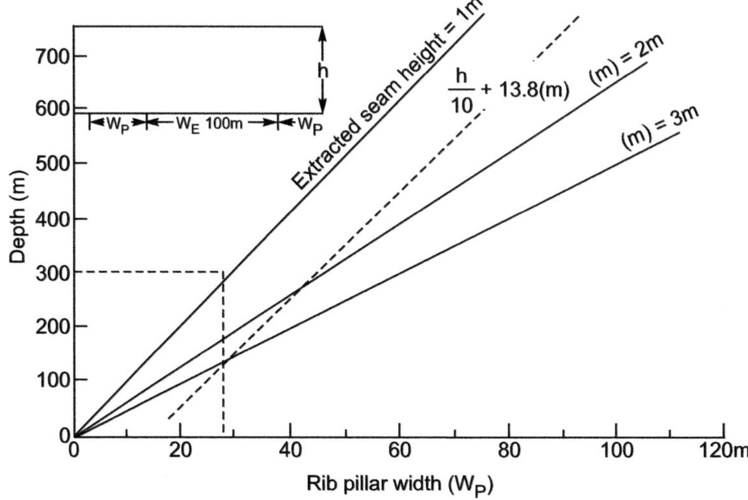

Figure 12.5(a) Rib pillar size design chart for 100 m wide longwall face

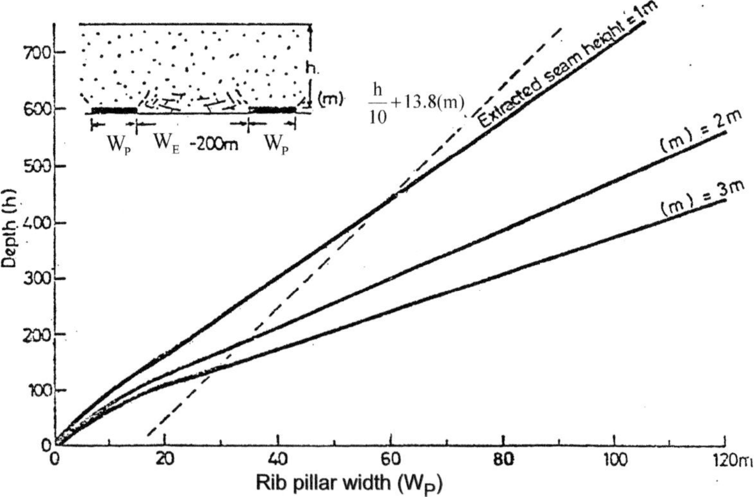

Figure 12.5(b) Rib pillar design chart for 200 m wide longwall faces

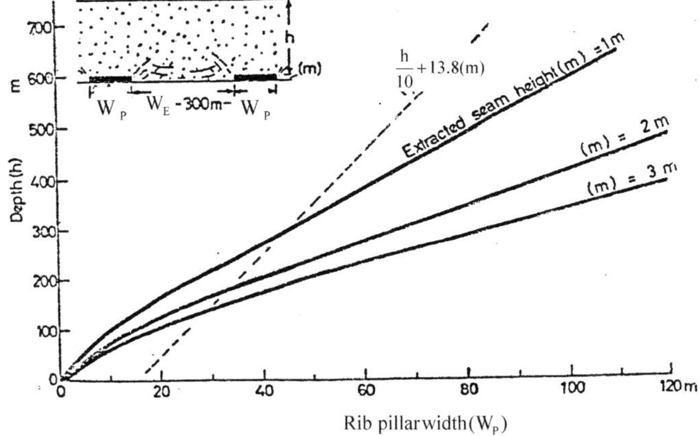

Figure 12.5(c) Rib pillar design chart for 300 m wide single entry longwall face (Whittaker and Singh, 1979)

Examination of Figures 12.5(a, b and c) indicates that keeping the other factors constant:

1. the size of rib pillar increases with increase in depth below the surface,
2. the size of rib pillar increases with increase in extracted seam height and
3. the size of rib pillar increases with increase in width of the face.

It can also be concluded that this method of pillar design underestimates the size of pillar for thin seam conditions for narrow panel and overestimates the size of rib pillar for thicker seam extraction and wider face widths.

12.5 Progressive Failure Theory

This method of pillar design was pioneered by Wilson and Ashwin (1973) at Mining Research and Development Establishment, UK. The stress distribution on a rib pillar surrounded by two extracted longwall faces at both sides of the pillar is given in Figure 12.6.

It can be seen that the stress distribution on the rib pillar is uneven due to the following factors.

1. Rib pillars are flanked by goaf on both sides (Figure 12.6).
2. A yield zone is formed around the pillars with a solid core at the centre.
3. Vertical stress is low at the pillar edge due to confining stress generated by the roadway support and increases rapidly to the peak stress at a distance 3 to 9 m from the rib.

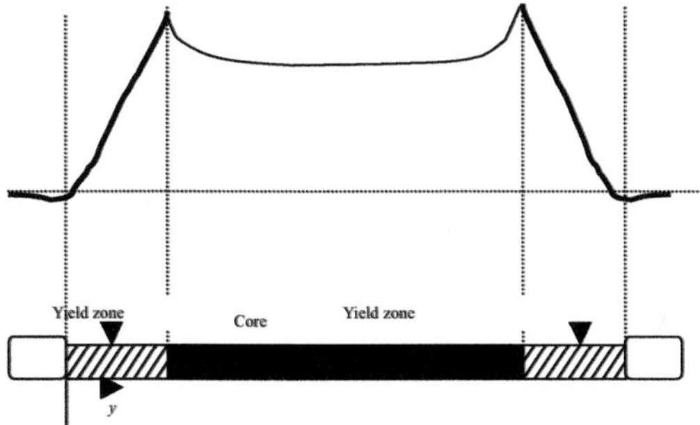

Figure 12.6 Stress distribution on the rib pillar

4. Strength of the core increases due to confining constraint due to the yield zone, given by the following relationships (Figure 12.7):

$$\sigma_1 = \sigma_0 + \sigma_3 \tan\beta \qquad \text{(peak strength)}$$

where

σ_1 = failure stress of coal
σ_0 = uniaxial compressive strength of coal pillar
σ_3 = confining stress offered by the broken rock to the coal core
$\tan\beta$ = triaxial stress factor of coal measures rock = K

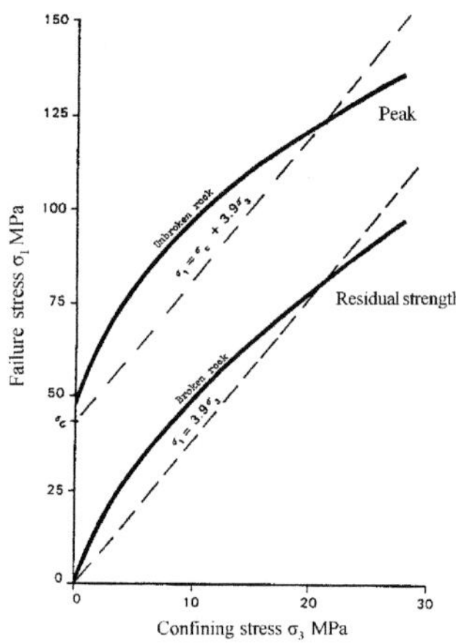

Figure 12.7 Pillar strength based on triaxial test (NCB, 1972)

Also

$$\sigma_1 = p' + K\sigma_3 \qquad \text{(residual strength)}$$

where

p' is taken as 0.1 MPa.

The abutment stress peak is located at the boundary of the yield zone and the elastic boundary of the pillar core. Its magnitude is given by the following equation:

$$\text{Abutment stress} = \hat{\sigma} = K\sigma_v$$

where

$$\sigma_v = \text{vertical stress due to depth} = 0.025 \times \text{depth}$$

Representative values of K are as follows:

Weak mudstone and uncemented rock	K = 2 to 3
The average coal measure rock and coal	K = 3 to 5
Cemented and angular sandstone	K = 4 to 6

Wilson and Ashwin (1973) considered the stress at equilibrium on an element δ_y in width, l in length, and m in thickness at the edge of the pillar as shown in Figure 12.8.

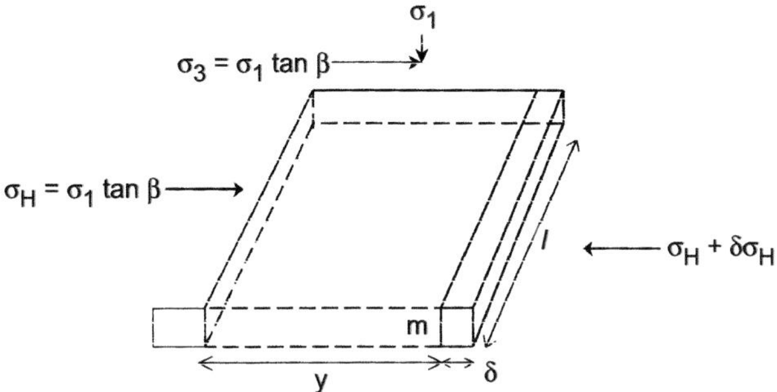

Figure 12.8 Pillar design principle based on the progressive failure concept (Wilson and Ashwin, 1973)

The stabilising force on the element is equal to the destabilising force due to the out of balance horizontal stress as follows:

$$\text{Frictional resistance} = \text{Horizontal force}$$

$$2\,\sigma_V \tan\beta \times 1 \times \delta_y = \delta\sigma_H\, m \times 1$$

Integrating:

$$\frac{y}{m} = \frac{1}{\sqrt{\tan\beta(\tan\beta - 1)}} \ln\frac{\sigma_H}{\sigma_o}$$

Based on the above analysis Wilson and Ashwin (1973) constructed two design charts for the design of pillars under various dimensional conditions (NCB, 1972).

Figure 12.9 shows the safe size of square pillars in room and pillar mining for five sizes of roadways from 3.6 m, 4.5 m, 5.4 m, 6.3 m and 7.2 m wide. It was assumed that the roof and floor comprised moderate strength coal measures rock with a triaxial strength factor of 4 and a factor of safety of 1.

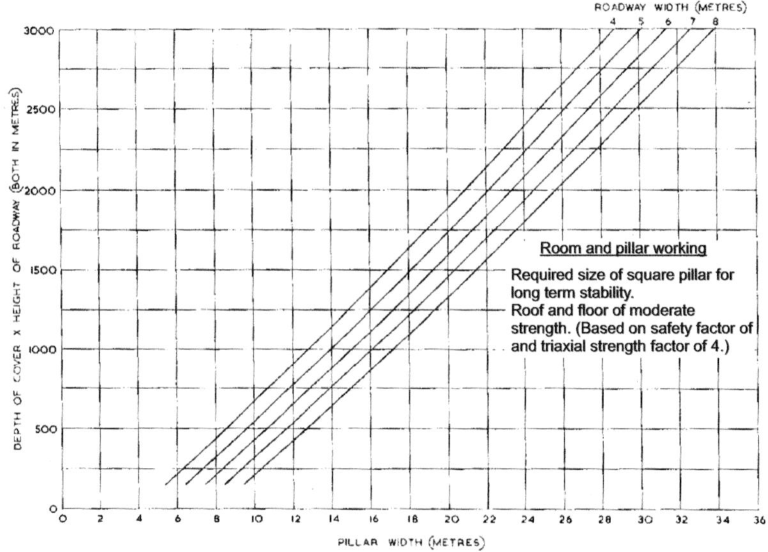

Figure 12.9 Sizing of pillar during development (NCB, 1972)

Figure 12.10 shows the design parameters of a rib pillar bounded by two 3 m wide gate roadways with moderately strong roof and floor strata and a factor of safety of 1. The design chart presents the relationship between the

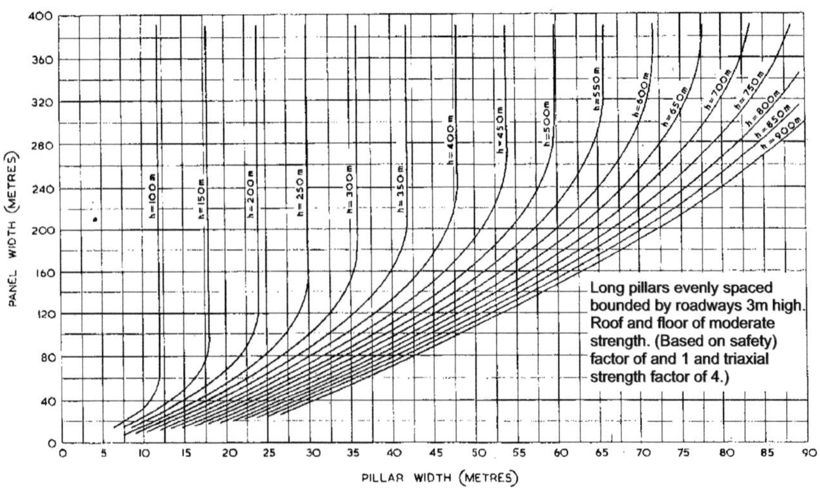

Figure 12.10 Pillar sizing in single entry longwall mining for various panel widths (NCB, 1972)

widths of a panel against the size of the rib pillar for various depths below the surface. It can be shown that if the rib pillar width exceeds 0.12 time the depth below the surface, then an infinitely wide panel can be mined safely on either side of the rib pillar. The effective width of the panel is 0.6 h, where the curves representing the width of pillars for a given depth become vertical with the point of flexure representing the stable pillar size. It may also be noted that the widths of the panels and that of the pillars depend upon the depth of cover and extracted seam height. Figure 12.10 shows that the width of the panels are between 0.18 h to 0.3 h and the size of the rib pillars is between 0.12 h and 0.25 h (NCB, 1972).

12.6 Stress Balance Method of Pillar Design

Developed by Wilson (1981), this method of analysis uses a stress balance method. Table 12.1 summarises the expressions in calculating the stress distribution on rib sides.

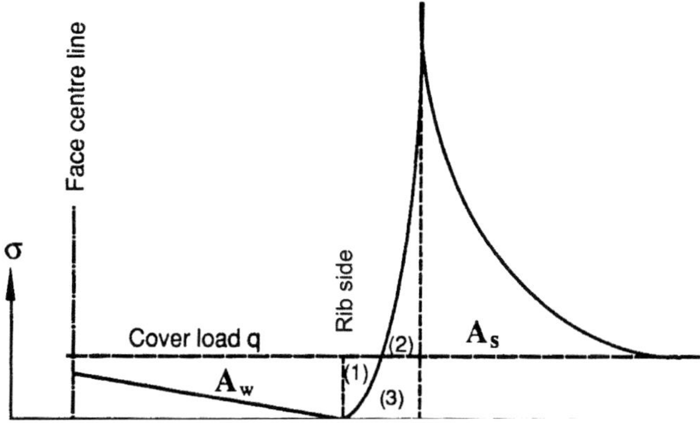

Figure 12.11 Idealised stress balance across rib side (Wilson, 1981)

12.6.1 Assumptions

1. After mining, the total vertical forces are equal to the pre-mining overburden forces.
2. Pre-mining load:
   $q = 0.025 \times h$ (MPa), where h is the depth below the surface.
3. Compared to total vertical force to be redistributed, the load transmitted by the support is minimal and may be ignored.
4. Distribution of stress in caved area is based on field observations. Stress reaches overburden stress q at a distance 0.3 h from the rib.

5. Distribution of stress on the rib-side within the yield zone is calculated by elasto-plastic analysis.
6. Resulting non-linear stress distribution can be approximated by a triangular distribution.

Table 12.1 Expression used in calculating the stress distribution on rib sides (Wilson, 1982)

Condition	*Weak seam and strong floor*	*Strength of roof, floor and seam*
Vertical Stress	$\sigma_v = K(p + p') \cdot e^{xF/m}$	$\sigma_V = K(p + p') \cdot \left(\frac{2x}{m} + 1\right)^{k-1}$
Peak abutment stress or yield stress	$\hat{\sigma} = \sigma_o + K \cdot q$	$\hat{\sigma} = \sigma_o + K \cdot q$
Width of yield zone	$X_b = \frac{m}{F} \ln \frac{q}{(p+p')}$	$X_b = \frac{m}{2}\left[\left(\frac{q^{\frac{1}{K-1}}}{(p+p')}\right) - 1\right]$
Vertical force carried on the yield	$A_b = \frac{m}{F} K(q-p'-p)$	$A_b = \frac{m}{2}(p+p')\left\{\left(\frac{q}{(p+p')}\right)^{k/k-1} - 1\right\}$

where $\quad p' = 0.1$

$$F = \frac{K-1}{\sqrt{K}}\left(1 + \frac{K-1}{\sqrt{K}} \tan^{-1}\sqrt{K}\right) = \text{a variable}$$

where, $\tan^{-1}\sqrt{K}$ is expressed in radians

$$\sigma_v - q = (\hat{\sigma} - q)\exp\frac{(X_b - x)}{C}$$

σ_v = post-mining vertical stress (MPa)
$\hat{\sigma}$ = peak abutment stress
q = pre-mining vertical stress (σ_h (MPa))
σ_o = uniaxial compressive strength
K = triaxial stress factor
m = height of extraction (m)
X_b = width of yield zone (m)
p' = unconfined compressive strength of broken rock (0.1 MPa)
p = support pressure (MPa)
C = distance of the abutment load decay on the pillar
F = a variable depending upon the triaxial stress factor

By equating the area under the curve:

$$A_w + (1) = A_s + (2)$$
$$A_w + (1) + (3) = A_s + (2) + (3)$$
$$A_w + X_b \times q = A_s + Ab$$
$$C = \frac{A_W + q \times X_B - A_b}{(\hat{\sigma} - q)}$$

A_w = load deficiency in goaf

For W > 0.6 h

$$A_w = 0.15\,\gamma h^2$$

For W< 0.6 h

$$A_w = \frac{1}{2} W\gamma\left(h - \frac{W}{1.2}\right)$$

where W = width of extracted area (m)

$$W_p = 2\,(X_b + C)$$

12.6.2 Worked Example

Stress distribution about rib side (Fig. 12.12)

This method of pillar design has been illustrated by Wilson (1981) from the actual field data for a typical coal mining condition in the Midlands Coalfield in UK.

It is given that:

$$h = 500\text{ m}, K = 3.7, \sigma_o = 4\text{ MPa}, p = 0.1\text{ MPa}, p' = 0.1\text{ MPa}$$
$$q = \gamma h = 0.025 \times 500 = 12.5\text{ MPa}$$

The abutment stress can be calculated as follows:

$$\hat{\sigma} = \sigma_o + K \times q = 4 + 3.7 \times 12.5 = 50.25\text{ MPa}$$

$$F = \frac{K-1}{\sqrt{K}}\left(1 + \frac{K-1}{\sqrt{K}} \tan^{-1}\sqrt{K}\right) = \frac{3.7-1}{\sqrt{3.7}}\left(1 + \frac{3.7-1}{\sqrt{3.7}} \tan^{-1}\sqrt{3.7}\right) = 3.55$$

Width of the yield zone is given by:

$$X_b = \frac{m}{F} \ln \frac{q}{(p+p')} = \frac{3}{3.5} \ln \frac{12.5}{(0.2)} = 3.5\text{ m}$$

$$A_w = 0.15\,\gamma h^2 = 0.15 \times 0.025 \times 500 \times 500 = 938\text{ MPa/m}$$

Distance of load decay is given as follows:

$$C = \frac{A_W + q \times X_B - A_b}{(\hat{\sigma} - q)};\; C = \frac{(938 + 12.3 \times 3.5 - 38.46)}{(50.25 - 12.5)} = 24.98$$

$$W_p = 2 \times X_b + 2C = 2 \times 3.5 + 2 \times 25 = 57\text{ m}$$

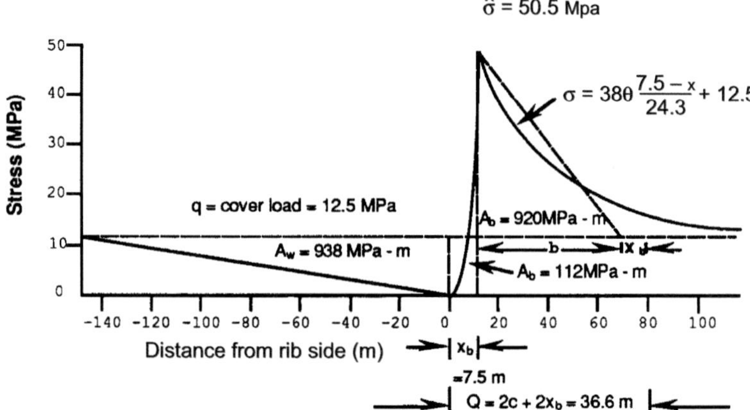

Figure 12.12 Stress distribution about rib side in a worked example (Wilson, 1983)

Empirical method

$$W_p = 0.1 \times \text{depth} + 13.8\ \text{m} = 63.8\ \text{m}$$

12.7 Chain Pillar Design in Longwall Mining

In many countries, it is a statutory obligation to use multi-entry roadways as a means of ingress and egress to a longwall face for safety reasons. Figure 12.2 shows a two-entry longwall face where as Figure 12.13 presents a three-entry longwall face. In both the above layouts chain pillars are provided to protect the gate entries from the caving rock mass in the adjoining goaf. Figure 12.13 shows three longwall faces. The first longwall face on the left has been finished, while the second face has retreated half way and the gate entries for the third face are being driven. The right side entries to the working face are the main gates which carry intake air to the face and the tail gate entries are return airways from the face. On the completion of the second longwall face, the newly developed entries for the new face will act as main gates while the present main entries will act as the return roadways for the new face. The sizing of the chain pillar is important from a ground control point of view and also for achieving fast development rates. The chain pillars in longwall mining undergo different loading during the progression of mining. Moreover, various pillars are subjected to different loading regimes as follows.

1. Barrier pillars for the protection of bleeder roadways and main entries.
   - The barrier pillars are subjected to development loads during development of the main entries and the face.

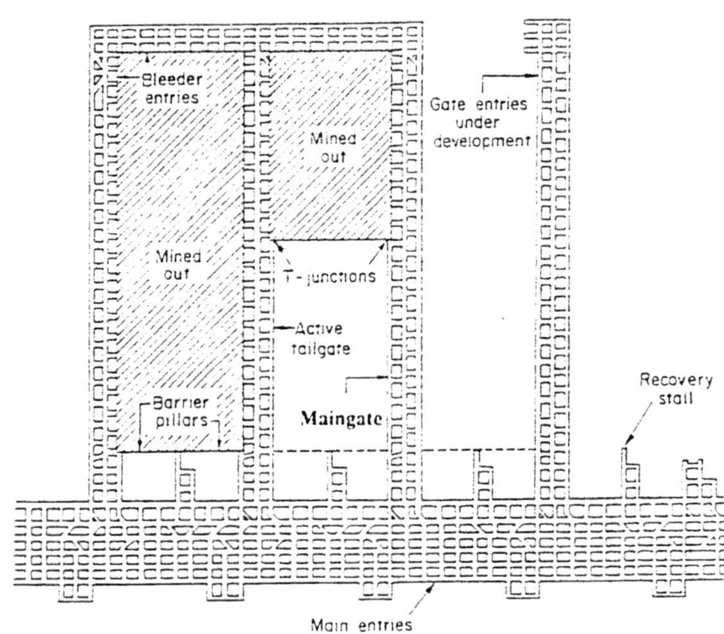

Figure 12.13 Layout of longwall retreat face (Mark, 1990)

- During the working life of the longwall face these pillars are loaded with front and back abutment loading during the start of the face. The barrier pillar is also subjected to the front abutment pressure during the finishing of the longwall retreat face. Thus, these pillars are subjected to increased abutment loading only once during the life of the face.

2. Chain pillars are subjected to side abutment loading

These pillars are loaded four times, first as a development load, second as front abutment followed by side abutment stress as a main gate of the former face and then as side abutment pressure of the second face as tail gate chain pillars. Therefore, the design of chain pillars involves loading at several stages of development and should be able to bear the maximum load.

There are two different basic philosophies which have been used for the design of chain pillars.

1. Conventional design approach involving the use of large pillars to carry the side abutment load.
2. Yielding pillar technique involving the use of small pillars which are located in relatively de-stressed ground to transfer the load onto the large abutment pillars and gate entries.

12.7.1 Conventional Design Approach

This design approach is based on calculation of the maximum abutment load on the chain pillar systems as consequence of longwall mining and calculates the strength of the pillar material and the factor of safety against failure. If the calculated factor of safety is not within the accepted range the size of the pillar is increased until the desirable factor of safety is achieved. This method is known as the Analysis of Longwall Pillar Stability or ALPS method (Mark, 1990).

The method consists of the following discrete steps.

1. Calculate the load on the pillars by using the following four steps.
   (a) Calculation of development load
   (b) Calculation of side abutment load
   (c) Calculation of front abutment load based on field work to determine β, F_m and F_T.
   (d) Calculation of load distribution on the pillar system
2. Calculate longwall pillar strength depending upon the bearing capacity of the longwall pillar system.
3. Calculate the stability factor.

Figure 12.14 shows a layout of a four-entry longwall retreat face together with the loading of barrier pillars. The loads applied on the longwall pillars may be calculated as the development loads and the abutment loads.

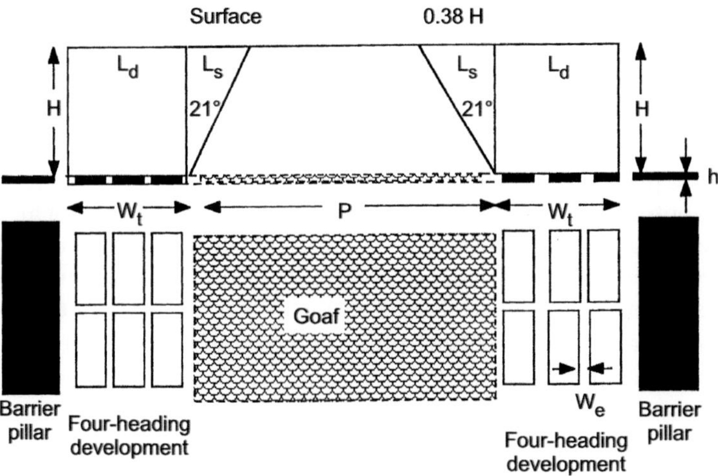

Figure 12.14 Chain pillar layout for four-entry longwall retreat face

(a) Calculation of development loads

The development load L_d, which is present before the commencement of longwall extraction, can be calculated as follows:

$$L_d = \gamma H W_t$$

$$W_t = \sum W + (n+1) W_e$$

where

H = depth of cover (m)
W_t = width of the pillar system (m)
W_e = width of the entry (m)
γ = unit weight of overburden
n = number of entries in gate entry system

(b) Calculation of abutment load

As described in Chapter 8, the abutment loads on the longwall face may be divided into two parts:

- Side abutment load
- Front abutment load at the face and T-junctions

(i) Calculation of side abutment In the past two concepts have been developed for the calculation of the side abutment load based on subsidence theory.

- Wilson (1973) concept
- King and Whittaker (1971) concept

Figure 12.15 shows Wilson's concept which assumes that the vertical stress on the pillar increases over the goaf linearly from zero at the rib to the original overburden pressure at a distance of 0.3 h from the edge of the pillar.

The approach by King and Whittaker (1971) is based on the assumption that a shear angle β of the overburden rock mass determines the loading of the chain pillar (Figure 12.16).

For critical and supercritical conditions (Whittaker and Pye, 1977):

$$P \geq 2 H \tan \beta$$

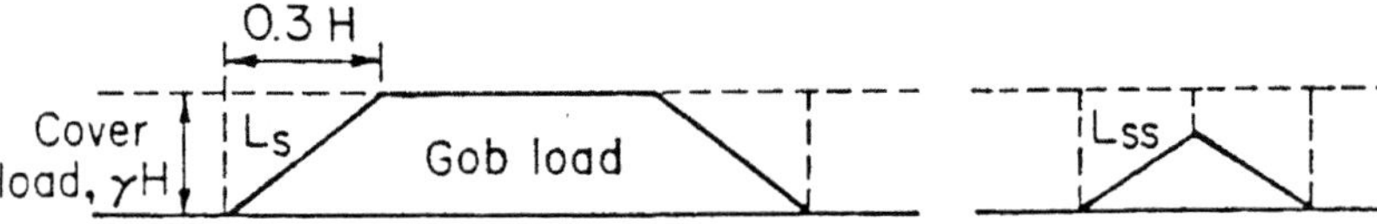

Figure 12.15 Abutment loading (Wilson, 1981)

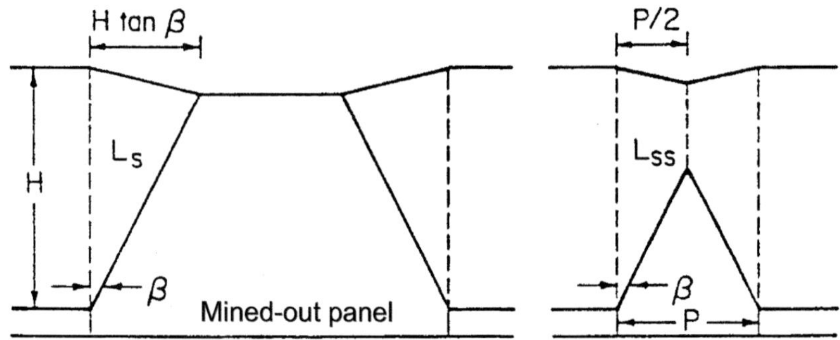

Figure 12.16 Abutment loading based on the subsidence theory by King and Whittaker (1971)

and

$$L_s = \frac{1}{2}\gamma H^2 \tan\beta$$

For sub-critical subsidence:

$$P < 2H\tan\beta$$

$$L_{ss} = \gamma\left[\frac{H \times P}{2} - \frac{P^2}{8\tan\beta}\right]$$

where

L_s = pillar loading for supercritical panel
L_{ss} = pillar loading for sub-critical panel width
H = depth of panel (m)
β = shear angle of the overburden material
γ = rock mass density 2300 kg/m^3
P = panel width (m)

(ii) Calculation of front abutment The front abutment is calculated as a fraction of the side abutment load as follows:

- Front abutment at the main gate *T*-junction

$$L_{fm} = F_m L_s$$

- Front abutment at the tail gate end

$$L_{ft} = F_t L_s$$

F_m and F_t are front abutment factors whose value depends upon local geology, with values less than 1.

(iii) Field values for the design parameters F_m, F_t and β Field work carried out in the United State in five mines at 16 sites (Mark, 1985; Scheurger, 1985;

Allwes 1985; Listak, et al. 1985) has indicated that the mean values for the various design parameters are as follows (Mark, 1990):

Abutment angle $\beta = 21°$

Main gate front abutment factor = $F_m = 0.5$

Tail gate front abutment factor = $F_t = 0.7$

(iv) Calculations for distribution of the abutment load on the pillar system

Distance of zone of influence

When a longwall face with a multi-entry gate roadway system is mined, the side abutment load is transferred onto the chain pillars and adjoining barrier pillar.

Peng and Chiang (1984) defined the width of the abutment influence zone D from the panel edge (Figure 12.17) as follows:

$$D = 5.09\, H^{0.5}$$

where D = distance from the panel edge

H = depth below surface (m)

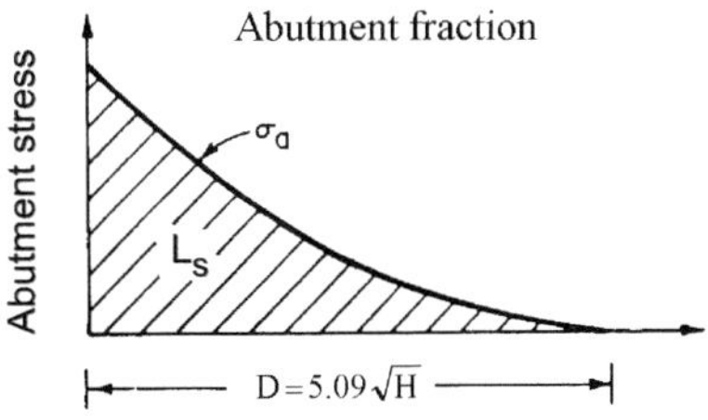

Figure 12.17 Abutment fraction (Peng and Chiang, 1984)

Distribution of the abutment pillar load

The stress distribution within the stress influence zone can be calculated in terms of the inverse square of the distance from the panel edge, Airey (1977). After calculating D, the abutment load width and knowing the width of the chain pillars A and B, the side abutment shared by the individual pillars and the barrier pillar BP can be calculated as shown in Figure 12.18. Assuming an inverse square stress decay function, the abutment stress distribution is given by the following equation (Airey, 1977):

$$\sigma_a = \left(\frac{3L_s}{D^3}\right)(D - x)^2$$

where

D = extent of the side abutment influence zone

W_t = width of chain pillar system (m)

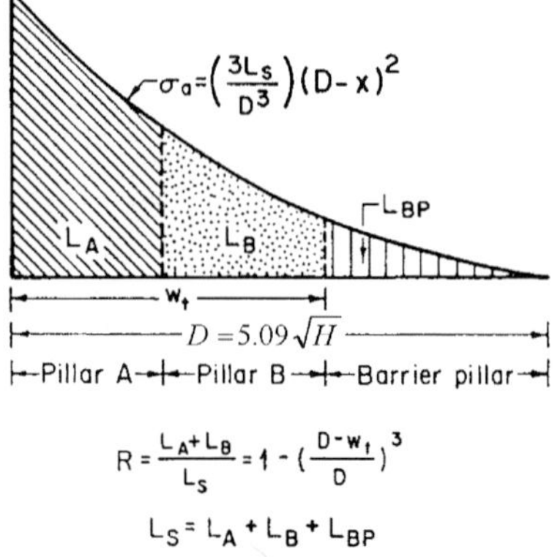

Figure 12.18 Proportional load on the pillar (Mark, 1985)

x = distance from the edge of the longwall panel (m)
σ_a = abutment stress distribution function
H = depth below surface
L_s = total side abutment load
L_A = side abutment load on pillar A
L_B = side abutment load on pillar B
L_{BP} = side abutment load on barrier pillar
R = abutment fraction

It can be shown that:

$$L_S = L_A + L_B + L_{BP}$$

and the fraction R of the total side abutment shared by the chain pillars is given by the following equation (Mark, 1990):

$$R = 1 - \left[\frac{D - W_t}{D}\right]^3$$

where $W_t < D$

It may be noted that where W_t is > D or where no adjacent panel is mined, then R = 1.

Figure 12.19 shows the comparison between the measured average front abutment stresses and calculated abutment pillar stresses using Airey's decay function (Mark, 1990).

Estimation of pillar strength

Estimation of the load bearing capacity of a longwall chain pillar system involves:

- determination of the individual pillar strength
- determination of load bearing capacity of the chain pillar system.

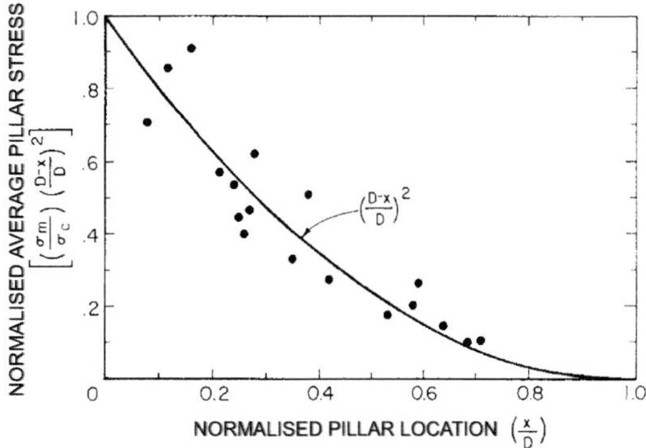

Figure 12.19 Measured and calculated front abutment stress (Mark, 1985)

Figure 12.20 shows the effect of size on the uniaxial compressive strength of coal indicating that the size effect reaches an asymptotic value at a specimen width of 1.5 m.

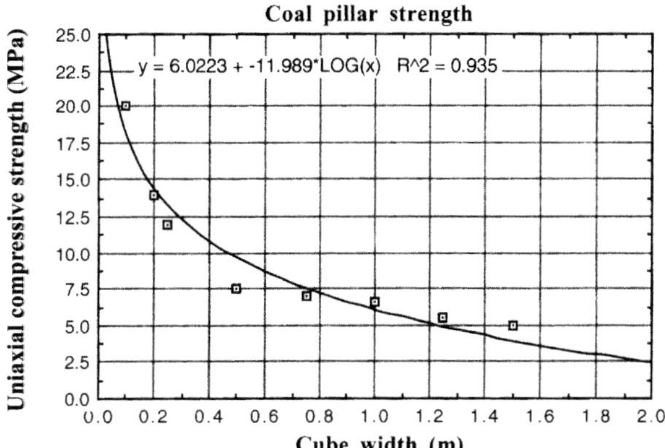

Figure 12.20 Effect of specimen size on compressive strength of coal

In the past, the following empirical relationships have been used to estimate the strength of coal pillars (Figure 12.21).

$$\sigma_p = S_1\left(0.78 + 0.22\,\frac{W}{h}\right) \qquad \text{(Holland-Gaddy formula)}$$

$$\sigma_p = S_1\left(0.64 + 0.22\,\frac{W}{h}\right) \qquad \text{(Modified Bieniawski formula)}$$

$$\sigma_p = 7.180\left(\frac{W^{0.46}}{h^{0.66}}\right) \qquad \text{(Salamon and Munro, 1967)}$$

where

S_i = *in situ* strength of coal (MPa)
W = width of the pillar (m)
h = height of the pillar (m)
σ_p = pillar strength (MPa)

Average value of *in situ* coal strength is taken as 6.21 Mpa.

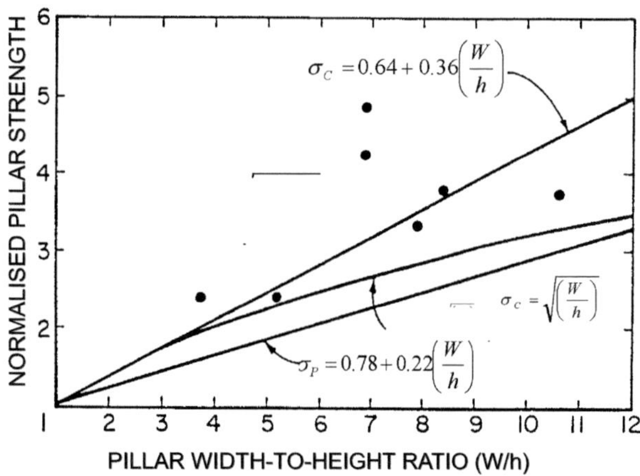

Figure 12.21 Effect of specimen size on compressive strength

Once the strength of the individual pillar has been estimated, it can be shown that the load bearing capacity of individual pillars is given by the following equation:

$$B_p = \frac{\sigma_p W l_p}{(l_p + W_e)}$$

The load bearing capacity of the pillar system is given by:

$$B = \sum B_p$$

Calculation of the stability factor

The stability factor for the pillar systems should be between 1.0 and 1.6 and this can be calculated as follows:

$$SF = \frac{B}{L}$$

where SF = stability factor (>1.0)
B = load bearing capacity of pillar system
L = designed pillar loading

In the above calculation one of the three pillar loading conditions may be used as follows:

(i) Main gate loading consisting of development loading together with the first front abutment loading is given by:

$$L_m = [L_d + (L_s \times F_m \times R)]$$

(ii) Bleeder pillars loading, comprising development load plus the first full side abutment load

$$L_B = [L_d + (Ls \times R)]$$

(iii) Most severe loading in tail gate loading comprising development load and two side abutments loads as follows

$$L_T = [L_d + (1 + F_t)\, L_S]$$

The value of the stability factor should be between 1.0 to 1.3 (Mark, 1990). A study has been carried in Australian coal mines to validate ALPS method for Australian conditions (Colwell, *et al.*, 1998)

12.7 Worked Example

A mine is developing its first two longwall panels in a block 600 m × 1200 m coal reserve. The seam is 1.8 m thick and the average depth of cover is 300 m. The face length is selected as 240 m because of the availability of current longwall equipment. The longwall face is served by a three entry gate system; entries being 5.4 m wide.

Because the ventilation requirement during the development phase, the length of the pillars has been selected as 33 m.

The proposed layout of the mining block is shown in Figure 12.22. Calculate the width of chain pillar if equal sized pillars are to be used (Mark, 1990).

Solution Each roadway will be subjected to different service and hence to different service loading. In order to minimise sterilising some of the coal reserves, the pillar will be sized for each gate system.

Input data

$$H = 300 \text{ m}$$
$$P = 240 \text{ m}$$
$$\gamma = 2300 \text{ kN/m}^3$$
$$W_e = 5.4 \text{ m}$$
$$l_p = 33 \text{ m}$$
$$h = 1.8 \text{ m}$$

1. Width of pillar system

$$W_t = 2W + 2\, W_e$$

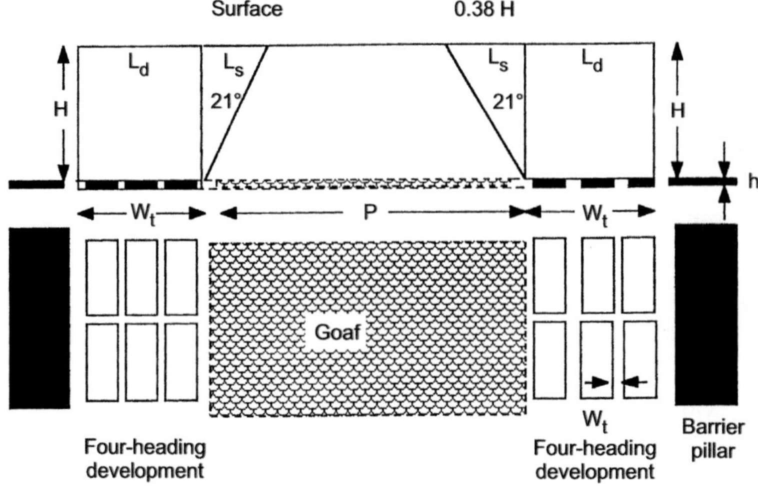

Figure 12.22 Longwall chain pillar layout

First estimate W is 21.5 m

$$W_t = 2W + 2W_e$$
$$= 2 \times 21.5 + 2 \times 5.4 = 54 \text{ m}$$
$$L_d = 2300 \times 981 \text{ H } W_t/1000000$$
$$L_d = 0.023 \text{ H } W_t = 0.023 \times 300 \times 54$$
$$= 372.60 \text{ MN/m}$$

2. $P_{crit} = 0.77 \text{ H} = 0.77 \times 300 = 231$ m

   As the face length is 240 m which is > P_{crit}, supercritical conditions exist.

3. Therefore, L_s is given by:

$$L_s = H^2 \tan\beta \times \frac{\gamma}{2}$$
$$= 300 \times 300 \times \tan 21 \times \frac{2300}{2}$$
$$= 397.3 \text{ MN/m}$$

4. Side abutment influence zone $= 5.09\,(H)^{0.5}$
$$= 5.09 \times (300)^{0.5}$$
$$= 88.16 \text{ m}$$

5. Abutment fraction $R = 1 - \left[\frac{D - W_t}{D}\right]^3$

$$= 1 - \left[\frac{(88.16 - 54)}{88.16}\right]^3$$

$$= 1 - 0.058 = 0.94$$

6. In order to estimate maximum pillar load

$$L_B = L_d + (R \times L_s)$$
$$= 372.6 + 0.94 \times 397.30 = 746.78 \text{ MN/m}$$
$$L_m = L_d + (R \times L_s F_m)$$
$$= 372.6 + 0.94 \times 397.3 \times 0.5$$
$$= 559.33 \text{ MN/m}$$
$$L_t = L_d + ((1 + F_t) \times L_s)$$
$$= 372.6 + (1 + 0.7)\,397.30 = 372.6 + 675.10 = 1047.7 \text{ MN/m}$$

7. *In situ* strength of pillar using Bienwiaski formula:

$$\sigma_p = 6.21\left[0.64 + 0.36 \times \frac{W}{h}\right] = .21\left[0.64 + 0.36 \times \frac{21.5}{1.8}\right] = 30.80 \text{ Mpa}$$

8. $B_p = \sigma_p \dfrac{W \times l_p}{(l_p + W_e)} = 30.80 \dfrac{21.5 \times 33}{(33 + 5.4)} = 571.2 \text{ MN/m}$

$$B = 2 \times B_p = 2 \times 571.2 = 1142.4 \text{ MN/m}$$

9. Factor of stability

$FS = \dfrac{B}{L_t} = \dfrac{1142.2}{1047.7} = 1.09$ (Acceptable which is within the range of 1.0 to 1.3)

12.8 Conclusions

Empirical methods of pillar design were based in the past on practical experience which was also included in the mining regulations of many countries. As the empirical methods were based on the experience of shallow mine workings, these methods are not relevant for the pillar design in deep mine workings where they tend to result in overdesign. This will result in sterilisation of large coal reserves.

This chapter has presented different approaches in the design of rib pillars based on subsidence theory, the yield zone theory or a combination of the two methods.

An empirical approach for the design of chain pillars based on the subsidence theory and practical observations in 16 longwall faces is also given. This approach has been validated in some Australian coal mines with limited success (Colwell *et al.*, 1998).

References

Airey, E.M. (1977). 'An analysis of stress around mine roadways: Stress in Failed Rock', *Appendix to a study of yield zone around roadways*, Publication of European Communities, Luxembourg, Rep. EUR 5815e, pp. 44–88.

Allwes, R.A., J.M. Listak, G.J. Chekan and D.R. Babich (1985). *The effect of retreating longwall on a three entry gate road system*, BuMines RI, 8996, pp. 19.

Colwell, M., R. Frith and C. Mark (1998). *Chain Pillar Design (Calibration of ALPS)*, ACARP C6036, Final Report, October, pp. 67.

Hustrulid, W.A. (1976). 'A review of Coal Pillar Strength Formulae', *Rock Mechanics*, Vol. 8, No. 2, pp. 114–145.

Husing, S.M. and S.S. Peng (1985). 'Chain pillar design for US longwall panels', *Mining Science and Technology*, Vol. 2, pp. 279–305.

Listak, J.M., J.L. Hill III and J.C. Zelanko (1986). *Direct measurement of longwall strata behaviour*, A case study, Bureau of Mines RI 9040, pp. 19.

King, H.J. and B.N. Whittaker (1971). 'A review of current knowledge of roadway behaviour', *Proceedings of the Symposium of Roadway Strata Control*, Institution of Mining and Metallurgy, London, pp. 73–87.

Madden, B.J. (1988). Squat pillar design, *Coal International*, January, pp. 6–9.

Mark, C. (1990). *Pillar design method for longwall mining*, United States Department of Interior, Bureau of Mines Information Circular, IC 9247, pp. 51.

Mark, C. (1985). 'Longwall pillar design—some recent developments', *Society of Mining Engineers*, AIME preprint, No. 89–103, 1989, pp. 9.

Mark, C. and Z.T. Bienwiaski (1986). 'An empirical method for the design of chain pillars in longwall mining', *Proceedings of 27th Rock Mechanics Symposium*, Tusacaloosa, Alabama, Society of Mining Engineers, AIME, pp. 415–422.

Mark, C., F.E. Chase and A.A. Campoli (1995). 'Analysis of retreat mining pillar stability', *Proceedings of the 14th Conference on Ground Control in Mining*, West Virginia University, 1, August pp. 63–71.

NCB (1972). *Design of mine layouts with reference to geological and geometrical factors*, National Coal Board Mining Department, Working Party Report, pp. 19–21.

Peng S.S. and H.S. Chiang (1984). *Longwall Mining*, Wiley, pp. 708.

Peng, S.S. (1986). *Coal Mine Ground Control*, Wiley, 2nd edition, pp. 243–245.

Salamon, M.D.G. and A.H. Munro (1967). 'A study of the strength of coal pillars', *Journal of the South African Institute of Mining and Metallurgy*, Vol. 68, No. 2, pp. 55–67.

Scheurger, M.G. (1985). 'An investigation of longwall pillar stress history', *4th Conference in Ground Control in Mining*, WV University, Morgantown, WV, pp. 41–50.

Whittaker, B.N. and J.H. Pye (1977). 'Design and layout aspects of longwall method of coal mining', *Proceedings, Design Methods in Rock Mechanics, 16th Symposium in Rock Mechanics*, Minneapolis, MN, 1975, American Society of Civil Engineers, pp. 303–314.

Whittaker, B.N., and R.N. Singh (1979). Design and stability of pillars in longwall mining, *Mining Engineer*, July, pp. 59–73.

Whittaker, B.N. and R.N. Singh (1981). 'Design and Stability of Pillars in Longwall Mining', *International Journal of Rock Mechanics and Mining Science*, Vol. 18, No. 4, pp. 331–335.

Wilson, A.H. (1983). 'The Stability of Underground Workings in Soft Rocks of the Coal Measures', *International Journal of Mining Engineering*, Vol. 1, pp. 91–187.

Wilson, A.H. and D.P. Ashwin (1973). 'A Hypothesis Concerning Pillar Stability', *Mining Engineer*, London, pp. 409–417.

Wilson, A.H. (1981). 'Stress and Stability in Coal Rib-sides and Pillars', *Int Proc. I[st] Annual Conference on Ground Control in Mining*, West Virginia University, pp. 1–12.

Wilson, A.H. (1982). 'Pillar stability in longwall mining', *State of the art of ground control in longwall mining and mining subsidence*, Editors Y.P. Chugh and M. Karmis, Ch. 5, pp. 85–95.

Structural Stability of Excavations in Jointed Rock

13
Chapter

13.1 Introduction

It is now common practice to use analytical methods to carry out the design of excavations in continuous and massive rock deposits. However, when the rock material is intercepted by naturally occurring geological features such as joints, bedding planes, folds, faults, dykes or shear zones, it becomes necessary to resort to an alternative design technique based on previous experience. In order to quantify this experience, a similar rock mass is considered subject to identical loading conditions which would have similar support requirements. This approach provides an empirical method for the design of structures in rock masses. This chapter describes various systems of classifying rock masses into different categories and using the resulting data for the design of structures in rock.

13.2 Terminology of Rock Structural Weaknesses

The most common discontinuities found in rock are joints, fractures, partings, faults, bedding planes, folds, dykes, shear zones and so on, as defined here.

Joints Joints are defined as breaks, of geological origin, in the continuity of the body of rock. They have the following characteristics.

- They can occur singly but they usually occur in sets or systems.
- There is little or no displacement.
- Joints are accompanied by alterations or depositions of products of decomposition on the joint surfaces.

Fractures Fractures can be defined as fresh breaks in the continuity of the body of rock. They have the following attributes.

- They have no displacement.

- There is no regular system.
- Fractures may be open or closed but not bonded.
- Fractures are often manmade.

Partings A parting is a thin layer of deposited or altered material separating beds in sedimentary or metamorphic rock. In sedimentary rock, the depositional layer may contain carboniferous material. It may be noted that the parting is nearly always unbonded.

Separation A separation is a relatively fresh break along a bedding plane or beds of sedimentary or metamorphic rock. A separation has the following characteristics.

- Partings often give rise to separation.
- Partings are usually manmade (as a consequence of mining).

Faults Faults are fractures in rock along which some movement has taken place. A weakened material known as fault gouge usually envelopes fault planes. Fault provides discontinuity in the rock mass.

Major joint systems The major joint systems can be traced for significant distance ranging from tens to hundreds of metres and are characterised by planes parallel to each other.

Cross-joints Cross-joints are sets of joints intersecting the major joint sets. They have the following attributes.

- They are usually less pronounced than major joint sets.
- They are likely to be curved and irregularly spaced.

Joint fillings Joints can either be open or filled with minerals such as calcite, dolomite, iron oxide or clay minerals. Open joints can be either dry or may contain water or may be subjected to water pressure.

Types of joints The intersection of joint sets can result in formation of either blocky or columnar structures as shown in Figure 13.1.

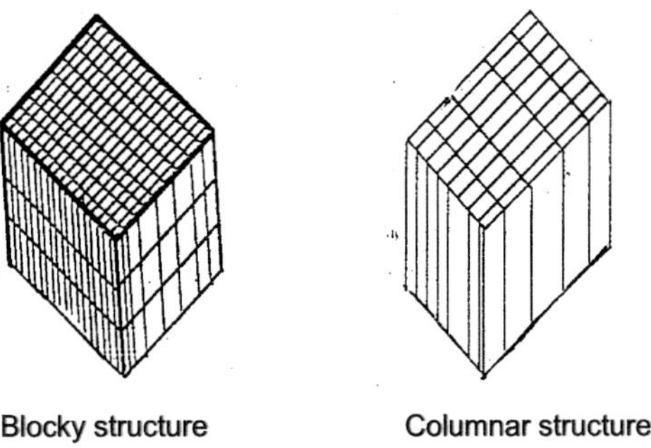

Figure 13.1 Types of joints

Joints and bedding planes can be classified based on their spacing and the thickness of the bedding planes as shown in Table 13.1

Table 13.1 Descriptive terminology of joint spacing and bedding thickness

Descriptive joint spacing	*Descriptive bedding thickness*	*Spacing, m*
Very close	Very thin	<0.05
Close	Thin	0.05–0.3
Moderately close	Medium	0.3–1.0
Wide	Thick	1.0–3.0
Very wide	Very thick	>3.0

13.3 Rock Mass Classification

13.3.1 Aims of Rock Mass Classification

The purpose of a rock mass classification system is to isolate in qualitative manner the categories of rock which under a particular set of engineering constraints will behave in similar ways. The principle on which the classification systems are based is the premise that a number of selected parameters can be determined quickly and easily by the field geologist or engineer to allow him to characterise the behaviour of the rock mass. For a particular rock, a numerical assessment is made within the designated limits of the quality of the rock as it relates to a particular parameter. These numbers are then summed or multiplied together to give an index number which can be related to the behaviour of the rock mass.

Thus, the aims of rock mass classification are as follows.

1. To divide a particular rock mass into groups of similar behaviour
2. To provide a basis for understanding the characteristics of each group
3. To yield quantitative data for engineering design
4. To provide a common basis for communication to aid design

13.3.2 Attributes of Rock Mass Classification Systems

The above-mentioned aims can be fulfilled by ensuring that a rock mass classification system has the following attributes

1. It is simple, easily remembered and understandable.
2. Each term is clear and the terminology used is widely accepted by engineers and geologists.
3. The most significant properties of the rock mass are included.
4. It is based on a rating system that can weigh the relative importance of the classification parameters.

5. It is functional by providing quantitative data for the design of rock support.

A rock mass classification is divided into two groups.

1 Intact rock classification
2 Rock mass classification

13.4 Intact Rock Classification

The major limitation of the intact rock classification is that it does not provide quantitative data for engineering design purposes. However, it provides for better communications relating to the qualitative behaviour of the intact rock.

Table 13.2 Strength classifications for intact rock based on ISRM (1979) classification

Range of value	*U.C.S. (MPa)*
Very low	1–5
Low strength	2–25
Moderate	25–50
Medium	50–100
High	100–250
Very high	250–700

Table 13.3 presents the strength classification of intact rock based on the ultimate uniaxial compressive strength (U.C.S.) of the rock and its qualitative description. The table also summarises various intact rock classifications developed by various professionals working in isolation. Obviously, these intact rock classification systems do not provide a common basis for communication to the professionals in different fields.

13.5 Rock Mass Classification

Many rock mass classification systems are in existence today. Some of these require special attention as they are very well known in the industry, and they include those proposed by Terzaghi (1946), Lauffer (1958), Deere and Miller (1966), Wickham, Tiedeman and Skinner (1974), Bieniawski (1973), Barton, Lien and Lunde (1974), Ghose and Raju (1981) and Molinda and Mark (1994) as shown Table 13.4. Three main systems of rock mass classification popular in the mining industry are presented in this chapter.

Table 13.3 Various strength classifications for intact rock

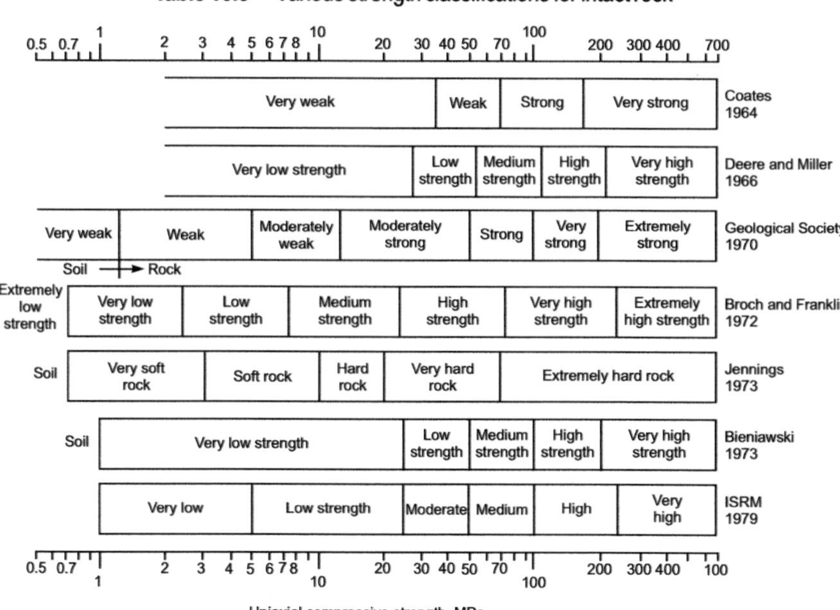

Table 13.4 Major rock mass classification systems

Name of classification	*Originator and date*	*Application*
Rock loads	Terzaghi, 1946	Tunnels with steel supports
Stand-up time	Lauffer, 1958	Tunnelling
Rock quality designation	Deere, 1964	Core logging and tunnelling
Intact rock strength	Deere and Miller, 1966	Communication
RSR concept	Wickham *et al.*, 1974	Tunnelling
Geomechanics classification (RMR system)	Bieniawski, 1973	Tunnels, mines, foundations
Q-system	Barton *et al.*, 1974	Tunnelling, large chambers
Rock mass weakening	Singh, 1986	Mining
Basic geotechnical classification	ISRM, 1981	General
Modified RMR system	Ghose and Raju, 1981	Roof bolting in coal mines
Coal mine roof rating	Molinda and Mark, 1994	Coal mines

Terzaghi's classification was the first practical rock mass classification introduced and it has been proved successful in the USA for the past 57 years for tunnelling with steel support. During construction of a tunnel, some relaxation of the rock mass will occur above and on the sides of the tunnel. The loosened rock will tend to move in towards the tunnel. Friction forces along the lateral boundaries will resist this movement and these friction forces transfer the major portion of the overburden weight on to the material on either side of the tunnel. The roof and sides of the tunnel are required to support the balance which is equivalent to a height of loosened block. Terzaghi's classification provides a practical means of finding the support loading in a variety of rock mass conditions.

13.6 Rock Quality Designation

Deere's classification introduced the rock quality designation (RQD) index, which is a simple and practical method of describing the quality of rock from a borehole core. The qualitative relationship between the RQD indices and the engineering quality of the rock was proposed by Deere (1964). The procedure for the measurement and calculation of RQD is given in Figure 13.2.

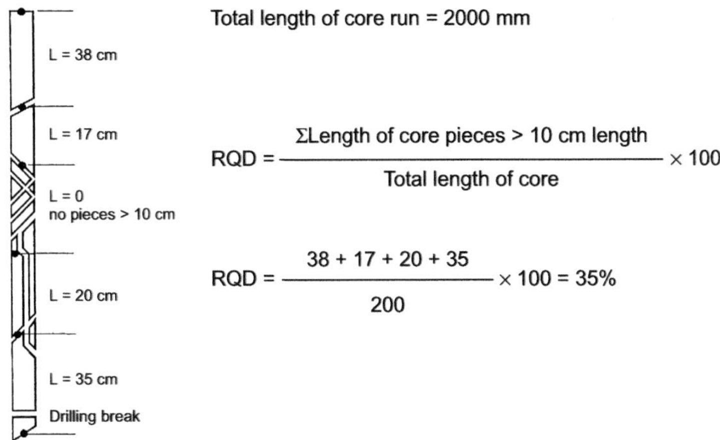

Figure 13.2 Procedure for measurement and calculation of RQD (after Deere, 1964)

RQD has limitations in that it does not provide information regarding joint orientation, tightness, and gouge (infilling material). It is therefore not sufficient in its own right to provide an adequate description of a rock mass. Also, it may be noted that:

$$RQD = 115 - 3.3 * Jv \tag{13.1}$$

where Jv is equal to the number of joint per cubic metre. Table 13.5 gives the RQD and its qualitative rock quality description of the rock.

Table 13.5 Rock quality designation

RQD	Rock quality
<25	Very poor
25–50	Poor
50–75	Fair
75–90	Good
90–100	Excellent

13.7 Geomechanics Classification System

This system was developed by Bieniawski in 1973 and utilises the following parameters, all of which are measurable in the field and can be obtained from borehole data.

- Uniaxial compressive strength (UCS) of intact rock material
- Rock quality designation (RQD)
- Spacing of discontinuities
- Condition of discontinuities
- Groundwater conditions
- Orientation of discontinuities

To apply this system, the rock mass along the tunnel route is divided into a number of representative structural zones having certain uniform geological features within each zone. Once the classification parameters are determined, the importance ratings are assigned to each parameter. In this respect, the typical rather than the worst conditions are evaluated. Futhermore, it should be noted that the importance ratings which are given for discontinuity spacing apply to rock masses having three sets of discontinuities. After the importance ratings of the classification parameters are established, the rating for the other parameters are then summed to yield the basic rock mass rating for the structural region under consideration. The RMR system provides guidelines for the selection of roof support to ensure long-term stability of the various rock mass classes. These guidelines depend on such factors as the depth below the surface, tunnel size, shape and the method of excavation.

This system has been used extensively in mining particularly in the USA and the Republic of South Africa where it is also applicable to rock foundations and slopes. Ghose and Raju (1981) and Venkateswaralu, Ghose and Raju (1989) modified Bieniawski's classification for coal mine roof support design in India which found wide industry acceptance.

Using case histories of roof falls, the roof span and stand-up time have been correlated for various rock mass ratings as shown in Figure 13.3.

Table 13.6 Geomechanical classification of rock mass (Bienwiaski, 1936)

A. CLASSIFICATION PARAMETERS AND THEIR RATINGS

	Parameter		*Range of values*						
	Strength of	Point-load strength index	> 10 MPa	4–10 MPa	2– 4 MPa	1– 2 MPa	For this low range-uniaxial compressive test is preferred		
1	intact rock material	Uniaxial comp. strength	> 250 MPa	100–250 MPa	50–100 MPa	25–100 MPa	5–25 MPa	1–5 MPa	< 1 MPa
	Rating		15	12	7	4	2	1	0
	Drill core quality RQD		90%–100%	75%–90%	50%–75%	25%–50%	< 25%		
2	Rating		20	17	13	8	3		
	Spacing of discontinuities		> 2 m	0.6–2 m	200–600 mm	60–200 mm	< 60 mm		
3	Rating		20	15	10	8	5		
4	Condition of discontinuities (see E)		Very rough surfaces Not continuous No separation Unweathered wall rock	Slightly rough surfaces Separation < 1 mm Slightly weathered walls	Slightly rough surfaces Separation < 1 mm Highly weathered walls	Slicken-sided surfaces or Gouge < 5 mm thick or Separation 1–5 mm Continuous	Soft gouge > 5 mm thick or Separation > 5 mm Continuous		
	Rating		30	25	20	10	0		
		Inflow per 10 m tunnel length (l/m)	None	< 10	10–25	25–125	> 125		

Table 13.6 Contd.

5	Ground water	(Joint water press)/(Major principal σ)	0	< 0.1	0.1–0.2	0.2–0.5	> 0.5
		General conditions	Completely dry	Damp	Wet	Dripping	Flowing
		Rating	15	10	7	4	0
B. RATING ADJUSTMENT FOR DISCONTINUITY ORIENTATIONS (See F)							
Strike and dip orientations			Very favourable	Favourable	Fair	Unfavourable	Very unfavourable
	Ratings	Tunnels and mines	0	–2	–5	–10	–12
		Foundations	0	–2	–7	–15	–25
		Slopes	0	–5	–25	–50	
C. ROCK MASS CLASSES DETERMINED FROM TOTAL RATINGS							
Rating			100 ← 81	80 ← 61	60 ← 41	40 ← 21	< 21
Class number			I	II	III	IV	V
Description			Very good rock	Good rock	Fair rock	Poor rock	Very poor rock
D. MEANING OF ROCK CLASSES							
Class number			I	II	III	IV	V
Average stand-up time			20 yrs for 15 m span	1 year for 10 m span	1 week for 5 m span	10 hrs for 2.5 m span	30 min for 1 m span
Cohesion of rock mass (kPa)			> 400	300–400	200–300	100–200	< 100
Friction angle of rock mass (deg)			> 45	35–45	25–35	15–25	< 15

Table 13.6 Contd.

E. GUIDELINES FOR CLASSIFICATION OF DISCONTINUITY CONDITIONS					
Discontinuity length (persistence)	< 1 m	1–3 m	3–10 m	10–20 m	> 20 m
Rating	6	4	2	1	0
Separation (aperture)	None	< 0.1 mm	0.1–1.0 mm	1–5 mm	> 5 mm
Rating	6	5	4	1	0
Roughness	Very rough	Rough	Slightly rough	Smooth	Slicken-sided
Rating	6	5	3	1	0
Infilling (gouge)	None	Hard filling < 5 mm	Hard filling > 5 mm	Soft filling < 5 mm	Soft filling > 5 mm
Rating	6	4	2	2	0
Weathering	Unweathered	Slightly weathered	Moderately weathered	Highly weathered	Decomposed
Ratings	6	5	3	1	0

F. EFFECT OF DISCONTINUITY STRIKE AND DIP ORIENTATION IN TUNNELLING			
Strike perpendicular to tunnel axis		Strike parallel to tunnel axis	
Drive with dip - Dip 45°– 90°	Drive with dip - Dip 20°– 45°	Dip 45°– 90°	Dip 20°– 45°
Very favourable	Fovourable	Very unfavourable	Fair
Drive against dip - Dip 45° - 90°	Drive against dip - Dip 20°– 45°	Dip 0°– 20° Irrespective of strike	
Fair	Unfavourable	Fair	

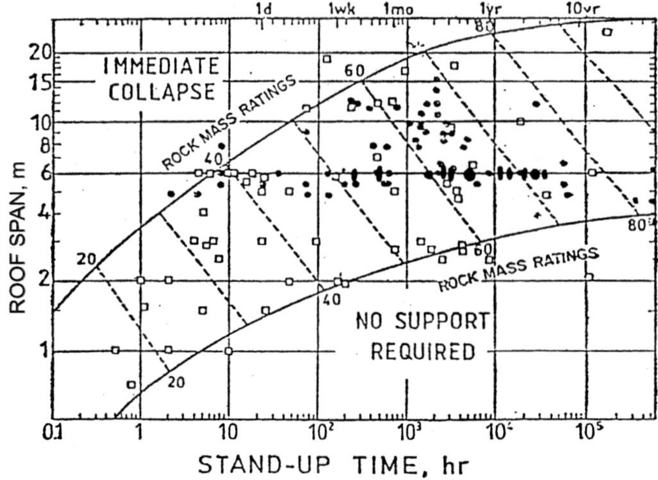

Figure 13.3 Geomechanical classification of rock mass for mining and tunnelling based on case histories of roof falls

The intercept of the RMR line with the desired tunnel span determines the stand-up time. Alternatively, the intercept of the RMR line with the top boundary line determines the maximum span possible in a given rock mass. Any larger span would result in immediate roof collapse. An intercept of the RMR line with the lower boundary line determines the maximum span that can stand unsupported indefinitely.

Example 1 Consider a mine roadway with a 5.4 m span which is to be driven in a gypsum mine. The following classification parameters have been determined, as shown in Table 13.7.

Table 13.7 Rock mass rating in a gypsum mine

	Item	*Value*	*Rating*
1.	Strength of rock material	34 MPa	4
2.	RQD	79%	17
3.	Spacing of discontinuities	0.2–0.6 m	10
4.	Condition of discontinuities; not continuous joint, no separation, unweathered rock wall	A	30
5.	Groundwater	<10 l/min	13
	Basic rock mass value		74
6.	Orientation of joints	Favourable	–2
	Final RMR		72

The rock mass is class II (from Table 13.6) and is classified as good rock. The stand-up time will be about 1 year (from Figure 13.3) and it is recommended to support the roadway using 3 m long rock bolts.

Example 2 The rock mass classification of the shale formation in an open pit iron ore mine in which a slope is to be excavated was obtained as follows (Gray, 1988).

Table 13.8 Rock mass classification for iron ore deposit

	Item	*Value*	*Rating*
1.	Strength of rock material	50 MPa	4
2.	RQD	40%	5
3.	Spacing of discontinuities	0.05–0.2 m	6
4.	Condition of discontinuities;		
	polished and moderately weathered	A	15
5.	Groundwater	Damp	10
	Basic rock mass value		40
6.	Orientation of joints	Favourable	–2
	Final RMR		38

The RMR is 38 which indicates that the rock mass is of poor quality.

13.8 Q-System

The **Q-system** of rock mass classification was developed in Norway by Barton, Lien, and Lunde (1974). Its development represented a major contribution to the subject of rock mass classification for a number of reasons. The system was produced on the basis of an analysis of some 200 tunnel case histories. It is a quantitative classification system and it is also an engineering system enabling the design of tunnel support.

The **Q-system** is based on a numerical assessment of the rock mass quality using six different parameters.

(1) RQD
(2) Number of joint sets
(3) Stress condition
(4) Roughness of the most unfavourable joint or discontinuity
(5) Degree of alteration
(6) Water inflow

The following six parameters are grouped into three quotients to give the overall rock mass quality Q as follows:

$$Q = \frac{RQD}{J_n} \times \frac{J_r}{J_a} \times \frac{J_w}{SRF} \tag{13.2}$$

where

RQD = rock quality designation
J_n = joint set number
J_r = joint roughness number
J_a = joint alteration number
J_w = joint water reduction number
SRF = stress reduction number

In Table 13.9, the numerical values of each of the above parameters are interpreted as follows.

The first two parameters represent the overall structure of the rock mass and their quotient is the relative measure of the block size. The quotient of the third and fourth parameters is said to be an indicator of the water pressure.

The fifth parameter is a measure of

(1) loosening load in the case of shear zones and clay bearing rock,
(2) rock stress in competent rock, and
(3) squeezing and swelling loads in plastic incompetent rock.

The sixth parameter is regarded as the parameter representing total stress.

The **Q** value is related to tunnel support requirements by defining the equivalent dimensions of the excavation. The equivalent dimension, which is a function of both the size and the purpose of the excavation, is obtained by dividing the span, diameter, or the wall height of the excavation by a quantity called the excavation support ratio (ESR), thus

$$\text{Equivalent dimension} = \frac{\text{Excavation span, diameter or height (m)}}{\text{ESR}}$$

The relationship between the index **Q** and the equivalent dimension of an excavation determines the appropriate support measures. The maximum unsupported span can be obtained as follows:

$$\text{Maximum span} = 2\,(\text{ESR}) \cdot Q^{0.4}$$

The Q value and the parameter of support pressure P_{roof} are calculated from the following equation:

$$P_{roof} = 2 \cdot Q^{-1/3} / Jr \qquad kg/cm^2 \tag{13.3}$$

If the number of joint sets is less than three,

$$P_{roof} = 2/3 \cdot \frac{Jn^{1/2} \cdot Q^{-1/3}}{Jr} \qquad kg/cm^2 \tag{13.4}$$

Table 13.9 Rock tunnelling quality index (after Barton *et al.*, 1974)

Description	*Value*		*Notes*
1. Rock quality designation	**RQD**		
A. Very poor	0–25		1. Where RQD is reported or measured as ≤ 10 (including 0), a nominal value of 10 is used to evaluate Q.
B. Poor	25–50		
C. Fair	50–75		
D. Good	75–90		2. RQD intervals of 5, i.e. 100, 95, 90 etc. are sufficiently accurate.
E. Excellent	90–100		
2. Joint set number	J_n		
A. Massive, no or few joints	0.5–1.0		
B. One joint set	2		
C. One joint set plus random	3		
D. Two joint sets	4		
E. Two joint sets plus random	6		
F. Three joint sets	9		1. For intersection use ($3.0 \times J_n$)
G. Three joint sets plus random	12		
H. Four or more joint sets, random, heavily jointed, 'sugar cube' etc.	15		2. For portals use ($2.0 \times J_n$)
J. Crushed rock, earthlike	20		
3. Joint roughness number	J_r		
a. Rock wall contact			
b. Rock wall contact before 10 cm shear			
A. Discontinuous joints	4		
B. Rough and irregular, undulating	3		
C. Smooth undulating	2		
D. Slicken-sided undulating	1.5		1. Add 1.0 if the mean spacing of the relevant joint set is greater than 3 m.
E. Rough or irregular, planar	1.5		
F. Smooth, planar	1.0		2. J_r = 0.5 can be used for planar, slicken-sided joints having lineations, provided that the lineations are oriented for minimum strength.
G. Slicken-sided, planar	0.5		
c. No rock wall contact when sheared			
H. Zones containing clay minerals thick enough to prevent rock wall contact	1.0 (nominal)		
J. Sandy, gravely or crushed zone thick enough to prevent rock wall contact	1.0 (nominal)		
4. Joint Alteration number	J_a	**or degrees (approx.)**	
a. Rock wall contact			
A. Tightly healed, hard, non-softening impermeable filling	0.75		1. Values of ϕr, the residual friction angle, are intended as an approximate guide to the mineralogical properties of the only alteration products, if present.
B. Unaltered joint walls, surface staining	1.0	25–35	
C. Slightly altered joint walls, non-softening mineral coatings, sandy particles, clay-free disintegrated rock, etc.	2.0	25–30	
D. Silty-, or sandy-clay coatings, small clay-fraction (non-softening)	3.0	20–25	
E. Softening or low-friction clay mineral coatings, i.e. kaolinite, mica. Also chlorite, talc, gypsum and graphite etc., and small quantities of swelling clays. (discontinuous coatings, 1–2 mm or less)	4.0	8–16	

Table 13.9 Contd.

Description	*Value*	*Notes*	
4. Joint alteration number	**J_a**	**ϕr degrees**	
b. Rock wall contact before 10 cm shear		**(approx.)**	
F. Sandy particles, clay-free, disintegrating rock etc.	4.0	25–30	
G. Strongly over-consolidated, non-softening clay mineral fillings (continuous < 5 mm thick)	6.0	16–24	
H. Medium or low over-consolidation, softening clay mineral fillings (continuous < 5 mm thick)			
J. Swelling clay fillings, i.e. montmorillonite (continuous < 5 mm thick). Values of J_a depend on percent of swelling clay-size particles, and access to water.	8.0–12.0	6–12	
c. No rock wall contact when sheared			
K. Zones or bands of disintegrated or crushed	6.0		
L. rock and clay (see G, H and J for clay	8.0		
M. conditions)	8.0–12.0	6–24	
N. Zones or bands of silty- or sandy-clay, small clay fraction, non-softening	5.0		
O. Thick continuous zones or bands of clay	10.0–13.0		
P. & R. (see G, H and J for clay conditions)	6.0–24.0		
5. Joint water reduction	**J_w**	**approx. water pressure (kgf/cm^2)**	
A. Dry excavation or minor inflow i.e. < 5 l/m locally	1.0	< 1.0	
B. Medium inflow or pressure, occasional outwash of joint fillings	0.56	1.0–2.5	
C. Large inflow or high pressure in competent rock with unfilled joints	0.5	2.5–10.0	1. Factors C to F are crude estimates: increase J_w if drainage installed
D. Large inflow or high pressure	0.33	2.5–10.0	
E. Exceptionally high inflow or pressure at blasting, decaying with time	0.2–0.1	> 10	2. Special problems caused by ice formation are not considered.
F. Exceptionally high inflow or pressure	0.1–0.05	> 10	
6. Stress reduction factor		**SRF**	
a. Weakness zones intersecting excavation, which may cause loosening of rock mass when tunnel is excavated			
A. Multiple occurrences of weakness zones containing clay or chemically disintegrated rock, very loose surrounding rock (any depth)		10.0	1. Reduce these values of SRF by 25 - 50% but only if the relevant shear zones influence do not intersect the excavation
B. Single weakness zones containing clay, or chemically disintegrated rock (excavation depth < 50 m)		5.0	
C. Single weakness zones containing clay, or chemically disintegrated rock (excavation depth > 50 m)		2.5	
D. Multiple shear zones in competent rock (clay free), loose surrounding rock (any depth)		7.5	
E. Single shear zone in competent rock (clay free). (depth of excavation < 50 m)		5.0	
F. Single shear zone in competent rock (clay free) (depth of excavation > 50 m)		2.5	
G. Loose open joints, heavily jointed or 'sugar cube' (any depth)		5.0	

Table 13.9 Contd.

Description		*value*		*Notes*
6. Stress reduction factor			SRF	
b. Competent rock, rock stress problems	σ_c/σ_1	$\sigma_t\sigma_1$		2. For strongly anisotropic virgin stress field
H. Low stress, near surface	> 200	> 13	2.5	(if measured): when $5 \le \sigma_1/\sigma_3 > 10$, reduce
J. Medium stress	200–10	13–0.66	1.0	σ_c to $0.8\sigma_c$ and σ_t to $0.8\sigma_t$. When $\sigma_1/\sigma_3 >$
K. High stress, very tight structure	10–15	0.66–0.33	0.5–2	10. Reduce σ_c and σ_t to $0.6\sigma_c$ and $0.6\sigma_t$,
(usually favourable to stability, may				where σ_c = unconfined compressive
be unfavourable to wall stability)				strength, and σ_t = tensile strength (point
L. Mild rockburst (massive rock)	5–2.5	0.33–0.16	5–10	load) and σ_1 and σ_3 are the major and
M. Heavy rockburst (massive rock)	< 2.5	< 0.16	10–20	minor principal stresses.
c. Squeezing rock, plastic flow of incompetent rock				3. Few case records available where depth
under influence of high rock pressure				of crown below surface is less than span
N. Mild squeezing rock pressure			5–10	width. Suggest SRF increase from 2.5 to
O. Heavy squeezing rock pressure			10–20	5 for such cases (see H).
d. Swelling rock, chemical swelling activity depending on presence of water				
P. Mild swelling rock pressure			5–10	
R. Heavy swelling rock pressure			10–15	

ADDITIONAL NOTES ON THE USE OF THESE TABLES

When making estimates of the rock mass quality (Q), the following guidelines should be followed in addition to the notes listed in the tables:

1. When borehole core is unavailable, RQD can be estimated from the number of joints per unit volume, in which the number of joints per metre for each joint set are added. A simple relationship can be used to convert this number to RQD for the case of clay free rock masses: RQD = 115–3.3 J_v (approx.), where J_v = total number of joints per m^3 ($0 < RQD < 100$ for $35 > J_v > 4.5$).
2. The parameter J_n representing the number of joint sets will often be affected by foliation, schistosity, slaty cleavage or bedding etc. If strongly developed, these parallel 'joints' should obviously be counted as a complete joint set. However, if there are few 'joints' visible, or if only occasional breaks in the core are due to these features, then it will be more appropriate to count them as 'random' joints when evaluating J_n.
3. The parameter J_r and J_a (representing shear strength) should be relevant to the weakest significant joint set or clay filled discontinuity in the given zone. However, if the joint set or discontinuity with the minimum value of J_r/J_a is favourably oriented for stability, then a second, less favourably oriented joint set or discontinuity may sometimes be more significant, and its higher value of J_r/J_a should be used when evaluating Q. The value of J_r/J_a should in fact relate to the surface most likely to allow failure to initiate.
4. When a rock mass contains clay, the factor SRF appropriate to loosening loads should be evaluated. In such cases the strength of the intact rock is of little interest. However, when jointing is minimal and clay is completely absent, the strength of the intact rock may become the weakest link, and the stability will then depend on the ratio rock-stress/rock-strength. A strongly anisotropic stress field is unfavourable for stability and is roughly accounted for as in note 2 in the table for stress reduction factor evaluation.
5. The compressive and tensile strengths (σ_c and σ_t) of the intact rock should be evaluated in the saturated condition if this is appropriate to the present and future *in situ* conditions. A very conservative estimate of the strength should be made for those rocks that deteriorate when exposed to moist or saturated conditions.

13.7.1 Support Categories Used for Tunnel Support

Based on the field work done by Barton *et al.* (1974), the tunnel support may be classified into the following groups.

(1) Unsupported
(2) Spot bolting
(3) Systematic bolting

(4) Systematic bolting with shotcrete
(5) Fibre reinforced concrete 50–90 mm and bolting
(6) Fibre reinforced concrete 90–120 mm and bolting
(7) Fibre reinforced concrete 120–150 mm and bolting
(8) Fibre shotcrete > 150 mm with rib of concrete bolting

Example 3 Consider a roadway tunnel of 9 m span in phyllite rock mass with the following known rock mass quality data:

Joint set 1:	Rough, undulating	$J_r = 3$
	Gypsum coating	$J_a = 4$
	18 joints per metre	
Joint set 2:	Smooth, undulating	$J_r = 2$
	Slightly altered walls	$J_a = 2$
	8 joints per metre	

thus

$$J_v = 18 + 8 = 26 \text{ and } RQD = 115 - 3.3{*}J_v \text{ (approx.)}$$

$$RQD = 30$$

$$J_n = 4$$

Minor water inflow	$J_w = 1$
UCS of phylite	$\sigma_c = 45$ MPa
Major principal stress	$\sigma_1 = 5$ MPa
Minor principal stress	$\sigma_3 = 2$ MPa

$$Q = \frac{30 \times 2 \times 1}{4 \times 4 \times 1.5} = 2.5 \text{ (poor rock)} \tag{16.4}$$

Hence $\sigma_1/\sigma_3 = 2.5$ and $\sigma_c/\sigma_1 = 9$ (high stress) SRF = 1
Support estimate B = 9 m, B/ESR = 9
For Q = 2.5; support category = 17
Permanent support = rock bolt spaced at 1 m, bolt length 3 m

13.9 Comparison of Classification Systems

A correlation has been provided between the RMR and Q value. A total of 120 cases were analysed which resulted in the following relationship:

$$RMR = 9\, L_n Q + 44 \tag{13.5}$$

Rutledge and Preston (1978) determined the following correlation between the two classification systems:

$$RMR = 13.6 \log Q + 43 \qquad (\text{S.D.} = 9.4) \tag{13.6}$$

A comparison of the stand-up time and the maximum unsupported span, as shown in Figure 13.4 reveals that the RMR is more conservative than the Q-system (Grimstad and Barton, 1993).

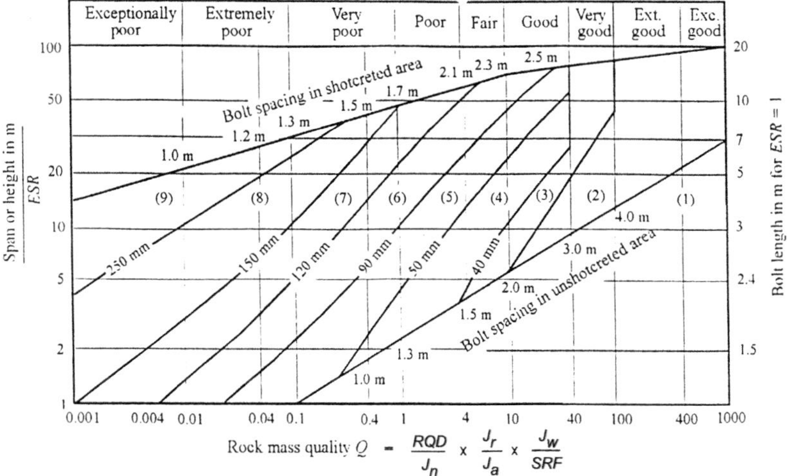

Figure 13.4 Estimated support categories based of rock quality index (Grimstad and Barton, 1993)

13.10 Rock Mass Weakening Coefficient for Strength Determination

Singh (1986) suggested a classification system for coal measure rock formations in which a scale factor, W_c is defined, so that the strength parameter of the rock mass may be designated as W_c, a fraction of the intact rock samples. This classification system includes the following terms.

i) Rock Quality Designation, RQD (R)

(i) Spacing index (K_1)

(ii) Joint index (K_3)

(iii) Joint surface index (K_2)

(iv) Joint filling

(v) Joint aperture index (K_4)

These jointing parameters are quantified as suggested in Table 13.10. The jointing coefficient is calculated as follows:

$$K = K_1 * K_2 * K_3 * K_4 \tag{13.7}$$

And the weakening coefficient is given by:-

$$W_C = R * K \tag{13.8}$$

Table 13.10 Weakening coefficient of coal measure formation (Singh, 1986)

	Rock classification			
Parameters	*Poor*	*Moderate*	*Good*	*Strong*
RQD (%)	40	40–60	60–80	80
index (R)	0.4	0.4–0.6	0.6–0.8	0.8
Joint spacing	Close (<0.3 m)	Moderate (0.3–1.0 m)	Wide (1.2–2.0)	Remote (>2.0 m)
index (K_1)	0.7	0.8	0.9	1.0
Joint surface	Polished	Smooth	Rough	Dorment
index (K_2)	0.7	0.8	0.9	1.0
Joint filling	Open	Soft filling	Tight filling	Asperite
index (K_3)	0.7	0.8	0.9	1.0
Joint aperture	5 mm	1–5 mm	0.1–1.0 mm	<0.1 mm
index (K_4)	0.6	0.7	0.8	0.9
Weakening coefficient	0.1	0.1–0.2	0.2–0.5	0.5–0.7

where

W_C = weakening coefficient of the rock mass.
R = RQD(%)

A correlation was made between RMR rating values and the corresponding weakening coefficients and the following relation was obtained (Figure 13.5).

$$W_C = 0.0248\, e^{(0.037\ \text{RMR})} \tag{13.9}$$

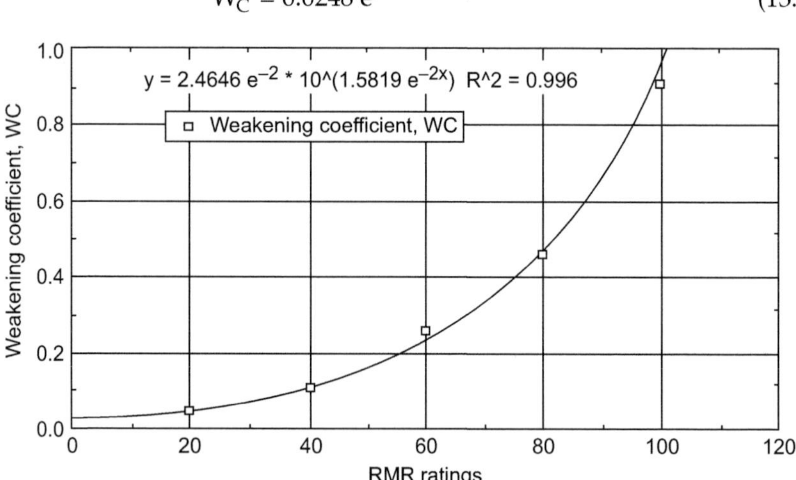

Figure 13.5 Correlation between weakening coefficient (W_C) and RMR rating values

13.11 Structural Stability of Excavations in Jointed Rock

Instability of hard rock excavations mainly depends upon the excavation geometry and the presence of structural discontinuities like joints, faults or bedding planes and their attributes. The most important intrinsic factors affecting the stability of surface and underground excavations in hard rocks include the following.

- Range and extent of discontinuities
- Properties of discontinuities (aperture, roughness, and type of filling material)
- Orientation of discontinuities
- Number of joint sets
- Lithological composition, the mineralogical constituents, strength, porosity, density and water content of the rock
- Various geotechnical parameters such as shear strength and density of the rock mass
- Rock mass classification used to estimate the behaviour of the rock mass at the excavation boundaries
- Size and shape of the excavations

One of the more inexpensive methods of stability analysis in a hard rock excavation is to carry out joint surveys within the excavation in the jointed rock mass and to interpret the results with the aid of stereographic projection techniques. This method can be used to predict the size and shape of the potentially unstable block together with its possible mode of failure by taking into account the shear strength of the natural discontinuities. This method of analysis can be applied to both surface as well as underground excavations. Use of this technique for the stability analysis of excavations is discussed in the following sections.

13.12 Geological Assessment of Jointed Rock

13.12.1 Joint Surveys

A joint survey is carried out using a clinometer for measuring the orientation and dip of the discontinuities and a tape to measure relative distances. The clinometer is placed on the joint plane against the folding lid of the instrument and the instrument is levelled before the dip and dip direction of the joint is recorded. The tape is used to measure the line along which the survey is taking place with respect to a suitable survey reference point. Dip and dip direction of all joints, faults and bedding planes intersecting the tape measure are recorded together with the distance along the line and the type of discontinuities and their attributes. The main attributes of the discontinuities which are recorded are the following:

- Discontinuity length
- Discontinuity orientation
- Discontinuity aperture
- Discontinuity curvature
- Discontinuity persistence
- Discontinuity infilling
- Discontinuity roughness
- Water flow through the discontinuities

13.12.2 Stereograph

Two types of strereographic projection methods are available.

1. The equal angle projection
2. The equal area projection

There is no specific advantage of using a particular method but the system chosen for a particular analysis should be consistently used throughout. A basic description of the method of graphical presentation of geologic data is given by Hoek and Brown (1980), Hoek and Bray (1981) and Brady and Brown (1985).

In essence, the method consists of plotting of the poles of the discontinuities recorded in the field onto the equal area stereonet. Different symbols are used to denote the various types of structural features and their continuity. Maximum pole concentrations on the stereographic-net are determined by using a counting net and the contours of equal pole concentration are plotted. This contour plot is transferred onto a meridianal stereographic-net and great circles representing the planes of discontinuities and their poles are marked. Intersections by three major joint sets may form wedges within the excavation of various shapes and sizes posing a range of instability problems discussed in a subsequent section.

13.13 Structural Stability of Underground Excavations

Figure 13.6 shows the stereographic representation of three rock wedges in the roof of a tunnel in a jointed rock showing stable condition, condition of gravity fall of the unstable wedge and sliding wedge from the roof of a tunnel.

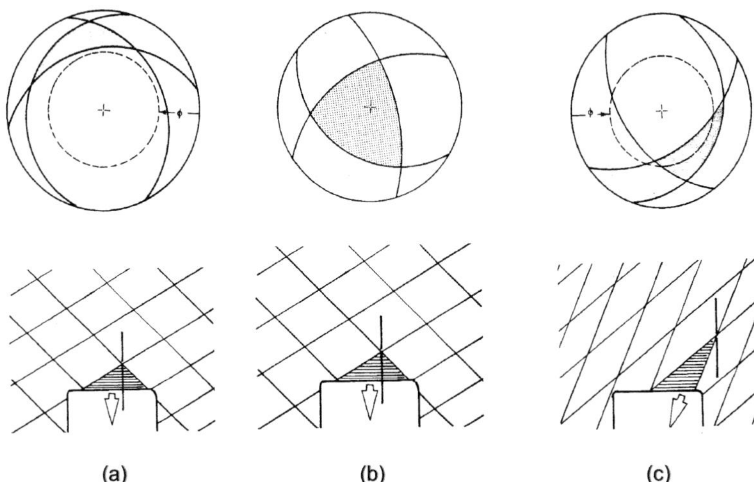

Figure 13.6 Stereographic presentation of various modes of instability in an excavation in jointed rock: (a) stable wedge condition, (b) gravity fall of roof wedge, (c) sliding wedge failure from the roof of a tunnel

13.13.1 Determination of the Condition of Instability of Roof and Side Wedges

The method of determining the size and shape of the wedges formed by the intersection of three major discontinuities is based on a graphical method presented by Goodman (1976). In Figure 13.6 the stereographic plot shows the base of three wedges formed by the intersection of three major sets of discontinuities.

If the stereographic plot indicates a vertical line through the apex of the wedge representing a centre point of the net is passing through the area of the base, the wedge is potentially unstable. In order for a rock wedge to fall freely into the tunnel it is necessary to examine the internal angle of friction of the joint planes. If a great circle representing the angle of friction ϕ is drawn on the stereographic net and the intersected rock wedge lies completely outside the friction circle then the wedge is stable as shown in Figure 13.6(a).

If in the stereographic projection, as shown in Figure 13.6(b) the vertical line passing through the base of the wedge and the area of the base lies inside the great circle representing the internal angle of friction of the joint then the mode of roof instability indicates a gravity fall of the roof wedge.

If three joints intersecting form a wedge but the vertical line through the apex of the wedge does not pass through the base of the wedge, then failure may occur by preferential sliding along one of the bounding planes.

A condition for sliding wedge failure indicating that the wedge forms a closed figure which does not contain the centre of the stereographic net is shown in Figure 13.6(c).

13.13.2 Determination of Geometry and Weight of the Block Intersected by the Set of Joints

The stereographic projection method can be used for a detailed estimation of the size and shape of potentially unstable wedges in both the roof and sidewalls of an excavation as follows.

(a) Roof failure analysis

Figure 13.7 shows a method of evaluation of the shape, size and volume of a wedge in the roof of a tunnel in jointed rock. Three planes determine the maximum pole concentrations of the joints and/or bedding planes and these are marked by three great circles designated as A, B and C in the stereonet. The strike lines of these great circles are labelled as a, b and c and the traces of the vertical planes through the centre of the stereonet and the great circle intersections are marked ab, ac and bc. If an underground roadway had a width W in Figure 13.7(b), the directions of the strike lines correspond to the traces of the planes A, B and C on the flat roof of the excavation. The intersection of these strike lines can result in the formation of a maximum size of wedge which can fall into the underground roadway. In plan, the apex of the wedge is defined by determining the point of intersection of the lines ab, ac and bc projected from the corners of the base of the triangular wedge shown in Figure 13.7(b). The height of the wedge can be

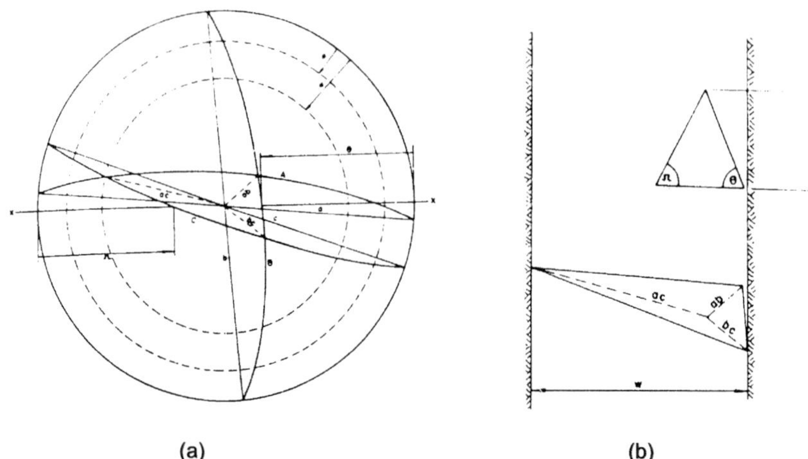

(a) (b)

Figure 13.7 Evaluation of shape and volume of a wedge in the roof of a tunnel determined by taking a section through the apex of the wedge normal to the plane of the roadway The apparent dips of the planes A and C are represented by the angles θ and Ω measured on the stereographic net along the line XX passing through the centre of the net Figure 13.7(a). Having determined the shape of the base of the wedge, its area can be calculated by a simple geometric relationship. The volume of the wedge can be calculated as ⅓ area of the base × height, and its weight determined if the bulk density of the rock is known. This method permits the evaluation of the maximum size of the unstable wedge which needs to be supported. In order to determine optimal spacing of support a statistical analysis of the distribution of joints or bedding planes is necessary.

(b) Sliding wedge failure of roof

The principle used to determine the size and shape of the unstable rock wedge prone to sliding movement is similar to that used in the falling wedge analysis. The method consists of determining the true plan view of the largest wedge by drawing lines parallel to the strike lines of the great circles A, B and C in Figure 13.8(a) and forming a triangle in Figure 13.8(b). The apex of the wedge in Figure 13.8(b) is determined by drawing lines parallel to lines bc, ac and ab which intersects at a point in Figure 13.8(b). However, the largest size of the wedge is not an important consideration as the span of the roadway is also not an important constraint against failure.

In order to determine the height of the wedge a projection of the points of intersections of the line XX with the base of the wedge is made at the direction at right angles to the line ab in Figure 13.8(b). The intersection of

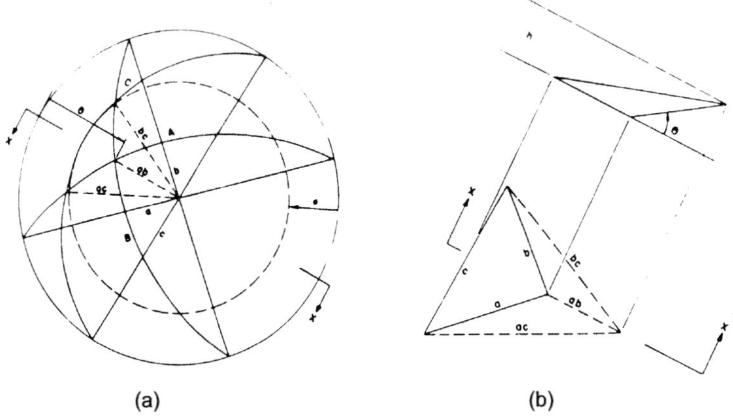

Figure 13.8 Geometrical evaluation of the height and volume of a wedge failed due to sliding of wedge from the roof of a tunnel

great circles A and B in the stereographic net in Figure 13.8(a) gives angle θ as the true dip of the intersection and Ω as the dip of great circle C which will enable determination of the height of the unstable wedge in Figure 13.8(b).

13.13.3 Case History Analysis of Instability of Underground Excavation in Jointed Rock Mass

The site of investigation was a drive in 1100 ore body in Mount Isa Mines in central Queensland, Australia. P575 scavenger drive on 19B sub-level was driven to provide access to the sill pillar below P575 stope draw point in order to obtain access to 26,000 tonnes of remnant copper ore. The scavenger drive was located in the area of a high to medium grade chalcopyrite ore containing >2% copper. The rock mass surrounding the drive had proliferation of joints dipping 70° north-east but contained no visible bedding planes. There were a number of small faults intersecting the drive. The chalcopyrite exhibited brittle behaviour without relieving any stress concentrations. Further to the south of the drive in the M62 drive the basement contact was greenstone which was highly sheared and showed little sign of foliation. The basement contact I in this area dipped in the direction north-east with a dip angle between 50° to 60°. The structural map of the 19B P575 scavenger drive is presented in Figure 13.9(i).

The stereographic analysis of instability is shown in Figure 13.9(ii) predicting that the shape of the unstable slabs in the roof will be triangular with a height of 1.5 m, and a base area of 3.3 m^2. The mode of failure was predicted as the sliding wedge failure of the roof. Visual observations within the scavenger drive confirmed the sliding mode of failure of slabs from the roof and sides.

13.14 Instability of Surface Mining Slopes

Instabilty analysis of hard rock slopes is essentially a two part process: (1) stereographic analysis of structural fabrics of the site for delineating the possible mode of structurally controlled instability and (2) limiting equilibrium stability analysis to determine the factor of safety against failure. Some of the most important modes of failure are described in the following section.

13.14.1 Stereographic Representation of Various Modes of Slope Failures

Figure 13.10 shows the three main types of structurally controlled slope failures common in hard rock quarrying along with stereographic plots of structural features likely to promote slope failures.

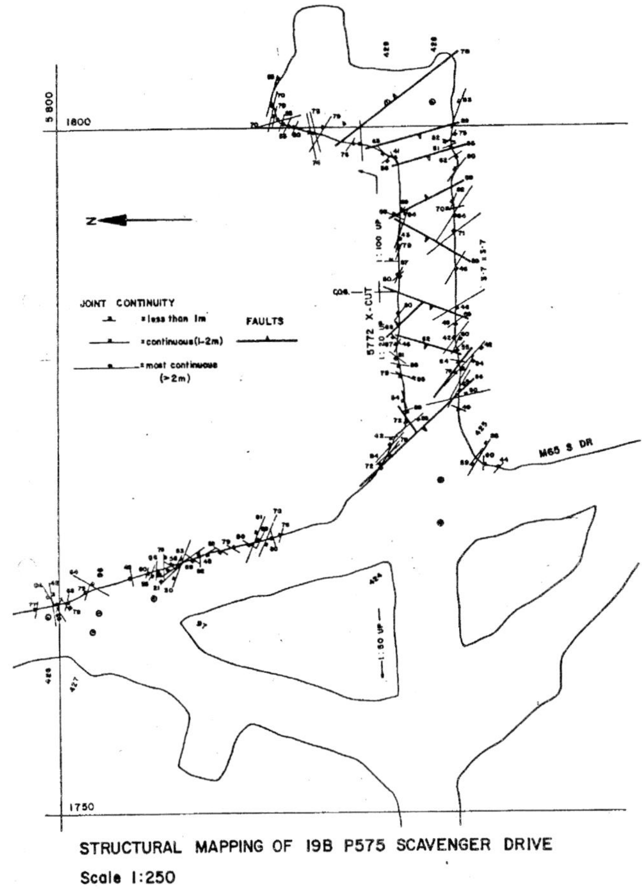

Figure 13.9(i) Structural map of 19B P575 scavenger drive in 1100 ore body in Mount Isa Mine

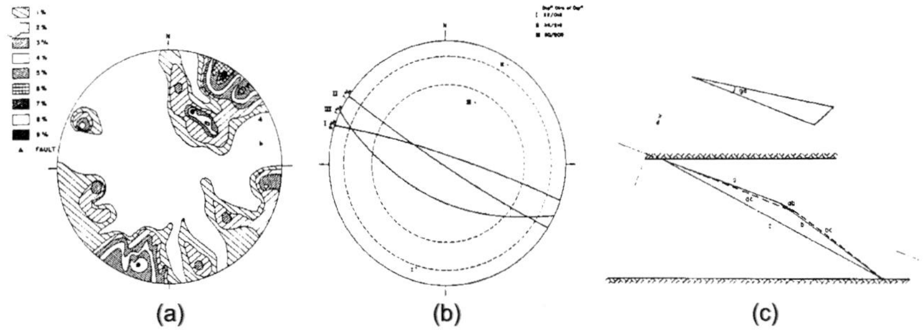

(a) Contour plot of structures, (b) three major structural planes, (c) sliding wedge failure

Figure 13.9(ii) Stereographic stability analysis of roadway instability in jointed rock mass

Figure 13.10(a) shows a planar failure in rock with highly developed structural features. The corresponding stereographic plot represents the great circle showing the slope face and the great circle representing the main structural features. The direction of sliding is at right angles to the strike line of the failure surface. For planar failure to take place the following four conditions should be fulfilled.

(1) The dip direction of the planar discontinuities must be within 20° of the dip direction of the slope face.
(2) The dip of the planar discontinuity should be less than the dip of the slope face.
(3) The dip of the planar discontinuity should be greater than the angle of friction of the failure surface.
(4) The lateral continuity of the potential failure surface must be broken by release surfaces.

Figure 13.10(b) indicates wedge failure bounded by two intersecting discontinuities. The corresponding stereograph shows the great circle indicating the slope face and that of two intersecting discontinuities. The line joining the centre of the stereo-net and the point of intersection of the two discontinuities indicates the direction of sliding either against both planes simultaneously or along the steeper of the two planes in the direction of maximum dip.

Figure 13.10(c) shows toppling failure in hard rock capable of forming columnar structure separated by steeply dipping discontinuities. The great

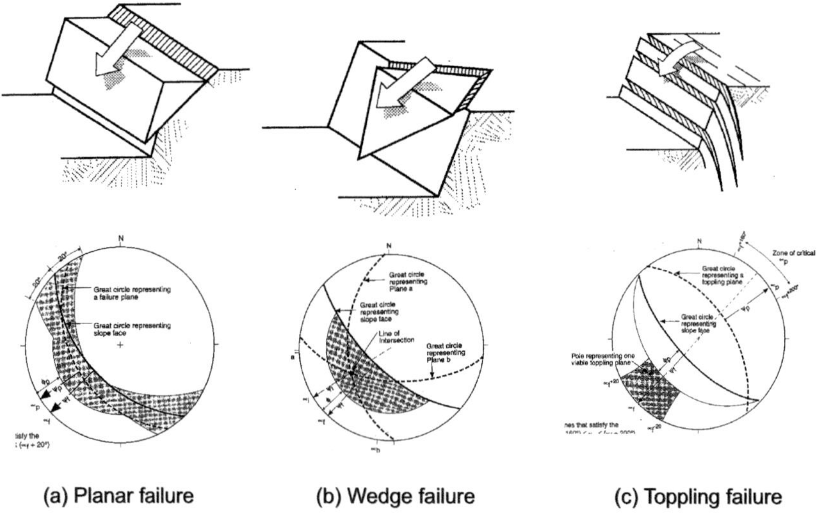

(a) Planar failure (b) Wedge failure (c) Toppling failure

Figure 13.10 Various modes of failures in surface mining circle representing the slope face and the great circle representing the discontinuity planes dip in opposite direction and steeply into the face.

A practical application of this approach for assessing the stability of a slope can be exemplified by a case study in the Marulan limestone quarry (Khanlari, 1996).

13.14.2 Face Instability at a Limestone Quarry, NSW, Australia

The site of this investigation was the Marulan quarry located in the Bungonia region, some 10 km east of the Hume Highway, south-east of the NSW capital, Sydney. Based on field observations and geological studies it was concluded that the limestone rock mass at the Marulan quarry is a recrystallised and medium-grained, light to dark grey limestone. It is a thickly bedded deposit of a high quality and forming a semi-weathered rock mass. Due to the presence of some major discontinuity sets and a few dolerite dykes that are mainly weathered, instability of the rock slope may manifest itself as discrete unstable blocks.

13.14.3 Methodology of Discontinuity Data Collection and Analysis

The method of data collection relating to the discontinuities of the slope faces in the Marulan quarry was based on the scan line method as suggested by Brown (1981). For stability assessment, the slope faces of the quarry were divided into two sections designated as section A and section B. Geological and geotechnical field investigations were carried out in order to study the main type of rocks and geological features present. Three-dimensional joint survey programmes were carried out in order to collect geotechnical data from the discontinuities in the slope faces. Some 768 readings were taken from the discontinuities that intersected the scan-line. Discontinuity data collected in this manner were analysed using statistical analysis techniques and were also used as input data for the evaluation of the rock mass quality using different rock mass classifications.

13.14.4 Discontinuity Data From Marulan Quarry

Table 13.12 shows the statistics derived from the discontinuity measurements and their attributes for section A and section B slopes for the Marulan quarry. The main parameters noted in the table were length, orientation, dip angle, aperture and the nature of the joint infill material. Other attributes of the discontinuities included in the table were discontinuity persistence, joint roughness, joint spacings, joint water inflow and joint compressive strength as derived from the Schmidt hardness rebound test.

Table 13.12 Statistics of discontinuity distribution at Marulan quarry (Khanlari, 1996)

Discontinuity parameters	*Section A*	*Section B*
Number of discontinuities	357	411
Orientation		
• Mean	179.24° ± 17.1 (56.6%)	197.5° ± (21.2%)
• Maximum concentration	180–200° (14.3%)	100– 120(21.2%)
Dip angle	70 ±17° (17%)	57°± (27.45%)
• Maximum	70–80° (33%)	80–90° (34.5%)
Aperture	2–6 mm (69%)	2–6 mm (72%)
Water inflow	98.6 % dry	78% dry
Discontinuity curvature		
• Stepped	46%	37%
• Undulating	11%	20%
• Planar	43%	43%
Discontinuity roughness		
• Roughness	83%	100%
Discontinuity persistence		
• Outcrop	69%	70%
• Rock	10%	10%
• Beyond exposure	21%	20%
Joint spacing		
• 0–0.15 m	20%	26%
• 0.215–0.25 m	36%	47%
• 0.25–0.3524%	15%	
• 0.35–0.45 m	12%	5%
• >0.45 m	8 %	7%
Joint compressive strength (Schmidt hammer hardness)	23±8	23±6
Joint length (m), range 0.5 to 3.5 m	1.58±0.97	1.56±0.98

The discontinuity data collected from the scan line survey and site investigation was classified to determine the frequency distribution of the discontinuities. The factor analysis of the discontinuity data indicated that the joint spacings, the number of joint sets and their orientation are the main factors contributing to the failure mode of the hard rock slopes. Figure 13.11 shows the distribution of joint spacings in section A of the Marulan quarry together with the Rose diagram of the discontinuity sets. It can be seen that there are five major sets of joints in section A.

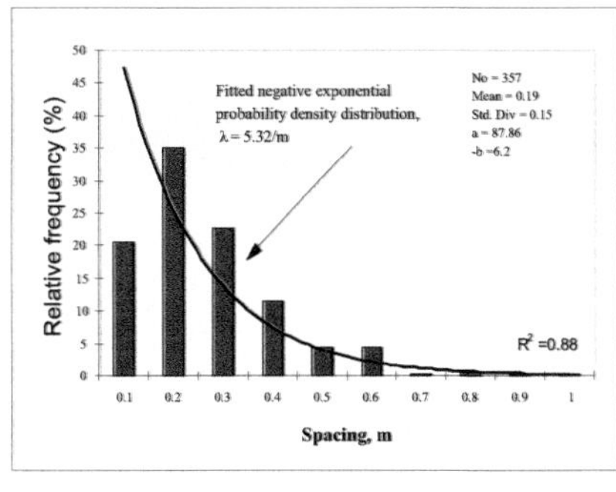

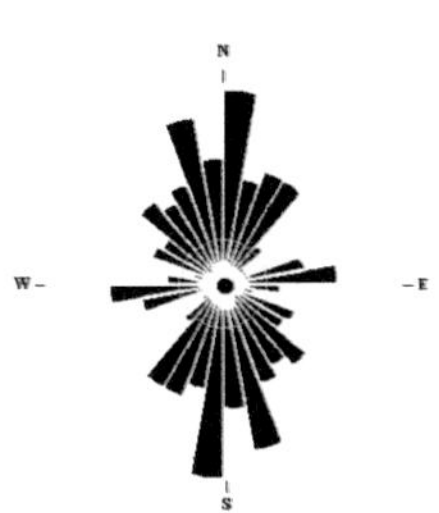

(a) Joint spacing (b) Rose diagram of discontinuities

Figure 13.11 Probability density distribution of joint spacing in section A (Khanlari, 1996)

Figure 13.12 shows the spacing distribution and Rose diagram of discontinuity sets in section B of the quarry indicating four major sets of discontinuities with joint spacing of 0.2 m.

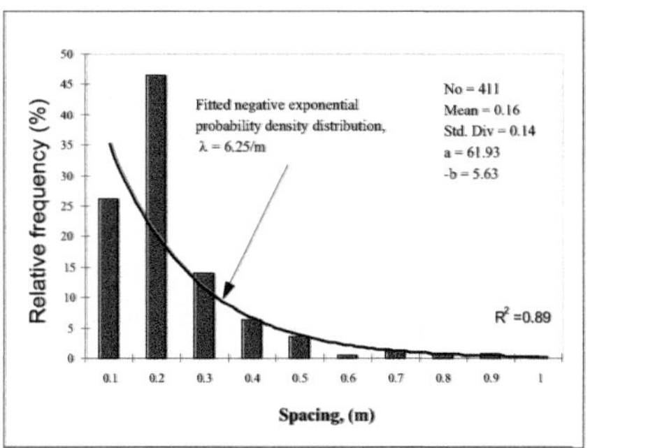

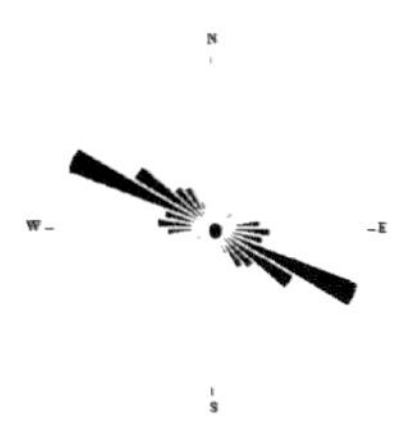

(a) Joint spacing at section B of the quarry (b) Rose diagram of discontinuities

Figure 13.12 Probability density distributions of joints and Rose diagram in section B (Khanlari, 1996)

Rock quality designation (RQD)

Different values of RQD for sections A and B were calculated from joint surveys data. Also the theoretical RQD* was calculated from Equation 13.10 and the results are presented in Table 13.13. As is clear from Table 13.12, the measured RQD is within 9% of the calculated RQD using only the average number of discontinuities per metre:

$$RQD^* = 100\,(0.1\,\lambda + 1)\,e^{-0.1\lambda} \tag{13.10}$$

Table 13.13 Comparison between measured and theoretical RQD in limestone rock mass

Source of data	*Total scan-line length (L (m))*	*Average discontinuity frequency (λ)*	*Measured RQD %*	*Theoretical RQD* %*	*Differences in RQD values (%)*
Section A	67.80	5.26	83.3	90.27	+6.97
Section B	65.79	6.25	78.32	87.4	+9.08
Total	133.59	5.75	80.68	82.11	+1.43

13.14.5 Qualitative Assessment of Rock Mass Properties

Table 13.14 shows the average discontinuity frequency in section A, section B and for the total scanned length together with the measured RQD for the quarry. The quality description charts enabled the estimation of Rock Quality Percentage (RQP) and also Rock Quality Risks (RQR) to be evaluated for the limestone rock mass on the basis of a technique developed by Sen (1990). It can be seen in Table 13.14 that there is an inverse relationship between (RQR) and the average number of fractures along a scan line for a given (RQD) value. It means that the smaller value of (RQR) represents a better quality rock mass. From Table 13.14, it may be concluded that the RQP and RQR systems of rock mass quality classifications provide relative frequency distribution of good quality rock in the same rock mass and also gives an indication of risk associated with the actual RQD.

Table 13.14 Different values of RQD, RQP, and RQR for the limestone rock mass

Source of data	*Average discontinuity frequency (λ)*	*RQD*	*RQP*	*RQR*
Section A	5.26	83.3	Good–Fair 0.04–0.96	0.96
Section B	6.25	78.32	Good–Fair 0.02–0.98	0.98
Total	5.75	80.68 0.05–0.95	Good–Fair	0.95

13.15 Rock Mass Characterisation at Marulan Quarry

For characterising the rock mass at the Marulan quarry the following distinct steps were carried out.

1. Determining engineering properties of the intact rock by laboratory testing
2. Estimating rock mass strength from the laboratory and field data and classification of the rock mass using (1) RMR system, (2) Q system and (3) Weakening coefficient system

Table 13.15 presents some of the important properties of the intact limestone samples tested in the laboratory. It should be noted that these engineering properties were used as input data for rock mass classification.

Table 13.15 Engineering properties of intact rock samples from Marulan quarry

Engineering properties of intact rock	*Structural regions A and B*
Uniaxial compressive strength (MPa)	77.84
Tensile strength (MPa)	3.81
UCS derived from diametral point load test (MPa))	77.37
UCS derived from axial point load (MPa)	86.13
Poisson's ratio	0.35
Elastic modulus (GPa)	18.45
Friction angle	45.10
Bulk density (kg/m^3)	2660.0

13.15.1 Application of Rock Mass Classification Systems

The data obtained in the field work was analysed in terms of four recent rock mass classification methods; the Geomechanics Classification (RMR) (Bieniawski, 1989), the Q Index (Barton, 1974; Barton and Choubey, 1977) as well as the Weakening coefficient system (W_C) (Singh, 1986; Singh and Gahrooee, 1990). The application of these classifications is described below.

13.15.2 Geomechanics Classification System (RMR)

The calculated RMR rating values based on the geotechnical and geological parameters for sections A and B of the limestone quarry are given in Table 13.16. Results indicate that the rock mass quality in section A is fair and in section B a very poor class rock.

Table 13.16 Geomechanics classification (RMR) of the rock mass at the Marulan quarry

Rock mass parameters	*Section A*		*Section B*	
	Value	*Rating*	*Value*	*Rating*
UCS (MPa)	77.84	7	77.84	7
RQD %	83.3	17	78.32	17
Discontinuity spacing (m)	0.19	8	0.16	8
Discontinuity condition	Class III	20	Class II	25
Groundwater condition	Dry	15	Dry	15
Orientation rating	Class III	25	Class V	–60
Total RMR rating		57		12

13.15.3 Application of Q System

The Q system was based on the measurement of six geotechnical parameters as described by Barton *et al.* (1976). The Q index was calculated by use of the following equation:

$$Q = \frac{RQD}{J_n} \times \frac{j_r}{j_a} \times \frac{J_w}{SRF} \tag{13.2}$$

The designation of rock mass parameters in equation 13.2 and the results of the Q system of classification for the limestone are given in Table 13.17.

Table 13.17 Engineering rock mass classification (Q system) for the Marulan quarry (Singh *et al.*, 2002)

Rock mass parameters	*Structural region (A)*		*Structural region (B)*	
	Value	*Rating*	*Value*	*Rating*
Rock quality designation (RQD) %	83.3	83.3	78.32	78.32
Joint set number (J_n)	5	15	4	15
Shear strength factor (J_r/J_a)	3/2	1.5	3/1	
Joint water reduction factor (J_w)	Dry	1	Dry	1
Rating adjustment factor (RAF)	5		5	
Total rating	1.65		3.13	
Description	Fair		Poor	

13.15.4 Application of Weakening Coefficient Classification System (W_c)

The W_C classification system was used by Singh (1986) for underground coal mining and in the design of rock slopes by Singh and Gahroohee (1990). The most important discontinuity parameters contributing to this system are RQD, joint spacing, joint surface index, joint filling materials and, finally, the discontinuity aperture. The W_C values for the limestone rock mass were quantified according to the discontinuity parameters and the appropriate values are presented in Table 13.18. It may be recalled that the relationship between these discontinuity parameters can be represented by the equation 13.7 as reproduced below:

$$\mathbf{K = K_1 \times K_2 \times K_3 \times K_4} \tag{13.7}$$

where K_1 = discontinuity aperture
K_2 = discontinuity spacing
K_3 = joint surface index
K_4 = joint filling index

The W_C can be calculated by Equation (13.8).

$$\mathbf{W_C = RQD \times K} \tag{13.8}$$

Table 13.18 Rock mass weakening coefficient for limestone in the Marulan quarry (Khanlari, 1996)

Rock mass parameters	*Section A*		*Section B*	
	Value	*Rating*	*Value*	*Rating*
RQD %	83.3	0.83	78.32	0.78
	0.19	0.7	0.16	0.7
Joint surface	Rough	0.9	Rough	0.9
Joint infilling	Open <5 mm	0.7	Open >5 mm	0.6
Discontinuity aperture	2 - 6 mm	0.7	2–6 mm	0.7
Weakening coefficient (WC)	0.256		0.206	
Orientation rating	0.29 × 0.37		0.13 × 0.1	
Total weakening coefficient (WC)	0.11		0.013	
Description	Moderate		Poor	

13.16 Appraisal of Face Stability at Marulan Quarry

In order to assess the stability of the slope faces at the quarry under consideration, the data collected from the joint surveys were analysed using statistical analysis methods to find the relationship between the various discontinuity parameters. The data from discontinuity orientations were used for the assessment of the potential mode of failure in two different parts of the limestone quarry using a stereographic projection technique as described in the following sections.

13.16.1 Stability Assessment of the Slope Face in Section A

Section A is located in the northern part of the quarry and contains six benches. A discontinuity survey program was carried out in order to assess the stability of the slope faces in this part of the quarry. Samples of naturally jointed rocks were collected during the site investigation. From direct shear tests a friction angle of 18.9 degrees and cohesion of 1.91 MPa were determined. The discontinuity dips in this section show that 32.77% of the discontinuities had a dip of between 70 and 80 degrees. A lower hemisphere equal area stereographic projection method was used for discontinuity data analysis.

Figure 13.13(a) shows poles of some 357 discontinuities measured from the slope faces of six benches. Figure 13.13(b) shows the contoured plot of the discontinuity poles in section A of the quarry plotted using the lower hemispherical stereographic net. It can be seen in the diagram that there are five discontinuity sets on this part of the quarry. Figure 13.14(a) represents a

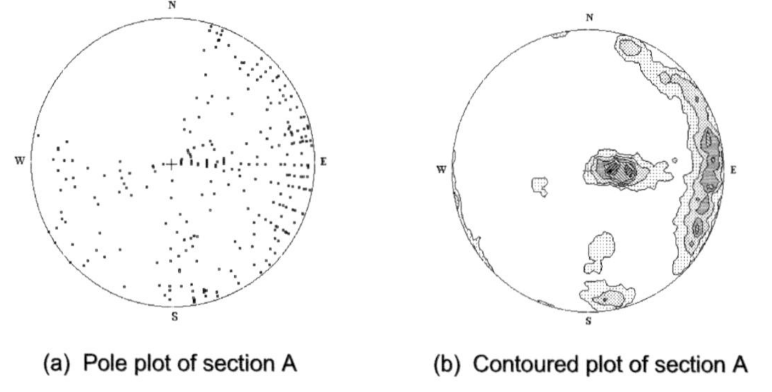

(a) Pole plot of section A (b) Contoured plot of section A

Figure 13.13 Analysis of discontinuity orientation data of section A from Marulan Quarry

contour diagram superimposed on a polar net and Figure 13.14(b) represents a stereographic projection of major discontinuity planes. It can be seen in Figure 13.14 (b) that five major joint sets were prominent, occurring at dip and dip directions of 17/267, 81/266, 78/297, 82/351 and 82/198 respectively. On this diagram, a great circle of slope face of 75 degrees with dip direction of 191 degrees was plotted.

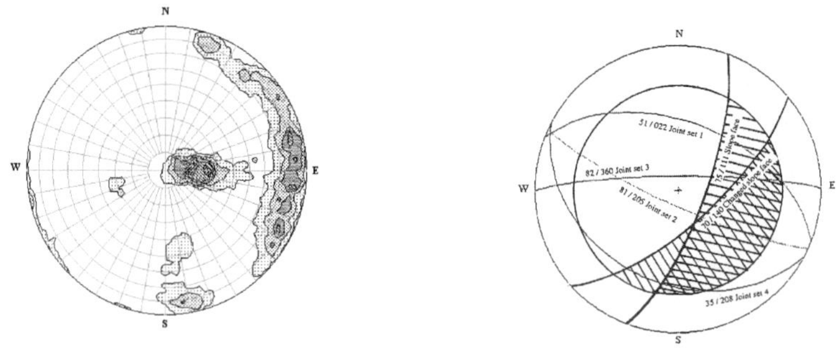

(a) Control plot of section A on polar net (b) Stereographic projections of major planes

Figure 13.14 Analysis of pole plot data of section A from Marulan Quarry

In addition, a circle representing an internal angle of friction of 18.9 degrees was superimposed on the stereo-net to delineate the unstable shaded area of potential wedge failure. It is recommended that the orientation of the slope face be changed to 68/214 degrees in order to eliminate the danger of potential instability.

13.16.2 Stability Assessment of the Slope Face in Section B

Section B is situated in the western section of the quarry. It contains some dolerite dykes intersecting the limestone rock mass which probably has an effect on the stability of the slope face in this part of the quarry. Results of the statistical analysis of the discontinuity data show that 34.55% of the discontinuities are dipping at 80 to 90 degrees in section B. It indicates that the joints in this section are sharply inclined in their dip. In section B of the quarry, the rock mass was subjected to a small-scale toppling failure which resulted from discontinuity and slope face orientations.

Figure 13.15(a) represents a plot of some 411 discontinuities on a lower hemispherical equal area stereo-net. Figure 13.16(b) shows a contour plot of these discontinuities showing the presence of four discontinuity sets in this part of the quarry, occurring at dip and dip directions of 51/022, 81/205, 82/360, and 35/208 respectively.

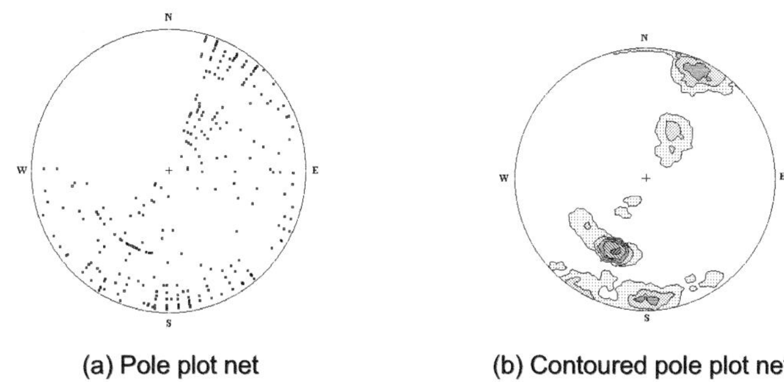

(a) Pole plot net (b) Contoured pole plot net

Figure 13.15 Discontinuity orientation data analysis of section B from the Marulan Quarry

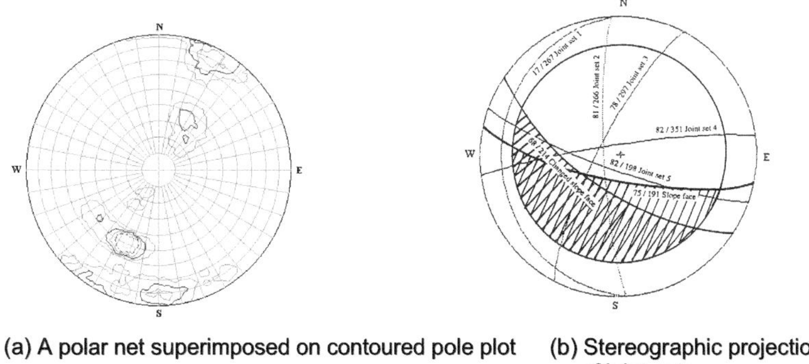

(a) A polar net superimposed on contoured pole plot (b) Stereographic projection of joint sets

Figure 13.16 Analysis of cluster poles of section B from the Marulan Quarry

In order to assess slope stability, a great circle of the quarry face at 75° degrees dip with a dip direction of 111 degrees and a friction cone at an angle of 18.9 degrees was superimposed on the stereo-net (Figure 13.16(b)).

The potential for possible toppling type of failure exists between planes one and four and also a potential for sliding is predicted in the direction of the intersecting lines I_{24}. In order to reduce the potential for instability, or to reduce risk, it is recommended that the orientation of the slope face be changed to 70/140 or that the slope angle be reduced to 65 degrees.

The orientations of dominant joint sets are given in Table 13.19 together with suggested remedial measures to reduce the danger of potential slope instability.

Table 13.19 Orientations of major discontinuity planes and slope faces

Structural region (Dip–dip direction)	*Section A*	*Section B*
First predominant joint set	17 / 267	51 / 022
Second predominant joint set	81 / 266	81 / 205
Third predominant joint set	78 / 297	82 / 360
Fourth predominant joint set	82 / 351	35 / 208
Fifth predominant joint set	82 / 198	—
Slope face	75 / 191	75 / 111
Changed slope face	68 / 214	70 / 140

13.17 Conclusions

A rock mass classification system is one of the most important methods employed to present the rock mass quality data. Assigning a numerical value to the selected parameters helps develop a holistic measure. This method provides a valid technique of classifying rock masses of similar properties and provides a successful comparison method from one site to another for rocks with similar rock mass properties. This approach provides a valid method of assessing

- a scale factor for reducing the laboratory properties of intact rock to the properties of the rock mass,
- aid in design of the support for an underground excavation, and
- a means to estimate the support pressure required for designing roof support.

It is also observed that the stereographic projection method incorporating a joint survey technique offers a very inexpensive and effective method of assessment of stability of mining excavations in jointed hard rock masses.

References

Barton, N., R. Lien and J. Lunde (1974). 'Engineering classification of rock masses for the design of tunnel support', *Rock Mechanics*, Vol. 6, pp. 189–239.

Barton, N. (1976). 'Recent experience with the Q-system of tunnel support design, ASTM STP984', American Society for Testing and Materials, Philadelphia, pp. 107–117.

Brady, B.H.D. and E.T. Brown (1985). *Rock Mechanics for Underground Mining*, George Allen and Unwin, London, ISBN 0-04-622004-6, pp. 484–488.

Barton, N. and Y. Choubey (1977). 'The shear strength of rock joints in theory and practice', *Rock Mechanics*, Vol. 12, pp. 1–54.

Bieniawski, Z.T. (1973). 'Engineering classification of jointed rock masses', Trans. South African Inst. of Civil Engineers, 15, No. 12, pp. 335–344.

Bieniawski, Z.T. (1989). *Engineering Rock Mass Classification*, John Wiley, New York, 251 pp.

Brown, E.T. (1981). *ISRM Suggested Methods*, Pergamon Press, Oxford, 211 pp.

Coates, D.F. (1964). 'Classification of rock for Rock Mechanics', Int. Jl. of Rock Mechanics and Mining Science, 1, pp. 421-429.

Deere, D.U. (1964). 'Technical description of rock cores for engineering purposes', *Rock Mechanics and Engineering Geology*, Vol. 1, No. 1, pp. 17–22.

Deere, D.U. and R.P. Miller (1966). *Engineering classification and index properties of intact rock*, US Air Force Weapons Laboratory Report AFWL-TR-65-16 Kirkland, New Mexico.

Ghose, A.K. (1996). 'Rock Mass Classification—A Design Tool for the Mining, Civil Engineering and Construction Industry', *Journal of Mines, Metals and Fuels*, February, pp. 63–76.

Ghose, A.K. and N.M. Raju (1982). 'Characterization of Rock Mass vis-à-vis Application of Rock Bolting–Modelling of Indian Coal Measures', *Proc. 22nd US Symposium on Rock Mechanics*, MIT, pp. 422–427.

Goodman, R.E. (1976). *Methods of geological engineering in discontinuous rocks*, West Publishing Company, New York, pp. 58–59.

Goodman, R.E. (1989). *Introduction to Rock Mechanics*, John Wiley and Sons, New York, pp. 141–177.

Gray, P. (1988). *Private communication*, BHP Engineering, Wollongong, Australia.

Grimstad, E., and N. Barton (1993). 'Updating the Q-system for NMT', Int. Symposium on sprayed concrete (Eds. Kompen, Opsahll, Berg.) Norwegian Concretc Association. Oslo.

Hoek, E. and J.W. Bray (1981). *Rock Slope Engineering*, Institution of Mining and Metallurgy, London, Third Edition, ISBN 0 900488 573, pp. 37–62.

Hoek, E. and E.T. Brown (1980). *Underground Excavation in Rock*, Institution of Mining and Metallurgy, London, Third Edition, ISBN 0 900488 573, pp. 61–86, 183–241.

Hiscock and Mitchell (1993). *Stone Artefact Quarries and Reduction Sites in Australia*, Australian Publishing Service, 94 pp.

Khanlari, G.R. (1996). 'Application of Rock Mass Characterisation to Slope Stability Problems', Ph.D. Thesis, Department of Civil and Mining Engineering, University of Wollongong, NSW, Australia, 490 pp.

Lauffer, H. (1958). Gebirgs Klassifiernng für den Stollenbau, *Geol Bauwesen*, 24(1), pp 46–51.

Molinda, G.M., and Mark, C. (1994). 'Coal Mine Roof Rating (CMRR): A practical rock mass classification for coal mines, U.S. Dept. of two Interior, Bureau of Mines, IC 9387.

Rutledge, J.C., and R.L. Preston, (1978). 'Experience with engineering classification of rock,' Proc. International Tunnelling Symposium, Tokyo, pp. A3. 1–A.3.71.

Sheorey, P.R. (1984). 'Use of rock classification to estimate roof caving span in oblong workings', *International Journal of Mining Engineering*, Vol. 2, pp. 133–140.

Sheorey, P.R. and B. Singh (1984). 'Application of rock mass classification to mining stability problems—some case studies', *Stability of Underground Mining*, USA, pp. 383–395.

Sen, Z. (1990), 'Technical Note: RQP, RQR and Fracture Spacings', *International Journal of Rock Mechanics and Mining Sciences and Geomechanics Extracts*, Pergamon Press, Vol. 27, pp. 84–94.

Singh, R.N., A.M. El'Mehrig and M. Sunu (1986). 'Application of rock mass characterisation to the stability assessment and blast design in hard rock surface mining excavations', 27th *US Rock Mechanics Symposium*, Alabama, pp. 471–478.

Singh, R.N., M. Eksi and A.G. Pathan (1987). 'Design evaluation of room and pillars workings for in gypsum mines', *Symposium on Underground Mining Methods and Technology*, Nottingham, England, pp. 403–414.

Singh, R.N. and M. Eksi (1987). 'Empirical design of pillars in gypsum mines using Rock Mass Classification Systems', *Journal of Mines, Metals and Fuels*, pp. 16–23 and 34.

Singh, R.N. and M. Eksi (1987). 'Rock characterisation of gypsum and marl', *Mining Science and Technology*, Amsterdam, pp. 105–112.

Sunu, M.Z. and R.N. Singh (1989). 'Application of modified rock mass classification systems to blasting assessment in surface mining operations', *Mining Science and Technology*, Vol. 8, pp. 285–296.

Singh, R.N. and D.R. Gahrooee (1990). 'Application of Weakening Coefficient for Stability Assessment of Slopes in Heavily Jointed Rock Mass', *International Journal of Surface Mining*, 3, p. 13.

Singh, R.N. and A.M. Elmherig (1985). 'Geotechnical investigations and appraisal of face instability in jointed rock in surface mining', *Proceedings of* 26th *US Symposium in Rock Mechanics*, Rapid City, South Dakota, 26–28 June 1985, pp. 31–39.

Singh, R.N., R. Khanlari, R.N. Chowdhury and J. Shonhardt (1996). 'A geotechnical investigation and instability assessments of quarry slopes in heavily jointed porphyry rock mass', *VIIth International Symposium on Landslides*, Balkema, Rotterdam, ISBN 909 5410 8185, pp. 1835–1841.

Singh T.N. (1986). 'Application of equivalent materials model in mining and tunnelling', *Proceedings on Workshop on Rock Mechanics and Problems in Tunnels and Mine Roadways*, Srinagar, pp. 42, 1–27.

Turner, K.A. and R.L. Shuster (1996). *Landslides Investigation, Mitigation*, Special report 247, TRB, National Research Council, ISBN 0309-049151-2, Chapter 15, Norrish, N.I. and D.C. Wyllie (Eds) pp. 391–425.

Venkateswaralu, V., A.K. Ghose and N.M. Raju (1989). 'Rock-Mass classification for design of roof supports. A statistical evaluation of parameters', *Mining Science and Technology*, 8, pp. 97–107.

Wickham, G.E., H.R. Tiedemann, and E.H. Skinner (1979). 'Ground support prediction model–RSR concept', Proc. Rapid Excavation and Tunnelling Conference, AIME. New York, pp. 691–707.

Supports in Mining and Tunnelling

14 Chapter

14.1 Introduction

The state of stress in rock mass can be characterised by the three stress components in three orthogonal directions namely vertical stress and the major and minor principal horizontal stresses. As an excavation is made in the rock mass, the *in situ* stresses readjust themselves and the strata surrounding the excavation is de-stressed, forming a pressure arch. The main purpose of support surrounding a mining excavation is to stabilise the de-stressed rock and make the excavation safe for entry by workmen. In the past, fall of ground in mines has been a major cause of accidents in underground mining especially due to roof falls at the coalface and in the access roadways to the mine and the coalface. With the advent of face powered supports and universal use the steel arch support, in the UK, this has resulted in the reduction of fatal accident rates from 0.27 to 0.04 per 100,000 man-shifts from 1927 to 1972 (Whittaker, 1975) and serious accidents from 0.94 to 0.22 per 100,000 man-shifts over the same period. Similar trends have been observed in the Australian coalmines too. Thus, the widespread use of face powered support and steel-arched support of roadway have virtually eliminated accidents due to fall of ground in underground coal mining. This chapter briefly reviews the methods and design of roadway supports and face powered supports in underground coalmines.

14.2 Roadway Supports

The two basic methods of support around a roadway can be defined as follows:

(1) Internal reinforcing supports increasing the strength properties of rocks
(2) Externally applied supports to the de-stressed rock surrounding the excavation

Examples of the above types of support are given in Figure 14.1 and Table 14.1. Rock reinforcement mobilises the frictional and a cohesive properties of the rock mass and creates a zone of active support whereas the external support provides resistance to the movement of rock as a consequence of de-stressing and creates a passive support zone. Both support types create a zone of rock with higher strength, which is indirectly supported by passively supported rock. Another classification of mine support is as follows.

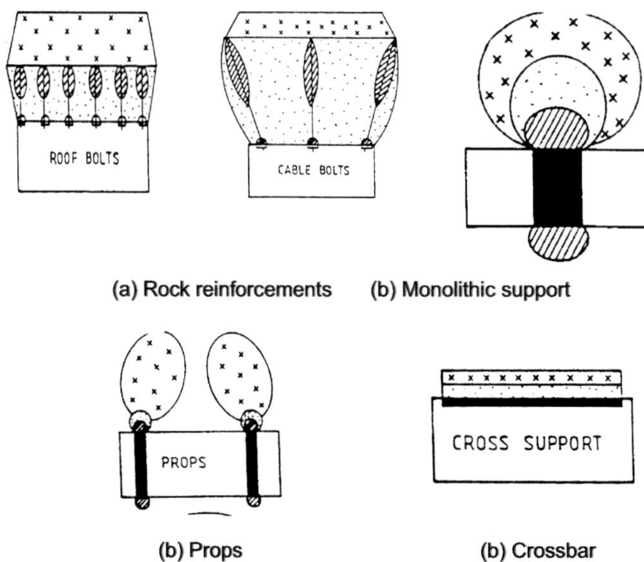

Figure 14.1 (a) Rock reinforcements, (b) externally applied supports

Active support is one that loads the geological mass in which it is placed immediately upon installation.

Passive support is one that is placed in the geological mass but does not play a role in ground support until the mass moves and subsequently loads the fixture.

Support layout and their material content are designed to prevent strata failure. In order to determine the support requirements of an excavation, an understanding of the mechanism of failure of roadways is necessary as presented in the following section.

14.3 Modes of Roadway Failures

The failure of an underground roadway in rock may take place due to a combination of factors such as inadequate strength and deformational

Table 14.1 Classifications of the roadway supports (Balusu, 1994)

Items	*Support type*	*Activity*
	Externally applied supports	
i) Timber	Props, chocks	Active
ii) Steel	Crossbars or channels	Passive
	Rib support	Passive
	W-strap	Active
iii) Friction	Square set	Passive
iv) Hydraulic	Props, screw jack	Active
	Chocks and powered supports	Active
v) Cement	Masonry blocks	Passive
	Shotcrete	Passive
	Internal reinforcing supports	
Full contact	Unheaded bolt	Passive
	Headed bolt	Passive
	Wooden dowel	Passive
	Cable bolts	Active
	Split set bolt	
Point anchors	Bolts in resin anchors	Passive
	Mechanical shell/ split wedge	Active
	Mechanical in resin	Active
Combination	Mechanical anchor + resin	Active
	Friction anchor and cement	Active
	Cable sling	Passive

properties of the host rocks, excessive rock stress in relation to the elements of roadway design, inappropriate shape and span of the roadway and inadequate application of the excavation support system.

If failure does take place then the remedial measure is either to strengthen the support or improve the design of the roadway such as changing the roadway width, shape or direction of the roadway's longitudinal axis with respect to the stress directions. Thus, the main cause of roadway failure is due to interaction of the following factors.

- Material properties of the rock
- State of stress
- Initial failure and application of the control measures
- Initial failure giving rise to different types of failure mechanism

The following types of roadway failures may occur (Richmond *et al.*, 1986):

- Span failure

- Skin failure
- Mid-span shear failure
- Guttering failure
- Cantilever failure
  - Wedge-shaped roof fall
  - Slicken-sided roof
  - Fault induced roof instability
- Rib failure

14.3.1 Span Failure

This type of failure occurs when the span of the roadway is too wide and the amount of roadway support is inadequate, promoting a sagging type of movement in the roof strata (Figure 14.2). As a consequence, tensile stresses due to gravitational loading initiate cracks to be developed on the roof beam at the centre of the roadway and also shear stresses at the roadway's edges. As the fracture planes interlink, the entire roof beam may fail. The span failure is generally associated with shallow workings in low horizontal stress regimes.

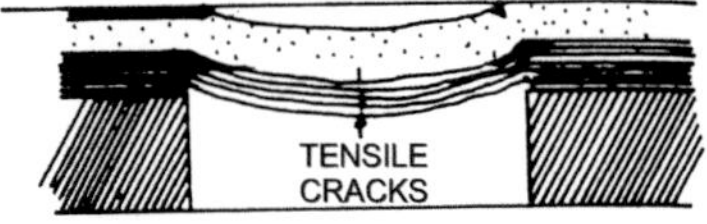

(a) Span failure—tensile mode

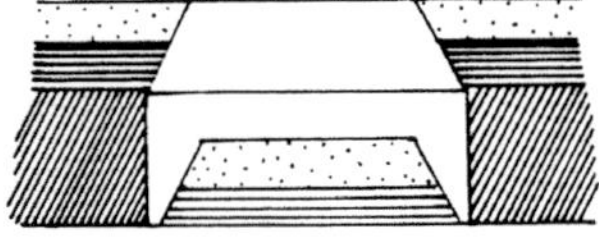

(b) Span failure—shear mode

Figure 14.2 Span failure of a roadway (Richmond *et al.*, 1986)

The remedial measures are as follows.

(1) Increasing the resistance of the roof to sag and reduction of the tensile stresses on the roof beam
(2) Increasing the roof bolt density to increase tensile strength of the roof beam.

14.3.2 Skin Failure

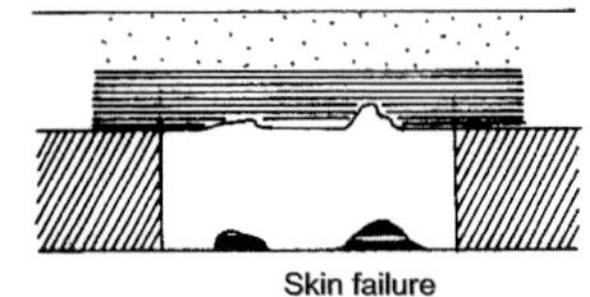

Figure 14.3 Skin failure of roof strata (Richmond *et al.*, 1986)

Figure 14.3 shows the skin type failure of the roof strata causing dislocation and caving of small fragments of rocks up to 0.01 m^3 – 0.25 m^3 size. The main reason for this type of failure can be attributed to inadequacies of rock structure and the rock fabric as exemplified by intensely bedded or friable roofs, cross-bedded strata units, slicken-sided strata units and low strength of the surrounding rock. The situation is accentuated if the roof consists of swelling claystone or seat-earth or fissile mudstones. The following remedial measures are adopted in order to control roof failure.

- Prompt installation of support at the face
- Increased coverage of roof by installing wire mesh in between the w-straps
- Selecting the roadway horizon in such a way so as to cut the weaker roof material
- Leave 0.3 m to 0.6 m thick coal layers against the friable roof in order to offer protection

14.3.3 Mid-span Shear Failure

Mid-span shearing can occur in the immediate roof beam or in a strata layer above the bolting horizon. This type of failure occurs in weak roof strata when bed separation has already weakened the roof beam and induced shear stresses in the mid-span due to lateral stresses (Figure 14.4). The remedial measure is to adopt closer bolting at the mid-span or installation of stiffer cross support.

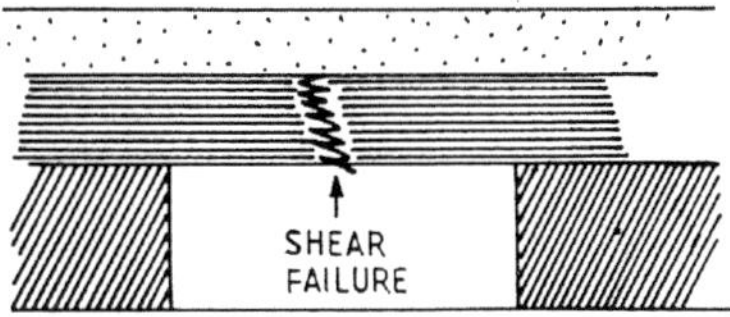

Figure 14.4 Mid-span shear failure (Richmond *et al.*, 1986)

14.3.4 Guttering Failure

Rib side shearing is the first stage of roof failure initiated in a square or rectangular opening due to high horizontal stresses and sharp corners. The presence of high horizontal stresses causes the development of the shear stress in the roof bed. If the shear stress exceeds the shear strength of the roof strata, the roof strata are generally dislodged and the fracture planes tend to extend at an angle over the heading (Fig. 14.5). Remedial action comprises

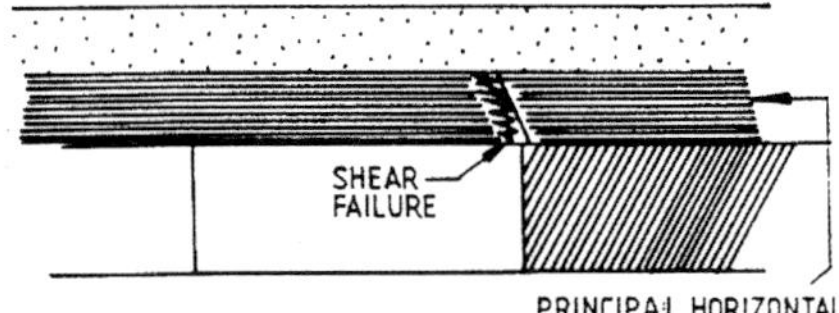

Figure 14.5 Shear failure of roof at the pillar edge (guttering) (Richmond *et al.*, 1986)

installation of roof bolts at an angle to the failure plane at this stage to arrest the further development of fractures. The development of guttering type of failure can be averted if the axis of the roadway is re-orientated at 45° to the direction of the principal horizontal stress.

14.3.5 Cantilever Failure

Figure 14.6 shows the cantilever type of failure of the roof beam, which occurs if the guttering fracture is not arrested by installation of effective roof support and allowed to propagate into the next stage of cantilever movement. This will develop compressive stresses at the fulcrum of the roof cantilever which may fracture the roof at this location leaning the roof across the roadway. Finally, the cantilever action continues forming a high arch profile by interconnecting the fracture systems. Thus, a cohesionless, arch-shaped rock mass will be formed which may culminate in a massive 1roof fall.

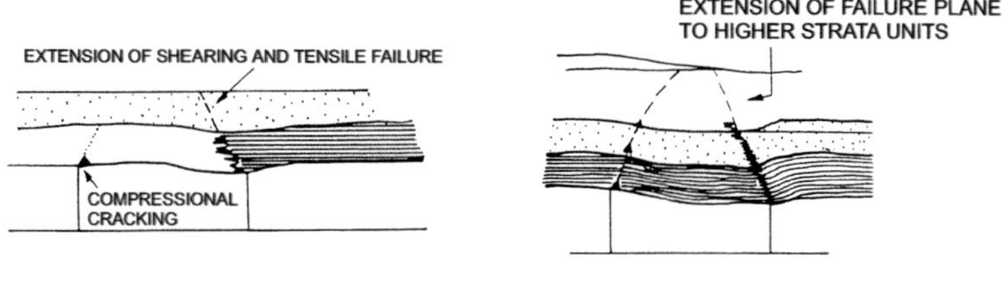

(a) Cantilever failure (b) Shear failure in higher strata units

Figure 14.6 Cantilever failure or shear failure of higher strata units

14.3.6 Failure Controlled by Geological Features

There are different types of geological structures which may cause the following types of instabilities.

(a) Wedge-shaped roof fall

This type of failure is caused by failure initiated by the presence of geological structure in the roof strata such as joints and faults.

The intersection of joint sets with roof strata destroys the continuity of the roof beam and creates broken wedges in the roof beam, which are held together in the roof due to the inherent frictional forces. These wedges tend to fall freely if the gravitational force acting on the unstable wedges exceeds the frictional force. A higher frequency of joints results in sagging of the roof beam which may be a major cause of roof fall. Thus, the orientation of the longitudinal axis of the roadway with respect to the strike line of the joint sets is an important consideration in the design of the mine layout.

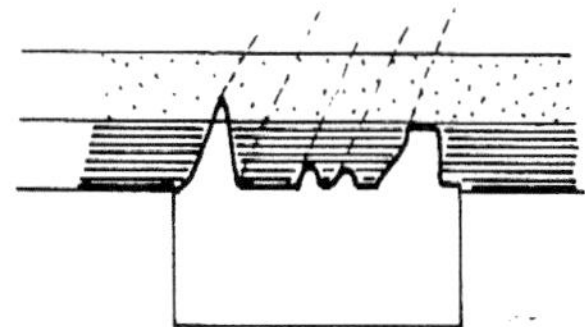

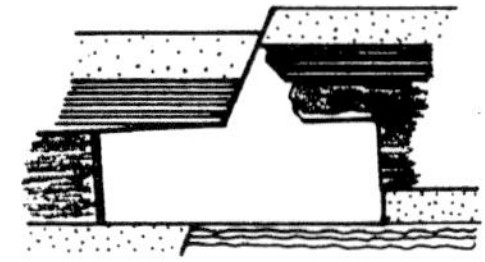

(a) Failure along minor joints (b) Failure induced by structural features

Figure 14.7 Roof failure induced by geological features (Richmond *et al.*, 1986)

(b) Slicken-sided roof

The presence of clay mylonite or other cohesion-less rock material in the roof permits rocks to dislodge unexpectedly from the main roof and thus cause major instabilities. Due to this feature, a slicken-sided roof is difficult to control especially during the roof support installation.

(c) Fault induced roof instability

Figure 14.7(b) shows roof instability caused by the presence of geological faults, which makes the surrounding rock weaker due to percolating water through the fault plane, and the presence of fault gouge. In addition, the rock surrounding the faults has higher frequencies of joints than normal. Therefore, additional supports are required to be installed in the vicinity of faults in order to avert the incidence of roof falls.

14.3.7 Failure of Roadway Sides or Rib Failure

Figure 14.8 shows two modes of pillar failure in coal caused by the presence of main cleat being parallel to the axis of the roadway and also the rib side buckling type failure.

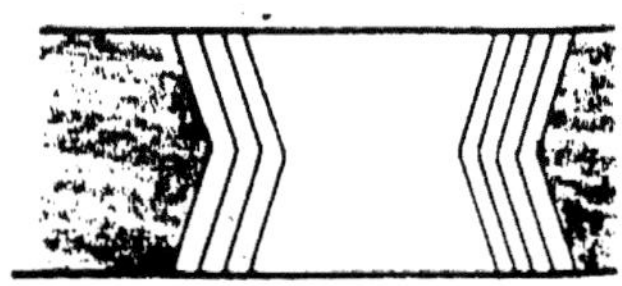

(a) Side failure due to presence of cleats (b) Rib side buckling failure

Figure 14.8 Mode of failure of roadway sides and floor (Richmond *et al.*, 1986)

The consequences of uncontrolled rib failure are very serious as indicated below.

- Buckling and falling of large blocks of coal up to 1 tonne in weight is a major hazard to the safety of the workmen.

- Side fall effectively increases the span of the roadway leading to span or beam failure of the roof. This problem is further increased by the presence of unsupported roof between the existing support and the edge of the degraded pillar.
- Progressive side spalling results in extension of the yield zone into the coal pillar, thus reducing the effective solid core of the pillar and impairing the overall structural stability of the mine.

14.4 Types of Roadway Supports

14.4.1 Props

The prop is the oldest and most versatile form of roof support used in mining and which still remains a main component of support even in modern mechanised systems of mining. In its simplest form it is an upright timber with a lid which supports the mine roof by tightening the prop against the roof by means of a wooden wedge. Thus, it is an active support providing a positive immediate support to the strata. Figure 14.9(a) shows a debarked tree trunk 150–230 mm in diameter being used as a prop to support the roof. Timber is a lightweight and versatile support material which can be easily transported and cut to the required length.

An inclined prop shown in Figure 14.9(b) is located on the floor by a steel hitch and set on to the roof mid-angle between the vertical line and a line normal to the seam and tightened in position by a wooden wedge. Figure 14.9(c) shows a prop with a pointed foot installed in position to give indication of the magnitude and direction of the loads and impending ground failure.

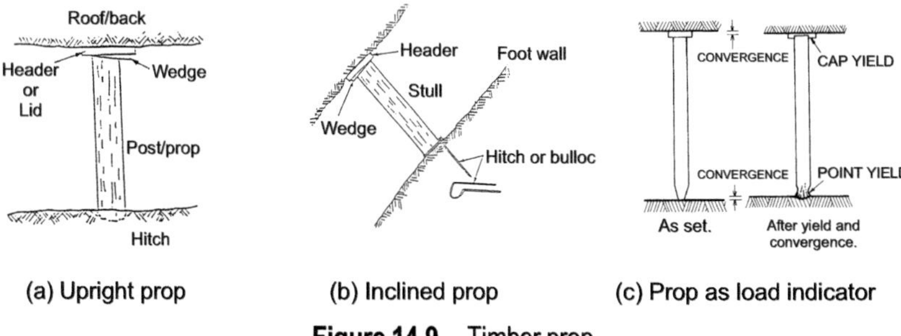

(a) Upright prop (b) Inclined prop (c) Prop as load indicator

Figure 14.9 Timber prop

14.4.2 Crossbars

A crossbar consists of a bar supported by two upright props tightened against the bar by wooden wedges. The stress on the roof is evenly distributed by installing cross bar, props and wooden laggings. Figure 14.10(b) shows a three-piece timber set with square section posts and crossbar used in metalliferous mines. Use of steel supports is common when the strata loading is relatively high and the life of the excavation is several years. Four-piece sets and stull timbering and also three-piece sets with inclined posts are used as support in metalliferous mines and need skilled timber men for their installation and maintenance.

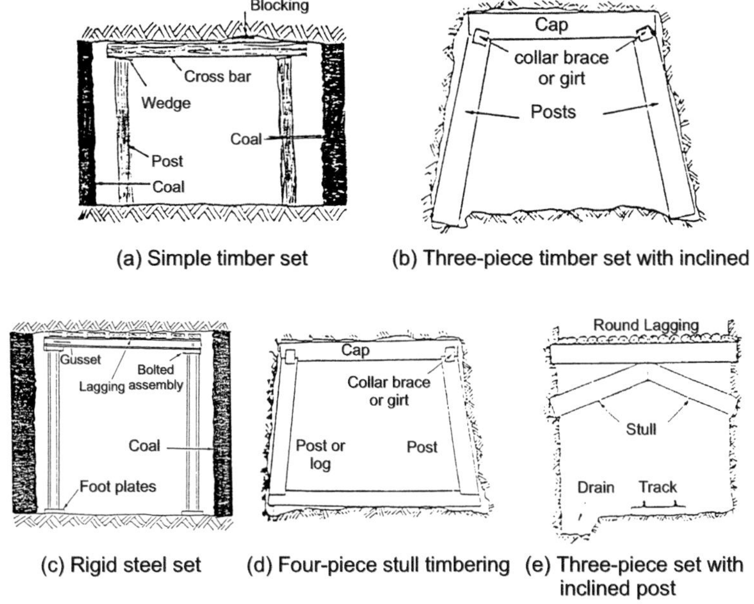

(a) Simple timber set (b) Three-piece timber set with inclined legs

(c) Rigid steel set (d) Four-piece stull timbering (e) Three-piece set with inclined post

Figure 14.10 Timber sets for roadway support

14.4.3 Chocks

Where the strata loading is excessive, chocks with greater load bearing capacity are used to support the roof and keep the excavations open. Timber chocks comprise sized timber 1.2 to 1.5 m long laid at right angles on top of each other and tightened against the roof by the use of wedges to support heavy loads. Higher load carrying capacities and longer life of chocks are achieved by replacing the timber with steel chocks, using timber props at the four corners of the chock or backfilling the chocks with broken rock aggregates. In the longwall face or in depillaring areas chocks are also used to promote caving along the goaf line.

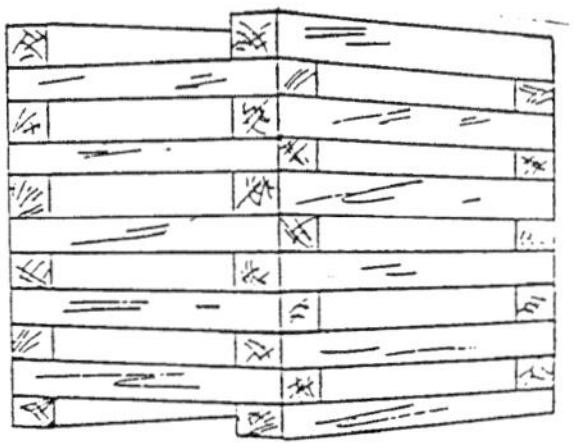

Figure 14.11 Chock support

14.4.4 Steel Rib Supports or Arches

Steel rib supports or arches are used in underground tunnelling operations where surrounding strata is unstable or loading is too severe for other supports or for longer life expectancy of the excavation such as in civil engineering tunnels requiring minimum maintenance. An arched shaped roadway closely conforms to the stress redistribution pattern within the de-stressed rock surrounding a roadway and therefore is a preferred shape. Figure 14.12(a) shows rigid arch type of supports offering greater load carrying capacity since the bending stresses encountered are reduced and transformed into compressive stresses (when an arch shape is used), and the load is evenly distributed all around the rib supports. Struts are used to equally space the arches along the axis of the roadway and lagging distributes strata load evenly onto the arch support. Similarly, floor plates are used at the foot of the arch to distribute stress on the weaker floor and also absorb roof convergence without deforming the arch supports. In the comparatively harder coal measures rock situation a rectangular roadway profile may be preferable using rigid steel rib supports. In weak rock conditions at depth in excess of 200–300 m it is possible for a gate roadway to undergo vertical closure in the order of 1–2 m either allowing the arch legs to penetrate the soft floor without distorting the arches or by using yieldable arches shown in Figure 14.12(b).

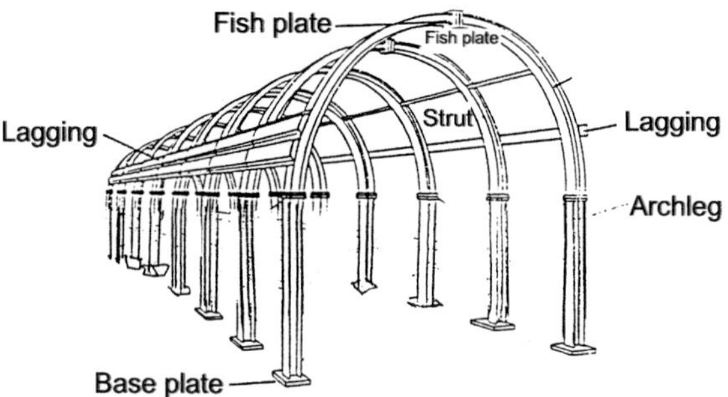

Figure 14.12(a) Rigid steel arches

Basically, the yielding arches are composed of three sections, with the top section sliding within the two side elements. Periodically, the U-bolts of the clamping bars are loosened to permit the arch sections to slide in order to relieve stress and as a consequence eliminate the arch distortion.

14.4.5 Concrete or Shotcrete

The use of concrete or shotcrete as a mining support element is very restricted due to the limited life span of mining excavations and also cost

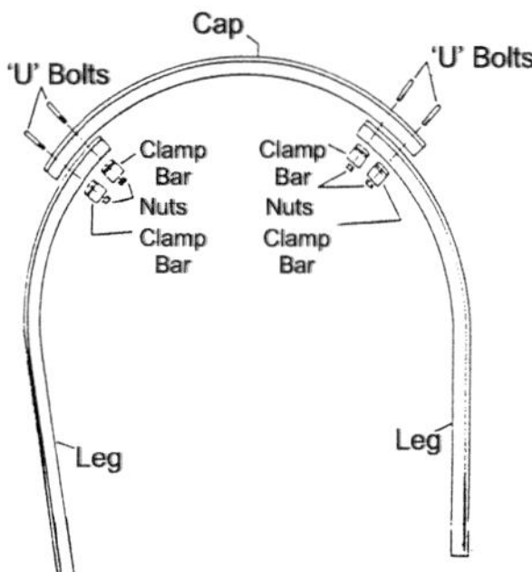

Figure14.12(b) Yieldable steel arches (Gregory, 1983)

considerations. While shotcrete or spray concrete is used to resist tensile movements on the crown of the excavation, concrete is designed to withstand compressive stresses. However, reinforced concrete is used to resist tensile, compressive and to a limited extent bending and shear stresses. Thus, concrete supports are mainly used to support permanent mining structures such as shafts, pit bottom excavations, pumping stations, tippler stations, water sumps, dams, and ventilation and fire stoppings.

14.4.6 Backfill

Backfilling of the voids created by mining is a well established mining practice for safe extraction of massive or thick ore deposits and also to limit the surface subsidence. Backfilling or stowing is also carried out for disposal of mining, quarrying or processing waste in an environmentally acceptable manner. If the filling operation is carried out immediately after the excavation, it not only reduces the strata movement, prevents cracking and deterioration of the supporting elements, prevents ventilation leakage but also controls the surface subsidence.

14.5 Roof Bolts

In many countries roof bolting has become a primary means of supporting mining excavations due to their cost benefits, ease in handling and for their ability to apply active support to the strata. This method of support offers a considerable amount of flexibility, causes minor disruption to the mining cycle and is considered to be a low volume support material reducing considerably the storage and handling requirements. This method of support is compact offering minimum obstruction to ventilation current and to moving machines and equipment in comparison to the standing supports. However, this system of support gives little indication of impending roof failure and rock bolts suitable for supporting very wide spans are impractically long. It is also difficult to ensure the quality of the rock bolting installations especially in a strain softening rock mass.

14.5.1 Theory of Roof Bolting

(1) Laminated beam theory

In bedded rock, two or more strata are bolted together, forming a stronger, laminated self-supporting beam that is capable of supporting the overlying rock. Rock in this condition offers the most favourable opportunity for improving the inherent strength of the rock by the use of bolting, Figure 14.13(a). Rock bolts also pre-stress the rock by working in the same manner as pre-stressed reinforced concrete.

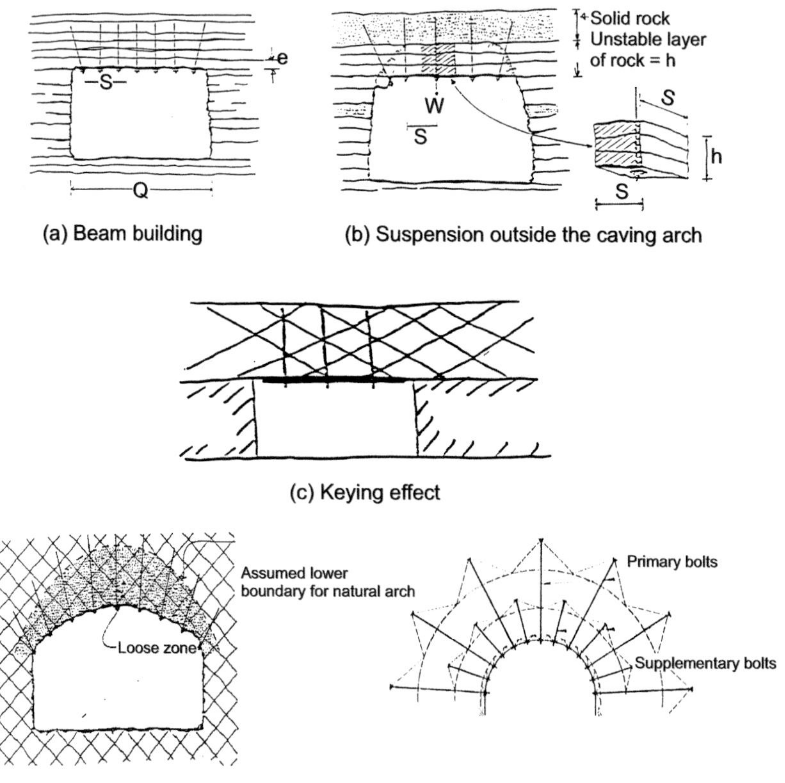

(a) Beam building (b) Suspension outside the caving arch

(c) Keying effect

(d) Arching effect in rectangular roadway (e) Formation of rock arch in arched roadway

Figure 14.13 Theory of rock bolting

(2) Suspension outside the caving arch

(a) It is well known that when the back of a drive fails, it will do so only for a short distance above the drive and the back will assume the shape of an arch.

(b) To prevent the back of the drive falling it is possible to anchor roof bolts into sound rock beyond the line of the natural arch.

(3) Consolidation of fragments

(1) Rock may be broken into numerous slabs, blocks or irregular interlocking fragments but without voids or open fractures.

(2) In such cases each rock bolt has the effect of clamping or squeezing together the rock fragments in a circular zone around it so that they tend to form a solid mass.

(3) A number of these competent/rigid blocks or key stones around the perimeter of a tunnel will act as a masonry arch and will support themselves.

Thus, tensioned rock bolts increase the internal friction of the rock and decrease the major to minor stress ratio, so that the incompetent rock may more readily form a ground arch and a competent rock is less liable to disintegrate.

(4) Surface keying effects

Rock bolts also perform the important function of compressing the rock surface and limiting the effect of rock expansion by preventing the opening of partings, cracks and joints which would be subject to attack by air, water or temperature changes.

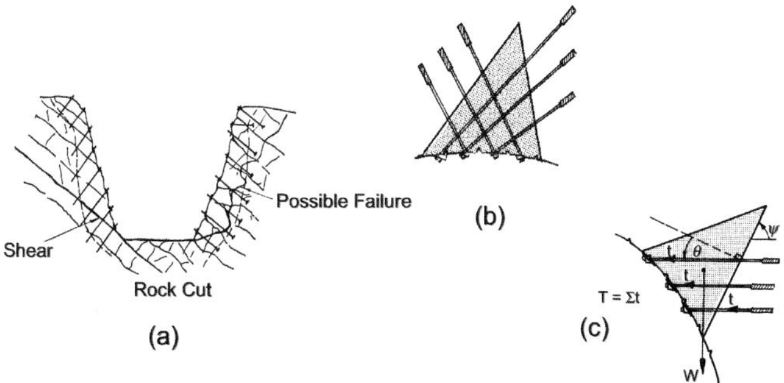

Support of (a) jointed rock in quarry, (b) free falling wedge, (c) sliding wedge

Figure 14.14 Rock bolt support in jointed rock

(5) Prevent structural failure in jointed rock mass

In a jointed rock mass, the rock bolts or cable bolts prevent the free fall of major unstable wedges due to gravitational loading in an underground excavation and they also reinforce an unstable rock wedge sliding along a single plane. In quarrying operations, rock bolts or cable bolts can reinforce the rock mass by keying effect.

(6) Catchment devices for holding rock fragments

Rock bolts are also used to hold in place wire mesh, slings or straps which prevent rock fragments from falling.

14.5.2 Types of Roof Bolts

Rock bolting forms a principal system of support in the mining industry. In the past, a large variety of anchoring devices have been used throughout the world. In general, rock bolts can be classified in the following categories.

(1) Mechanically anchored tensioned rock bolts
(2) Fully grouted steel or wooden dowels
(3) Grouted tensioned rock bolts
(4) Friction anchors or split set
(5) Wooden dowels

(a) Partially anchored tensioned rock bolts

The primary element of a mechanically anchored rock bolt is a steel rod inserted into a borehole drilled up to a predetermined depth in the rock. The remote end of the rock bolt contains an anchoring device which permits it to be firmly anchored to the far end of the borehole. The surface end of the rock bolt is equipped with a plate and nut that allows the bolt to be tensioned against the mouth of the borehole and permit compressive forces to be induced in the rock between the two ends of the bolts. Figure 14.15(a) shows a slot and wedge type anchor on a 25 mm diameter, 1.8 m long, rock bolt which can be tensioned to 200–300 Nm torque by tightening the nut at the outer end of the rock bolt. Figure 14.15(b) shows the expansion shell type of rock bolt and Figures 14.15(d) to (h) show a range of mechanical anchors from 18 mm to 24 mm in diameter and 80 mm to 180 mm in length. The mechanical type point load anchors were very popular in 1960s but they generate high stress concentration near the point anchor. However, the seismic shocks from blasting, etc. may cause the anchor to be loosened and become inoperative or exhibit excessive yield under heavy loading conditions. Figure 14.15(c) shows resin grouted and mechanically anchored rock bolts with the shell imbedded in resin. This system of support provides good active loading and excellent long-term anchorage. The bolt head is

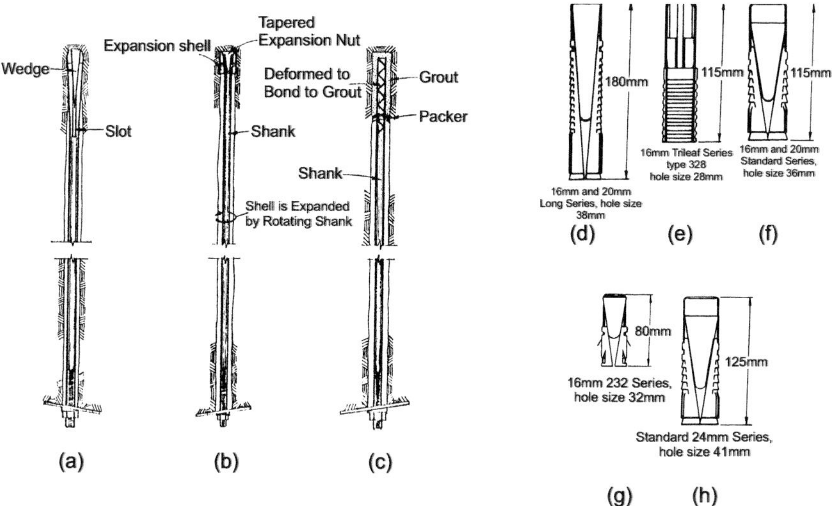

Figure 14.15 Mechanically anchored bolts and various anchoring devices

spun into the resin to anchor it and then, after setting, has a plate and nut placed on it for tensioning.

(b) Fully grouted and tensioned rock bolts

Although the point anchored mechanical bolts provide an immediate active support to the rock they suffer from a long term problem in that the tensioned bolts lose their effective tension over a period as follows.

- The attack of aggressive acidic groundwater on both the anchorage and the tensioning end would cause loss of tension of the bolt
- The possible damage to the tensioning end of the bolt by construction machines
- Loss of tension of the bolt due to slippage of the anchor
- The seismic loads due to blasting or creep effects or strata movement along the joints or bedding plane may cause spalling of the rock. This will result in loss of tension in the bolt due to spalling around and under the bearing plate of the bolt.

Some of the above-mentioned difficulties can be reduced by providing the following remedial actions.

(i) Using corrosion resistant rock bolts

(ii) Filling the bolted hole completely with resin or cement grout in order to retain effective tension in the rock bolt by preventing the slippage of anchorage and loss of tension in the bearing plate due to spalling

A typical arrangement of a grouted solid expansion anchorage bolt and a hollow core grouted rock bolt is shown in Figure 14.16.

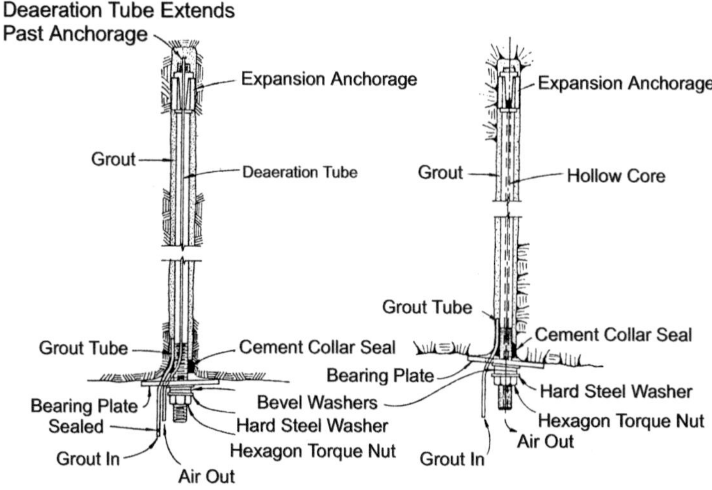

Figure 14.16 Fully anchored tensioned rock bolts

(iii) The fully grouted tensioned bolts offer shear resistance to lateral movements and prevent the bolt shank being pinched

(c) Fully resin grouted un-tensioned steel dowels

A full column contact resin grouted bolt consists of a fully deformed steel reinforcing bar with a threaded end. Resin cartridges containing the resin and the catalyst separated in the plastic container, are pushed into the borehole with a loading stick. The rebar is then inserted in the borehole and is spun. The resin cartridge breaks and thoroughly mixes with the accelerator and activates the resin to set in 20 seconds to 2 minutes. When the resin is set, it obtains high anchorage strength per unit length. The grouted rebar provides a passive support to the rock mass and provides axial resistance to rock movement as well as to shear movements along a joint plane or a bedding plane as shown in Figure 14.17. Thus the function of a rock dowel is analogous to ordinary reinforced concrete providing passive resistance to tensile and shear stresses and stabilising the rock mass.

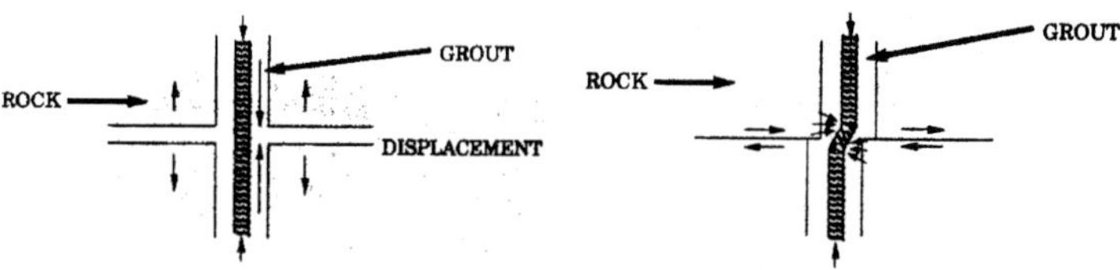

Figure 14.17 Mechanism of support for fully grouted dowels

As the resin cartridges are expensive and their shelf life is limited in high temperature environments, cement anchorages using high alumina cement have been developed which provide heavy hardening properties to the cement. Full column cement grouted dowels have the capacity for achieving loads up to 100 KN two hours after final set.

This kind of anchorage has been used as rib bolts, wooden dowels as well as steel roof bolts out bye in broken strata.

(d) Friction rock stabilisers/split set support system

The friction rock stabiliser is a hollow, compressible, high-strength steel tube with a split extending its full length of the bolt, a taper at the upper end and a retaining ring at the bottom. The split set tube is driven into a bore hole some 3 mm of smaller diameter than the tube as shown in Figure 14.18. The friction rock stabiliser anchors itself by the outward expansion force of the tube against the borehole wall. The anchoring force is distributed throughout the entire length of the borehole tube which develops resistance to axial movement of about 4 to 5 tonnes per metre length. The friction rock stabiliser exerts triaxial loading on the rock in the bolted zone which in turn increases the rock strength greatly.

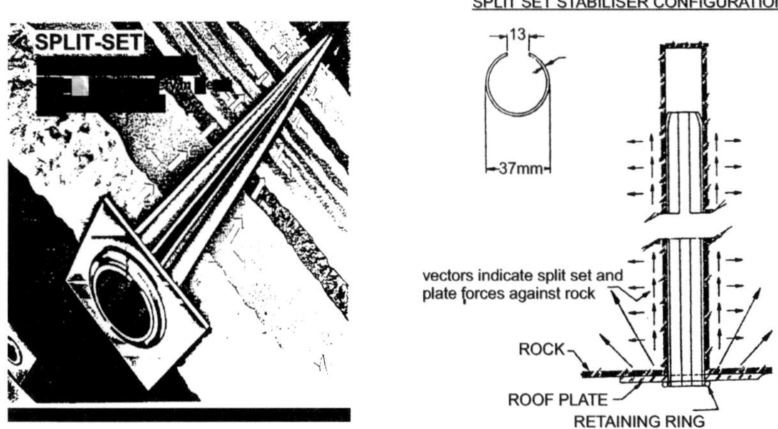

Figure 14.18 Split set friction rock stabiliser

The split set offers the advantage of providing resistance to ground movements caused by dynamic loading when the tube tends to distort and set more firmly within the hole, thus arresting the rock movement. Split sets can also withstand greater shearing stresses from lateral shifting of the bedding planes.

(e) Combination and special type rock bolts

Fibre glass bolts

Fibre glass is extremely strong in tension but easily cuttable by machine. It can therefore be used for longwall gate roadways and longwall face supports to prevent side and face spalling. The fibre glass dowel comprises a bunch of fibre glass strands inserted in the borehole and grouted in place using a polyester resin together with a hardener (Fig. 14.19(a)). As the fibre glass strand sets in the borehole the strand is cut to the desired length after installation. This bolting system offers the advantage of being highly flexible and can be inserted in curved and variable length bore holes. However, it is limited in application due to competitive advantages offered by other bolting systems.

Swellex Bolt

The swellex bolting system is another innovative friction rock stabiliser which has the advantage of requiring less head room for installation and simple installation equipment. It is essentially a 42 mm diameter and 2 mm thick steel tube which is reshaped to 29 mm diameter and reinforced by two welded short sleeves at both ends (Fig. 14.19(b)). The bolt is inserted into a 32 mm diameter and 1.8 m long borehole and expanded by injecting high pressure water in the bolt through a hole in a flanged sleeve through a detachable chuck. On expansion the bolt conforms the shape of the borehole and exerts radial pressure on the borehole wall. This bolting system has a lower ultimate strength, the ability of maintaining high anchorage resistance in spite of considerable rock deformation and therefore it is suitable for application in soft coal measures rock. This system offers the added advantage of being suitable for installation in water bearing ground and does not require special installation equipment other than a high pressure water hose.

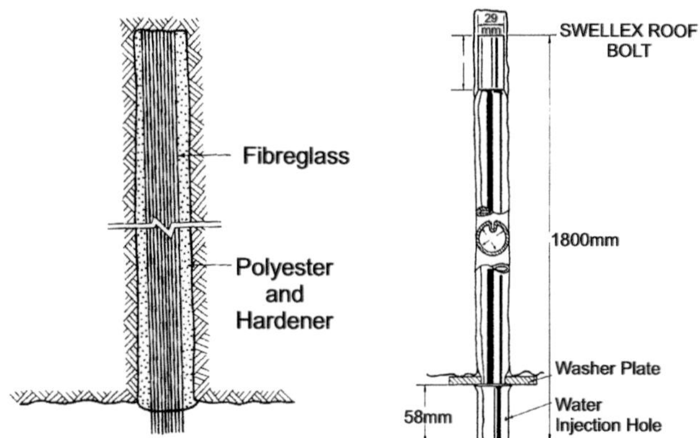

Figure 14.19 (a) Fibre glass bolt (b) Swellex bolt

Mechanical and resin anchored bolts

In the past many special types of rock bolts and combination of support systems have been used to surmount certain difficulties or for special applications. In soft coal measures rock, a mechanical anchorage may induce localised high stress at the point load anchor which may cause localised rock failure and slippage of the bolt. This can be prevented by use of the additional anchorage provided by a resin cartridge.

Cable sling/Roof stitching

The cable sling is an innovative roof control system providing continuous support to the full width of the roadway similar to that provided by a crossbar. Two 2.4 m long inclined boreholes are drilled over the pillars to the desired depth as shown in Figure 14.20. Slings are cut to correct lengths, inserted in the boreholes, split sets half inserted in the bore holes and the sling ends grouted at the borehole end with cement/resin. Two to three roof laggings are placed over two slings at specified intervals to serve as load transfer laggings for the mine roof. At this juncture the bearing plates are inserted on the split set before fully tensioning the support system and inserting the split to the full borehole end. Thus, the system provides full friction support all along the roof surface. This support system can also be used in a variety of ways to stabilise irregular rock surfaces such as pillar sides, ribs etc (Ghose and Raju, 1980).

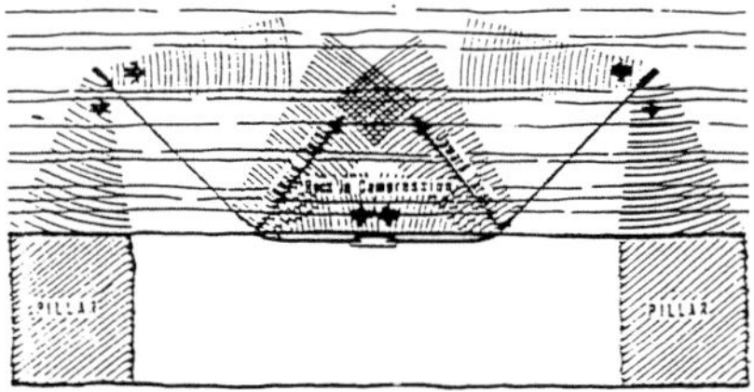

Figure 14.20 Cable sling as a roadway support

14.5.4 Design of Bolting System

Design of a rock bolting system for a mine requires the following considerations before finalising the system to be used.

(i) Size/width of the excavation to be supported since the length of bolt selected depends on the span of the opening.

(ii) Estimation of the depth of the loosened destressed zone as the length of bolt depends on this factor.

(iii) Depth of anchorage necessary depends upon the estimated weight of rock to be supported by each anchor.

(iv) The period for which support is required depends upon the estimated life of the opening, environmental conditions and whether the atmosphere is dry or wet or corrosive as the selection of the final bolt depends upon the expected life for which the bolt is required. For example, under dry conditions bolts will last up to five years while in wet and acidic mine environments the bolt may be corroded in three months, requiring the bolt to be fully grouted to increase its corrosion resistance.

(v) Ground conditions and geological structure have a major bearing on the selection of the type of the bolt to be used. For example in broken blocky ground, a tensioned rock bolt may be necessary while in massive solid rock an un-tensioned rock is all that is needed to be installed. In weak or leached rock a full column resin grouted anchor will be necessary to prevent the anchor becoming de-bonded due to strata deterioration.

(vi) Expected strata pressure, duration and frequency of the abutment pressures will determine if other standing supports or permanent supports are required.

(vii) The orientation of the opening with respect to the direction of the structural features and fracture planes will determine the rock bolt pattern and orientation.

Length and spacing of the bolt

The span of the opening is one of the important parameters in determining the length of bolt required for a given situation. Representative formulae derived by various workers are given below:

(a) For hard rock mining practice in Scandinavia;

$$L = 1.40 + 0.184 \times W \tag{14.1a}$$

where,

W = width of the opening (m)

L = length of bolt (m)

S = spacing of the bolt (m)

(b) Experience in the Snowy Mountain project,

$$L = 1.0 + 0.003 \times W^2 \tag{14.1b}$$

(c) Austrian tunnelling method, Rabcewicz (1955)

$$L = 0.5\,W \tag{14.1c}$$

(d) Based on the experience of Mount Isa mines

$$L = 0.5\,W \tag{14.1d}$$

(e) Based on Hoek and Brown (1980)

$$L = \text{between } \frac{W}{3} \text{ and } \frac{W}{2} \tag{14.1e}$$

The bolting density to be used with above formulae is generally assumed to be

$$S = 0.5 \times L \tag{14.1f}$$

Support load evaluation

Basically there are two approaches which can be used for the estimation of support loading (Daws, 1980).

(1) Estimation of the dead weight of the de-stressed material acting on the support system due to formation of the roadway. The active load due to volume expansion can be calculated as follows:

$$P_a = P\left[1 + \frac{1 - \sin\phi}{1 + \sin\phi}\right] \tag{14.2}$$

where

P = rock load on support system (MN)

P_a = active load on the support system (MN)

ϕ = angle of internal friction of rock

Table 14.2 gives the load on the support system calculated by various formulae for a 5.5 m × 4.3 m high arched-shaped roadway. It may be noted that the results vary considerably but the method used by Daws (1980) and Whittaker and Singh (1980) are of the same order.

(2) The second approach calculates the width of the yield zone surrounding the excavation using Wilson's formula discussed earlier:

$$X_b = \frac{H}{2}\left[\left(\frac{q}{p' + p_1}\right)^{\frac{1}{K-1}} - 1\right] \tag{14.3}$$

where

X_b = width of the yield zone (m)

H = equivalent diameter of the opening

q = strata stress due to depth (MPa)

K = triaxial stress factor of rock

p' = support pressure

p_1 = residual adhesion of rock in yield zone

Calculation of rock bolting design parameters

The design of a rock bolting system is based on the concept that systematic rock bolting creates a reinforced arching zone which provides a uniform zone of compression. In soft coal measure rock this rock arch provides an effective resistance to the strata surrounding the excavation so as to prevent the extension of the yield zone (Fig. 14.21).

Table 14.2 Load on support system calculated by different formulae for an arched-shaped roadway-size (5.5 × 4.3 m)

Author	*Assumptions*	*Height of envelope*	*Area of broken ground/m length, m²*	*Dead load on support tonnes/m*	*Active support load tonnes/m*
Airey (1974)	Broken Envelope Triangular.	$L/2 \text{ Cot } \phi = 4.05$ m	$\frac{L^2}{4} \text{ Cot } \phi = 11.10$	$L^2 \text{ Cot } \phi\gamma = 27.8$	35.6
Dejean and Raffoux (1980)	Parabolic Envelope	$L/4 \text{ Tan}\left(\frac{\pi}{4}+\frac{\phi}{2}\right) = 2.58$ m	$\frac{L^2}{6} \text{ Tan}\left(\frac{\pi}{4}+\frac{\phi}{2}\right) = 9.45$	$\frac{L^2}{6} \text{ Tan}\left(\frac{\pi}{4}+\frac{\phi}{2}\right)\gamma = 23.6$	30.3
Whittaker & Singh (1980)	Broken Envelope is arched-shaped. Can approx. to a rectangle.	H = 4.3 m	L.H = 23.65	L.H.γ. = 61.5	78.7
Daws (1980)	Triangular Envelope	1.5 L	$0.75\, L^2$	$0.75\, L^2.r$ = 57.8 tonnes/m	74.0
Terzaghi	Elliptical Envelope	Calculated by	*Vertical Stress*	*Horizontal Stress*	
(1943)		Wilson's Formula	$\sigma_v = \frac{B}{2K_1 \text{ Tan } \phi}\left(\frac{\gamma - 2c}{B}\right) \times$	$\sigma_H = 0.3\left(\sigma_v + \frac{1}{2}\gamma h\right)$	100.60 Vertical
		$\frac{H}{2}\left[\frac{(q)^{\frac{1}{K-1}}}{(p+p')} - 1\right]$	$\left(1 - e^{-K_1 \frac{\text{Tan } \phi\, 2D}{B}}\right)$	= 6.27 tonnes/m³ or	35.00 Horizontal
			= 15.38 Tonnes/m³ or 84.7 Tonnes/m	= 27.09 tonnes/m	

(*Source:* Daws, 1980)

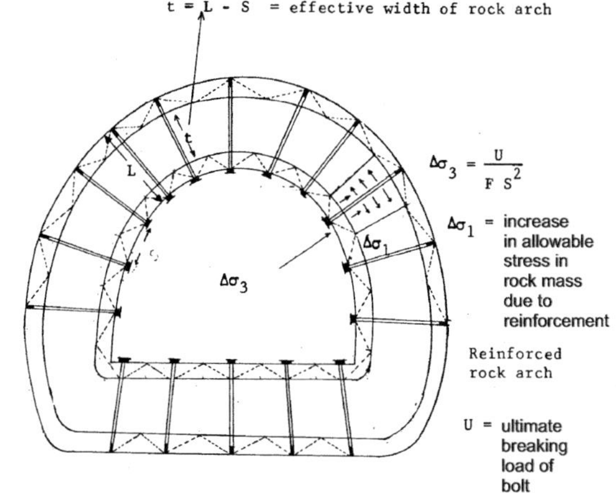

F—factor of safety, S—bolt spacing (m), L—length of rock bolt

Figure 14.21 Mechanism of roadway reinforcement

The installation of bolts effectively increases the confining stress in the rock mass within the reinforced arch by $\Delta\sigma_3$ thereby supporting the overlying rock surrounding the roadway by $\Delta\sigma_1$. The rock load supported by the bolt is given by:

$$P_a = \Delta\sigma_1 \times t \tag{14.4a}$$

$$P_a = \Delta\sigma_1 (L - S) \tag{14.4b}$$

$$\Delta\sigma_1 = \Delta\sigma_3 \times \tan^2\left(\frac{\pi}{4} + \frac{\phi}{2}\right) \tag{14.4c}$$

$$\Delta\sigma_3 = P \tan^2\left(\frac{\pi}{4} + \frac{\phi}{2}\right) \tag{14.4d}$$

Combining all the above equations we have the following quadratic equation:

$$P_a^2 FS^2 - U\,L\,\tan^2\left(45^\circ + \frac{\phi}{2}\right) + U\,S\,\tan^2\left(45^\circ + \frac{\phi}{2}\right) = 0 \tag{14.4e}$$

where t = effective thickness of the reinforced arch (m) = (L–S)
L = bolt length (m)
S = bolt spacing (m)
P_a = active rock load on the support (tonnes/m)
P = dead load acting on the support (tonnes/m)
U = ultimate strength of bolt in tonnes
F = factor of safety
ϕ = internal angle of friction of rock

Worked Example

A roadway 5.4 m wide and 4.3 m high is to be supported by rock bolts 22 mm in diameter. The internal angle of friction of the rock mass is 34.1°, the yield strength of the bolt is 16.5 tonnes and factor of safety is 1.3. Calculate the bolt spacing for the roof, floor and sides of the roadway.

Solution Taking the length of the bolt as 1/3 that of the width of the roadway (Hoek and Brown, 1980) = 1.8 m

Diameter of borehole = 25 mm

U = 16.5 tonnes

F = 1.3

P = 78.7 tonnes (Table 14.2)

Uniaxial compressive strength of rock = 36 MPa

Rock unit weight $\gamma = \tilde{22}$ kN/m^3

Modulus of Elasticity = 24.2 GPa

Poisson's ratio = 0.12

Triaxial stress factor $K = \tan^2 (45° + \phi/2) = 3.55$

Roof and floor

Substituting the above values in Equation (14.4e) :

$$78.6 \times 1.3 \times S^2 - 3.55 \times 16.5 \times 1.8 + 3.55 \times 16.5 \times S = 0$$

$$102.3\, S^2 + 58.7\, S - 105.43 = 0$$

$$S = 0.85 \text{ m}$$

Sides

$$\text{Horizontal load} = \sigma_H = 0.3\left[\sigma_V + \frac{\gamma}{2} H\right]$$

$$= 0.3\left[78.7 + \frac{2.6}{2}\, 4.2\right] = 25.25 \text{ tonnes/m}$$

Bolt spacing on the side of the roadway can be found as follows:

$$25.25 \times 1.3 \times S^2 + 3.55 \times 16.5 \times S - 105.4 = 0$$

$$S = 1.18 \text{ m}$$

14.5.5 Monitoring of Rock Bolts and Pull Tests

In order to test the effectiveness of an individual rock bolt and also the efficacy of the rock bolting system a range of tests and monitoring programmes are used. These monitoring systems can be divided into following categories.

(i) Anchorage capacity test or pull out test on individual bolts

(ii) Rock bolt monitoring using load cells or instrumented rock bolts

(iii) Measurement of bed separation using single point or multi-point borehole extensometers
(iv) Roadway deformation survey to monitor overall efficacy of the support system

Anchorage capacity test or pull-out test on individual bolts

Figure 14.22 shows a typical pull-out test arrangement for a threaded rock bolt. An extension rod is coupled to rock bolt to be tested by a double threaded extension sleeve. This enables a deflection dial gauge, a reaction chair or a reaction bridge and a hollow jack to be installed on to the rock bolt for the test. The purpose of the reaction bridge to transmit reaction pressure of the jack remote from the rock being supported by the bolt. The pumping pressure of the hollow jack can be related to the load on the bolt while the bolt extension can be measured by a dial gauge to an accuracy of ±0.01 mm (Figure 14.22). The jack calibration accuracy should be within ±1% as measured from two gauges. Figure 14.23 and Table 14.3 show typical results of pull out tests on the most common type of rock bolts used in coal mining.

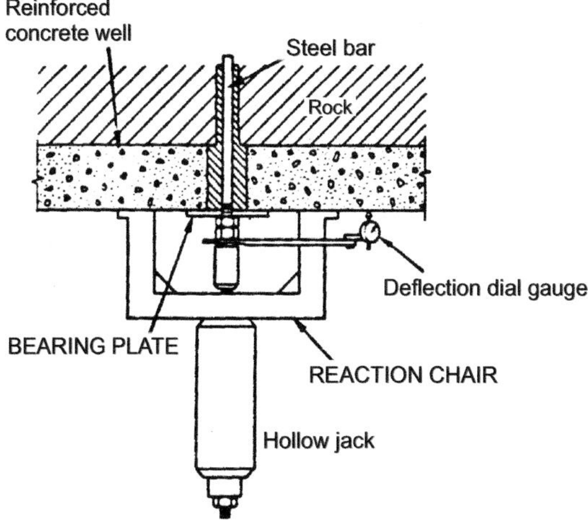

Figure 14.22 Pull-out test equipment using a hollow hydraulic jack

Table 14.3 Typical test results on common rock blots

Bolt type	*Average yield load (tonnes)*	*Average failure load (tonnes)*
High tensile bolt	15	26.0
Mild steel bolt	12	18.5
Swellex bolt	9	12.0
Split set	9	13.0

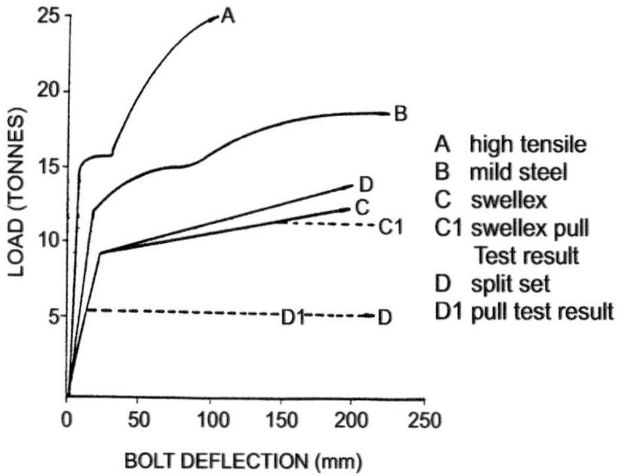

Figure 14.23 Pull-out test on rock bolt (Richmond *et al.*, 1980)

Rock bolt monitoring using load cells or instrumented rock bolts

In order to monitor the behaviour of a rock bolt *in situ* and to evolve design criteria, it is advisable to measure the load generated in the rock bolt and the magnitude of strata displacement at various points in the roof. A variety of rock bolt load cells with capacities of up to 400 KN are available which can be axially placed on the collar of the rock bolt and are designed to be used in underground coalmines (Norris and Yearby, 1980; Peng, 1986). Strain gauged bolts first used by Dunham (1976) are installed for the measurement of the amount and direction of strata movement in the bolted roof strata. The strain measurement devices consist of two precise diametric strain gauges installed at nine locations along the length of the bolt. This enables the measurement of forces and displacements developed at various levels in to the roof of a bolted roadway.

Routine measurement of bed separation

Figure 14.24 shows a monitoring scheme adopted by Cerchar in France for measuring strata movements in fully bolted mine roadways. The single point extensometer comprising a wooden plug with a strain wire is anchored 100 mm above the bolting horizon in a borehole located in the mine roof. A tapered wooden plug with a centre hole acts as a reference plate with respect to which the change of length of a marker on the strain wire is measured. The extensometer wire is tensioned by a standard weight and a simple tape measure is used for recording roof convergence to an accuracy of ±1 mm. These single point extensometers are installed every 10–15 m distance along a roadway and convergence data is recorded and processed every day on a computer located in the mine control room. The results

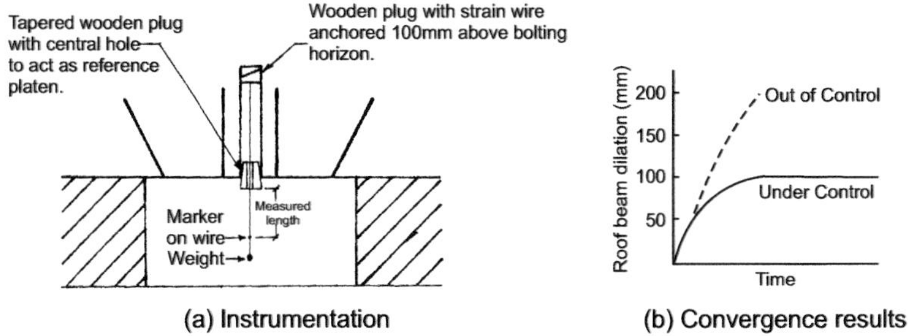

(a) Instrumentation (b) Convergence results

Figure 14.24 Extensometer for monitoring stability of rock bolted roadway

are plotted in convergence time graph shown in Figure 14.24(b). If the convergence exceeds the critical bed separation for a given coalfield, standing supports or additional supports are installed to improve the roadway stability (Daws, 1987)

Multi-position borehole extensometers

Figure 14.25 shows two types of multi-position extensometers used to carry out roof stability measurements in 6–8 m long boreholes placed vertically in the roof strata of a roadway. Displacement and load measurements are taken for each advance cycle of the heading or at more closely spaced intervals.

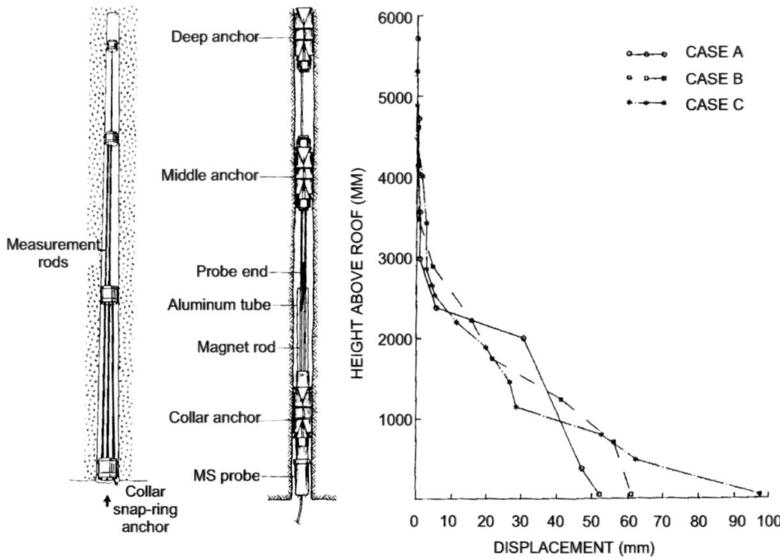

Figure 14.25 Multi-position borehole extensometers for measuring axial displacements at various heights of roof section

The displacement results in Figure 14.23 indicates (1) the depth of loosened zone in the roof, (2) bed separation at various anchors and (3) the condition of the rock above the opening. The slope of the profile section is indicative of the properties of rock in that section. For example, the shallow slope section represents fractured and weakened roof beds. Plotting of the height versus strain data can show points of high strains which are indicators of major partings, lack of rock integrity and development of increased strata loading.

In addition, instrumented roof bolts are installed on the roof in order to measure the forces generated in the encapsulated length of the bolt. Figure 14.26(a) shows a typical load development with respect to time of a short anchored bolt. Figure 14.26(b) shows load (KN) on a 2.4 m long fully encapsulated resin bonded bolt at nine measuring points along the length of the bolt showing that load is generated at two points A and B. Thus, the strata monitoring schemes are used to investigate the following parameters.

- Anchorage failure of roof bolts
- Overloading of bolts
- Excessive bed separation and requirement for standing supports and a remedial measure
- Establish long-term stability of the roadway

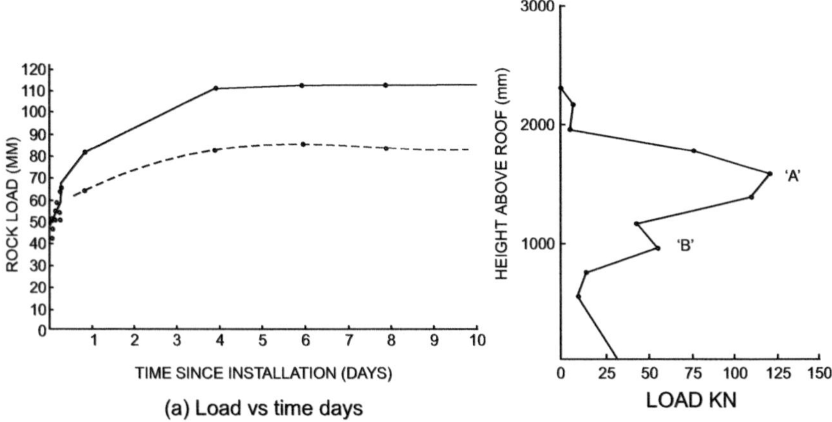

Figure 14.26 (a) Rock bolt load on 0.8 m encapsulated bolt and (b) load distribution along a 2.4 m-long fully encapsulated bolt

14.5.6 Roof Stability Performance for the Roadway Supported by Conventional Supports and Rock Bolted Roadway (Table 14.4–A Comparison)

Site conditions

Roadway width and height = 5.5 m width × 2.4 m height
Length of bolts = 1.8 m

Bolts per w-strap = 6 bolts per strap
Spacing between straps = 1.0 m

Support cost per metre length of the bolted roadway

Resin capsules (6)	=	\$19.32
Bolts (6)	=	\$31.74
Dome plates	=	\$5.04
Hemispherical seat	=	\$2.52
W-strap	=	\$16.94
Sub-total	=	\$75.50

Other cost (drilling, capital and running cost) = \$75.00

Total cost per metre	=	\$150.00
Cost per tonne	=	\$8.00 to \$9.50

Table 14.4 Roadway support characteristics for conventional and bolted roadway

Stability criteria	*Steel support Steel legs and crossbars*	*Rock bolt support*
Roof displacement	>100 mm	<40 mm
Height of failure	2.5 m	<1.9 m
Roof strength achieved in Immediate 2 m roof	1.75 MPa	6.0 MPa
Cost per tonne	\$26.6/t–\$31.64/t	\$8.00–\$9.50/t

14.6 Hydraulic Supports

14.6.1 Hydraulic Props

The hydraulic prop was the first step towards the mechanisation of roof support systems. It is virtually a self-contained upright hydraulic ram providing an active preload pressure to the roof strata by extending the ram and providing support resistance up to the designed capacity.

Light duty hydraulic props may have a capacity of 5 tonnes to 15 tonnes whereas high capacity props have a rated capacity of 20 tonnes to 45 tonnes and a height of 1.4 m to 4 m (Figure 14.27). The support is operated by setting it upright and operating the release valve in order to extend the prop against the roof. The operating valve is manually operated until load of 5 tonnes is achieved. The load on the prop continues to be built up as the strata converges until the design loading capacity of the prop is reached. Thus, the hydraulic prop is an active support pre-stressing the strata up to a designed load.

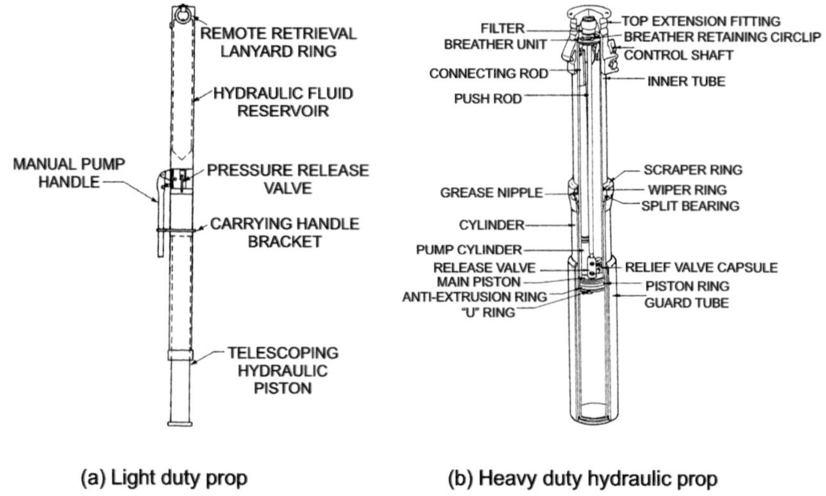

(a) Light duty prop (b) Heavy duty hydraulic prop

Figure 14.27 Hydraulic props

14.6.2 Breaker Line Support

This is a crawler mounted self-contained and self-propelled hydraulic chock support designed in South Africa for safe extraction of pillars in room and pillar operations. The breaker line hydraulic support shown in Figure 14.28 was first introduced in Australia in 1987 for extraction of pillars in the Wongawilli system of mining

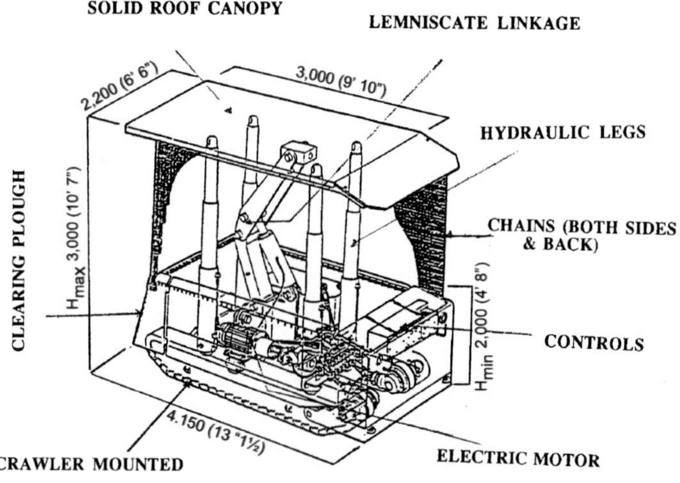

Figure 14.28 Breaker line powered support

The breaker line support consists of four vertical hydraulic props connecting the base of the support to a 5 m long × 2.5 m wide solid roof canopy and has a leminscate link interconnecting the base of the support to the roof canopy. This linkage comprises a geometrically designed four bar system with floating support cylinders and a lemniscate guided rear canopy to provide near vertical movement to the roof canopy throughout the support height range.

The yield load of the BLS is 540 tonnes and the working height of the support is between 2 m and 3.8 m. It has a length of 4.15 m, width of 2.2 m and weighs 21 tonnes. The power of the traction motor is 30 kW and the flitting speed 12 m/min. The support can be operated by a remote control and also by manually operated control valves located in the front of the support. Thus, the breaker line support is a completely self-contained support having its own hydraulic system requiring an electric power supply. The support is equipped with chain types of screens at both sides to protect it from accumulation of the roof flushings. A hydraulically operated plough located in front of the breaker line powered support keeps the floor in the vicinity of the BLS free from loose debris.

Since the introduction of the breaker line support, major advances in the safety against roof falls during depillaring operations have been achieved. Use of breaker line supports permits lifting of the pillar at both sides of the split, thus reducing the number of splits per pillar and hence reducing the cost of development during depillaring operations.

The following precautions are to be taken during the use of the breaker line support:

- Advance the goaf side breaker line support first.
- The minimum distance between two adjacent BLS units should be 0.6 m to 1.5 m.
- If the fender is < 10 m in width, then it should be completely extracted and then the breaker line support should be repositioned for the next sequence.
- BLS gauges are to be checked before moving the chock. If the BLS gauge indicates at or near yield loads, then the support should not be moved until caving or yielding of the support has taken place.
- If there is a danger of the BLS becoming jammed in the closed position, timber chocks should be erected in close proximity to share the load.
- Under heavy strata loading conditions the breaker line support should be moved only half of its length at a time.
- The electric cable supplying power to the BLS should be properly supported and protected.
- Additional care should be taken during the withdrawal of the breaker line support.

14.6.3 Longwall Powered Supports

Development of longwall face powered support is one the major contributory factors for reducing drastically accident rates at the longwall face due to roof fall. Basically two types of powered supports are available.

(1) Conventional types of powered support for supporting weak to moderate strength roof strata where the strata loading is predominantly vertical and the roof is easily amenable to caving behind the powered support.

(2) Shield support which is a high strength, high capacity powered support designed to work in strata offering a high lateral loading on the support and the roof is difficult to cave behind the powered support.

(a) Conventional chock-type powered support

This type of powered support has been successfully used for medium to thin seam condition especially in weak strata conditions. Figure 14.29(a) shows a four-leg conventional powered support comprising four-telescopic legs housed in the self-centring leg housings within the base structure of the chock. The powered support applies load to the roof strata through a full width rigid steel canopy to which an anti-flushing back shield is hinged.

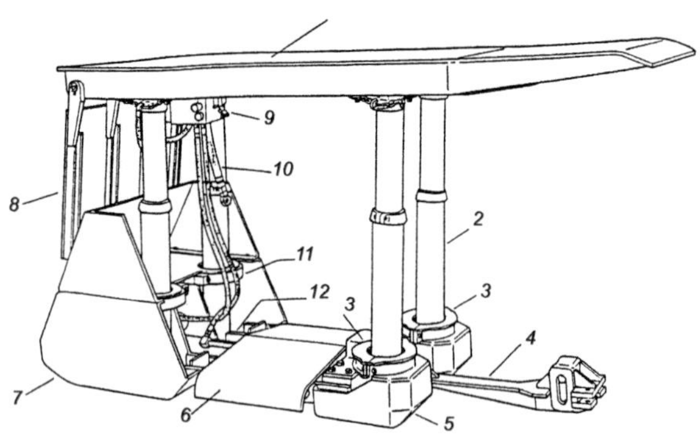

1. Full width rigid canopy	**2. Leg**
3. Self-centring leg housing	**4. Double acting ram**
5. Front base structure	**6. Walkway floor cover**
7. Rear base structure	**8. Anti-flushing shield**
9. Hydraulic control valve	**10. Hydraulic hoses**
11. Stabiliser	**12. Frame bars**

Figure 14.29(a) Chock-type powered support

A double acting ram connects the base of the powered support with the pan of the Armoured Face Conveyor (AFC). The space between the front and back legs of the support provides a good travelling way which may be obstructed with the flushing from the caved waste.

The conventional chock type powered support is a lightweight support which can be easily transported and assembled underground. It is also a fairly inexpensive support which can be easily repaired underground and easily maintained. The main disadvantages of this type of support is that lateral strata loading may cause the legs to lean towards the goaf especially in thick seam conditions and may cause major damage to the support if not properly dealt with. Problems can also arise in massive roof conditions (e.g. sandstone) where the strata does not easily cave due to insufficient strength of support. Thus, the support is incapable of withstanding high lateral, vertical and sudden loads.

(b) Chock shield type of support

Figure 14.29(b) shows a four-leg chock shield type of powered support that is a stable support suitable for operating in the strongest roof conditions. This type of support is suitable for thick seam conditions and gives excellent stability against high lateral thrusts. This support also provides high vertical thrust to the roof strata necessary for inducing effective caving in strong roof conditions. Being a heavy duty chock, it is capable of withstanding sudden roof weighting in strong massive roof conditions. This support offers a good travelling way for face workers and relatively easily access to face machinery.

This support system is applicable to medium to thick seams in flat and steep seam conditions and is regarded as superior to all other longwall face supports being able to deal with almost any mining and geological conditions.

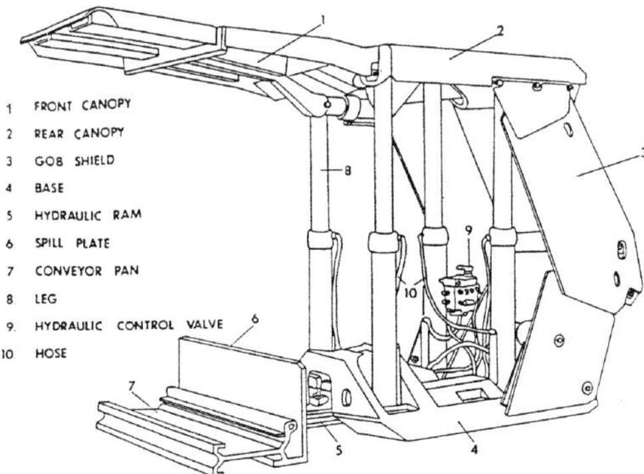

Figure 14.29(b) Chock shield type support

Figure 14.29(c) shows an inclined leg chock shield suitable for thick seam conditions.

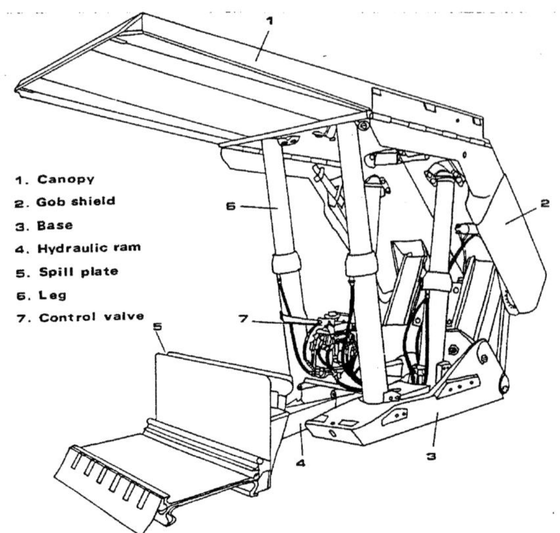

Figure 14.29(c) Inclined leg thick seam support

(c) Two-leg shield support

Figure 14.29(d) shows a two-leg lemniscate link type shield support for thick seam mining conditions. It consists of two inclined support legs connecting the base with the solid roof canopy. The lemniscate linkage assembly

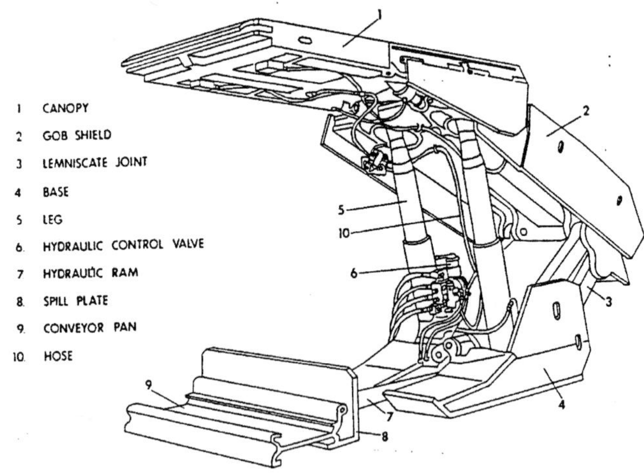

Figure 14.29(d) Two-leg leminscate thick seam shield support

connects the base plate of the support with a heavy duty gob shield and the roof canopy. This linkage permits virtually a vertical movement of the canopy throughout the height range of the powered support without imparting lateral thrust and twisting action to the hydraulic legs of the support. Moreover, the yield load of the support is independent of friction owing to absence of lateral movement of roof canopy. This is a heavy-duty support suitable for strong roof conditions but it needs a strong floor. It has excellent capabilities to cave strong roofs and for waste flush control.

(d) Two-leg calliper shield

Figure 14.29(e) shows a two-leg hinge link shield type of support for thin seam mining conditions. The calliper type of shield support offers lateral restraint to lateral movement of the roof. This type of support is particularly suitable for thin to medium seams less than 1.5 m owing to its ability to close up to a compact size for transport into mine and subsequent face transfers. As the prop free front for this support between the support legs and the face is relatively large, this type of support is applicable to good roof conditions. The calliper type of support offers a fairly free walkway for operators and a good access way for materials and machinery. However, for thick seam conditions spalling of coal is possible as it is not possible to incorporate powered sprags and fore poles on the powered supports. The hinge type of connection between the base plate and the goaf shield permits the roof canopy to move along an arc, thus, changing the distance between the

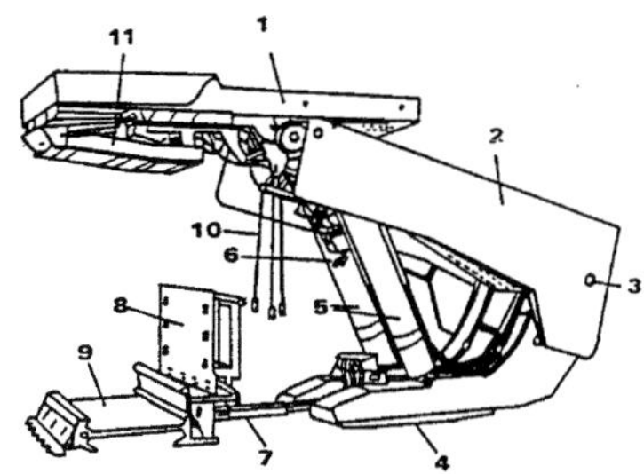

1. Canopy
2. Gob shield
3. Hinge
4. Base
5. Legs
6. Hydraulic control valve
7. Hydraulic ram
8. Spill plate
9. Conveyor pan
10. Hose
11. Antispalling plate

Figure 14.29(e) Two-leg shield support

canopy tip and the face throughout the height range of the support. This generates frictional forces within the linkage causing equipment damage and lateral movement of the roof canopy towards the goaf during support setting operations. Frictional force also reduces the supporting capacity of the shield support.

14.7 Powered Support Design

14.7.1 Determination of Caving Height

The design of the loading capacity of a powered support is based on the assumption that a detached block of thickness H, fractures and rests on the face powered support (Wilson, 1975; Ashwin *et al.*, 1970).

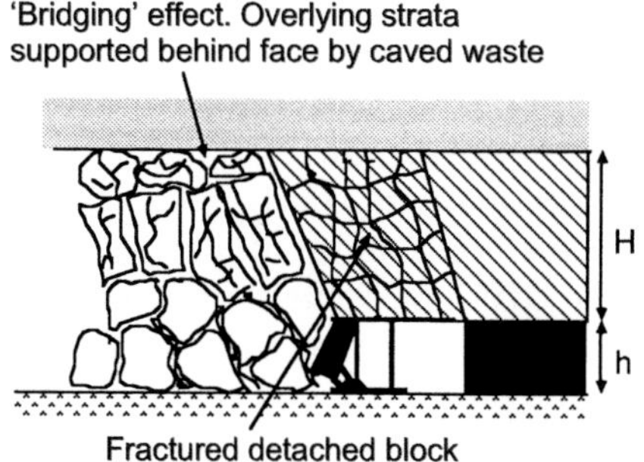

Figure 14.30 Detached block theory

As the powered support advances the detached block caves and bulks up to the level of the main roof and fills the void behind the support. Thus, the main roof is supported in front by solid coal and on the goaf side by the caved waste material. In ideal conditions, the height of the detached block can be calculated by the following equation:

$$(c+h)\rho_b = H\rho \qquad (14.5a)$$

$$\rho_b \times b_f = \rho$$

$$c = \frac{\rho_b}{(\rho - \rho_b)} \times h\ ; \qquad c = \frac{h}{b_f - 1} \qquad (14.5b)$$

where

H = height of detached block (m)
h = extracted seam height (m)
ρ_b = density of caved strata (t/m^3)
ρ = density of intact roof strata (t/m^3)

c = caving height (m)

b_f = bulking factor

If the bulking factor $\quad b_f = 1.5, \; c = 2\,h$

and if $\quad b_f = 1.25, \; c = 4\,h$

If the caving height (c) density of the intact rock and the support spacing are known, the minimum weight of each powered support setting load can be calculated.

This is multiplied by a factor of 1.5 will allow for lowering of the adjacent support. Thus, the setting and the yield load of the powered supports can be calculated as follows:

$$\text{Setting load} = 1.5 \times \text{block load}$$

$$\text{Yield load} = 2 \times \text{setting load}$$

14.7.2 British Support Requirement

Setting load and yield load can be related to the following parameters:

- Mining height 2 h
- Bulking factor 1.5.
- Caving coincides with the rating of the support canopy.

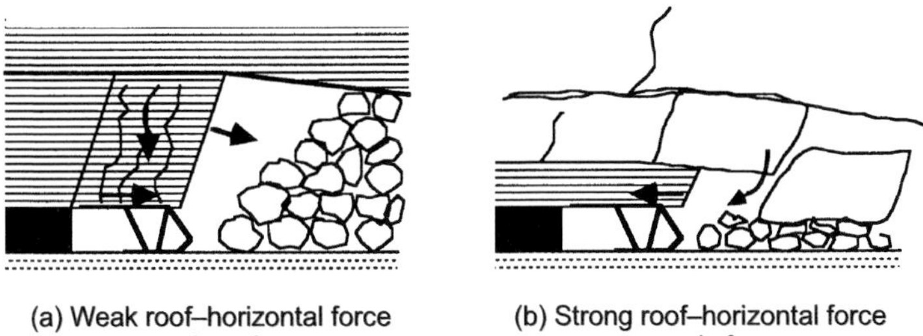

(a) Weak roof–horizontal force away from face

(b) Strong roof–horizontal force towards face

Figure 14.31 Loading on powered supports due to lateral strata movement

14.7.3 Setting Resistance

The setting resistance of the powered support can be calculated as follows:

$$\text{Weight of strata (t)} = \text{Area of room (m}^2) \times \text{Height of caved strata} \times \text{Strata density}$$

$$\text{Setting resistance} = \frac{\text{Weight of strata}}{\text{Area of roof}}$$

$$= \frac{s(s+d) \times 2h \times 2.5}{s(1+d_1)} = 5\,h\,(t/m^2) \qquad (14.6)$$

where

$$\begin{aligned} h &= \text{extracted seam height (m)} \\ s &= \text{support spacing (m)} \\ l &= \text{overall length of the powered support} \\ d_1 &= \text{maximum beam tip to face distance (m)} \\ w_1 &= \text{web depth (m)} \end{aligned}$$

This multiplied by a factor of 1.5 will allow for lowering of the adjacent support.

Minimum setting load $= 1.5 \times 5\,h\,(t/m^2) = 7.5\,h\,(t/m^2)$

Yield resistance $= 7.5\,h \times 2\,(t/m^2) = 15\,h\,(t/m^2)$

14.7.4 The 8 m Formula of Germany

In German mines, an empirical formula is used to select the ratings for a powered support for caved longwall faces. According to this formula, the minimum support resistance for the face powered support is taken as support in MPa equal to eight times the extracted thickness of the coal seam.

Assumptions

- Rock bulking factor of 1.5
- Specific weight of rock is $2.5\,t/m^3$
- Factor of safety of 1.6
- No strong beds occur in the immediate roof vicinity.

$$P = (5 + 0.15\,E)\,h \tag{14.7}$$

$$\begin{aligned} P &= \text{support resistance } (t/m^2) \\ E &= \text{inclination of seam (gon)} \\ h &= \text{seam thickness (m)} \end{aligned}$$

British and German approaches for support capacity give comparable values.

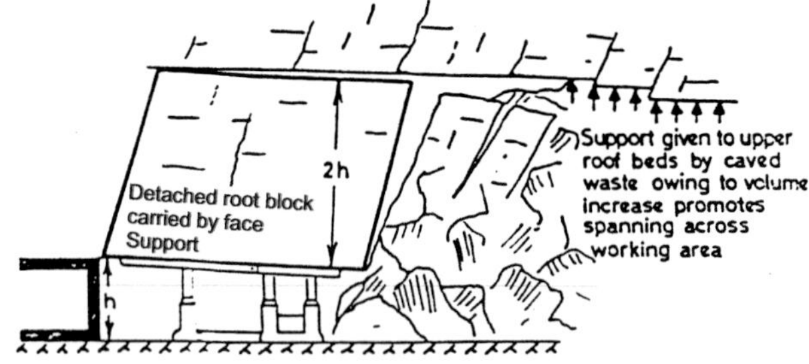

Figure 14.32 Loading face support by a roof block (Whittaker, 1974)

14.8 Development of Detached Block Theory

14.8.1 Support Rating in Level Seam

The weight of the detached block supported by the powered support depends upon the angle of break 'θ', depending upon the internal angle of friction of rock (Whittaker, 1975). For simplicity, the mean value of the angle of break 'θ' is taken as = 25° to 35°.

The powered support is represented by the resultant thrust by the front (R_f) and back legs (R_b) of the support acting at a distance D and (D+d) from the coal face. Three basic conditions of loading may arise in a level seam.

1. Centre of gravity of the block is between the front and back legs
2. Centre of gravity of block is in between the face and the front leg
3. Centre of gravity of the block is behind the back leg

(a) Centre of mass of roof block between front and back row of the support (Fig. 14.33)

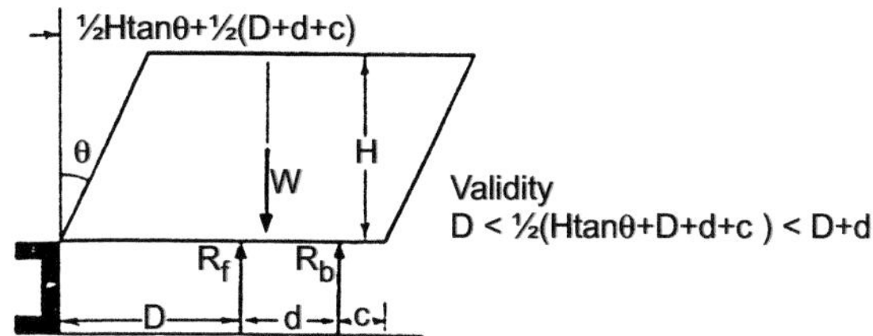

Figure 14.33 Weight of the block between front and back legs

For equilibrium

$$D < \frac{1}{2}(H \tan\theta + D + d + c) < D + d \text{ and}$$

$$W = R_f + R_b \tag{14.8a}$$

$$\frac{1}{2} W (H \tan\theta + D + d + c) = D R_f + (D + d) R_b$$

It may be noted that because of the location of the centre of gravity, the roof block is self-stabilising.

Taking moment about R_f (front leg)

$$R_b = \frac{W}{2d}(H \tan\theta - D + d + c) \tag{14.8b}$$

Taking moment about the back leg (R_b)

$$R_f = \frac{W}{2d}(D + d - H \tan\theta - c)$$

(b) Centre of gravity of the roof block is between the front leg and the face (Fig. 14.34)

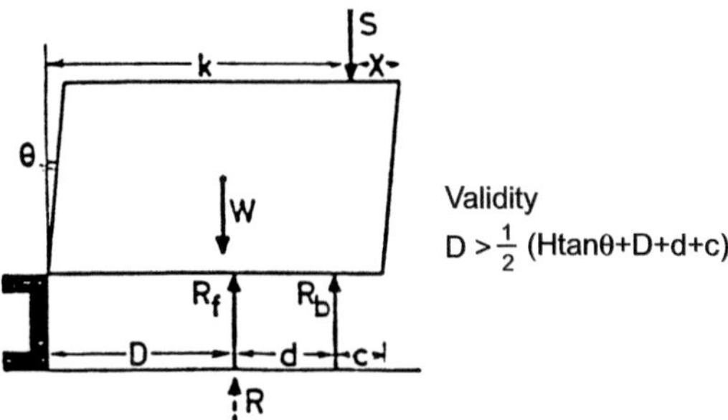

Figure 14.34 Centre of gravity between the face and front leg (Whittaker, 1975)

Condition of equilibrium

$$D > \frac{1}{2}(H\tan\theta + D + d + c) \text{ and}$$

$$W + S = R = R_f + R_b \tag{14.9a}$$

For minimum load condition $R_b = 0$ and $R_f = R$ and moment about the position of S.

$$R_f = \frac{W}{k - D}\left[k - \frac{1}{2}(H\tan\theta + D + d + c)\right] \text{ and } R_b = 0 \tag{14.9b}$$

and k is the position of S and is given by

$$k = H\tan\theta + D + d + c - X \tag{14.9c}$$

(c) Centre of gravity of roof block behind the back leg (Fig. 14.35)

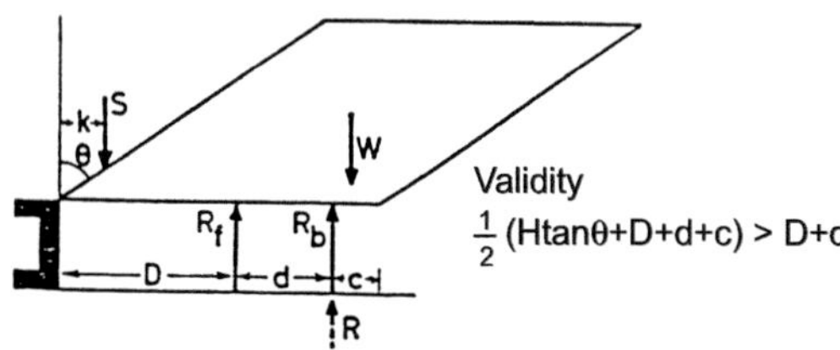

Figure 14.35 Centre of gravity behind the back leg (Whittaker, 1976)

This condition is defined as

$$\frac{1}{2}(H\tan\theta + D + d + c) > D + d \tag{14.9d}$$

For equilibrium in this case the following conditions must be met

$$W + S = R = R_b + R_f$$

$$(D+d)\times R = k\times S+\frac{1}{2}[H\tan\theta + D+d+c]\times W \quad (14.9e)$$

For minimum loading conditions

$$R_b = \frac{W}{(D+d-k)}\left[\frac{1}{2}(H\tan\theta + D+d+c)-k\right] \quad (14.9f)$$

and

$$R_f = 0$$

Based on the above analysis, Figure 14.36 shows the variation of loading of front and back legs.

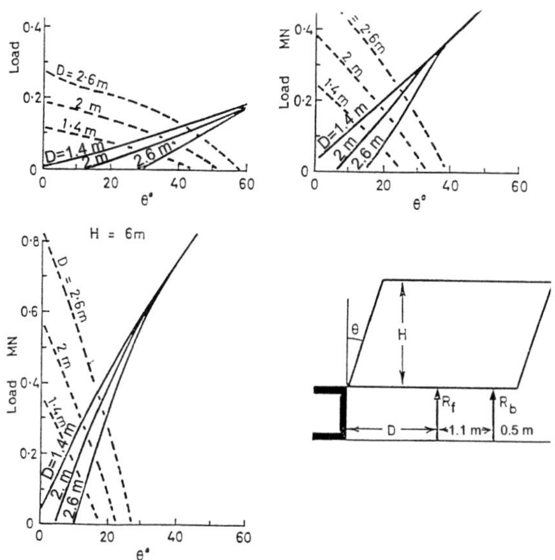

Figure 14.36 Variation of loading of front and back legs (Whittaker, 1975)

14.8.2 Steep Seam Working

Three alternative layouts of face advances are possible.

1 Working along the strike
2 Working towards the rise
3 Working towards the dip

In each case the forces acting on the roof block can be resolved along the plane giving:

$$W\sin\alpha = \mu S \qquad \text{and} \qquad W\cos\alpha + S = R$$

$$R = W\left[\cos\alpha + \frac{\sin\alpha}{\mu}\right] \quad (14.10)$$

According to Whittaker (1975) the above equations are valid irrespective of the direction of the face advance. Figure 14.37 shows a plot of the support load per unit thickness of the coal seam per metre length against the seam gradient from a level seam to a vertical seam. Support loads have been calculated for the prop-free front distance of 1.4 m, 2.0 m and 2.6 m.

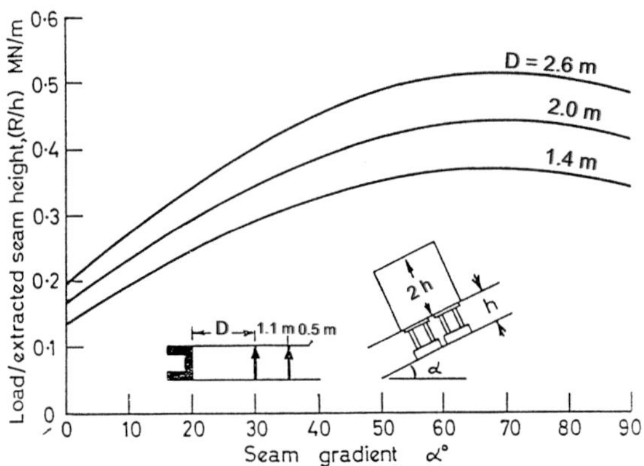

Figure 14.37 Variation of support loading with seam gradient compared to flat seam thrust (Whittaker, 1975)

The results indicate that the support loading per unit seam thickness increases considerably from level seam conditions to a seam gradient of 65° and then decreases slightly. Thus, the support load density for an inclined seam is considerably higher that that for a level seam.

14.8.3 Comparative Study of Various Formulae for Estimating Mean Support Densities of Powered Supports

Figure 14.38 shows that the suggested mean support density in terms of extracted seam heights as proposed in different countries is based on their local experience. It can be seen that the support loadings are minimum for the British mining conditions characterised by thin seam mining and soft rock conditions. In the Soviet Republic mining conditions, the support ratings are highest due to high lateral stresses on the roof support and also due to thick seams and soft rock conditions.

In the USA, it is customary to use high support load densities of the order of 70–100 tonnes/m^2 with a maximum load density of 120 tonnes/m^2. It is suggested that most of the powered support installations are overly conservative rarely attaining the yield load but irregular caving may create conditions of dynamic loading resulting in cessation or destruction of the powered support installations. It is further suggested that the powered

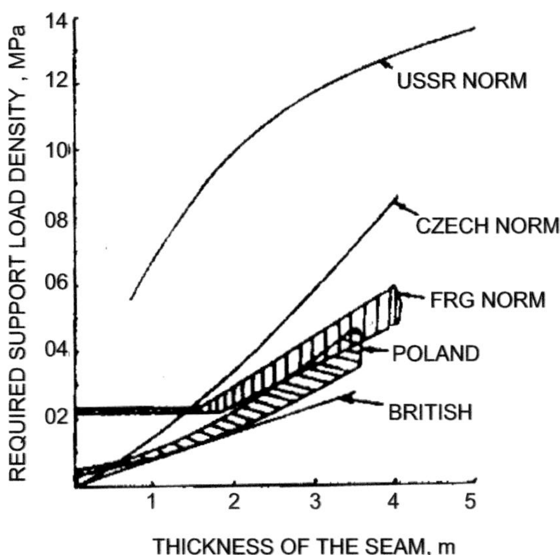

Figure 14.38 Suggested mean support density in various countries (Ghose, 1985)

support design for a specific locale should take into considerations the possibility of the dynamic load on the supports due to the following main factors (Ghose, 1985).

- Sudden caving of the bridging beam
- Rockburst
- Dynamic impact of a caving block of strata and
- Vibration of roof and floor induced by seismic waves.

14.9 Conclusions

This chapter briefly reviews methods of support in mining both during mine development and also longwall powered support. The mine supports can be sub-divided into two categories as active support and passive supports. The role of active support is to inhibit inherent strata movement in the de-stressed zone around an excavation and improve the strength of rock material by activating the inherent cohesion and internal friction. It has been shown that with the advent of mechanised powered support, breaker line supports and roof dowelling the accident rate due to fall of ground in underground mining has been considerably reduced.

The basic role of the face powered support is to ensure adequate strata control by preventing excessive convergence, bed separation and fall of ground in front of the roof canopy. The face support should provide safe access to workmen on the face and should prevent flushing of the caved material into the working area of the face.

A range of formulae has been developed to calculate the support load density of the powered support mainly based on local practical experience using an empirical approach. These formulae lack in basic understanding of the support strata interaction, they either do not directly take into consideration the dynamic loading of the supports by main and periodic fall of the main roof or they are rather too complicated to apply to the relevant field conditions. The supports are generally over-designed either by taking into consideration the dead weight of the strata as the support loading or using a very high factor of safety.

References

Aireys, E.M. (1974). 'The derivation and numerical solutions of equations relating to stresses around mining roadways', Ph.D. Thesis, Department of Mathematics, University of Surrey, July.

Ashwin, D.P., S.G. Campbell, J.D. Kibble, J.D. Haskayne, J.I.A. Moore and R. Shepherd (1970). 'Some fundamental aspects of face powered supports design', *Mining Engineer*, August, pp. 659–671

Balusu, S.R. (1994). 'Lecture notes on mine support', Private communication, 15 pp.

Bates, J.J. (1978). 'An analysis of powered support behaviour', *Mining Engineer*, 137, pp. 681–694.

Bates, J.J., J.W. Butler, G. Smith and B. Waring (1975). 'Design and analysis of powered support hydraulic supply and distribution system, suggestion for standard of system and components', Report No. 56, Mining Research and Development Establishment, NCB, *Mining Engineer*, Vol. 137, pp. 665–678.

Biron, C. and E. Ariouglu (1983). *Design of supports in mines*, A Wiley-Interscience Publication, John Wiley and Sons, New York, ISBN 0-471-86726-8, 237 pp.

Brabbins, M.W. (1991). 'A decade of roof support progress', Presidential address to the Manchester Geological and Mining Society, *Mining Engineer*, March, pp. 295–301.

Daws, G. (1978). 'An introduction to the design of resin anchors', *Civil Engineering*, December, pp. 61–63.

Daws, G. (1979). 'Resin rock bolting', *Civil Engineering*, November, pp. 39–43.

Daws, G. (1977). 'A method of designing a rock bolting scheme which is structurally equivalent to standing support system', *Seltite-Selfix Technical Service*, 6 pp.

Daws, G. (1980). 'Resin anchors in concrete', *The Civil Engineering*, April, pp. 35–41.

Daws, G. (1980). 'Calculation for rock bolting schemes, Bulcliffe Wood Colliery, Celcite (Selfix)', A special report to NCB (Unpublished).

Daws, G. (1987). 'A method of designing a rock bolting scheme which is structurally equilivalent to a standing support system', *Selfix Geotechnical Service*, 6 pp.

Dejean, M. and J.F. Raffoux (1980). 'The determining of roof bolting parameters', in *Rock Bolting*, Edition de le Revue Industrie Minrerale, Vol. 19, Rue du Grand-Moulin, 42029 Saint-Etienne Cedex, 18\980, 142 pp.

Dunham, R.K. (1976). 'Anchorage tests on strain gauged resin bonded bolts', *Tunnels and Tunelling*, September, pp. 73–76.

Ghose, A.K. (1985). 'Estimation of support resistance at longwall faces', *Journal of Mines, Metals and Fuels, Longwall Face Supports Developments*, Special Number, March.

Ghose, A.K., and N.M. Raju (1980), 'Roof truss for coal mine roof control – Laboratory and field Evaluation, Proc. 3rd Canadian Rock Mechanics Symposium, Toronto, May.

Gregory, C.E. (1983). *Rudiments of mining practice*, Trans Tech Publications, Clausthal-Zellerfeld, Germany, SBN 0-87201-783-4 (Chapter on Mine Support, pp. 79–86.)

Gupta, R.N. (1982). 'The influence of setting pressure and strata behaviour on longwall faces', Ph.D thesis, University of Newcastle-upon-Tyne.

Gupta, R.N. and I.W. Farmer (1983). 'Strata deformation and support performance at a longwall face', *Proc. 5th Congress Int. Soc. Rock Mechanics*, Melbourne.

Gupta, R.N. and I.W. Farmer (1982). 'Relation between strata deformation and support performance at longwall face', In: *Strata Mechanics* (Ed. I.W. Farmer), Elsevier, Amsterdam, pp. 74–81.

Hess, H. (1972). 'Roof control by powered support in West German coal mining industry', *International Conference on Strata Control*, London, Paper 1.

Hoek, E. and E.T. Brown (1980). *Underground excavation in rock*, IMM Publications.

ISRM (1974). 'Suggested methods for rock bolting testing', *International Society of Rock Mechanics*, Doc. 2, March.

Kenny, P. and A.H. Wilson (1963). 'Strata movement on faces equipped with conventional and powered supports', *Mining Engineer*, Vol. 122, pp. 524–38.

Lang, T.J. (1972). 'Rock reinforcement', *Bulletin of the Association of Engineering Geologists*, Vol. IX, No. 3, pp. 215–239.

Morgan, D. (1982). 'Longwall coal face roof supports for 1980s', *Mining Engineer*, Vol. 141, pp. 25–29.

Mukherjee, S.N. (1993). *Longwall machinery and mechanization*, Vol. 1 (Powered supports), A.M. Publishers, P.O. Indian School of Mines, 431 pp.

Norris, C. and M. Yearby (1980). 'The choice or the support available', *Proceedings of AusIMM, Melbourne, Roof Support Colloquium*, September, 4.1–4.17.

Peng, S.S. (1986). Coal Mine Ground Control, John Wiley & Sons., 491 pp.

Rabcewicz, I. (1955). 'Bolted supports for tunnels', *Mine and Quarrying Engineering*, Vol. 21.

Richmond, A. *et al.* (1986). 'Roadway Support', Chapter 18, *Australasian Coal Mining Practice*, Monograph 12, Editor C.H. Martin, pp. 263–293.

Roberts, B.M. (1988). *Notes on longwall powered supports*, Nottingham University, UK.

Sarkar S.K. and B. Singh (1987). 'Indian experience on Longwall Mining with caving under massive sandstone', *Underground Mining Methods and Technology*, Editors A.B. Swiliski and M.J. Richards, Elsevier Science Publishers B.V., Amsterdam, pp. 261–270.

Sheppard, R. (1972). 'Powered roof supports, today and tomorrow', *Mine and Quarry*, Vol. 8, pp. 17–26.

Singh, R.N, A. Heidereh-Zadeh and G. Daws (1980). 'Roadway support trials using rock bolts and grout-filled envelopes at Deep Hard Manriding Roadway, Annesley Colliery, South Nottinghamshire Area', A joint research report to NCB MRDE Tunnelling Branch, University of Nottingham and Celtite Selfix Ltd., November, 18 pp.

Singh, R.N., I. Porter and J. Hematian (1998). 'Stability Analysis of Four-way Intersections of Coal Mining Roadways', *International Conference on Geomechanics/Ground Control in Mining and Underground Construction*, 14–17 July, pp. 555–572.

Singh, R.N., I. Porter and J. Hemetian (2001). 'Finite element analysis of three-way roadway junctions in longwall mining', *Journal of Coal Geology*, Elsevier Science Publication, Vol. 45, pp. 115–125.

Smart, B.G.D., A.K. Isaac and C.G. Hinde (1980) Investigation into the relationship between powered support performance and working environment with particular reference to strata convergence, *Proceedings of the 21st U.S. Symposium in Rock Mechanics*, University of Missouri, Rolla, pp. 332–44.

Smart, B.G.D., P.W.H. Olsen and K. Metcalfe (1992). 'Consideration of lateral forces generated by powered support', *Mining Engineer*, London, pp. 189–196.

Smart, B.G.D., K. Mertcalfe and A. Redfern (1987). 'The prediction of powered support reaction under massive sandstone roofs', *Underground Mining Methods and Technology*, Editors A.B. Swiliski and M.J. Richards, Elsevier Science Publishers B.V., Amsterdam, pp. 297–308.

Scott, J.J. (1980). 'A new innovation in rock support—friction rock stabilizers', *CIM Bulletin*, February, pp. 65–69.

Thomas, L.J. (1978). *An introduction to mining*, revised edition, METHUEN, Sydney.

Terzaghi, K. (1943). *Theoretical Soil Mechanics*, John Wiley, New York, pp. 71–73.

Varcoe, R.A.B. (1987). 'An innovative approach to powered support system design with particular reference to control of lateral strata movement', Ph.D thesis, University of Strathclyde.

Wheeler, A. (1974). 'Support on the face', *Proceedings of Symposium Mining Methods*, Harrogate, Institution of Mining Engineers, London, pp. 65–68.

Whittaker, B.N. (1974). 'An appraisal strata control practice', *Mining Engineer*, Vol. 134, pp. 9–24.

Whittaker, B.N. (1975). 'A review of contribution made by powered roof supports on longwall mining', *University of Nottingham Mining Department Magazine*, Vol. 28, 1976.

Whittaker, B.N. and R.N. Singh (1980). 'The estimation of strata loading at a ripping lip', Special report to NCB, South Notts Area, February.

Wilson, A.H. (1975). 'Support load requirements on longwall faces', *Mining Engineer*, Vol. 134, pp. 479–91.

Wilson, A.H. (1978). 'Various aspects of longwall roof support', *Colliery Guardian International*, April, pp. 50–55.

Wilson A.H. (1980). 'The stability of underground workings in soft rocks of the coal measure', Ph.D Thesis, *International Journal of Mining Engineering*, Vol. 1, pp. 91–181.

Wilson, A.H. (1986). 'The problems of strong roof beds and water bearing strata in the control of longwall faces', *Proceedings Aus IMM, Illawarra Branch, Symposium Related to Ground Movement and Strata Control Related to Coal Mining*, August, pp. 1–8.

Mining Subsidence

15
Chapter

15.1 Introduction

Underground mining operations or abstraction of water from underground aquifers or extraction of petroleum from oil and gas reservoirs inevitably causes redistribution of strata stresses and possibly collapse of the overlying strata which may manifest itself at the surface as subsidence. In order for mining subsidence to take place some form of underground mining should have taken place followed by uncontrolled caving of the overlying strata to a substantial degree. Factors affecting the extent of surface subsidence due to mining are as follows.

- Type of ore deposits which can range from tabular deposits, stratified flat seams to steeply dipping irregular ore deposits or lenticular or massive ore bodies
- Method of mining ranging from partial extraction, the relation of caving methods to mining methods using backfilling or stowing techniques
- Nature of mineral deposits and that of the overlying strata including size, shape and inclination of the ore bodies, geology, hydrogeology and other characteristics defining the ground behaviour
- Geological and structural factors, which may modify the surface subsidence pattern

The subsidence process may cause vast destruction to sub-surface and surface rock formations resulting in damage to surface structures and infrastructures, and may influence hydrological, hydro-geological and biological environments on the surface.

Mining causes redistribution of strata stresses around an excavation resulting in rock movement in the proximity of the excavation which when allowed to continue unabated may culminate in surface subsidence. These processes are described in the subsequent sections.

15.2 Ground Movements Associated with Mining

15.2.1 Stress Distribution Around an Underground Roadway

Figure 15.1 shows the stress distribution around an underground roadway in a tabular deposit. As mineral is excavated from the roadway the overlying strata sags and detaches itself from the upper beds. This induces bending stresses in the roof which are tensile on the lower surface and compressive on the upper surface of the beam. Shear forces are developed on the roof at the pillar edges and also horizontal stress F_H on the end of the upper beams stabilising the roof. As the overlying beds expand toward the excavation a dome shaped relaxation zone is formed above the roadway and the strata stress in the overlying rock is transferred onto the abutment zone of the surrounding pillars. At the roadway sides, due to the interaction of vertical and horizontal stresses, shear stresses are developed on a triangular zone on the side of the pillar. If the rock surrounding the roadway is not supported the pressure arch expands forming a larger relaxed zone.

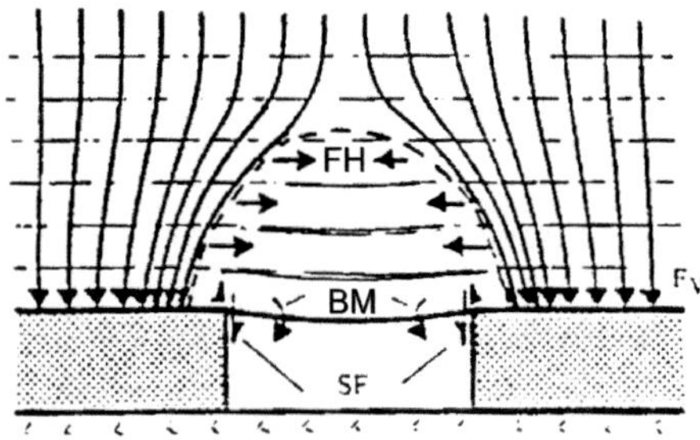

F_V = vertical component of force
F_H = horizontal compression
BM = bending moment of face
SF = shear force

Figure 15.1 Stress distribution in the vicinity of a rectangular narrow excavation (Fritzsche and Potts, 1954)

15.2.2 Ground Movement Above a Roadway at Shallow Depth

The mechanism of ground movement around a roadway depends upon the type of strata being negotiated by the roadway, namely:

(1) Unconsolidated overburden and
(2) Consolidated overburden.

(a) Shallow mine roadway overlain by unconsolidated overburden

The collapse process in unconsolidated formations near the surface is characterised by the following features (Figure 15.2a).

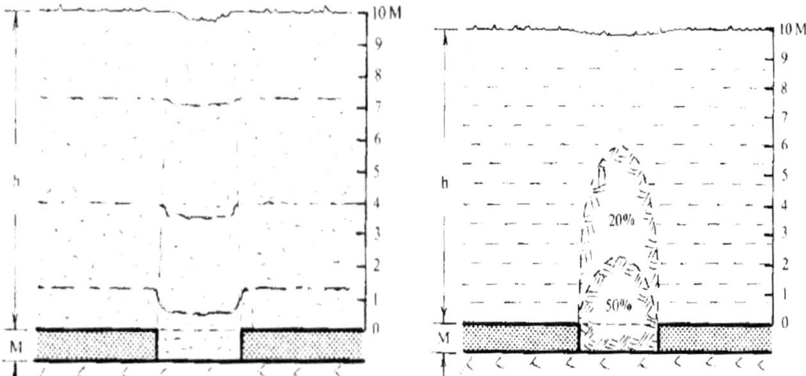

Figure 15.2 Subsidence above a roadway in (a) unconsolidated ground, (b) consolidated ground (Whittaker and Reddish, 1989)

(1) Shearing above the roadway sides.
(2) Relatively confined column of material slumping downwards towards the excavation.
(3) The resulting subsidence zone may extend up to a height 10 times that of the thickness of extraction.
(4) The width of the excavation controls the height of the subsiding dome.
(5) The strength and stratified nature of roof beds also have a major role defining the deformation pattern.
(6) Narrow extraction allows the development of a well defined dome while a wide extraction may modify the shape of the subsiding dome.

(b) Consolidated overburden

A roadway in a consolidated rock formation undergoes strata movement and caving after the removal of support (Figure 15.2b). This is followed by bulking of the fractured rock which will completely or partially fill the void. In most massive or sedimentary rocks, a dome-shaped collapse zone will be formed in the surrounding region of an excavation. The height of the unstable dome is governed by the following factors.

(1) Strength of the immediate roof as the roof strata has to resist the combination of tensile, bending, compressive and shear stresses to

maintain the integrity of the roof beds. For example, mudstone is weak and breaks and bulks easily while sandstone is a comparatively stronger rock liable to lesser degree of breaking and bulking.

(2) Bulking properties of the fractured rock within the collapsed dome is responsible for the subsidence being less than the extracted seam height. Broken walls will eventually fill up the void created by mining and thereby result in restoration of equilibrium. Figure 15.2(b) shows that if the bulking factor is 50% then the caving height is equal to twice the thickness of extraction and if bulking factor is 20% then the caving height is equal to five times the extracted seam height.

(3) Width of the excavation; wider excavations result in the formation of higher collapse zones.

(4) State of stress is an important parameter as in deeper excavations high strata stress may allow the surrounding rock to cave readily. High horizontal stress inhibits the development of the height of the collapsed zone.

(5) Presence of inherently weak beds, slip planes, jointing, bedding planes and faulting help the caving process.

(6) Presence of massive sandstone or other hard strata may assist in inhibiting the caving process.

(7) Domes of increased height do not tend to increase in width in relation to the original width of the roadway.

(8) Generally, the magnitude of surface subsidence depends upon the depth of the excavation below the surface.

(9) Small amounts of subsidence also occur due to general sagging of the overlying beds.

15.2.3 Ground Movement in Proximity of a Wider Excavation

Wider mining excavations like longwall faces results in increasing the size of the pressure arch surrounding the excavation. Thus, stress patterns and the following features mainly characterise displacements around a longwall face.

1. The natural arching process shields the longwall face from the full effect of the strata pressure due to the cover load.
2. Support of the roof on a longwall face is mainly based on the assumption that a detached roof block is held in place by the support system.
3. Caved waste is assumed to generate self-supporting resistance to the main roof by virtue of its bulking characteristics.
4. A bulking factor of 50% makes the height of the detached block twice the thickness of coal seam being extracted.

5. The variation of geological character of the roof strata of a coal seam has necessitated the design of powered supports of much higher ratings than indicated by the $2 \times M$ formula.
6. Figure 15.3 shows the presence of shear planes and fractures in coal measure rocks hading at 26° to the vertical in the flat seam in longwall mining with caving (Whittaker, 1974). Ground movement involving roof beds behind the face line is shown to be associated with differential slipping between sheared blocks coupled with the downward flexing of the beds. Immediately behind the support the caving process commences in the weak strata as immediate collapse while in strong strata as an overhanging cantilever which subsequently breaks due to excessive span. The caved rocks tend to act as a cushion for the overlying caving rocks in the affected zone.

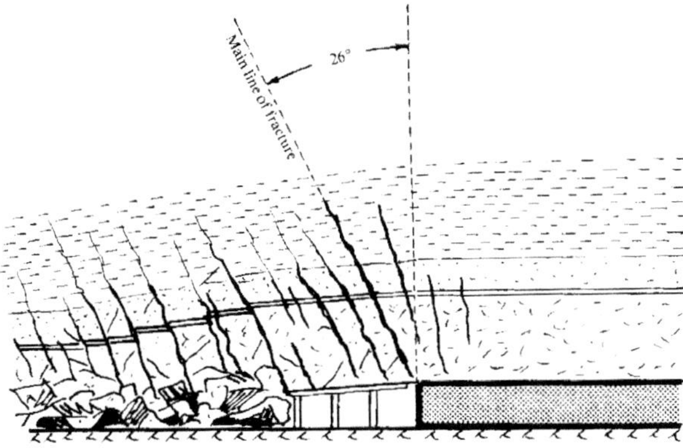

Figure 15.3 Caving process behind a longwall face (Whittaker, 1974)

15.3 Influence of Type of Mining Operation on the Resulting Surface Subsidence Pattern

Surface subsidence due to mining can be classified into the following two categories.

1 Discontinuous subsidence
2 Trough subsidence

Discontinuous subsidences are caused by mining a range of ore deposits of variable geometry over a limited area and are characterised by large step-wise surface movements following either sudden or progressive collapse. The discontinuous type of subsidence is difficult to predict with precision and difficult to control. On the other hand, the trough type of subsidence involves mining of tabular deposits over an extended area resulting in the formation of a smooth surface profile. This type of subsidence has been monitored or studied over a long period of time in many coal mining countries and can be predicted with reasonable accuracy.

The following factors determine the method of mining to be adopted for a given ore deposit and also the type of subsidence pattern.

1. Character of the mining deposit
2. Geological environment
3. Geotechnical characteristics
4. Time and place for mining
5. Value and distribution of the value of the minerals within the deposit
6. Strength properties of minerals and the host rock
7. Strength of surrounding rocks
8. Existence of special hazards
9. Location and geometry of deposit, dip and its depth below surface
10. The inclusion of support within the mined out areas and particularly as a permanent feature within the working stope forms a principal component of the mining method

15.3.1 Discontinuous Subsidence

The choice of the method of mining usually depends upon the type and geometry of the ore deposits and nature of the host rock. The potential problem of surface subsidence associated with some of the mining methods is given in the following sections.

(a) Solution mining

This method has been used for extraction of salt and Figure 15.4 shows that the mode of subsidence is associated with collapse features. The main characteristics of this system of mining are as follows.

1. Size and extent of solution cavity is unknown.
2. Potential for the subsidence of overlying rocks depends upon the strength properties and thickness of those overlying rocks.
3. Establishment of a groundwater flow-net at the pre-mined horizon is a prerequisite for this system of mining. This is followed by the collapse of the mining cavity. Further subsidence can be anticipated in the form of a deepening depression.

Figure 15.4 shows the chimney type of subsidence due to solution mining resulting in a sinkhole type of collapse migrating towards the surface. The subsidence pattern at the surface can either be

(i) saucer-shaped or
(ii) conical

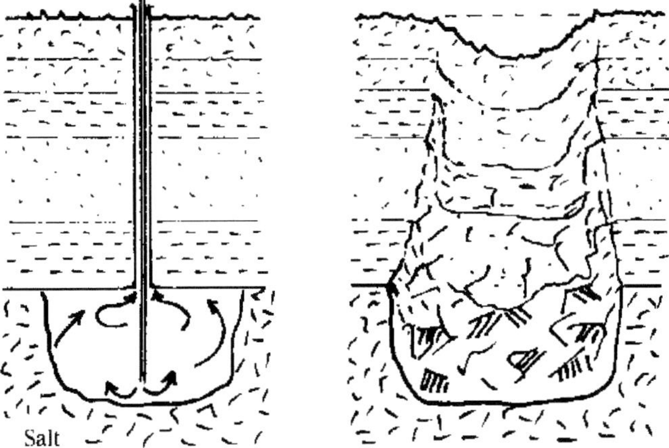

Figure 15.4 Subsidence due to solution mining (Whittaker and Reddish, 1989)

A sudden appearance of a surface subsidence cavity can occur resulting from a sudden collapse of the capping bed spanning a migrating collapsed chimney.

(b) Room and pillar mining

It is one of the most important methods of mining used for mining coal, gypsum, iron ore, rock salt, trona and many other minerals. Many abandoned old coal mine workings exist worldwide which may result in the following types of subsidence.

(1) Sinkhole type of subsidence caused by collapse of roadway junctions
(2) Widely spread saucer type subsidence due to pillar extraction or wide spread pillar collapse

The main features of surface subsidence associated with room and pillar mining are as follows (Figure 15.5).

1. Localised sinkholes are common types of subsidence associated with shallow workings and weak overburden.
2. Subsidence can be saucer-shaped if pillars are extracted or if large-scale pillar failure occurs.
3. Potential for trough type subsidence exists as rooms are left unsupported in a thick seam environment over a long term causing domino-type pillar failures.

(c) Subsidence in caving systems of mining

Mining of thick and inclined ore deposits can be carried out by using mining systems depending upon relative strength the hanging wall, footwall and the ore body.

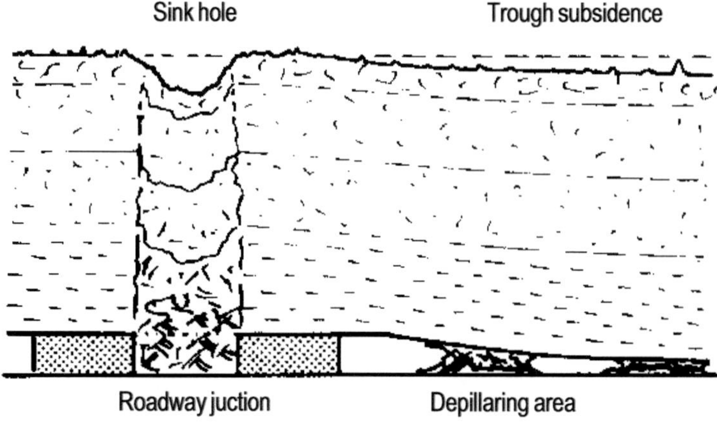

Figure 15.5 Collapse subsidence caused by room and pillar mining (Whittaker and Reddish, 1989)

There are three main caving systems that are applicable to mining thick and inclined ore bodies.

(1) Top slicing
(2) Sub-level caving
(3) Block caving

Figure 15.6 shows a cross-section of ore bodies using top slicing and sub-level caving methods of mining.

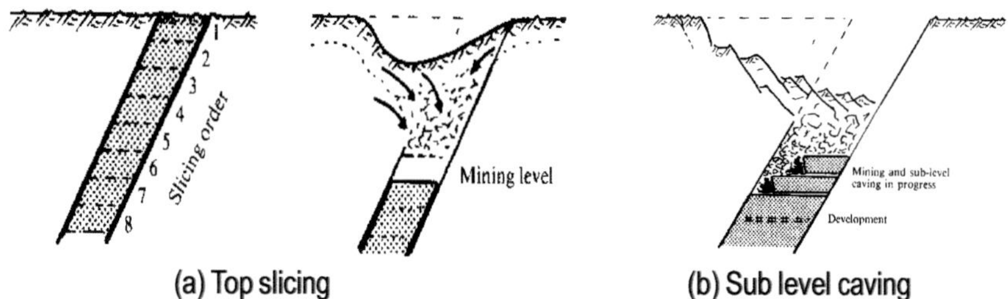

Figure 15.6 Subsidence due to caving systems of mining for thick and steep deposits

(a) Top slicing

The top slicing method is designed to extract successive slices of equal thickness in descending order. The development of the slice is carried out by driving a hanging wall and a footwall drive at the bottom layer of the slice while leaving a layer of mineral on the roof of the slice to support the broken ground in the previous slice. Once the boundary of the panel is reached, a level joins the footwall and hanging wall drives forming a level face. As the face retreats, the ground above the face which is supported by timber, is allowed to cave with timber above the next slice, thus forming a timber mat at the interface of the caved ground and the next slice. In this manner, a number of slices are extracted in descending order.

The subsidence pattern around the top slicing method of workings are characterised by the following factors.

- The surface subsidence is controlled by the position of footwall, thickness of the ore body, the length of each slice and depth of workings.
- Subsidence tends to be gradual with a surface depression developing as mining proceeds deeper.
- Surface extent of the subsidence depression depends upon the nature of the surrounding rocks and overlying superficials.
- The subsidence trough is step-wise and discontinuous.

(b) Sub-level caving method

The sub-level caving method of mining is applicable to steeply inclined thick ore bodies with strong ore, stable footwall and weak hanging wall. The ore is won by drilling and blasting, and a mechanised loading technique and the ground control is carried out by caving of the hanging wall. Development of sub-level, drilling, blasting and loading of mineral can be carried out at several sub-levels at a time. Thus, it is a mass production method, achieving production and productivities similar to that of some open cut mining operations. Some underground mines have produced 11 million tonnes of ore per year from sub-level caving operations.

The subsidence pattern of sub-level caving operations is shown in Figure 15.6(b) and can be summarised as follows.

- Caving of the hanging wall is initiated by loading of the blasted ore in sub-levels. Thus the caving operation is slow and progressive.
- The main form of subsidence is slumping of the hanging wall into the mined out area.
- The subsidence pattern is controlled by the footwall as footwall failures are comparatively rare.
- The subsidence pattern is characterised by stepped, discontinuous subsidence accompanied by large depressions and surface fissures.

(c) Block caving method

The block caving system of mining is applicable to a large, massive ore deposit, hundreds of square metres in cross-section (in plan) and extending to several hundreds of metres in depth. The host rock containing the ore body and the rock at the base of the ore body should be fairly competent. The ore body itself should be of such strength and should be intercepted with joints and other geological discontinuities so as to make it a readily cavable rock mass. Undercut levels and transport drifts are located at the bottom of the ore body in a competent host rock. The bottom of the under cut levels contain cone shape chutes connected to the underlying loading points in the transport drifts via finger raises. Once the caving process in the ore body is initiated by providing a free face and an undercut and an initial blast, the caving of ore body should continue under its own weight (Fig. 15.7).

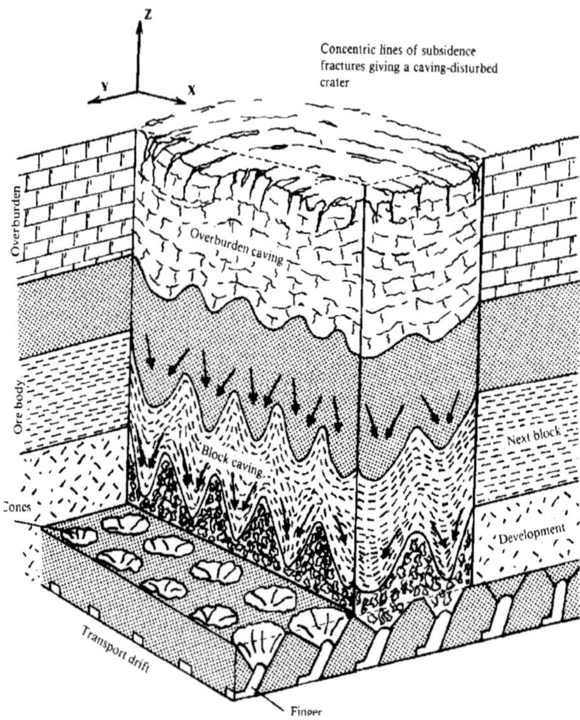

Figure 15.7 Surface subsidence development in block caving (Whittaker and Reddish, 1989)

Block caving mainly transmits its effects to the surface vertically giving rise to a disturbed crater with concentric rings at the surface.The effect of subsidence can be transmitted horizontally if the characteristics of the surrounding rocks allow the development of appreciable draw. Thus, the subsidence pattern is characterised by the following main features.

(a) Crater type of subsidence trough with concentric rings
(b) Mainly vertical depression in the mined-out area of the ore body

(c) A horizontal extension of the subsidence trough occurs laterally depending upon the angle of draw and also depending upon the nature and thickness of the overburden rocks

15.3.2 Trough Type of Mining Subsidence

Extraction of coal in a relatively wide excavation results in caving of the strata immediately surrounding the coal seam. This strata movement, when continued on a large scale, manifests itself at the surface as mining subsidence.

Prediction of mining subsidence is therefore important when coal mining is being conducted under the following conditions.

- Urban environments
- Major infrastructures
- Tidal water and estuaries
- Dams and water reservoirs

The rest of the chapter deals with the surface and sub-surface subsidence around a wide excavation, prediction of mining subsidence due to longwall mining and methods of controlling caving and subsidence.

Generalised continuous surface subsidence

Figure 15.8 shows the elements of a generalised subsidence curve around a wide excavation resulting in the development of a trough type of complete

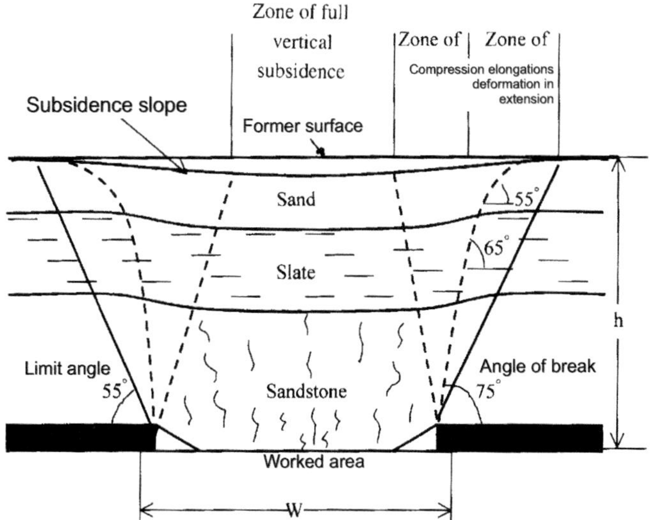

Figure 15.8 Components of mining subsidence

subsidence profile at the surface. It can be seen that the width of the surface subsidence zone is much larger than the width of extraction of the coal seam. Lines drawn from the edges of the extracted area at a limiting angle of 55° to the horizontal (also known as angle of draw γ from the vertical line) and intersecting the surface can delineate this surface subsidence zone. It has been observed that outside this surface subsidence zone, no ground movement due to subsidence takes place. A zone between a vertical line from the edge of the workings and the limiting line intersecting the surface forms a zone of elongation. In this zone the ground between two survey pegs elongates as a consequence of mining subsidence. This tensile zone results in the development of tensile cracks or the opening of the pre-existing joints as a consequence of mining. The main movement in this zone is characterised by a combination of horizontal displacement and vertical subsidence.

A line drawn from the edge of the worked out area over the goaf at a limiting angle and the vertical line intersecting the surface forms a zone of compression. In this zone the relative distance between two points on a survey line closes as a consequence of mining. This zone is characterised by an increase in slope towards the centre of the subsidence trough and relatively destructive compressive movement.

In the centre, there is a zone bounded by limiting lines where the ground movement due to mining is solely vertical subsidence. As mining proceeds the width of the zone of subsidence increases correspondingly.

15.4 Surface Subsidence and Sub-surface Subsidence Pattern

15.4.1 Depiction of Stress Zones Around Underground Mining Excavation

Figure 15.9 shows a schematic representation of sub-surface subsidence associated with a total extraction mining system that is characterised by three distinct zones of strata stresses.

The zone of vertical compression lies immediately above the coalface at the edge of the pillar where the vertical closure at the seam horizon is minimal due to the resistance to roof movement provided by the pillar edge and partial vertical subsidence at the surface. The zone of vertical extension lies immediately above the centre of the worked out area and is caused by the caving of waste as a consequence of mining and volume expansion of the caved material. However, the vertical compaction in the zone of full subsidence is 10–40% less than the extracted seam height resulting in overall extension of the overlying rock.

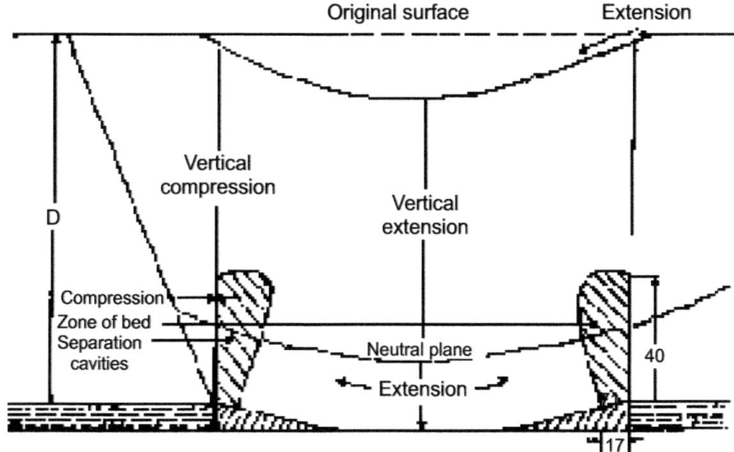

Figure 15.9 Depiction of stress zones in underground mining

In the mining horizon there is a zone of incomplete convergence characterised by full convergence at the centre of the caved area and incomplete convergence near the pillar edges. Directly above this zone is a region of incomplete convergence that contains bed separation cavities. This zone is 8 to 10 m wide and extends upwards to a height of 40 times the extracted seam height. This tension zone contains a large volume of bed separation cavities capable of storing copious amounts of water.

15.4.2 Idealised Model of *in situ* Permeability Changes Around a Total Extraction Mining Layout

Figure 15.10 is a generalised presentation of *in situ* permeability changes around a longwall panel. Three distinct zones of marked permeability changes are outlined.

(1) *Fracture zone* extending up to 30 to 58 times the extracted seam height above the seam horizon. A zone extending up to 3 to 5 times the extracted seam height exists directly above the extracted coal seam.

(2) *Aquiclude zone* is characterised by the constrained strata with no appreciable changes in the strata permeability. This zone extends from 30 times the extracted seam height above the extracted coal seam to 15 m from the surface. 30 t to h (15), where 'h' is overburden thickness, and 't' is the thickness of extracted seam height).

(3) *Zone of surface cracking* extends up to 15 m depth from the surface.

A study of the permeability changes in the roof above the rib pillar towards the centre of the panel indicated that permeability changes as a consequence of mining are 60 to 80 times the intact permeability of the over-

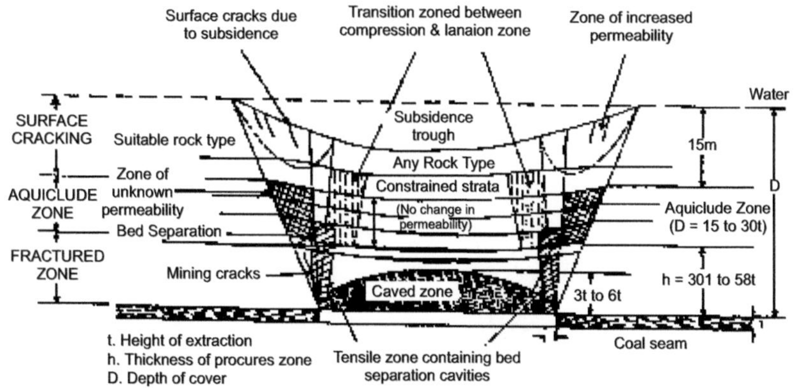

Figure 15.10 Strata behaviour with total extraction mining (Singh and Kendorski, 1979)

lying rocks (Figure 15.3). The disturbance at the surface manifests itself as subsidence and/surface cracks, depending upon the depth to seam thickness ratio, and other factors. With regard to the underground disturbance, there are the following distinct zones.

- Immediate roof caving into the void created by mining
- The layer above the caved zone is subjected to compressive and tensile stresses.
- The tensile strain is generally in the vertical direction which gives rise to bed separation, whereas in the horizontal direction it tends to open up joints.

The permeability changes above the longwall extraction are affected by the following factors:

1. Zone of crack formation
2. Orientation of fractures
3. Individual bed thickness
4. Rock mass flow characteristics
5. Lithology

15.4.3 Influence of the Area of Working on Surface Subsidence

The amount of surface subsidence depends upon the area of extraction in relation to the depth of workings.

Critical area of extraction is defined as the mining area, the working of which causes the complete vertical subsidence of one point on the surface (Figure 15.11).

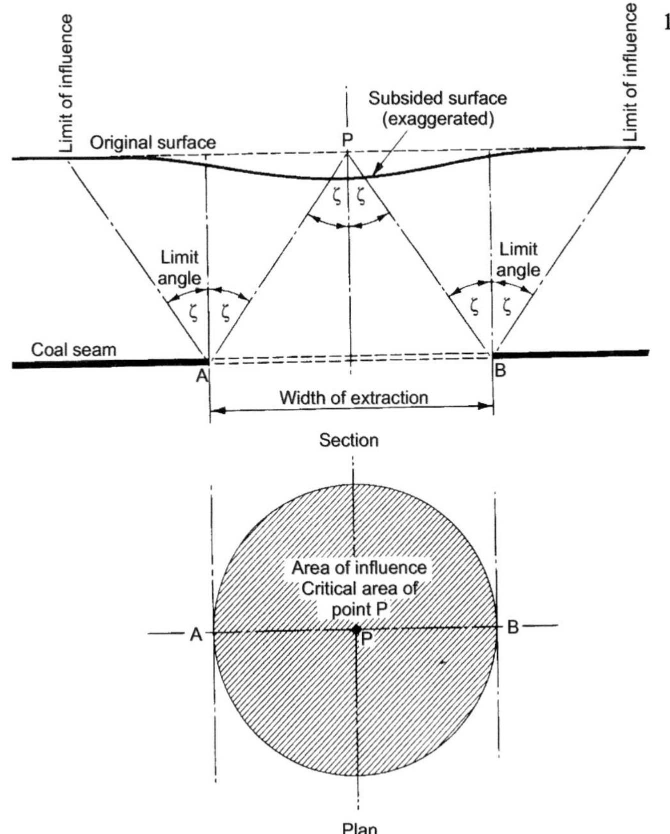

Figure 15.11 Effect of area of influence on the surface subsidence (Shadbolt, 1975)

Sub-critical area is defined as an area of working smaller than the critical area. When a sub-critical area is being worked, the point on the surface under consideration undergoes incomplete subsidence in relation to the thickness of extraction (Figure 15.12).

Super-critical area is an area of working greater than the critical area resulting in the area on the surface undergoing complete subsidence (Figure 15.12).

15.4.3 Effect of Width of Extraction to Depth Ratio on the Subsidence Pattern

Figure 15.13(a) shows the subsidence pattern, tensile strains, compressive strain and percentage slope at the surface as a consequence of mining a sub-critical area of extraction. If the width of extraction is equal to 1.4 × the depth

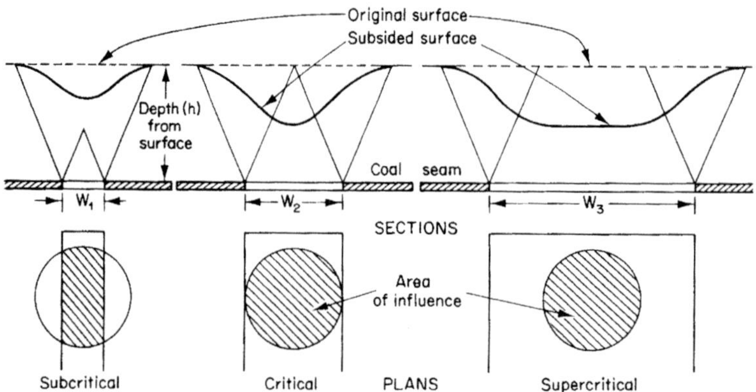

Figure 15.12 Changes in the surface subsidence by variation of width of working

below the surface, then only one point on the centre of the subsidence trough undergoes full vertical movement (Figure 15.13b).

The surface subsidence pattern as indicated in Figure 15.13(c) develops when the width of extraction is wide, exceeding 1.4 times the depth below the surface. This subsidence pattern is characterised by the development of vertical subsidence in the centre of the subsidence trough, and is typical of shallow workings.

15.4.4 Strata Pressure and Subsidence Along Longwall Face

Figure 15.14 shows the strata stress distribution along a transverse section of a longwall face, a pressure arch delineating the envelope of broken strata surrounding the longwall face and its interrelationship with the surface subsidence (Whittaker and Pye, 1977). Figure 15.14(a) shows the deep mining situation where the w/h ratio is less than 1.4 indicating a sub-critical condition. The stress distribution across the transverse section of the face indicates that there is a de-stressed zone in the extracted area whereas the rib sides are zones of high abutment pressure. At the edge of the rib pillars the abutment pressure will reach a peak stress within 1–3 m from the pillar edge and reduces in negative exponential manner to the cover load some 30–40 m from the pillar edge. Under these conditions the pressure arch does not intersect the surface and a surface point immediately above the centre line of the face does not undergo full subsidence.

Figure 15.14(b) shows that when the w/h ratio is > 1.4 and the pressure arch intercepts the surface, a part of the surface zone undergoes maximum surface subsidence and the corresponding part of the goaf is compacted so as to acquire the cover load.

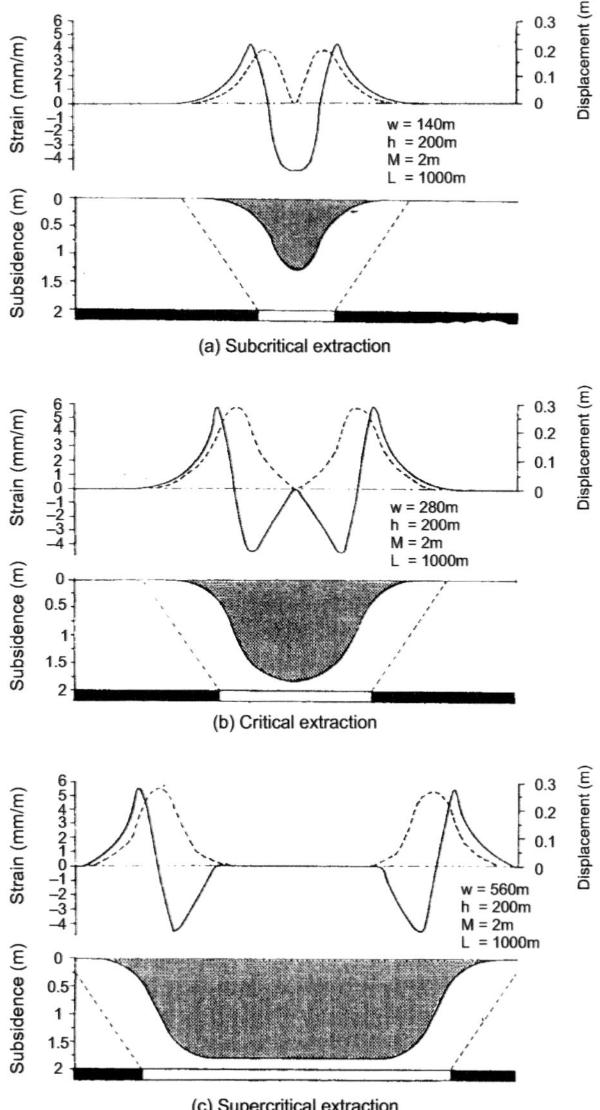

Figure 15.13 Comparison of subcritical, critical and supercritical mining subsidence situations

15.5 Prediction of Mining Subsidence

Surface subsidence due to longwall mining in flat coal seams has been well studied in the United Kingdom by the National Coal Board, based on

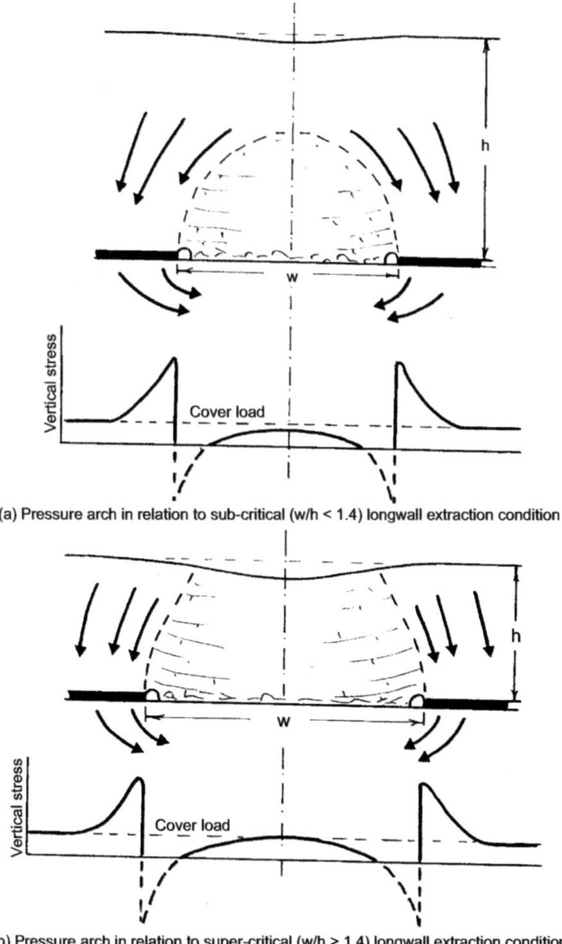

Figure 15.14 Illustrating arching of strata pressure across a longwall extraction in relation to subsidence according to Whittaker and Pye (1977)

principles derived from 299 case studies involving a wide range of mining and surface geological conditions (NCB, 1975). Some of the terms and representative values used in the handbooks are discussed below.

1. *Maximum subsidence* (S) can be evaluated as a function of the ratio of extraction width (w) to the depth of the longwall face (h) and the workable thickness of the coal seam (M).

$$\frac{S}{M} = f\left(\frac{w}{h}\right) \tag{15.1}$$

Whittaker and Breeds (1977) composed a table (Table 15.1) based on selected data taken from the *Subsidence Engineers Handbook* to demonstrate how S/M is related to w/h. The point of maximum subsidence is located directly above the centre of the excavation, Figure 15.15 and its magnitude can be simply calculated as follows:

Longwall width	w = 200 m
Longwall depth	h = 400 m
Extracted seam height	M = 2 m

Since $$w/h = \frac{200}{400} = \frac{1}{2}$$

Maximum subsidence S = 45% (2 m) = 0.90 m. The maximum subsidence magnitude at the centre line is of primary importance for the construction of the subsidence profile.

2. *Surface ground strain and tilt* are also illustrated in Figure 15.15, where subsidence produces regions of ground extension and compression by virtue of relative displacements towards the centre of the trough. The generalised relationship between the principal variables which represents the magnitude of the ground strains is expressed by Equation 15.2.

$$E = f\left(\frac{S}{h}\right) \tag{15.2}$$

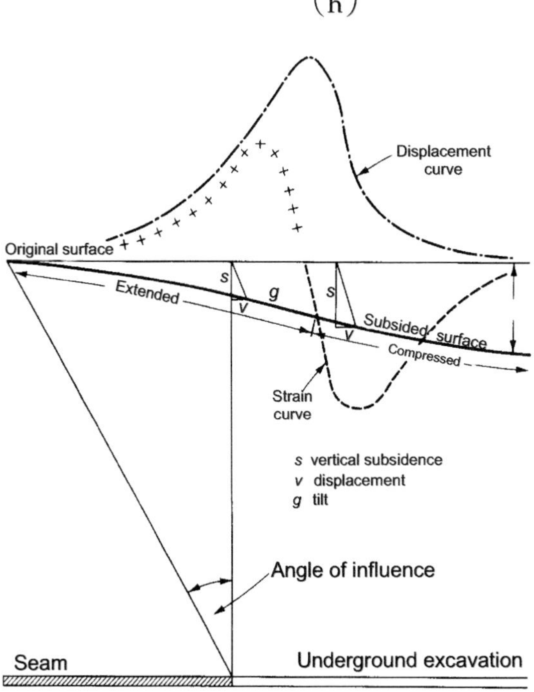

Figure 15.15 Elements ground movements due to extractions

The maximum tensile strain (–E) is located directly above or near to, but inside the edge of the rib and maximum compressive strain (+E) is either at the center or near to but inside the extraction area.

15.5.2 Worked Example of Subsidence Predictions

Table 15.1 gives a range of practical data for magnitude of –E and +E together with their respective positions in relation to the centre line of the longwall extraction area. The strain can be simply calculated. For the same longwall case in the previous paragraph, the calculated strain values are:

Table 15.1 Maximum subsidence, surface ground strain and tilt due to longwall mining (Whittaker and Breeds, 1977; NCB, 1975)

Longwall width/depth (w/h)	*0.2*	*0.25*	*0.33*	*0.5*	*0.75*	*1.0*	*1.4*
Max. subsidence/ extracted seam height S/M; caved wastes	*8%*	*12%*	*22%*	*45%*	*70%*	*84%*	*90% (max)*
Coefficients for deducing magnitude and position of maximum ground strains and tilt							
Max. strain due to compression (–E)	2.2 S/h	2.15 S/h	1.9 S/h	1.35 S/h	0.75 S/h	0.55 S/h	0.5 S/h
Position of –E from centre line (–Ex)	0	0	0	0.02 h	0.10 h	0.20 h	0.39 h
Max. strain due to extension (+E)	0.5 S/h	0.65 S/h	0.75 S/h	0.8 S/h	0.65 S/h	0.65 S/h	0.65 S/h
Position of +E from centre line (+Ex)	0.49 h	0.42 h	0.34 h	0.32 h	0.40 h	0.51 h	0.70 h
Max. ground tilt (G) (at transition pt.)	2.2 S/h	2.6 S/h	3.15 S/h	3.35 S/h	2.85 S/h	2.75 S/h	2.75 S/h
Position of G from centre line (Gx)	0.3 h	0.27 h	0.22 h	0.21 h	0.26 h	0.37 h	0.56 h

Maximum strain due to compression:

$$+\mathrm{E} = 1.35\frac{\mathrm{S}}{\mathrm{h}} = 1.35\frac{0.9}{400} = 0.0030 \text{ or } 3\ \mathrm{mm/m} \text{ at } 8\ \mathrm{m} \text{ from the centre line}$$

Maximum strain due to extension:

$$-\mathrm{E} = 0.8\frac{\mathrm{S}}{\mathrm{h}} = 0.8\frac{0.9}{400} = 0.0018 \text{ or } 1.8\ \mathrm{mm/m} \text{ at } 128\ \mathrm{m} \text{ from the centre line}$$

The maximum ground tilt as a consequence of subsidence occurs at the transition point between –E and +E and its magnitude is also a function of v/h. Whittaker and Breeds (1977) included values in Table 15.1 to calculate maximum tilt (G) for different relationships between w/h. The calculated magnitude of maximum tilt for a longwall case example is given below:

$$G = 3.35\frac{S}{h} = 3.35\frac{0.9}{400} = 0.0075 \text{ or } 7.5 \text{ mm/m at } 84 \text{ m from the centre line}$$

From an engineering point of view, the essential data for mining subsidence evaluation are the maximum subsidence (S), the maximum compressive (+E) and tensile (–E) strains and the maximum tilt (G).

3. *The angle of draw* represents the limiting plane of subsidence and is drawn from the edge of the extracted area in the seam to the point where subsidence is just measurable on surface. The angle between the vertical and the line of the draw is known as the 'angle of draw' and it varies from coalfield to coalfield and from one lithological unit to another. As an example, the angle of draw is approximately 35° in Great Britain. It is of interest that in Continental European and North American terminology the angle between the line of draw and the horizontal is known as the 'angle of caving' or limiting angle.

4. *Subsidence as a function of time* is also an important phenomenon in stability analyses. The main factor in subsidence duration is the depth of mine working. Subsidence as a function of time is summarised below (Fig 15.16).

(a) In the case of deep coal seams, the first phase of subsidence is not evident immediately after the excavation. It can only be measured after several years. However, in the case of shallow mining it can be noticeable after a short period.
(b) The second phase of subsidence is characterised by higher settlement rates and can be considered as accelerated settlement.
(c) The third phase of subsidence is abrupt and most of the total displacement occurs at this stage.

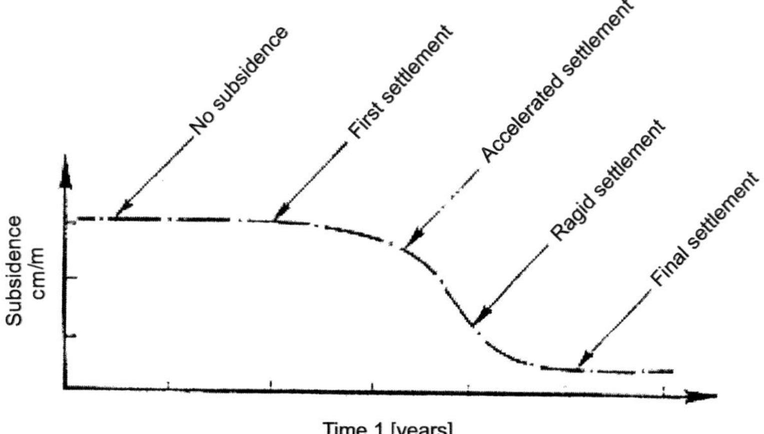

Figure 15.16 Time-dependent subsidence above a longwall panel

(d) Post-subsidence settlement is the final activity and is associated with very small displacement. It may take several years for the ground to come to rest.

Figure 15.16 is the curve of a subsidence profile and shows many similarities to a generalised creep curve. There may be inferred three stages of creep deformation, with rapid settlement corresponding to the failure deformation stage. Time-dependent subsidence can be expressed by the following equation:

$$\frac{S_t}{S} = 1 - e^{-ct} \tag{15.3}$$

where

S_t = subsidence at time of observation
S = maximum subsidence
t = time of observation of subsidence in progress
c = coefficient of velocity of strata displacement (dependent on the lithlogical units and geological structural conditions)

Subsidence as a function of time is of a primary concern in heavily populated areas.

15.5.3 Practical Examples of Sub-critical and Super-critical Subsidence

The classification of subsidence is defined by the method of mining and mine depth. Longwall mining is carried out in two geological settings. The first is shallow depth and weak rock strata, with the average working depth 50–100 m, length of the longwall face up to 100 m and the face advance of about 500 m during its life. The second, in appreciable depth and strong rock strata, with an average working depth 350–750 m, length of the longwall normally 200 m and the face advances up to 1000 m during its life.

The ratio between the extraction width to the depth of the longwall face and the angle of draw determine the subsidence category as follows.

(a) Super-critical subsidence is characteristic of very shallow mining and affects the ground surface substantially. The large width/depth ratios of the excavated area develop full subsidence commonly in the shape of flat-bottomed troughs at the surface. An important aspect arising from this is the high probability of appreciable surface fissuring occurring mainly along the longitudinal flanks of the subsidence trough. Such fissuring is usually time dependent where pseudo-plastic rocks and glacial till occur at the surface.

The advancing front of the subsidence trough rarely allows appreciably wide fissuring to develop parallel with the face. If the surface rocks are strong and have a joint or other natural fissure patterns, then the development of surface fissuring with subsidence may be governed more by geological characteristics than coincidence with maximum surface tensile strains. Consequently the geology at the surface needs careful study.

Figure 15.17(a) shows the general characteristics of surface subsidence to be expected in shallow workings. The relatively narrow surface strain zones are important areas, especially the location of maximum tensile strain which is a risk zone for the occurrence of significant surface fissuring and/or marked stepping in the subsidence profile.

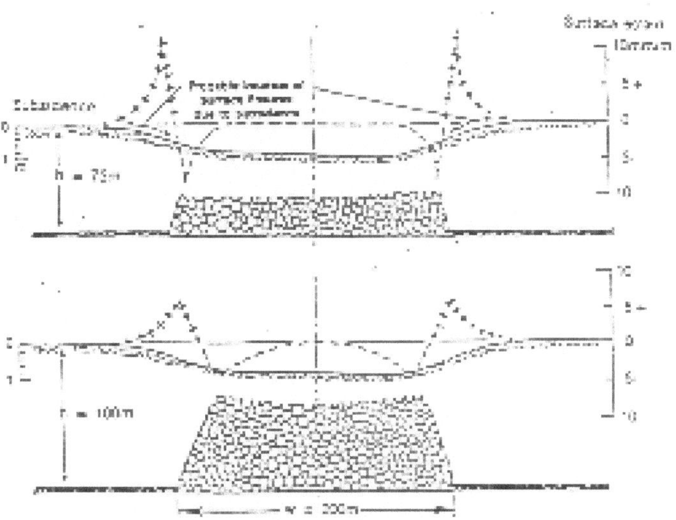

Figure 15.17(a) General characteristics of surface subsidence for super-critical extraction

(b) Critical subsidence (Figure 15.17b) is also related to shallow coal mining, and it has similar characteristics to sub-critical subsidence.

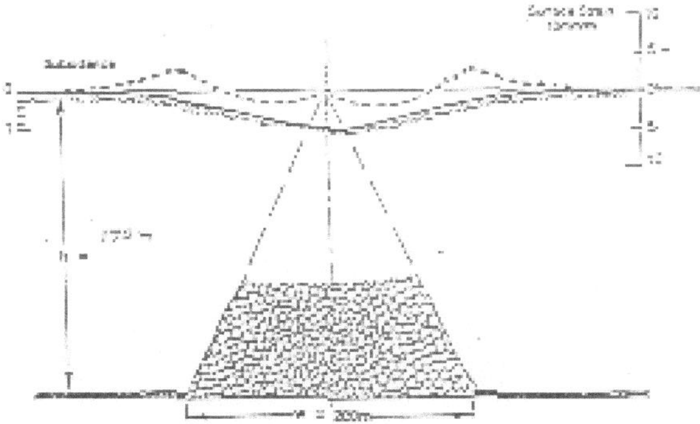

Figure 15.17(b) General characteristics of surface subsidence for critical extraction

The geological structural features of the rock strata are also important with respect to surface ground deformation as discussed in the previous category. The rock lithology does not appear to have a marked influence on subsidence in shallow mining.

(c) Sub-critical subsidence is typical of deeper coal mining, where the gob area is located below the ground surface because the draw planes do not intersect the ground surface (Figure 15.17c). This type of subsidence does not cause damaging deformations and fissuring of the ground surface because the ground surface gradually subsides without collapse or caving. The subsidence of deep mines continues after mining ceases and in the case of very deep coal mines, for example over 600 m, it can continue up to 10 years after completion of mining.

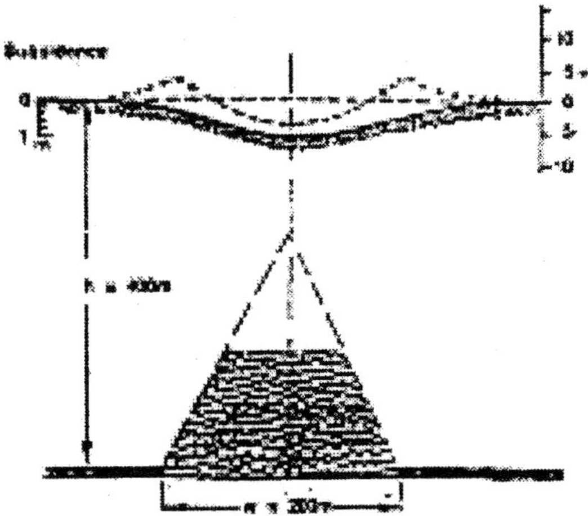

Figure 15.17(c) General characteristics of surface subsidence for sub-critical extraction area

The type of subsidence briefly described above considers only flat coal deposits which subside with a symmetrical profile about a vertical plane passing through the centre of the opening. If the coal seam is nearly flat and of a consistent thickness, the complete subsidence profile can be predicted by the use of a factor developed by the National Coal Board (1967).

Figure 15.18 shows the development of a subsidence trough and surface strains due to subsidence around a longwall face. A transverse section through the longwall panel indicates a zone of extension ahead of the longwall face and a compression zone behind the face. It can also be seen that there is an extension zone on both sides of the longwall panel over the solid coal whereas a compression zone is within the goaf area of the panel.

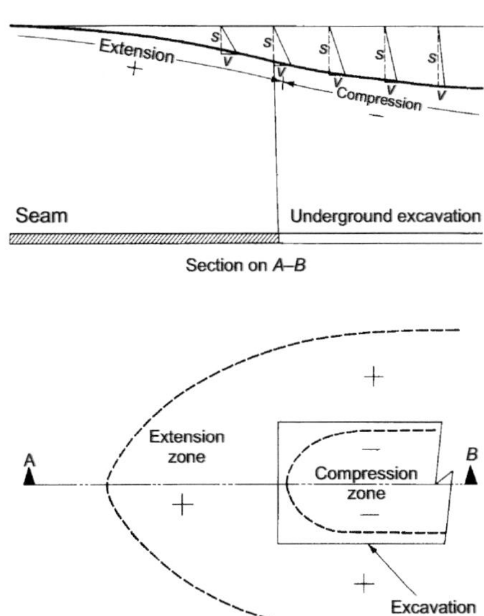

Figure 15.18 Development of strain zones on the surface due to longwall mining (Shadbolt, 1975)

Figure 15.19 shows the graphical relationship between the multiplying factors for predicting the maximum slope and strain for various width to depth ratios of the panel. It can be seen that by comparing the two strain curves in Figure 15.19 that the intensity of extension is considerably less than that of compression.

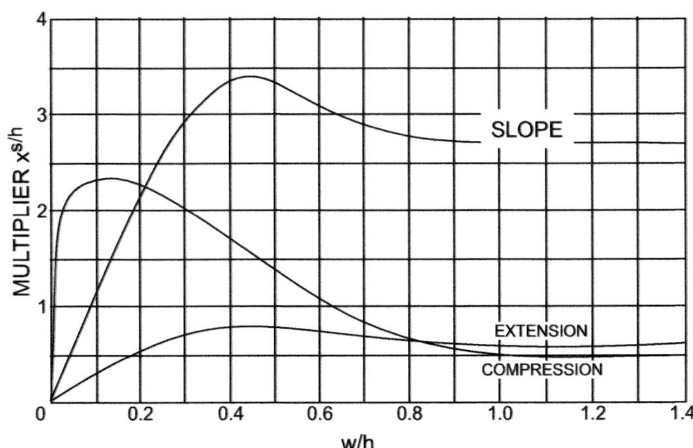

Figure 15.19 Relationship of slope and strain to the w/h ratio of workings (NCB, 1975)

15.6 Controlled Caving and Subsidence

The main role of controlled caving and subsidence in longwall mining is to reduce strata movement and caving and by this means to minimise ground surface effects and damage to surface structures. In order to achieve this, several methods are employed in coal fields over the world, as discussed below.

15.6.1 Stowing

Stowing of the excavated void can be by pneumatic or hydraulic means. The type of stowing governs the magnitude of subsidence as given by the equation below:

$$S_{max} = at \qquad (15.4)$$

where

S_{max} = maximum subsidence
t = workable thickness of the seam
a = subsidence factor dependent on the type of mining and stowing

Values for factor 'a' have been developed from surveying data of surface subsidence as follows:

Strata caving system	a = 0.75
Solid stowing (pack)	a = 0.50
Pneumatic and mechanical stowing	a = 0.30
Partial extraction (50%) and hydraulic stowing	a = 0.15

Successful stowing is achieved where the roof does not sag before filling has been introduced into the void.

However, over 95% of longwall production units have a gob created behind the face by total caving of the roof strata. For example, in Great Britain pneumatic stowing was used to eliminate caving and reduce surface subsidence. However, it is now no longer considered practical since more effective techniques can be used to give greater protection.

As a general rule, the maximum subsidence has been found to be 45% of the workable seam thickness for w/h ratios greater than 1.2 and with pneumatic stowing. With no stowing, the maximum subsidence is 90% for the same conditions (Figure 15.20).

15.6.2 Methods of Subsidence Control in Same Seam

One of the most common method of controlling the subsidence in the UK is by limiting the width of longwall faces and using intervening rib pillars so that surface subsidence can be reduced as follows.

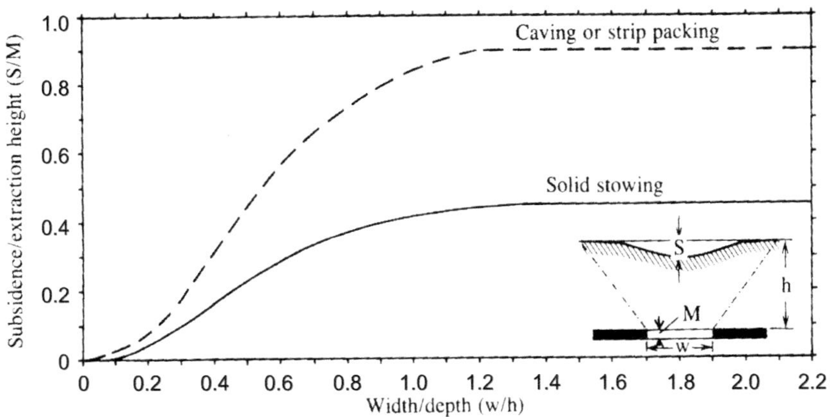

Figure 15.20 Effect of method of goaf strata control on subsidence (NCB, 1965)

1. Subsidence of the surface is gradual and spread over an area exceeding the mined-out area.
2. Surface subsidence is predictable with considerable accuracy in most mining situations.
3. The duration of subsidence is also predictable and it is relatively short lived by comparison to other mining methods.

Figure 15.21 shows the interactive effect of surface subsidence induced by two longwall faces resulting in a smooth overall subsidence profile.

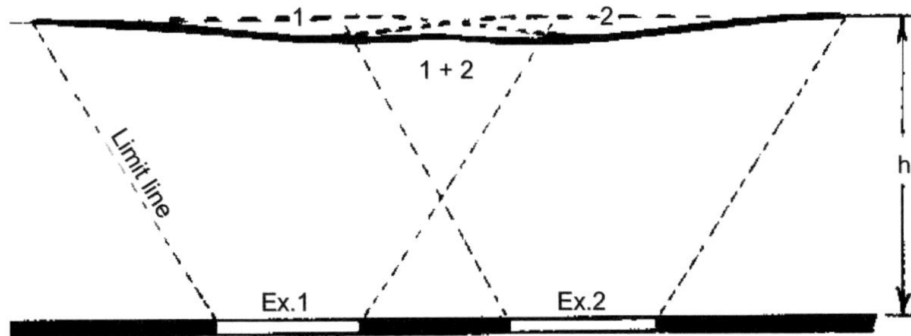

Figure 15.21 Interaction of surface subsidence profiles from extraction of same seam

15.6.3 Interaction of Surface Profiles from Extraction of Different Seams

Figure 15.22 shows a method of subsidence control using a multi-seam mining technique which involves staggering longwall faces in different seams and using different widths of rib pillars. The considerations for phasing multi-seam longwall faces for subsidence control are as follows.

1. Staggering of pillars under extraction can produce a smoothing effect to the resultant subsidence profile.
2. The example pre-supposes extraction A to have been completed before B_1 and B_2 commence.
3. Surface ground strain is less due to resulting smooth subsidence profile.
4. When multi-seam mining is carried out, the upper seam should be mined first and then the lower seams.

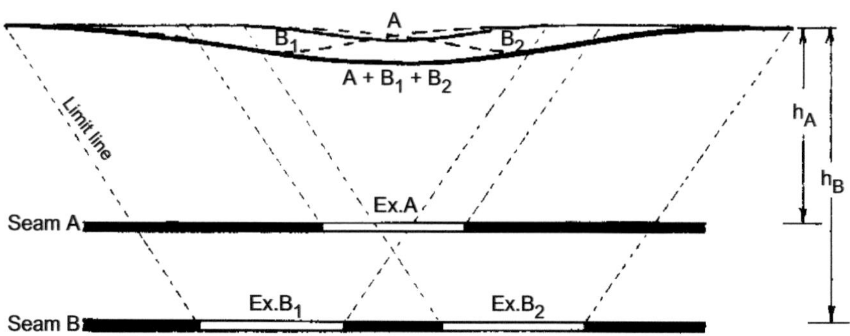

Figure 15.22 Subsidence control by multiple seam workings

In multiple seam situations careful planning and phasing of layouts are necessary in order to (1) minimise subsidence and (2) reduce strata pressure interaction between workings.

A careful extraction phase plan is required if the upper seam is to continue to remain workable.

15.6.4 Harmonic Mining

Harmonic extraction usually requires wide longwall faces. Simultaneous advancement of individual extraction faces should be in harmony with each other and two principal cases are considered.

(1) Multiple seam mining should have the spacing of the longwall faces shown in Figure 15.23. Mining of two or more seams has to be simultaneous to permit the compressive lateral strains, +E, from one mining horizon to offset the tensile strains, –E, from the lower mining horizon. Prevention is by summing strains with opposite signs (Figure 15.23). It is necessary to emphasise that advancing of both faces should be continuous and harmonious.

(2) Single seam mining faces should not be at an angle to each other ('two-winged') faces because of the danger of summing stresses with

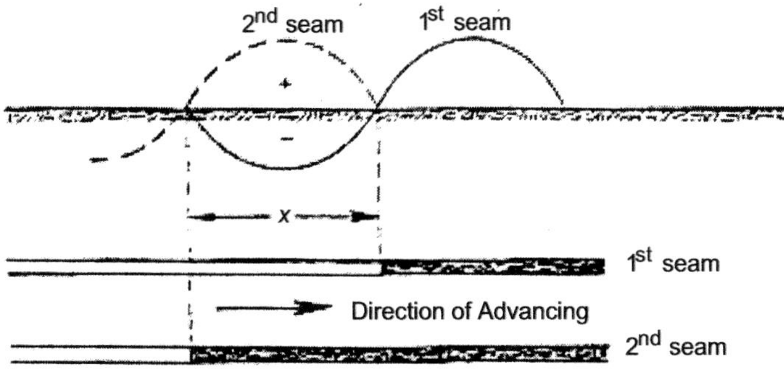

Figure 15.23 Harmonic mining as a method of subsidence control

the same signs. Mining should be with staggered faces, advancing in a stepped sequence, Figure 15.24. The tensile strain travelling ahead of the face for the right side panel is relatively small while the compressive strain behind the face will to some extent be compensated by the travelling tensile strain from the left side panel. With this sequence, the strain induced at a surface structure is maintained as a minimum.

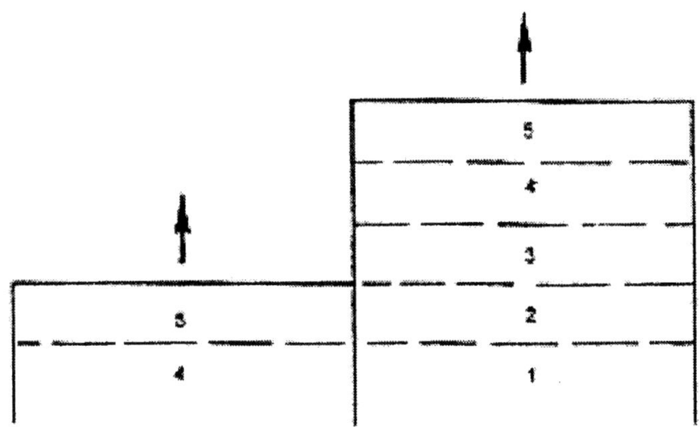

Figure 15.24 Mining in the same seam for controlling subsidence

15.6.5 Rapid Rate of Advance

Rapid extraction results in a fast moving active subsidence zone similar to a wave. The maximum tensile strain is less than the final obtainable strain from slower extraction and it may be at a permissible level (Figure 15.25).

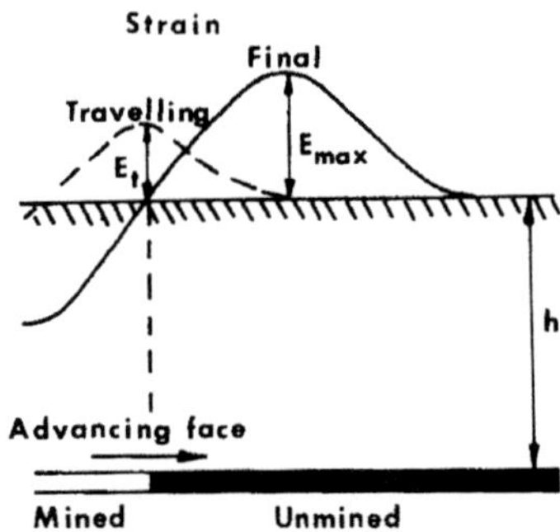

Figure 15.25 Rapid extraction as a means for controlling subsidence

The mechanics of controlled caving and subsidence is important due to the requirement for large, continuous and rapid coal extraction rates. This trend, in the near future, will be further increased with the need for the development of larger underground mining operations.

15.7 Effect of Mining on Surface Structures

A number of factors due to surface subsidence affect the stability of surface features. These are:

- Subsidence
- Tilt
- Tensile (stress)
- Compressive (stress)

Under normal conditions in coal measures rocks the mining subsidence can be predicted in terms of width to depth ratio and thickness of extraction. However, the following parameters affect the damage to surface structures due to mining subsidence (Figure 15.26).

1. Geology of overlying strata
2. Presence of newer measures, superficials and soils
3. Normal and abnormal structural factors and length

The most important factors are the length of the structure and the amount of surface strain.

Figure 15.27 indicates a method of evaluating the severity of damage to the length of building based on the horizontal strain.

The method of calculation of the surface strain is shown in the next section.

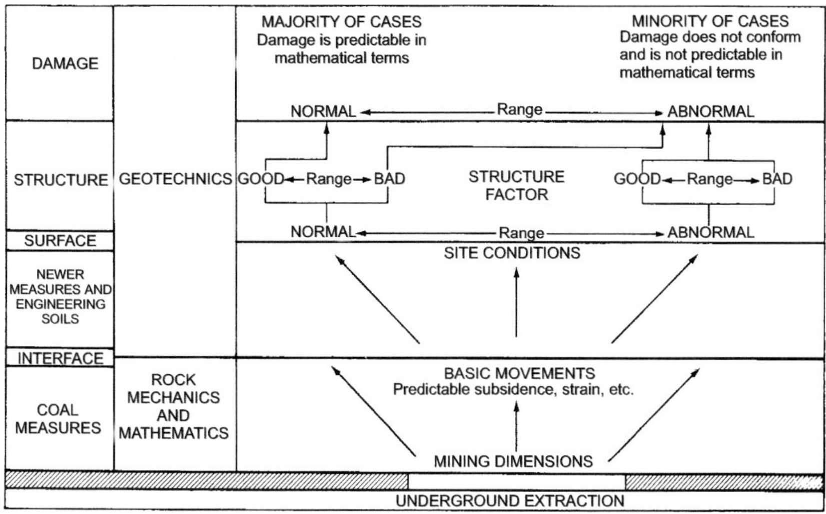

Figure 15.26 Factors effecting damage to the surface structure due to mining subsidence (Shadbolt, 1975)

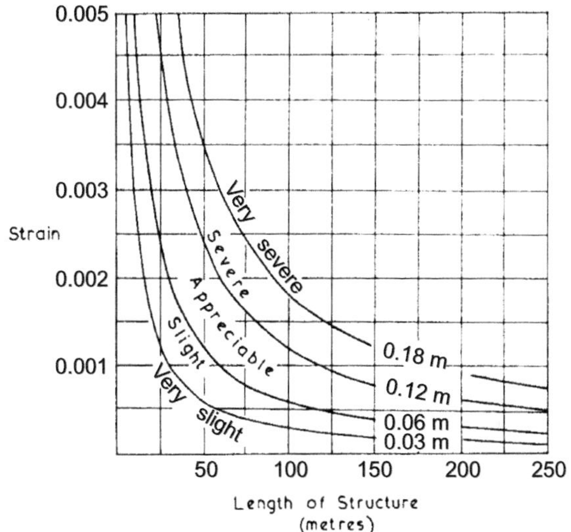

Figure 15.27 Length of structure and horizontal strain (after NCB, 1975)

15.7.2 Method of Calculating Surface Strain due to Longwall Mining

The maximum tensile strain at the bottom of an aquifer or a surface accumulation of water can be estimated by the following equation:

$$E = \frac{S}{h} = \frac{0.9tCv}{h}$$

where

E = surface tensile strain
= 0.9 × extracted seam height t (m)
v = maximum surface subsidence
S = subsidence = C × v
h = depth below surface (m)
C = disturbance factor at the bottom of the aquifer
t = extracted seam height (m)

Figure 15.28 shows the relationship between the width of workings and the depth below the surface and the disturbance factor for coal measures rock. It can be shown that for an extraction width of 200 m at a depth of 300 m below an aquifer with a total extraction of 2.0 m, the disturbance factor is 0.71 m and the maximum strain is 4.26.

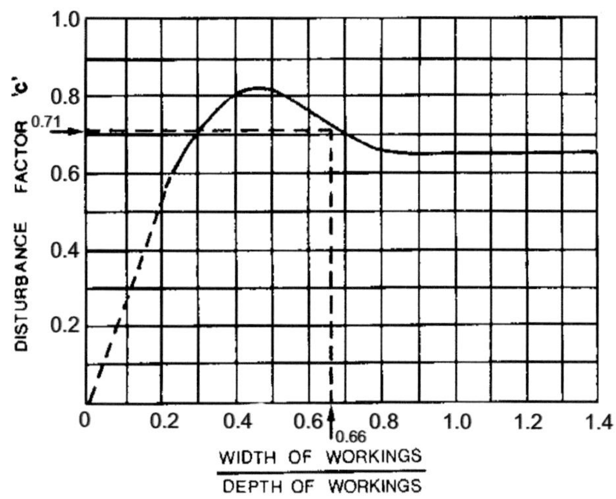

Figure 15.28 Disturbance factor for various w/h ratios (Watson, 1980)

15.7.3 Subsidence Prediction in Australia

Figure 15.29 shows the surface subsidence development curves of longwall mines in the Sydney basin and the Newcastle Coalfield as compared to British data from the *Subsidence Engineers Handbook* (1975).

The Australian experience of deep longwall mining beneath massive sandstone is as follows.

- In the Southern coalfields, the subsidence is 30% of the extracted seam height at Appin Colliery as compared to 90% in the British coalfields.

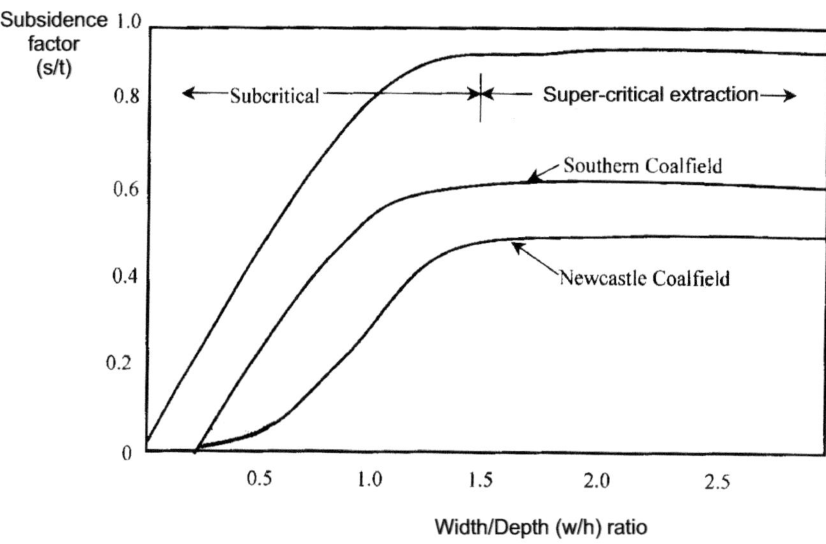

Figure 15.29 Difference in subsidence pattern in the UK Southern and Newcastle Coalfields

- In the Newcastle Coalfield, the surface subsidence is 26% of extracted seam height at Ellalong Colliery which is substantially less than the British norms.
- Maximum tensile strain is 0.5 mm/m as compared to 10 mm/m in UK conditions.
- Maximum compressive strain 1.2 mm/m compared with those predicted for the similar UK mining geometries.

The reasons for this anomaly in Australia are as follows.

- Greater mining depths in the Southern Coalfield incorporating thicker overburdens.
- Strong overburden units.
- Supercritical extraction width only achieved after mining several longwall panels.

15.7.4 The Differences in Subsidence Pattern between Newcastle Coalfield and the Southern Coalfield

(a) The Southern Coalfield

The main differences in the surface subsidence with respect to predictions carried out in the *Subsidence Engineering Handbook* are as follows.

- Rapid subsidence starts when w/h ratio exceeds 0.3.

- The sandstone units in the Southern Coalfield are less effective in bridging the strata across the goaf.
- Depth of cover is between 200–500 m resulting in more sag for equal w/h ratio.

(b) Newcastle Coalfield

The main subsidence characteristics in Newcastle Coalfield are as follows.

- The perceptible subsidence starts at w/h ratios of 0.6 as shown in Figure 15.29.
- The conglomerate strata units overlying the coal seams provide the bridging effect above the goaves and inhibit the subsidence.
- Depth of cover in Newcastle Coalfield is < 200 m.

15.8 Environmental Effects of Mining Subsidence

Development of subsidence around mine workings may induce fissures in overlying brittle rocks connecting the mine workings to the surface and create a range of problems, as follows.

- Formation of sinkholes in limestone and dolomite rocks.
- Infiltration of ground water from aquifers to the underground workings resulting in depletion of the water supply.
- Aquifers outcropping at the surface causing floods in agricultural fields.
- Breathing of air through the subsidence cracks and increasing the risk of underground fire due to spontaneous combustion.
- Subsidence movement rupturing the aquifers.
- Contamination of mine water through introduction of surface water.

15.8.1 Formation of Sinkholes

Mining beneath fissured limestone or dolomites may result in modification of ground water inflow pattern to the mining horizon causing enlargement of the existing fissures. The mining process also changes the hydraulic gradient from the aquifer to the mine workings and increased pumping rates also increase the groundwater flow through the porous and pervious ground in the mine roof. This is followed by enlargement of fissures, caving of the cap rock into the existing cavern and forming a sub-surface hole. This progressive failure of overlying rocks continues resulting in the formation of sinkholes extending up to the surface as illustrated in Figure 15.30.

Some of these sink holes are tens of metres in diameter and extend down to a depth of several metres.

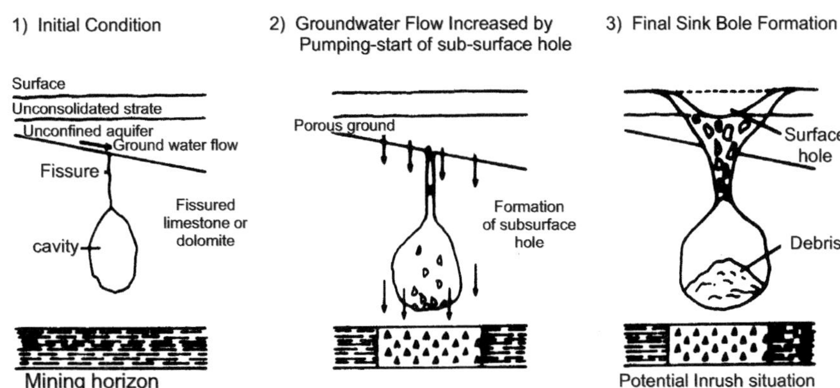

Figure 15.30 Formation of sinkholes in limestone and dolomites

15.8.2 Mining Subsidence Changing Surface Flow Pattern and Water Table

(a) Thick plastic impermeable beds underneath the aquifer

Figure 15.31 shows a coal seam overlain by bedded layers including a thick plastic aquiclude layer below an unconfined aquifer. A wide excavation in the coal seam will create disturbance of the overlying strata resulting in forming a subsidence trough without disrupting the plastic beds and forming water-conducting fractures. This will result in outcropping of the water table at the surface. The surface water due to rain or melting snow will aggravate this situation. The example of this form of flooding of aquifer water was observed at Annesley Colliery in the South Nottinghamshire area of the East Midlands Coalfields, UK.

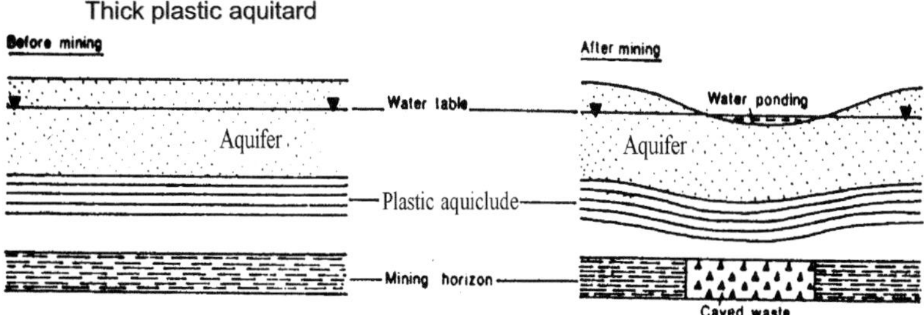

Figure 15.31 Surface subsidence causing outcropping of aquifer water at the surface

The remedial measure is to sink absorbing wells into a suitable permeable rock formation to drain the surface water (Figure 15.32a). If the underlying strata are impermeable, then the dewatering technique using the absorption well is unsuitable. Under theses circumstances absorbing drains assisted by horizontal mole drains or French drains at regular intervals should be used to discharge the flood water to a suitable surface watercourse as shown in Figure 15.32(b).

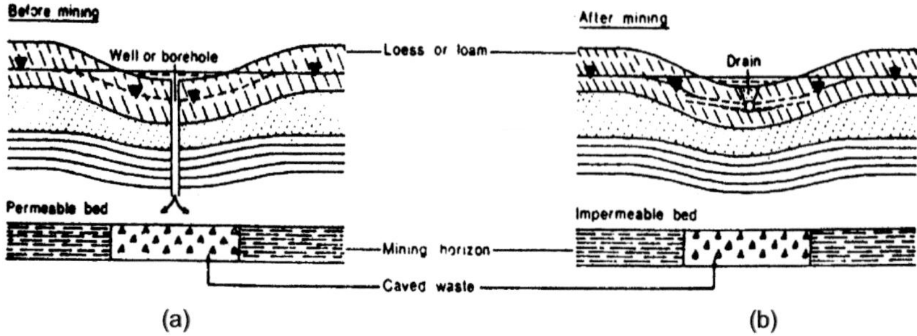

Figure 15.32 Remedial measures for preventing surface flooding by an aquifer: (a) Absorbing well discharging into a permeable bed; (b) Absorbing drains discharging into watercourse

(b) Solid brittle strata above the tabular deposit

Figure 15.33 shows a stratified ore deposit overlain by bedded overburden and confined and unconfined aquifers. The effect of a wide mining excavation is to form a subsidence trough above the mining horizon. This will result in the development of fractures in the brittle rock mass, change the hydraulic gradient and induce the flow of water towards the excavation. This will be followed by localised lowering of the piezometric surface of the aquifer, thus affecting the water supply of the wells with consequential environmental and ecological effects.

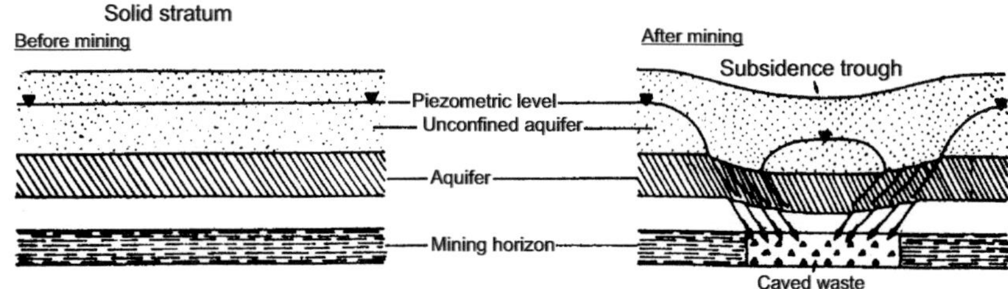

Figure 15.33 Surface subsidence causing the rupture of the overlying aquifer

15.8.3 Subsidence Increasing Potential Risk of Underground Fire

Subsidence has an effect on the initiation and progression of underground fires in coal seams liable to spontaneous combustion. The cracks developed as a result of mining subsidence over shallow mine workings allow circulation of air to the workings through the fissures. If coal left in the goaf is exposed to air, it may catch fire due to the inherent nature of the coal and may develop into a serious incident if left unattended. Underground fires in abandoned shallow mine workings in Jharia Coalfield in India originated from the large-scale collapse of a number of pillars inducing surface subsidence (Figure 15.34). Perhaps unsystematic mining of thick seams at shallow depths without paying due attention to the stability of slender pillars resulted in the development of such unprecedented fires in the underground workings in the Jharia Coalfield (Saxena, 1990).

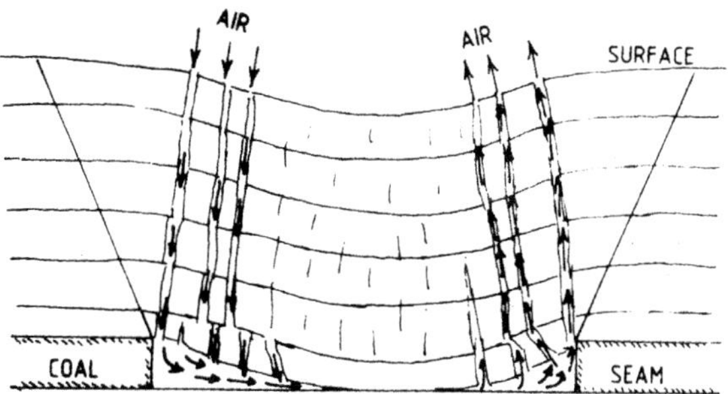

Figure 15.34 Subsidence increasing potential risk of underground fire in shallow room and pillar mine workings (Saxena, 1990)

15.8.4 Subsidence Rupturing Aquifers

Figure 15.35 shows the effect of subsidence due to shallow and thick mine workings which culminate in developing cracks and fissures in the overlying rock mass. If any thick aquifers are present, this subsidence process will inevitably result in increased inter-granular and pervious flow between the aquifers and the mine workings and also between the two

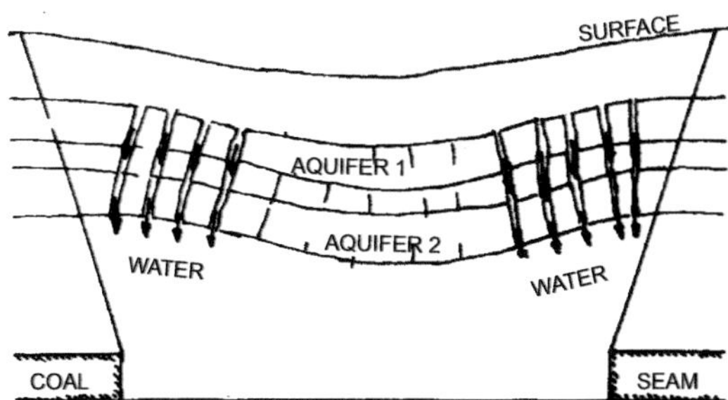

Figure 15.35 Mining subsidence interconnecting aquifers and disturbing groundwater chemistry (Saxena, 1990)

aquifers. This will result in loss of water, loss of yield of underground wells and mixing of water between different aquifers (Saxena, 1990).

15.8.5 Surface Water Contaminating Underground Water due to Subsidence (Saxena, 1990)

The development of open fissures in the subsided strata overlying a mine working is a typical characteristics of a brittle rock not normally experienced with the longwall system or the panel system of mining. Figure 15.36 shows subsidence cracks in the overlying rock mass above the mine workings which may conduct contaminated surface water from sewers to aquifers and underground workings (Saxena, 1990).

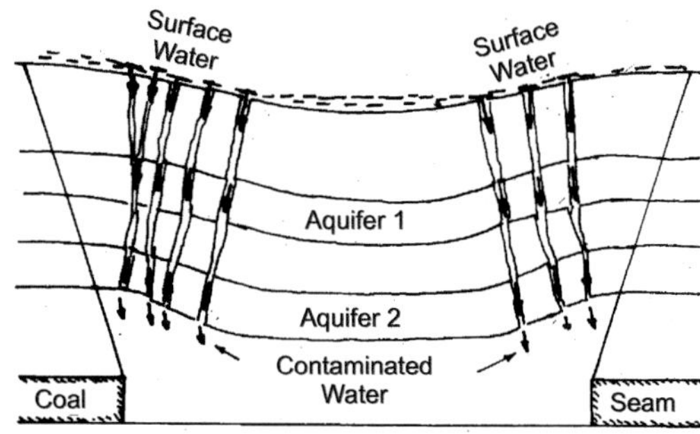

Figure 15.36 Contamination of underground water through subsidence cracks (Saxena, 1990)

15.9 Conclusions

Large-scale underground mining causes surface subsidence resulting in the destruction of sub-surface and surface rock formations around the mine workings causing a wide range of damage to surface structures and infrastructures. There are two types of surface subsidence: the discontinuous type associated with hard rock or in room and pillar mining characterised by step type deep subsidence depressions. The second type of subsidence is the trough type which is associated with longwall coal mining and is characterised by the formation of smooth subsidence profiles.

The trough type of subsidence can be predicted with accuracy with regard to magnitude, effects and durations and can be controlled by a careful and phased programme of mining. A brief account of the effects of mining subsidence on surface structural damage is given together with the design and precautionary measures.

Mining subsidence also modifies the hydraulic gradients and in certain circumstances results in the formation of sink holes, causing surface inundation, drying of wells, mixing of aquifers and contamination of aquifers with surface waste water. Carefully planned mining operations should be carried out to prevent the subsidence problems associated with underground mining.

References

Brady, B.H.G. and E.T. Brown (1985). 'Mining induced surface subsidence', Chapter 16, *Rock Mechanics for Underground Mining*, George Allen and Unwin, London, 525 pp.

Craft, J.L. and T.M. Crandell. (1988). *Mine configuration and its relationship to surface subsidence*, US Bureau of Mines, IC 9184, pp. 373–381.

Down, C.G. and J. Stock (1978). 'Environmental Impact of Mining', Chapter 12, *Mining Subsidence*, Applied Science Publishers Ltd., London, ISBN 085 334716 6, pp. 311–335.

Elbert, J. and L. Guernsey (1988). *An analysis of extremely dangerous problems associated with subsidence of abandoned coal mine lands in Southeren Indiana*, US Bureau of Mines, IC 9184, pp. 383–389.

Fritzche, C.H. and E.L.J. Potts (1954). 'Horizon Mining', Chapter 2, *Strata control and subsidence*, George Allen and Unwin Ltd., London, pp. 94–136.

Fawcett, R.J., S. Hibbert and R.N. Singh (1986). 'Analytical calculations of hydraulic conductivities above longwall coal faces. *International Journal of Mine Water*, Vol. 5, No. 1, March, pp. 45–60.

Hsiung, S.M., P.M. Lin and S.S. Peng (1988). *Structures and ground damages over abandoned mine lands*, US Bureau of Mines, IC 9184, pp. 362–368.

King, H.J., B.N. Whittaker, and C.H. Shadbolt (1974). 'Effects of mining subsidence on surface structures', in: *Minerals and the Environment Symposium*, Institution of Mining and Metallurgy, London, pp. 617–42.

Peng S.S. (1986). 'Coal Mining Ground Control', Second Edition, Chapter 13, *Surface Subsidence*, A Wiley Science Publication, John Wiley and Sons, New York, ISBN 0-471-82171-3, pp. 420–461.

NCB (1975). *Subsidence Engineering Handbook*, National Coal Board, Mining Department, 111 pp.

Orchard, R.J. (1964). 'Partial extraction and subsidence', *Mining Engineer*, vol. 123, No. 43, pp. 417–430.

Orchard, R.J. and W.S. Allen (1974). 'Time dependence in mining subsidence', *Minerals and the Environment* IMM, London, pp. 643–659.

Sengupta, M. (1993). 'Environmental Impacts of Mining, Monitoring, Restoration, and Control', Chapter 10, *Mining subsidence*, ISBN 0-87371, Lewis Publishers, Boca Raton, pp. 431–440.

Shadbolt, C.H. (1975). *Mining subsidence, in site investigations in areas of mining subsidence*, Editor F.G. Bell, Newnes-Butterworths, London, pp. 109–148.

Singh, R.N. and A.S. Atkins (1982). 'Design considerations for mine workings under accumulation of water', *International Journal of Mine Water*, Vol. 1, No. 4, pp. 35–56.

Singh, R.N. (1986). 'Mine Inundations', *International Journal of Mine Water*, Vol. 5, No. 2, pp. 1–28.

Singh, M.M. and F.S. Kendorski. (1979). 'Strata disturbance prediction for mining beneath surface water and waste impoundments', *Proceedings of I[st] Annual Conference in Ground Control in Mining*, University of West Virginia, July 1981.

Saxena, N.C. (1990). 'Environmental impacts of underground mining', Chapter 9, *Environmental Management of Mining Operations*, Editor B.B. Dhar, ISBN 81-7024-273-8, pp. 208–261.

Wardell, K. and P. Egnon. (1968). 'Structural concepts of strata control and mine design', *Transaction of Institution of Mining Engineers*, vol. 127, No. 95, pp. 633–656.

Watson, H.F. (1979). 'Undersea coal mining in south east of England', *Proceedings of 10th World Mining Congress*, Istanbul, Turkey, Vol III.

Whetton, J. and H.J. King (1959). *Aspects of subsidence and related problems*, Transaction of Institution of Mining Engineers, vol. 118, pp. 663–676.

Whittaker, B.N., (1974). An appraisal of strata control practice, Trans. Institution of Mining Engineers, Vol. 134, pp 9–24.

Whittaker, B.N. and D.J. Reddish. (1989). *Subsidence, Occurrence, Prediction and Control*, Elsevier, Amsterdam, ISBN 0-444-87274-4, Vol. 56, 528 pp.

Wilson, A.H. (1986). 'The problem of strong roof beds and water bearing strata in the control of longwall faces', *The AusIMM, Illawarra Branch, Ground Movement and Control Related to Coal Mining*, Wollongong, August, pp. 1–12.

Stability Analysis of Surface Mining Slopes

16
Chapter

16.1 Introduction

One of the important issues in surface mining is the stability of the slopes surrounding the mining excavations. Unexpected slope instabilities may impair the safety of the personnel or impose the danger of burying mining equipment and resulting loss of production. Steeper slopes result in higher ore to overburden ratios and therefore higher profitability of the mining operations. Four types of slopes encountered in surface mining are soil slopes, spoil slopes, rock slopes and slopes in jointed or bedded rocks. In this chapter, slope stability analysis techniques associated with soil, spoil and rock slopes using a limiting equilibrium method are discussed together with the slope stabilising techniques. Mining engineers also have to deal with the stability of tips and lagoons using the same underlying design principles as with slopes.

16.2 Factors Affecting Slope Instabilities in Surface Mining

In surface mining the major constraints are those of a geotechnical nature while the main investment decisions must be made on sound economic bases (Atkinson, 1989). Slope stability problems in excavated strata or spoils within a mining lease can be related to alignment, direction of advance, method of mining, scale of operations and geotechnical constraints. Some of the major geotechnical parameters are:

- material properties including homogeneity of soil or rock
- rock structure in jointed or bedded rock
- groundwater regime
- external factors such as surcharge loading or presence of old underground mine workings

16.2.1 Material Properties of Soils, Spoils and Rocks

In surface mining, the most important mechanism of failure of slopes in surface mining which can occur is sliding of rock, spoil or soil along a preferred slip surface (Figure 16.1). In the case of spoil and soil slopes the failure surface may have the configuration of a part of circle along which rotational failure may take place. In the case of rock slopes the failure may take place along a single failure surface (Figure 16.1b) or along several failure surfaces as shown in Figures 16.1(c) and (d). In order for a slope to fail it is necessary that the shear stress acting on the slope mass exceeds the shear strength of the slope material along the failure plane. Therefore, shear strength parameters of soil and rock material are important properties for stability evaluation of the surface mining slopes. The shear strength of soil or rock is usually expressed by the Coulomb equation which is a linear combination of both cohesion and friction as follows:

$$\text{Shear strength} = \text{cohesion} + \text{friction, or}$$

$$\tau = c + \sigma_n \tan \phi \tag{16.1}$$

where

τ = shear strength (MPa)
σ_n = normal stress (MPa)

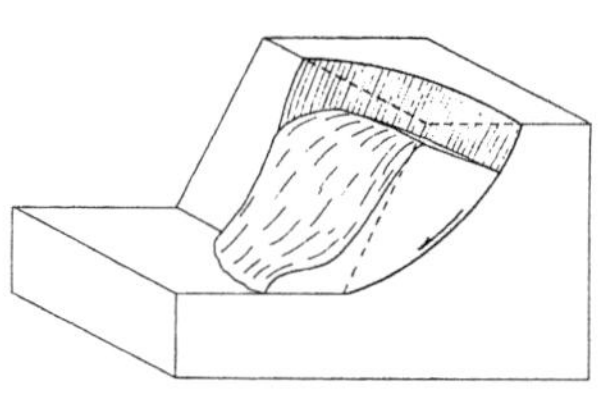

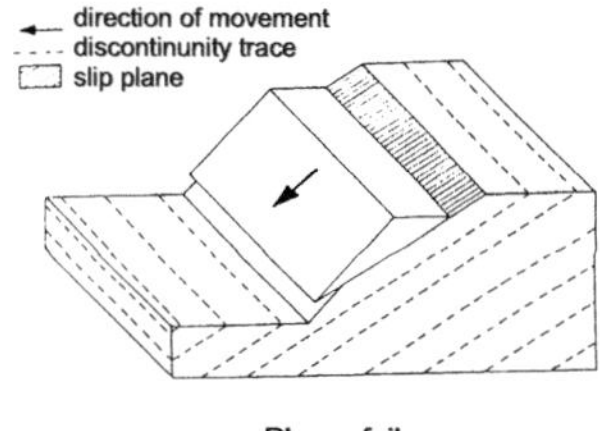

(a) Soil or spoil slope failure by rotational slip

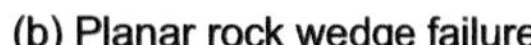

(b) Planar rock wedge failure

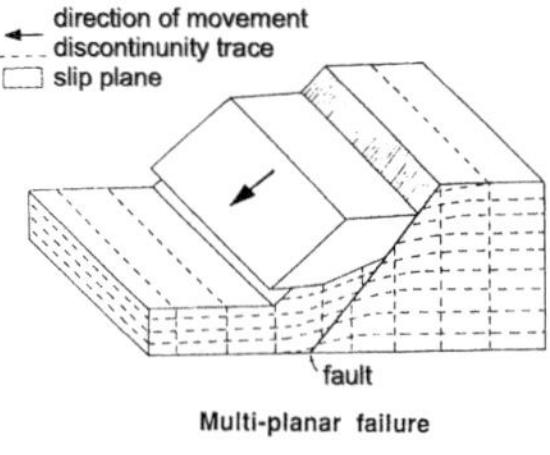

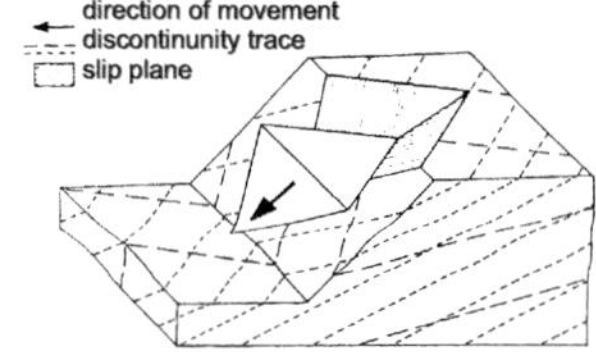

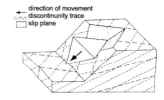

(c) Multi-planar failure

(d) Wedge failure along two failure surfaces

Figure 16.1 Failure surface in soil, spoil or rock slopes

ϕ = internal angle of friction
ρ = unit weight of rock or soil (kg/m^3)
c = cohesive strength (MPa)

The shear strength parameters of soil, spoil, rock material and rock joints are somewhat different. The shear strength of soil in Figure 16.2 shows the relationship between the shear stress and the normal stress on the shear surface. The normal stress on the shear plane is due to weight of the overlying soil given by $\rho \times h$. In general, rock or clayey soil possesses both cohesion and friction and the shear strength increases with an increase in the normal stress on the failure plane. A cohesive soil or rock has finite shear strength at zero normal stress. Most sands possess high friction and negligible cohesion whereas clays have comparatively low friction and some finite cohesion at zero normal stress. If sand is completely saturated it looses both cohesion and friction (viz. $c = 0, \phi = 0$) while saturated clay has a finite cohesion and virtually no friction. It can also be seen that loose sand has a lower angle of friction compared to dense sand.

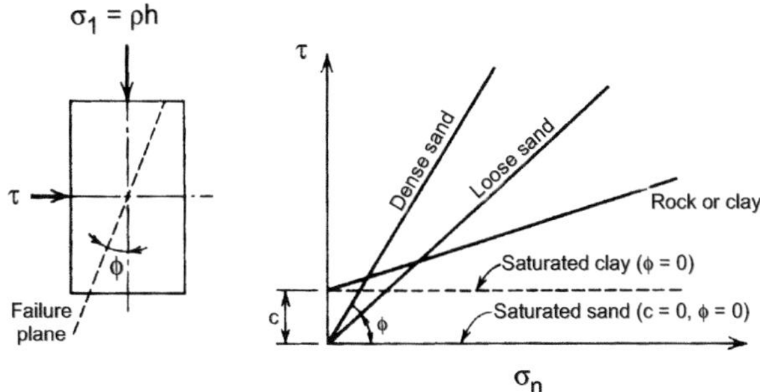

Figure 16.2 Shear strength of soil

Figure 16.3(a) shows the relationship between shear stress and normal stress for a rock discontinuity with adhesion at zero normal stress and a finite friction. Figure 16.3(b) shows the relationship between shear stress and shear displacement of a rock mass for increasing normal stresses. At a pre-failure regime the relationship between the shear stress and shear displacement is normally linear which becomes slightly non-linear before the peak stress is reached when the specimen fails and a residual strength is reached.

A plot of peak shear stress against the normal stress gives a linear peak strength envelope and a plot of residual strength against normal stress gives the residual strength envelope as shown in Figure 16.4(a). The use of the linear strength criterion is acceptable in situations representing low levels of

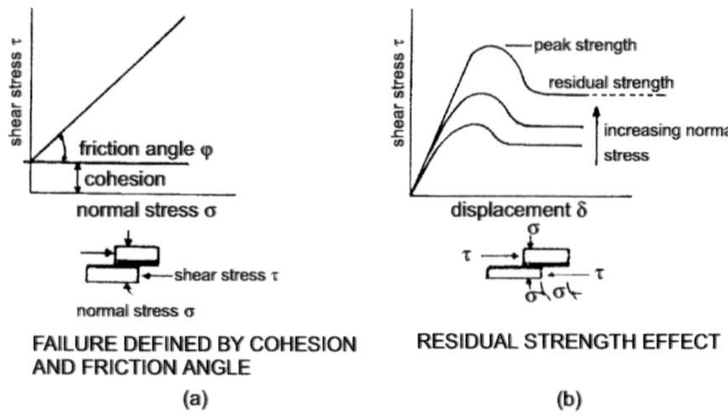

Figure 16.3 Shear strength of rock joint

constant stress along the failure plane. If there is a significant stress gradient along the failure surface it is then desirable to use a non-linear shear stress criterion shown in Figure 16.4(b). Denby (1983) proposed a power law representing the shear strength envelope of the coal measures rock of the following form:

$$\tau = A\,\sigma^{B} \tag{16.2}$$

where

τ = shear strength
σ = normal stress
A = constant
B = constant

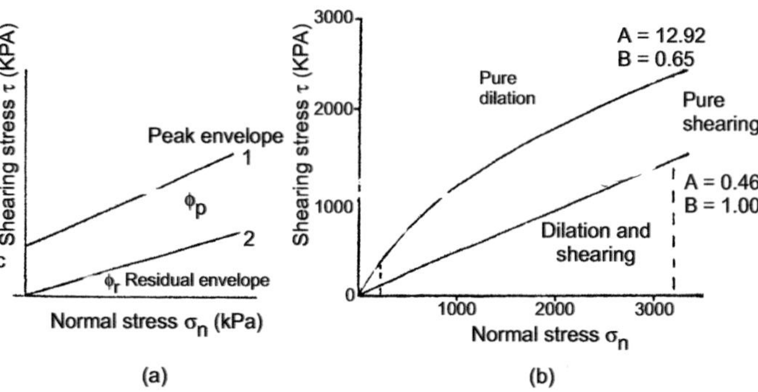

Figure 16.4 Linear and non-linear shear strength envelopes of soft rocks (Denby, 1983)

The properties of backfill material differ greatly from *in situ* rock as the mining process breaks up the strata into a wide range of particle sizes, alters porosity and permeability and causes weathering of the backfill material to varying degree, thus creating a soil-like material. Consequently, the slopes in the backfill material should be designed based on soil mechanics principles.

In some soils the inter-particle bonding in the soil is lost, possibly due to some form of leaching process whilst the slope retains the original structure. At some stage when most of the inter-particle bonding has been loosened an external shock load may induce complete collapse of the soil structure. This may result in creation of fluid-like material with low viscosity (solifluction) which can flow at high velocity to a considerable distance. Sites backfilled with a washed colliery discard may be susceptible to this type of failure and pose a danger over a considerable lateral extent (Denby, 1983).

16.2.2 Role of Geological Structures on Slope Stability

In harder rocks, the material strength may by far exceed the stress level imposed by the mining excavation. Failure may take place in the form of catastrophic sliding or by some amount of movement along pre-existing planes of weakness or discontinuities such as faults, bedding planes or joints or a combination of these geological structures. All structural features have a variety of properties as follows.

- Orientation and continuity of the structures.
- Shear strength of discontinuities including roughness, tightness and frequency.

Bedding planes have low strength but high continuity and form the most important structural feature in coal measures rock. Continuous joints are not very common in weaker coal measures rock but are relatively common in sandstones and in massive channels. Faults may have high continuity, low frequency and low shear strength. Minor structural features such as channels, washouts, swilley-type structures, and the occasional fossil slip surface along ancient stream channels may be of local importance and may give rise to local high angle discontinuities. Such strata are difficult to excavate to form conventional slopes. Superficial structures resulting from natural landslides which are relatively shallow should be avoided for the disposal of spoil to prevent landslides encroaching the excavated areas (Walton and Atkinson, 1978).

16.2.3 Effect of Groundwater

Resistance to sliding along a slip plane not only depends upon the shear strength of soil or rock material but also on the normal stress acting across the sliding surface. Groundwater affects the slope stability in a number of ways.

- It lubricates the clay and shale surfaces so that the cohesive strength lowers and the internal angle of friction ϕ decreases.
- Normal stress acting on the sliding surface reduces.
- Dead weight acting on the slope increases.

The reduction in normal stress is shown by the modified Coulomb equation as follows:

$$\tau = c + (\sigma_n - u) \tan \phi \qquad (16.3)$$

where 'u' is the pore water pressure. It is therefore necessary to allow the soil or rock slopes to drain freely so as to reduce the pore pressure and thereby increase the shear strength of the slope and consequently decrease the risk of slope failure. However, the low permeability of most soil types results in delay in both the dissipation of excessive pore pressures during the draining process and migration of water pressure within the ground mass.

16.2.4 External Factors Affecting Slope Instability

The external factors that control the stability of an excavated slope in surface mining operations are as follows.

- Mining method and slope geometry
- Old underground mine workings
- The effect of seismic shock
- Time effect
- Surcharge loading due to mining equipment or mining structures

(a) Mining methods

The method of surface mining controls the slope orientation and its geometry which in turn influences the slope stability. In general, flatter overall slopes are more stable than steeper slopes and this is achieved by benching since excavating machines and their mode of operation produce somewhat inflexible slope angles as outlined in Table 16.1. These slopes apply to individual benches.

Slope design in strip and terrace coal mining depends upon the cut orientation and the method of dewatering and only minor changes to slope angles are possible except when benching forms an integral part of the overburden removal operation. In general, in an open cut coal mine a highwall angle of 70–80° is maintained and the highwall bench is usually pre-blasted to facilitate drainage of the highwall to reduce pore water pressure and keep the highwall stable over a mining time span of a few days to few weeks (Thomas, 1979). In the case of unconsolidated overburden and wet ground conditions the slope angle is reduced to the natural angle of repose of 35–45° of the overburden and the extra material removed does not represent an economical disadvantage except at the limits of reach of the stripping equipment and at the ultimate cut. In highly inclined and deep-seated metalliferous mining operations the relatively steeper mining slopes offer economic advantages (Thomas, 1979).

Table 16.1 Typical bench slopes produced by excavating equipment in surface mining (Atkinson, 1989)

Equipment type and mode of operation	*Slope angle*	*Equipment type and mode of operation*	*Slope angle*
Dragline		**Bucket wheel excavator**	
With key cut	60–80°	Dropcut, digging face only	90°
Without key cut	70–80°	Dropcut, highwall	45–70°
Forecut	70°	Terrace cut, digging face	60°
Crosschop with key cut	60–80°	Terrace cut, highwall	50–70°
Crosschop without key cut	90°		
Stripping shovel	70–80°	**Loading shovel, standard crowd**	60–80°
Hydraulic excavator		**Tractor-scraper**	<30°
Shovel	45–90°	Conventional	<40°
Backhoe	30–90°	Tandem powered	<40°
		with rip dozer at toe	<90°

(b) Effect of blasting

Blasting is a common method of ground preparation for hard overburden in surface mining operations. Blasting induces artificial seismic effects due to the development of short wavelength high frequency vibrations due to the propagation of a detonation wave through the surrounding rock. Shock loads due to blasting may impose dynamic loads on the soil or rock slope and decrease its factor of safety leading to instability (Anon., 1988).

(c) Effect of time

Most of the factors affecting the stability of the surface mining slopes are time-dependent. Many strip mines advance rapidly but delays in the deeper excavations generally increases the likelihood of possible failure due to a decrease in peak strengths with time. Also the rate of change of pore water pressure in the excavated slope has an important bearing on the stability of slope. Groundwater recovery in the excavated surface mining slope also has an effect on the pore pressure development and the consequential effective stress on the slope. The development of weathering on the slope and backfill material is also time-dependent (Rao, 1980; Atkinson, 1989; Denby, 1989 in Singh and Denby, 1989)

(d) Surface mining above collapsed voids

Presence of old underground mine-workings, man-made caverns and natural voids within a surface mining lease can promote slope instability as shown in Figure 16.5 (Walton, 1988). Collapse of the overlying rocks above

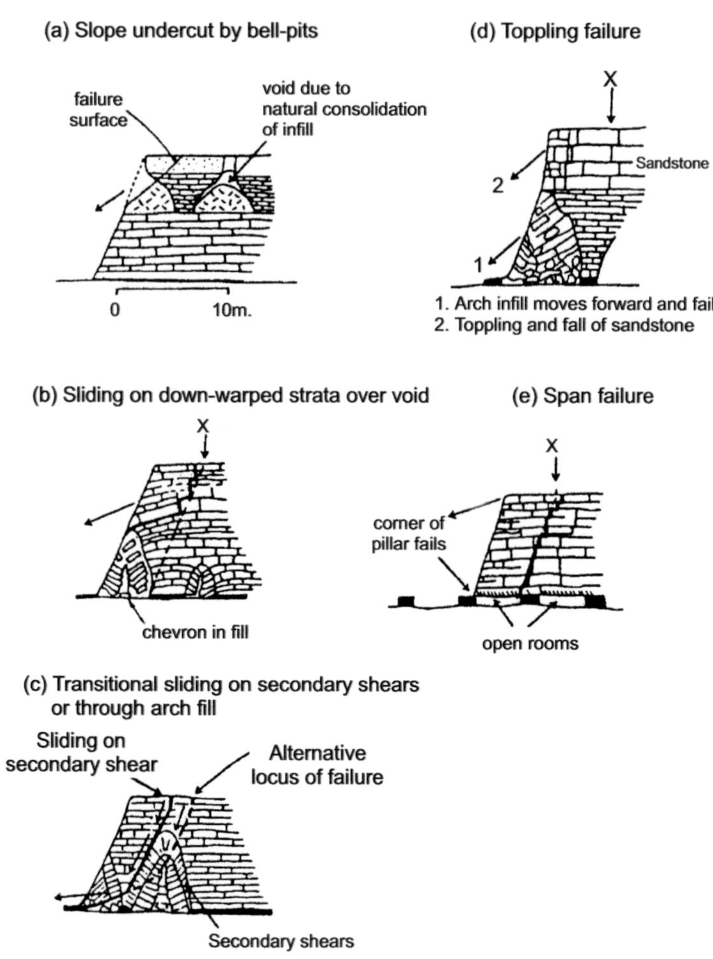

Figure 16.5 Instability of surface mining slopes on old mine workings (Walton and Taylor, 1978; Walton, 1988)

an old underground mining excavation causes caving of the strata which leads to deterioration of the slope structure. Types of instability encountered under these circumstances include translational sliding, toppling failure and rock falls and these types of instabilities usually affect a single bench slope (Walton and Taylor, 1978; Denby, 1983).

16.2.5 Effect of Adverse Surcharge Loading Conditions on Slope Stability

Figure 16.6 shows the surcharge loading by static and dynamic loading of the surface mining slopes due to traffic of heavy mining machinery. The

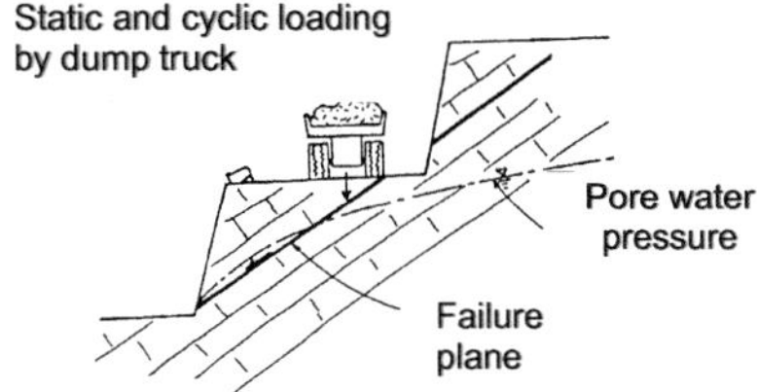

(a) Static and dynamic loading of highwall slope on haul roads

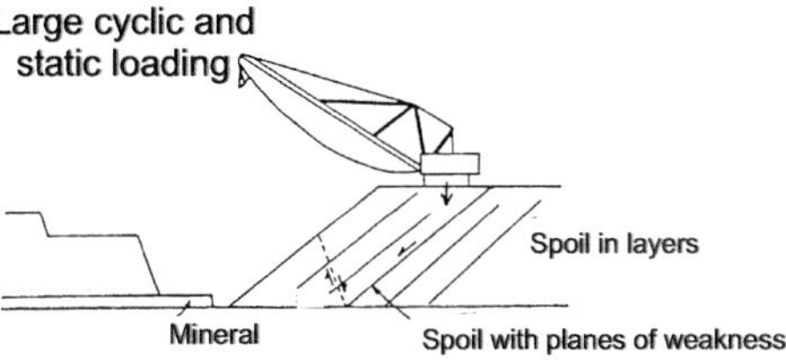

(b) Cyclic and static loading of spoil layers by large draglines

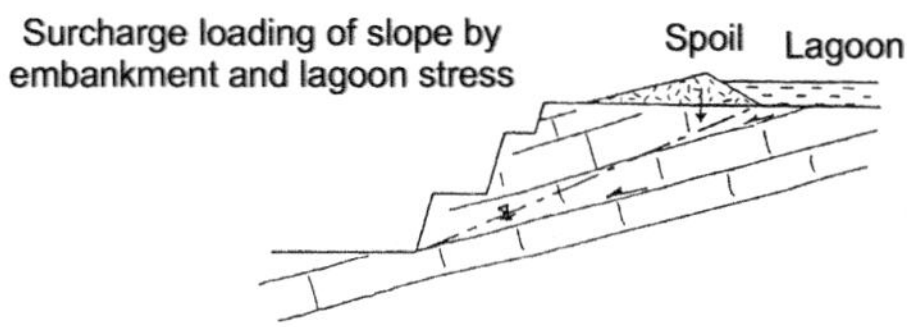

(c) Slope loading by embankment and lagoon

Figure 16.6 Loading of a mining slope (Denby, 1983)

dynamic load of the moving traffic may reduce the factor of safety of the marginally stable surface mining slope and promote sliding along a plane of potential failure specially where the stability condition is aggravated by pore water pressure. In a spoil slope, placement of spoils in layers creates the plane of instability where the excessive static and dynamic loading by large dragline located on the spoil pile may create the conditions leading to slope failure. In many mine-sites the surface land is so restricted that the boundary of lagoons and surface impoundments may lie in the vicinity of the excavated rock and soil slopes. The weight of the dam and waste water in lagoon not only provides the surcharge loading of the slope but also waste water leaking from the lagoon onto the mining slope may increase the pore water pressure and impair the overall stability of the excavated slope.

16.3 Stability Evaluation of Soil and Spoil Slopes

Before assessing the stability of a slope in a mining operation it is essential to classify the slope taking into consideration the main material, geological structure or type of instability present. If the slope consists of purely a soil slope or a spoil slope behaving like a soil slope then the method of analysis is based on the concepts of soil mechanics. In general, two types of failures may occur in the soil or spoil slopes:

1. Circular failure of slope between the crest and toe of the slope (Figure 16.7(a)
2. Base failure culminating at a rock ledge (Figure 16.7(b))

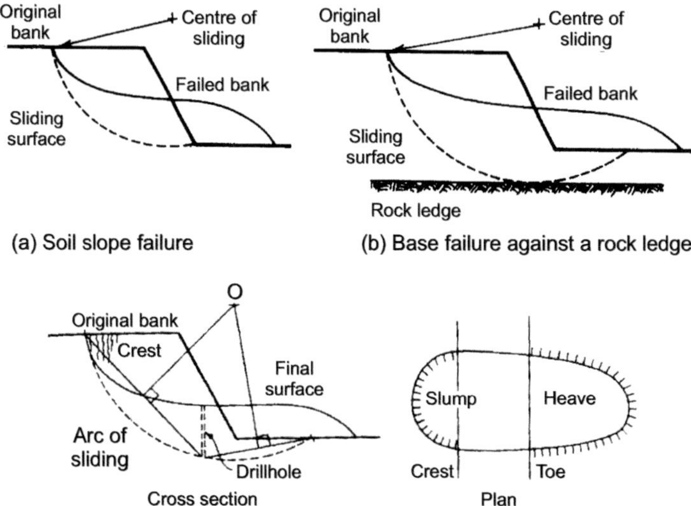

Figure 16.7 Types of soil and spoil slope failures due to rotational slip (Hartman,1987)

A shallow type of toe failure occurs in soils with low internal angle of friction in steep slopes (slope angle exceeding 55°). The sliding surface is circular in shape and invariably passes through the toe of the slope and some times joins the tension crack on the crest of the slope. In the case of a deep seated circular failure the instability is initiated at a hard plane of weakness usually below the toe of the slope. The slope failure is characterised by a massive flow resulting in the slumping of the crest of the slope and heaving of the toe of the slope (Figure 16.7a and b). In practice, the base of a circular slope surface in the failed soil can be located by drilling an exploratory drill hole on the failed slope and the centre of the circular arc 'O' can be located by constructing two perpendicular bisectors as shown in Figure 16.7(c) (Hutchinson, 1983). Impending slope instability is usually indicated by the presence of tension cracks at the crest of the slope some distance away from the slope edge.

16.3.1 Analytical Procedures for Slope Stability Analysis

The method of slope stability analysis is based on the limiting equilibrium method which comprises the following discrete steps.

(1) Selection of a critical failure surface
(2) Calculation of the destabilising forces acting on the slope
(3) Location of the centre of slip circle and the line of gravity acting on the slope segments
(4) Calculation of total stabilising forces acting on the slope
(5) Calculation of the factor of safety as ratio of stabilising force and destabilising forces. A factor of safety between 1.0 and 1.5 indicates a stable slope while 1.25 is an acceptable value for temporary slopes.

In practice two basic methods of slope stability analysis are available as follows.

(1) Total stress analysis method incorporating opposing slope weight
(2) Method incorporating the Swedish method of slices

16.3.2 Total Stress Analysis Method

This method of stability analysis assumes that the base of the failure surface is an arc of a circle with the centre at O with radius 'r' and tangential to the plane of weakness below the slope. A vertical line from the centre of the circle O divides the slope into two parts, the sector at the left of the line has weight W_1 destabilising the slope and the sector on the right has weight W_2 developing a rotational movement stabilising the slope. In addition the shear strength acting along the length 'L' of the sliding arch tends to stabilise the slope. A diagram of a deep circular type of failure representing a base failure is shown in Figure 16.8. It can be shown that:

$$\text{Moment of destabilising force} = W_1 \times \ell_1$$
$$\text{Moment of stabilising force} = W_2 \times \ell_2$$

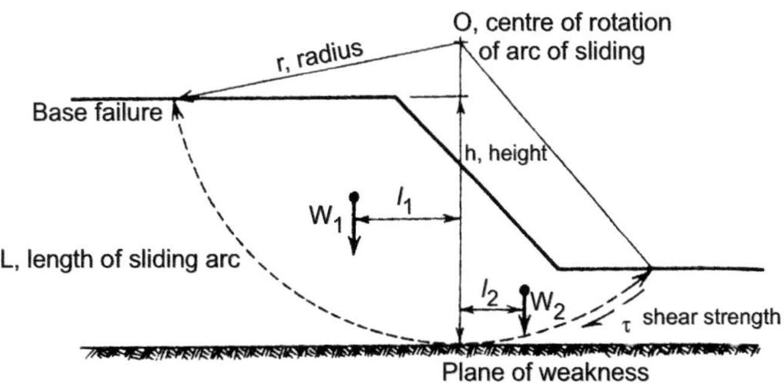

Figure 16.8 Total stress analysis method

Moment of the shear strength acting along the sliding arc $= \tau \times L \times r$

$$\text{Factor of safety against slope failure} = F = \frac{W_2 \times \ell_2 + \tau r L_r}{W_1 \times \ell_1} \quad (16.4)$$

$$\tau = c + \gamma h' \tan\phi$$

where

c = cohesive strength of soil
γ = unit weight of soil
ϕ = internal angle of friction of soil
τ = shear strength of soil
h' = mean height of the slope
W_1, W_2 = weight of sectors of the soil slope
ℓ_1, ℓ_2 = length of lever arms of each sectors of the soil slope
L = length of the sliding arc = $r\,\theta$
θ = angle subtended by the sliding arc on the centre of rotation of the arc

Measuring the area on a scaled diagram and multiplying it by the unit weight of the soil determines the weight of each sector, W_1 and W_2. The lengths of the lever arms ℓ_1 and ℓ_2 can be easily determined by constructing a geometric replica of the slope sectors on card board and locating their centres by suspending it freely on a pin. Location of the rotational centre of the sliding arc is determined by trial and error and is discussed in Section 16.3.5.

16.3.3 Effect of Tension Cracks

In cohesive soil slopes tension cracks may appear on the top of the slope crest as shown in Figure 16.9 indicating impending instability due to loss of continuity of the slope. In wet climates these tension cracks can be filled with water exerting hydraulic pressure on the slope. In a purely cohesive soil the depth of the tension crack, h_c, can be calculated by the following formula (Smith, 1973):

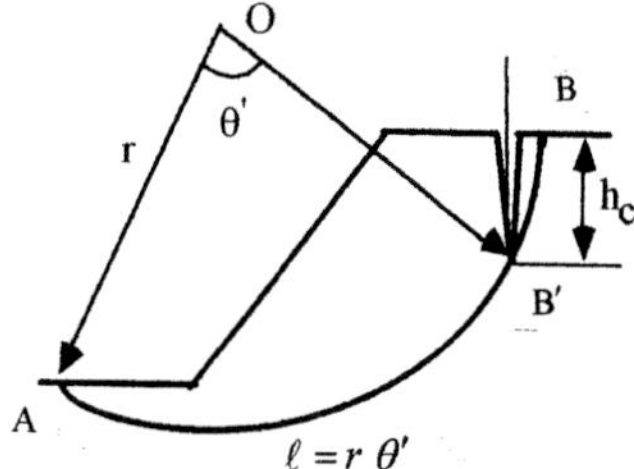

Figure 16.9 Tension crack in a cohesive soil (Smith, 1973)

$$h_c = \frac{2c}{\gamma} \tag{16.5}$$

where

h_c = depth of the tension crack
γ = soil density (kN/m^3)

The effect of the development of tension crack is to reduce effective length of the arc of the sliding surface from AB to AB′ thereby changing the angle of the arc of contact from θ to θ' in the formula for factor of safety F. However, the full weight W of the sector is used in calculations to compensate for any water pressure that may be exerted if the crack fills with rainwater.

16.3.4 The Swedish Method of Slices Analysis

In order to provide better mathematical accuracy to total stress analysis method utilising two segmental weights, the Swedish method of slices can be used utilising many segments of equal width. The soil in this method of analysis is considered to have both cohesion 'c' and an internal angle of friction ϕ.

The method is based on dividing the sector into a suitable number of vertical slices. The stability of one such slice is considered in Figure 16.10(b) where the lateral reactions on the two vertical sides of the wedge, L_1 and L_2 are assumed to be equal. At the base of the slice a triangle of forces is drawn with the weight of the slice W as vertical to a scale, a normal line (N) from the centre of the slice and a tangential component from the bottom of vertical line is drawn to complete the triangle of forces. Thus the magnitude and direction of the tangential component T and normal component N of the weight are determined.

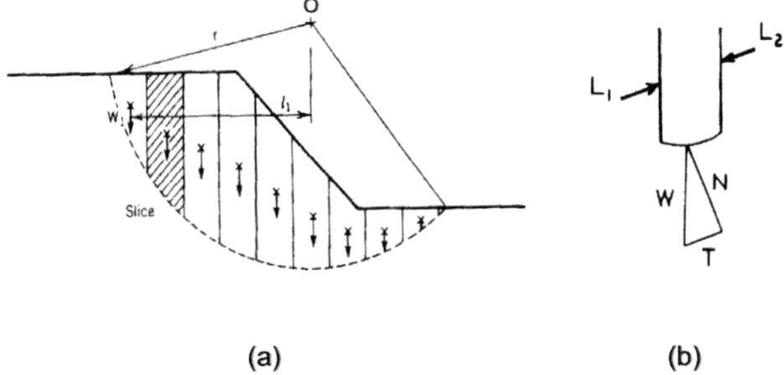

Figure 16.10 Slope stability analyses using the Swedish method of slices

It can be shown that by taking moments about the centre of rotation O:

Disturbing moment $= r\,\Sigma T$

Restraining moment $= r\,(c\,r\,\theta + \Sigma N\, \tan\phi)$

Hence $$F = \frac{\text{Restraining moment}}{\text{Disturbing moment}} = \frac{c\,r\,\theta + \Sigma N \tan\phi}{\Sigma T} \quad (16.6)$$

In this case the effect of a tension crack is given by:

$$h_c = \frac{2C}{\gamma} \tan\left(45° + \frac{\phi}{2}\right) \quad (16.7)$$

16.3.5 Location of the Most Critical Circle

The centre of the most critical slip circle is determined by trial and error method by plotting the centre of various trial slip circles being analysed with the corresponding factor of safety (Figure 16.11).

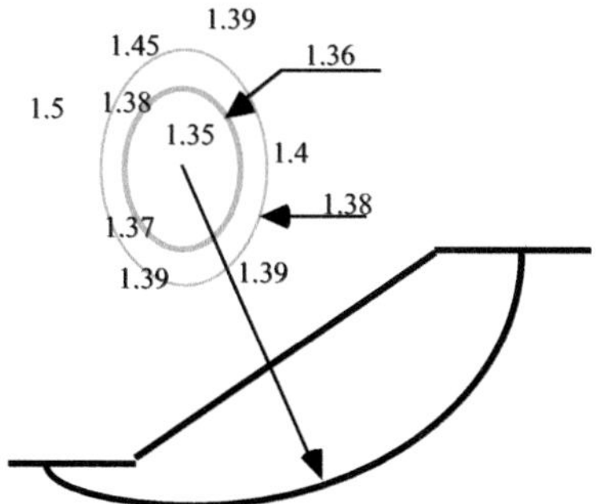

Figure 16.11 Method of determining centre of critical circle (Smith, 1973)

After plotting several centres of trial slip circles it is possible to draw 'contours' of F values, which are roughly elliptical so that the centre indicates the position of the centre of the critical circle. It may be noted that the value of the factor of safety is more sensitive to horizontal moments of the circle's centre than to vertical moments.

The mode of slope failure in homogenous soil primarily depends upon the soil properties and the steepness of the slope (Figure 16.12). In the case of soil slopes with slope angles greater than 53°, the critical slip circle invariably intersects through the toe for any value of friction angle exceeding 3°. In the case where there is a layer of relatively stiff material at the base of the slope, then the slip circle passes tangentially to this layer intersecting the slope above the toe (Figure 16.12b). For cohesive soils with a small angle of friction the slip circle tends to go deeper below the toe of the slope and extends in front of the toe (Figure 16.12c). The position of the deep slip circle is also controlled by the position of a layer of stiff material below the slope (Figure 16.12d).

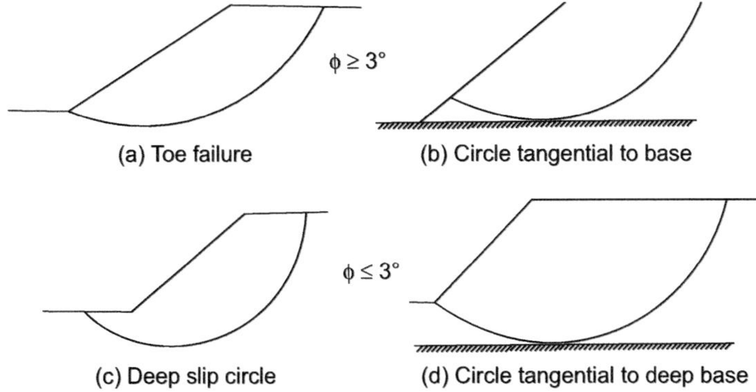

Figure 16.12 Modes of circular slip failures for homogeneous soils (Smith, 1973)

In the case of a slope made out of homogeneous cohesive soil it is possible to determine directly the centre of the critical circle by a method proposed by Fellenius (1936) (Figure 16.13). The centre of the circle can be obtained by the intersection of two lines set off from the bottom and top of the slope at angles α and β respectively as given in the Table 16.2.

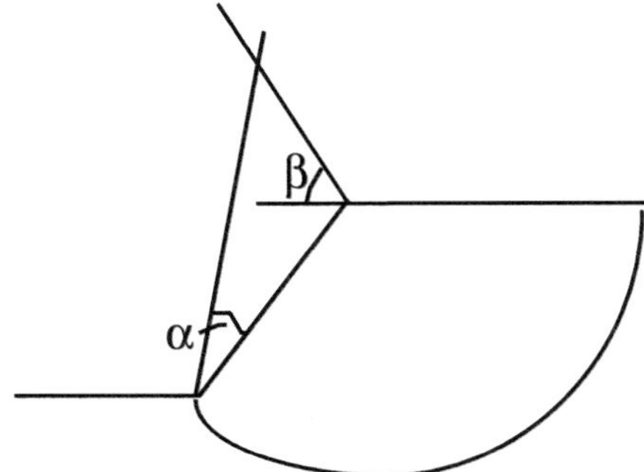

Figure 16.13 Fellenius's construction for centre of critical circle (Fellenius, 1936)

Table 16.2 Angles α and β proposed by (Fellenius 1936)

Slope	*Angle of slope*	*Angle α*	*Angle β*
1 : 0.58	60°	29°	40°
1 : 1	45°	28°	37°
1 : 1.5	33.79°	26°	35°
1 : 2	26.57°	25°	35°
1 : 3	18.43°	25°	35°
1 : 5	11.32°	25°	37°

The limitation of this method of is that it is applicable for obtaining a set of trial slip circles in average conditions and is not applicable when the slope is irregular or in the presence of pore water pressure within the slope or when the soil is not homogeneous.

16.3.6 Example

The analytical procedure for analysing the stability of a slope is demonstrated by the following numerical examples.

Example 16.1 Analyse the bank shown in cross section in Figure 16.14, with the worst possible case for the centre of sliding as shown. The soil has physical properties as follows:

$$\gamma = 1900\ \mathrm{kg/m^3}$$
$$c = 5.50\ \mathrm{kPa}$$
$$\phi = 30°$$

Analysis for two cases: (1) saturated and (2) dry.

Solution (1) Assume soil is saturated. Then

$$c = 0, \phi = 0, \text{and } \tau = 0$$

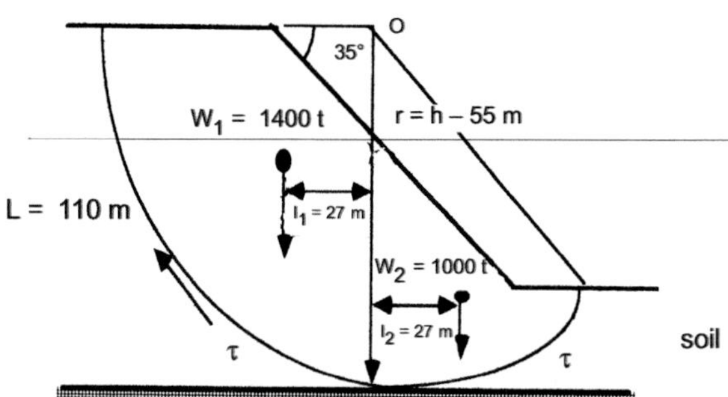

Figure 16.14 Analysis of a base failure by slip circle technique—Example 16.1

Writing the moment equation per unit width of slope, using units of tonnes-metre:

$$F \times W_1 \ell_1 = W_2 \times \ell_2 + \tau L r$$

$$F = \frac{W_2 \times \ell_2 + \tau r L_r}{W_1 \times \ell_1} \tag{16.4}$$

$$F = \frac{0 \times 55 \times 110 + 1000 \times 27}{1400 \times 20} = \frac{27000}{28000} = 0.96, \text{ therefore the slope is unsafe}$$

(2) Assume soil is dry (or only moist). Calculate shear strength by:

$$h' = h/2 = 27.5 \text{ m}$$

$$\gamma = 1900 \text{ kg/m}^3 = \frac{1900 \times 9.81}{1000} \text{ kN/m}^3 = 18.64 \text{ kN/m}^3$$

$$\tau = c + \gamma h' \tan\phi = 5.50 + 27.5 \times 18.64 \times \tan 30° = 30.15 \text{ tonnes/m}^2$$

$$\tau L r = 30.15 \times 110 \times 55 = 182407.5 \text{ tonnes-m}$$

Summing moments and finding the factor of safety:

$$F = \frac{27000 + 182407.5}{28000} = 7.48$$

The slope is safe, by a considerable margin.

Example 16.2 A soil slope consisting of frictionless soil has a cohesive strength of 30 kN/m^2 and unit weight of 18 kN/m^3. The centre of the trial circle is 12 m vertically above the toe of the slope. The weight of the soil slope is 582 kN at an eccentricity of 6.65 m from the centre of the rotation. Calculate the factor of safety (a) for the slope and (b) the factor of safety when a 2 m wide slice of soil as shown as shaded portion in Figure 16.15 is removed from the slope. Assume that no tension cracks were developed on the slope in both the above cases.

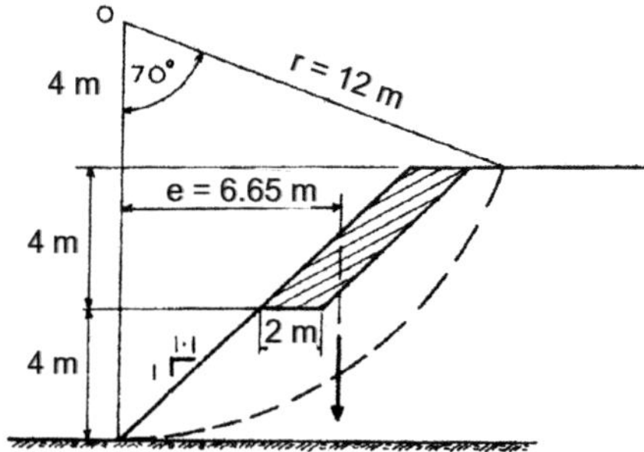

Figure 16.15 Cross-section through the soil slope

Solution

(a) Disturbing moment $= 582 \times 6.65 = 3870.3$ kN-m

$$\text{Restraining moment} = c\, r^2\, \theta = 30 \times 12^2 \times \frac{70}{180}\pi = 5280 \text{ kN-m}$$

$$F = \frac{\text{Restraining moment}}{\text{Disturbing moment}} = \frac{5280}{3870.3} = 1.36$$

Area of the slope undercut $= 2 \times 4 = 8 \text{ m}^2$

Weight of the slope removed $= 8 \times 18 = 144$ kN

$$\text{New eccentricity of the slope} = 1.1 \times 4 + \left(\frac{4.4+2}{2}\right) = 7.6 \text{ m}$$

$$\text{Relief of disturbing moment} = 144 \times 7.6 = 1094.4 \text{ kN-m}$$

$$F = \frac{5280}{3870.2 - 1094.4} = \frac{5280}{2775.8} = 1.902$$

16.4 Stability of Rock Slopes

If the potential sliding surface for an unstable rock slope is inclined at angle α and the slope face angle is β, the normal stress acting on the potential sliding surface is given by (CANMET, 1977):

$$\sigma_n = \frac{W}{A} \cos\alpha$$

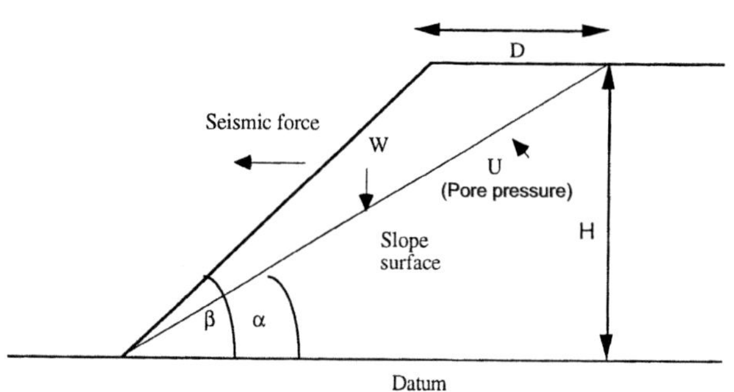

Figure 16.16 Stability of rock slopes

and the sliding stress is given by

$$\frac{W}{A} \sin\alpha$$

The shear strength of rock is given by:

$\tau = c + \sigma_n \tan\phi$
τ = shear force per unit area
c = cohesive strength
σ = normal stress
ϕ = internal angle of friction
γ = unit weight of slope material
W = weight of the slope
α = slope of the instability surface to horizontal
β = angle of the slope face to horizontal

$$k = \tan \varepsilon = \text{seismic force}$$
$$\varepsilon = \tan^{-1} k$$

Using limiting equilibrium equation

$$F = \frac{\text{Stabilising force}}{\text{Disturbing force}} = \frac{\text{Shear strength per unit area}}{\text{Disturbing force}}$$

It can be shown that the stability of planar rock slope with pore water pressure and seismic loading is given by (CANMET, 1977; Baron, Coates and Gyenge, 1970):

$$F = \frac{[cH \operatorname{cosec} \alpha + \{W \cos(\alpha + \varepsilon) - U\} \tan \phi]}{W \sin(\alpha + \varepsilon)} \tag{16.8}$$

where

$$W = \frac{1}{2} \gamma H^2 (\cot \alpha - \cot \beta)$$
$$k = \text{seismic load} = \tan \varepsilon$$
$$u = \text{water pressure} = \rho g h \times \frac{1}{2} H \operatorname{cosec} \alpha$$
$$h = \text{head of pore water pressure}$$

Stability of planar slope with pore pressure is given by:

$$F = \frac{[cH \operatorname{cosec} \alpha + \{W \cos(\alpha) - U\} \tan \phi]}{W \sin(\alpha)}$$

Stability of planar and dry slope is given by (CANMET, 1977, Baron, Coates and Gyenge, 1970):

$$F = \frac{[cH \operatorname{cosec} \alpha + W \cos \alpha \tan \phi]}{W \sin(\alpha)}$$

Example 16.3 A rock slope has a planar instability surface which is inclined to the horizontal at an angle of 40°. The height of the slope is 27 m and the angle of the slope face is 50°. The cohesive strength of the rock is 35 kPa and the internal angle of friction is 39°. The lateral force due to seismic effects may be taken as 0.12 times the weight of the unstable rock mass. The mean unit weight of rock is 2850 kg/m^3 and groundwater pressure is 1.6 m head of water. Calculate the factor of safety against sliding for the following conditions: (i) when slope is dry, (ii) when rock mass is saturated and exerting a hydraulic pressure on the slope and (iii) when the groundwater and seismic load are active simultaneously.

Solution

$$W = \frac{1}{2} \gamma H^2 (\cot \alpha - \cot \beta)$$

$$W = \frac{1}{2} 2850 (27^2) (\cot 40 - \cot 50) \frac{9.81}{1000}$$

$$= \frac{1}{2} 2850 (27^2)(1.19 - 0.84) \frac{9.81}{1000} = 3.28 \text{ MN/m}$$

$$k = \text{Seismic force} = \tan \varepsilon = 0.12$$

$$\varepsilon = 6.84°$$

$$u = \text{Water pressure} = \rho g h \times \frac{1}{2} H \operatorname{cosec} \alpha$$

$$= 1000 \times 9.81 \times 1.6 \times 0.5 \times 27 \times \operatorname{cosec} 40°$$

$$= 211896 \times 1.55 = 328 \text{ kN}$$

Stability of planar and dry slope

In case of dry slope u = 0 and the factor of safety is given by:

$$F = \frac{[cH \operatorname{cosec} \alpha + W \cos \alpha \tan \phi]}{W \sin \alpha} \tag{16.8}$$

$$F = \frac{1000 \times 35 \times 27 \times \operatorname{cosec} 40° + 3.59 \times 10^6 \cos 40° \tan 39°}{3.59 \times 10^6 \times \sin 40°}$$

$$= \frac{1.39 + 2.23}{2.307} = 1.56$$

Stability of planar slope with pore pressure

$$F = \frac{[cH \operatorname{cosec} \alpha + \{W \cos \alpha - U\} \tan \phi}{W \sin \alpha}$$

$$F = \frac{1000 \times 35 \times 27 \times 1.55 + (3.59 \times 0.77 - 0.33) \tan 39° \times 10^6}{3.59 \times 0.64 \times 10^6}$$

$$= \frac{1.47 + 2.43}{2.59} = 1.49$$

Stability of planar rock slope with pore pressure and seismic load.

The equation of factor of safety is as follows (Coates and Gyenge, 1965):

$$F = \frac{cH \operatorname{cosec} \alpha + \{W \cos(\alpha + \varepsilon) - u\} \tan \phi}{W \sin(\alpha + \varepsilon)} \tag{16.10}$$

$$F = \frac{1000 \times 35 \times 27 \times \operatorname{cosec} 40° + \{W \cos(\alpha + \varepsilon) - u\} \tan 39° \times 10^6}{3.59 \times \sin 40° \times 10^6}$$

$$= \frac{1.46 \times 10^6 + \{3.59 \times \cos(40 + 6.84) - 0.33\} \tan 39° \times 10^6}{2.308 \times 10^6}$$

$$= \frac{1.46 + 1.716}{2.308} = 1.38$$

16.5 Slope Stabilisation and Remedial Work

16.5.1 Introduction

In surface mining, the role of a geotechnical engineer is to manage the stability of the slopes in and around the excavation and to ensure stability. One approach for achieving these objectives is to carry out a slope stability analysis of all major soil and rock slopes using routine analytical techniques. If the analysis indicates that the slope is potentially unstable or a slope shows sign of instability during mining operation it is prudent to take the following steps.

- Monitor the stability of the slope
- Identify the causes of slope instability
- Redesign the slope to improve the factor of safety against failure.
- Use supports to increase the strength of slopes for ensuring long-term stability

Design and selection of support requires the causes of instability to be identified. The causes may be summarised as follows.

(i) Slope is too high or too steep.
(ii) The material from which the slope is made is too weak to maintain the slope in its present profile.
(iii) The pore water pressure is too high and thus adversely affecting the soil or spoil strength.
(iv) Slope is affected by external influences as follows:
  - Torrential rains
  - Seismic loading
  - Applied load from the earth-moving equipment, structures or surcharge loading.

These solutions will always be easiest and cheapest if slope is considered in isolation from its surroundings. Slides are prevented or controlled by

(i) flattening the slope,
(ii) proper drainage,
(iii) placing a retaining wall or ballast at the toe, or use of a buttress
(iv) driving piling,
(v) reinforcing the bank with rock bolts or bolts and wire mesh or
(vi) closely monitoring the surface for tensile cracks near the slope crest

16.5.2 Slope Investigation Programme

A systematic study of rock slope or spoil slope stabilisation techniques requires the following discrete steps:

- Identification of mode of instability
- Quantitative assessment of instability
- Theoretical and practical knowledge of possible supporting techniques
- Techniques for design and installation of mechanical supports
- Instrumentation, equipment and facilities for the fabrication, installation and monitoring
- Engineering and economic analyses to demonstrate feasibility and viability

Support is expensive and therefore it should only be used where benefits can be clearly identified and economically justified. In practice it is necessary to establish that:

(a) the slope is critical to the operation, that is instability will close the access, damage plant, equipment or machinery or bury the ore reserves;
(b) the slope is unstable and that a slide is imminent;
(c) alternative methods of slope stabilisation such as cut and fill method or drainage of the groundwater are not feasible or are more expensive.

16.5.3 Stabilising Mining Slopes by Re-grading

The most common methods of slope stabilisation in surface mining is to regrade the slopes particularly highwall and sidewall slopes in unconsolidated soils and rocks as well as spoil slopes as earth moving equipment is readily available.

Figure 16.17 shows some of the most common ways of dealing with slope instability in surface mining. There are three alternative procedures available as follows:

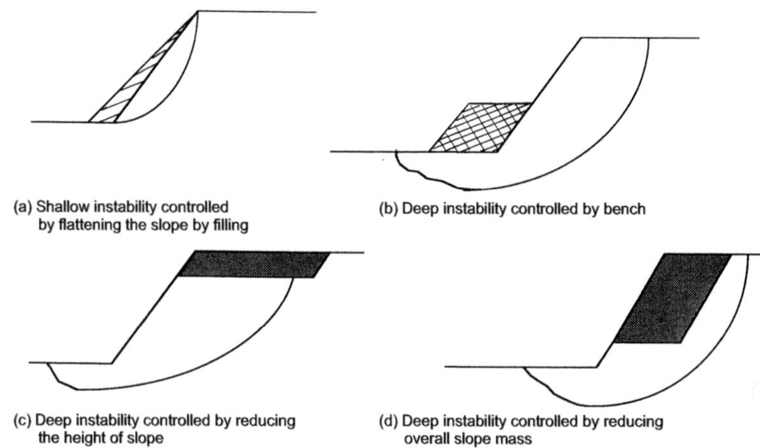

(a) Shallow instability controlled by flattening the slope by filling
(b) Deep instability controlled by bench
(c) Deep instability controlled by reducing the height of slope
(d) Deep instability controlled by reducing overall slope mass

Figure 16.17 Remedial action for slope instability by re-grading

(i) Grade to a uniform flatter angle

This method is most effective when dealing with instability in spoil heaps or shallow instability in which the movement is confined to soil layers close to the ground surface.

(ii) Concentrate filling the toe of the spoil slope until a step is created

For a deep-seated instability in soil or spoil dumps where a concentrated load of spoil on the toe of the slope can have a much better stabilising effect (Figure 16.17b).

(iii) Reduce overall slope height while retaining the slope profile (Figure 16.17c)

The same considerations apply to making cuttings or benches with the aim of reducing the slope height or reducing the overall angle of the slope.

(iv) Reducing overall slope angle by benching, (Figure 16.17d)

Slope flattening is relatively a better way for treating shallow mode failures than forming benches.The latter is more effective for deep-seated slope instabilities (Figure 16.17d).

16.6 Types of Support in Surface Mining

The types of support used in surface mining are as follows.

- Rock anchors and cable bolts or cable dowels
- Buttress wall
- Retaining walls

16.6.1 Rock Anchors, Rock Bolts, Cable Bolts and Cable Dowels

The most effective method of rock slope stabilisation is by improving the inherent shear strength of discontinuities, joints, bedding planes, and faults. In sedimentary rock, the bedding planes are the most significant discontinuity and instability occurs through movement along the discontinuity, and by separation and opening of the plane of the discontinuity together with the fracture of intact rock between the discontinuities. The purpose of the rock bolt or cable bolt is to increase the normal stress on the discontinuity surface by providing external tension and to provide resistance to sliding.

Friction is an important component of the shear properties of the sliding surface and of the normal force across the sliding surfaces. An effective way of increasing rock mass strength is to increase normal stress on the discontinuities associated with instability. This increase can be achieved by using rock anchors or cable dowelling (Figure 16.18(a) and (b)).

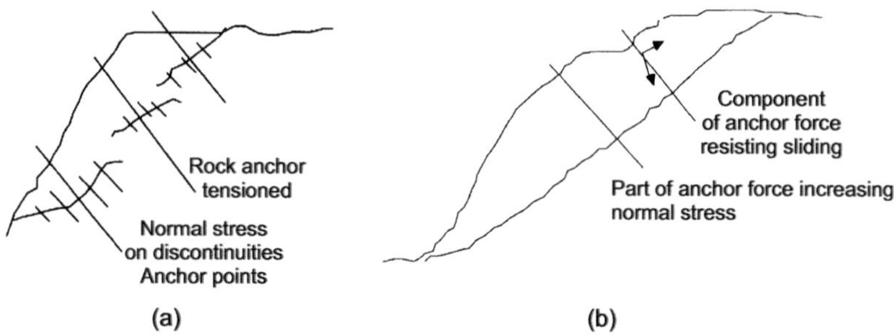

Figure 16.18 Rock anchorage to stabilise (a) discontinuities (b) planar rock slide (CANMET, 1977)

Rock anchors also control instability by directly resisting sliding. The force that can be realistically exerted by anchors in most cases is small compared with the weight of the rock mass. However, most instability in surface mining is borderline between sliding or remaining stable. A small increase in the resistance force to sliding may often be sufficient to ensure stability indefinitely.

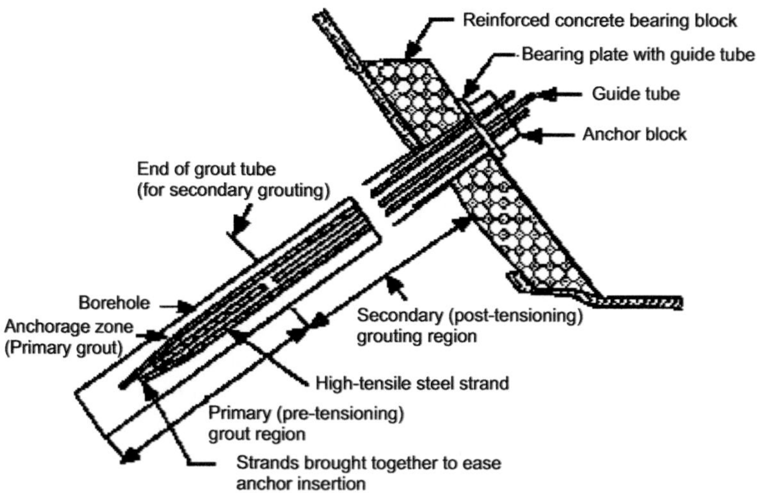

Figure 16.19 Post-tensioned grouted anchorage for stabilising rock slopes (Baron, Coates and Gyenge, 1970)

A rock anchor is a high strength cable or a bar, tensioned inside a borehole to 60–70% of its breaking load. Tension in the anchor is transferred to the surrounding rock mass by the anchor points at the ends. The length of the rock anchors used can be from 10–100 m. The tension in a rock anchor is selected between 20 tonnes to 220 tonnes occasionally reaching up to 450 tonnes (Fig. 16.19). The rock anchors have the most important application in stabilising deep-seated instability modes in which sliding or separation on a discontinuity is an inherent characteristic.

The material and techniques of ground anchorage has been adopted from pre-stressed concrete technology. Especially designed pre-stressing cable should be used for rock anchors rather than discarded mining ropes which suffer disadvantage of losing the anchor tension due to the strands slipping over each other. Mining ropes also contain fibre-cores that allow cable to compress, affecting grip of the anchoring wedges (Coates and Gyenge, 1965).

Rock anchors are placed at an angle to the discontinuities that form the instability (Figure 16.20). A component of the anchor force then acts directly to oppose sliding. The anchorage within the rock mass is formed by grouting the end of anchor for a length of about 16 m, however the actual length of anchorage is designed from the rock/grout bond strength. A conventional cement grout is often used but additives are also used to accelerate the setting time.

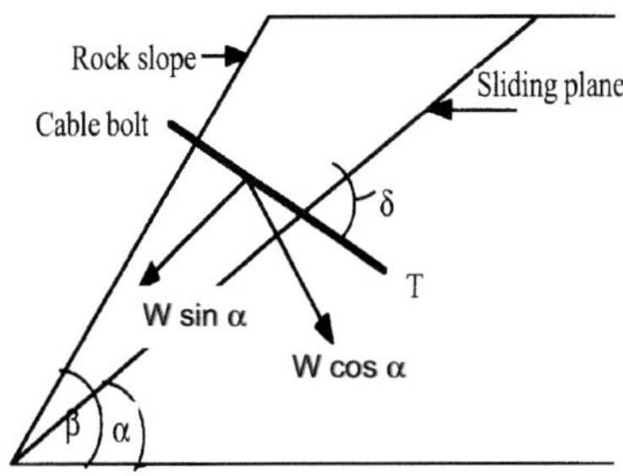

Figure 16.20 Typical rock anchor installation

16.6.2 Buttress

A buttress is a massive structure which provides support in the following ways (Figure 16.21).

- By providing lateral restraint through deadweight
- By increasing the strength of material below the buttress by increasing the normal stress

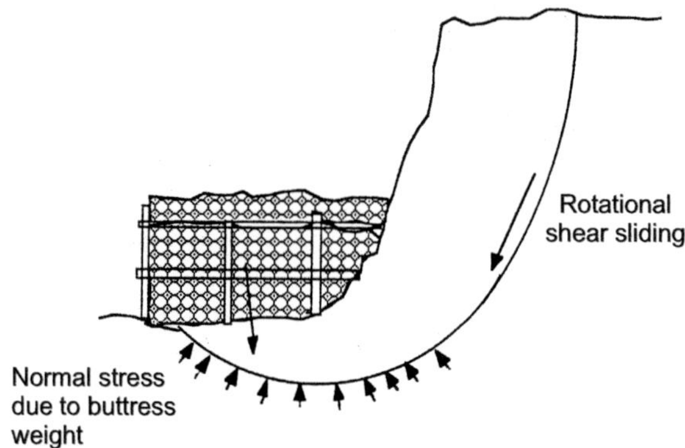

Figure 16.21 A buttress wall or crib with dead load of rock provides normal stress in the toe region and provides resistance to sliding (Baron *et al.*, 1970; Canmet, 1977)

A buttress usually consists of a timber brace or a concrete wall or crib to retain rock fill. An important requirement of such fill is that it should be self-draining to ensure that no build-up of groundwater pressure takes place so as to negate the strengthening effects of a buttress.

16.6.3 Retaining Walls

A retaining wall is a timber or a reinforced concrete structure for restraining loose material or to provide direct support to slope. These walls are shown in Figure 16.22. Timber braces are best suited to meet short term requirements whereas reinforced concrete is a durable support but requires more time and expense to install. Retaining walls are relatively costly to design and construct. They are used only for special applications such as to protect and support haul roads threatened with closure by the collapse of the slope and by falling materials.

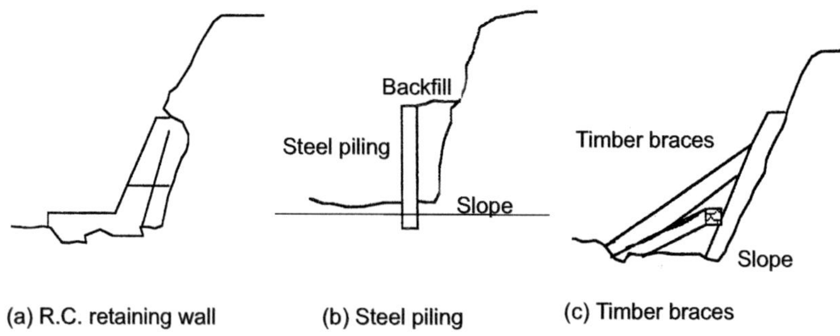

Figure 16.22 Retaining walls (Canmet, 1977)

16.7 Design of Surface Mine Supports

16.7.1 Design of Anchors

The design process of a support installation in a surface mine involves the following steps.

(1) Identification of a real or potential instability which will require the following information.
   - Mode and type of instability
   - Structural geology information
   - Groundwater condition and pore pressure
   - Mechanical properties of rock mass

(2) Quantitative engineering analysis of stability both with and without support

(3) Economic appraisal of support

Simple Plane Shear

The failure mode is considered plane shear when a discontinuity strikes approximately parallel to the slope at a shallow angle. The extent of the instability parallel to the slope is assumed to be large. The following assumptions are made.

- It is assumed that the plane shear mode is stabilised with rock anchors.
- Analysis is carried out in two dimensions only.

(a) Static plane shear in dry conditions with anchor

Considering that the disturbing force is W sin α, the component of the weight of the rock and the corresponding resisting forces are τ, the shear strength on the sliding surface, and T, the force due to the anchor. Resolving the forces parallel to the sliding surface:

$$F \times (W \sin\alpha) = \tau + T\cos(\alpha + \delta) \tag{16.11}$$

The factor of safety F against sliding is defined as follows:

$$F = \frac{\text{Resisting force}}{\text{Disturbing force}} = \frac{\tau + T\cos(\alpha + \delta)}{W \sin\alpha}$$

If resistance to sliding is represented by Coulomb's law

$$\tau = c + \sigma_n \tan\phi \tag{16.1}$$

where

τ = shear strength per unit area

T = total force due to anchors acting on the sliding surface
c = cohesive strength per unit area
σ_n = normal stress
ϕ = internal angle of friction

If the normal stress is assumed to be uniformly distributed over the sliding surface, the factor of safety against failure is given by the following equation (Baron, Coates and Gyenge, 1970):

$$F = \frac{[cH \operatorname{cosec} \alpha + \{W \cos \alpha + T \sin(\alpha + \delta)\} \tan \phi + \{T \cos(\alpha + \delta)\}]}{W \sin \alpha} \qquad (16.10)$$

where $W = \frac{1}{2} \gamma H^2 (\cot \alpha - \cot \beta)$

γ = average density of slope material

The Equation 16.10 indicates that the factor of safety against sliding increases as T increases provided $(\alpha + \delta)$ is within the range of 0 to 90 degrees.

The optimum direction of anchors can be calculated by maximising Equation 16.11 (differentiating with respect to δ and equating to zero)

$$\tan(\alpha + \delta) = \tan \phi \qquad (16.12)$$

(b) Pore pressure, seismic and blasting load

The effect of groundwater is to reduce resistance to sliding by transmitting part of the normal stress through water. In addition, seismic and blasting loads may both result in promoting ground movements with serious consequences.

The effect of blasting is not likely to cause the long period waves that result from earthquakes, but ground acceleration should be estimated and an equivalent load included in the analysis. The forces involved in both groundwater and seismic loads are very complex but a simplified analysis using equivalent loads is given here:

- Groundwater force, 'U' is assumed to be uniformly distributed on the sliding surface.
- Earthquake force or blasting can be assumed to be a horizontal equivalent force given by $k \cdot W$.

W = weight of the sliding surface
k = constant which has value less than 0.25

Inclined resultant force due to earthquake effect may be given by Equation 16.13.

$$R = W \sqrt{(1 + k^2)}$$

For $k < 0.25 \qquad R \approx W \qquad (16.13)$

By including the pore water pressure and earthquake forces, Equation 16.10 can be rewritten as follows (Baron, *et al.*, 1970).

$$F = \frac{[cH \operatorname{cosec} \alpha + \{W \cos(\alpha + \varepsilon) - U + T \sin(\alpha + \delta)\} \tan \phi + \{T \cos(\alpha + \delta)\}]}{W \sin(\alpha + \varepsilon)} \tag{16.14}$$

where,

$$\tan \varepsilon = k$$

The optimum inclination remains as far as the dry case is concerned. In equation 16.14, the term T occurs twice as follows:

Increase in normal stress $= T \sin(\alpha + \delta)$

Component directly resisting sliding $= T \cos(\alpha + \delta)$

Sometimes Equation 16.14 is rewritten as follows (Baron, Coates and Gyenge, 1970):

$$F = \frac{[cH \operatorname{cosec} \alpha + \{W \cos(\alpha + \varepsilon) - U + T \sin(\alpha + \delta)\} \tan \phi]}{W \sin(\alpha + \varepsilon) - T \cos(\alpha + \delta)}$$

In Equation 6.14, the term $T \cos(\alpha + \delta)$ is directly reducing the effect of the disturbing force $W \sin \alpha$.

(c) Worked example

A rock slope has a planar instability surface which is inclined to horizontal on an angle of 45°. The height of the slope is 31 m and the angle of the slope face is 60°. The cohesive strength of the rock is 35 kPa and the internal angle of friction is 35°. The lateral force due to seismic effect may be taken as 0.10 times the weight of the unstable rock mass. The mean unit weight of rock is 2650 kg/m^3 and groundwater pressure is 1.50 m head of water. Calculate the factor of safety against sliding for the following conditions:

(1) when slope is dry
(2) when the rock mass is saturated and exerting a hydraulic pressure on the slope
(3) when the ground water and seismic load are active simultaneously (Canmet, 1977; Baron, Coates and Gyenge, 1970)

Solution

$$W = \frac{1}{2}\gamma H^2(\cot \alpha - \cot \beta)$$

$$W = \frac{1}{2} 2650\,(31^2)\,(\cot 45 - \cot 60)\frac{9.81}{1000}$$

$$= \frac{1}{2} 2650\,(31^2)\,(1 - 0.577)\frac{9.81}{1000} = 5.275 \text{ MN/m}$$

$$k = \text{seismic force} = \tan \varepsilon = 0.10$$

$$\varepsilon = 5.7°$$

$$u = \text{water pressure} = \rho gh \times \frac{1}{2} H \operatorname{cosec} \alpha$$
$$= 1000 \times 9.81 \times 1.50 \times 0.5 \times 31 \times \operatorname{cosec} 45°$$
$$= 223668 \times 1.414 = 322.5 \text{ kN}$$

Stability of planar and dry slope

$$F = \frac{[cH \operatorname{cosec} \alpha + W \cos\alpha \tan\phi]}{W \sin\alpha}$$

$$= \frac{1000 \times 35 \times 31 \times \operatorname{cosec} 45° + 5.275 \times 10^6 \times \cos 45° \tan 35°}{5.275 \times 10^6 \times \sin 45°}$$

$$= \frac{1.534 + 2.61}{3.73} = \frac{4.14}{3.73} = 1.11$$

Stability of planar slope with pore pressure

$$F = \frac{cH \operatorname{cosec} \alpha + \{W \cos\alpha - u\} \tan\phi}{W \sin\alpha}$$

$$= \frac{1000 \times 35 \times 31 \times \operatorname{cosec} 45° + 10^6 \times \left\{5.275 \times \cos 45° - 0.3225\right\} \tan 35°}{5.275 \times 10^6 \times \sin 45°}$$

$$= 1.08$$

Stability of planar rock slope with pore pressure and seismic load

$$F = \frac{cH \operatorname{cosec} \alpha + \{W \cos(\alpha + \varepsilon) - u\} \tan\phi}{W \sin(\alpha + \varepsilon)}$$

$$= \frac{1000 \times 35 \times 31 \times \operatorname{cosec} 45° + 10^6 \times \left\{5.275 \times \cos(45 + 5.7) - 0.3225\right\} \tan 35°}{5.275 \times 10^6 \times \sin(45° + 5.7°)}$$

$$= \frac{1.534 + 2.114}{4.08} = 0.89$$

Anchor force

For maximum factor of safety $\tan(\alpha + \delta) = \tan\phi$

$$\tan(45° + \delta) = \tan 35°$$
$$\delta = -10°$$

$$F = \frac{[cH \operatorname{cosec} \alpha + \{W \cos(\alpha + \varepsilon) - u + T \sin(\alpha + \delta)\} \tan\phi]}{W \sin(\alpha + \varepsilon) - T \cos(\alpha + \delta)}$$

$$\frac{\left[{}^{35}\!/_{1000} \times 31 \times \text{cosec}45 + \{5.275 \times \cos(45+5.7) - 0.3225 + \text{T}\sin(45-10)\}\tan 35\right]}{5.275\sin(45+5.7) - \text{T}\cos(45-10)}$$

$$= \frac{1.534 + \{3.34 - 0.3225 + 0.57\,\text{T}\} \times 0.7]}{4.039 - \text{T} \times 0.819}$$

$$\text{F} = \frac{3.647 + 0.399\,\text{T}}{4.082 - 0.819\,\text{T}}$$

for $\text{F} = 1, \text{T} = 357\ \text{kN/m}$

$\text{F} = 1.1\ \text{T} = 649\ \text{kN/m}$

16.7.2 Plane Shear Instability Supported by Buttress

The use of a buttress to stabilise a simple plane shear instability is shown in Figure 16.23. The analysis of stability assumes that the buttress has a dead weight, V, and a resulting horizontal sliding of μV.

where

μ = Coefficient of friction between the buttress and the ground

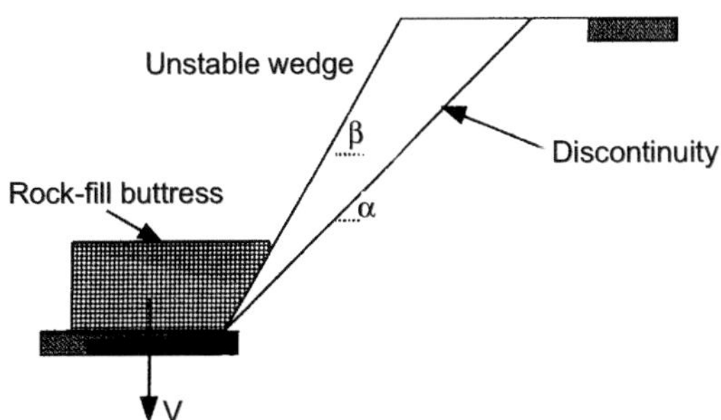

Figure 16.23 Principle of buttress support to a plane shear instability (Canmet, 1977; Baron, *et al.*, 1970)

Slope stability analysis indicated in Equation 16.16 can be used to analyse the effect of a buttress wall on stability. If the additional resisting force parallel to the plane of discontinuity due to sliding resistance of the buttress is given by $\mu V \cos\alpha$ then the factor of safety against failure is given by:

$$\text{F} = \frac{cH\,\text{cosec}\,\alpha + \mu V\cos\alpha + \{W\cos(\alpha+\varepsilon) - u\}\tan\phi}{W\sin(\alpha+\varepsilon)} \qquad (16.16)$$

16.7.3 Worked Example

Given that the planar shear instability case is as follows:

Height of the slope	= 31 m
Slope angle of unstable plane (α)	= 45°
Angle of the face (β)	= 60°
Lateral force due to seismic effect	= 0.1 W = kW
Cohesive strength (c)	= 35 kPa
Internal angle of friction (ϕ)	= 35°
Rock density	= 2650 k/m^3
Groundwater pressure	= 1.50 m head

Calculate the weight of the buttress wall required to raise the factor of safety from 0.88 to 1.25.

Solution

$$W = \frac{1}{2}\gamma H^2(\cot\alpha - \cot\beta)$$

$$W = \frac{1}{2}2650\ (31^2)\ (\cot 45 - \cot 60)\frac{9.81}{1000}$$

$$= \frac{1}{2}2650\ (31^2)\ (1 - 0.577)\frac{9.81}{1000} = 5.275\ \text{MN/m}$$

$$k = \text{seismic force} = \tan\varepsilon = 0.10$$

$$\varepsilon = 5.7°$$

$$u = \text{water pressure} = \rho g h \times \frac{1}{2}H\ \text{cosec}\,\alpha$$

$$= 1000 \times 9.81 \times 1.50 \times 0.5 \times 31 \times \text{cosec}\ 45°$$

$$= 223668 \times 1.414 = 322.5\ \text{kN}$$

$$\mu V \cos\alpha = (1.25 - FS_1)\ W \sin(\alpha + \varepsilon)$$

$$V = \frac{(1.25 - 0.88) \times 5.275 \times \sin(50.7)}{0.7 \times \cos 35°} = \frac{1.515}{0.5734}$$

$$= 2.64\ \text{kN/m of wall}$$

16.8 Conclusions

Geotechnical investigations and assessment of the stability of slopes around surface mining excavations are essential requirements for ensuring stability and safety of these operations. The use of steeper surface mining slopes results in increasing overburden ratios, thus affecting economy in deep-seated, steep open pit metalliferous mining operations. However, in strip and terrace mining the earth-moving equipment and mode of operation hardly permit readjustment of the slope angle as a means of improving economy or stability of the mining operations. Desktop studies involving simple slope stability analyses provide a means of identifying future instabilities of the surface mining slopes leading to ensuring the stability and safety of the operations. Stabilities of slopes in soils, spoil or in tips and embankments can be assessed by using soil mechanics principles. Stability of jointed and bedded rock slopes are assessed by applying stereographic projection techniques together with the limiting equilibrium analyses as discussed in Chapter 13.

Methods of re-grading surface mining slopes are applicable to soil, spoil slopes and rock slopes in shallow situations where heavy earth-moving machinery is readily available. Another alternative measure would be to change the orientation of the cut. However in deep surface mining conditions other remedial measures including the use of rock anchorage, slope dewatering techniques, buttress walls or retaining walls should be considered.

References

Atkinson, T. (1989). 'Economic and geotechnical factors in slope design', Lecture 1, Part II, Editors, R.N. Singh and B. Denby. Geotechnical Engineering Course, University of Nottingham, Department of Mining Engineering, 4 volumes, pp. 1.2 to 1.18.

Anon. (1985). Technical review of the stability and hydrogeology of minerals workings for the Department of Environment, UK, Geoffery Walton Consulting Mining and Engineering Geologists, and University of Nottingham, Department of Mining Engineering.

Anon. (1988). Geoffery Walton Practice and University of Nottingham, Department of Mining Engineering, *Technical Review of Stability and Hydrogeology of Minerals Workings*, Final Open File Report to the Department of Environment, London, HMSO.

Anon. (1986). Geoffery Walton Practice and University of Nottingham, Department of Mining Engineering, *Review of Current Geotechnical Practice in British Quarries and Related Interests and Requirements of Mine Planning Authorities and Other Statutory Bodies*, Open File Report, Department of Environment, London, U.K.

Anon. (1991). Geoffery Walton Practice and University of Nottingham, Department of Mining Engineering, *Handbook on the Design of Tips and Related Structures*, Department of Environment, London HMSO, ISBN 011 752535 1, 132 pp.

Baron, K., D.F. Coates and M. Gyenge (1970). 'Artificial support of rock slopes', Canadian Department of Energy, Mines and Resources, Mining Research Centre, Canadian Department of Energy, Mines and Resources, Ottawa, Final Report, 145 pp.

Chowdhury, R.N. (1978). 'Slope analysis', *Development of Geotechnical Engineering'* Vol. 22, 423 pp.

Cobb, Q. (1981). 'Slope Stability in British Surface Coal Mines', Ph.D Thesis, Nottingham University.

Canadian Department of Energy, Mines and Resources (1971). *Design Guides for Mine Waste Embankments in Canada*, Mining Research Centre, Canadian Department of Energy, Mines and Resources, Ottawa, Final Report, May, 185 pp.

CANMET (1977). *Pit Slope Manual*, 10 Chapters, Canadian Centre for Minerals and Energy Technology, Energy Mines and Resources, Canada, Report No. 77–13.

Coates, D.F. and T.S. Cochrane (1970). 'Development of design specifications for rock bolting from research in Canadian Mines', *International Mining Congress.*

Coates, D.F., M. Gyenge and J.B. Stubbins (1965). 'Slope stability studies in Knob Lake', *Proceedings of Rock Mechanics Symposium*, Toronto, pp. 35–46.

Denby, B. (1983). 'Shear strength assessment in mine slope design', Ph.D. Thesis, University of Nottingham, Mining Engineering Department.

Hartman, H.L. (1987). 'Introductory Mining Engineering', Chapter 6, *Surface Mining: Mechanical Extraction Methods*, John Wiley and Sons, New York, pp. 177–230.

Hoek, E. and J. Bray (1981). *Rock Slope Engineering*, The Institution of Mining and Metallurgy, London, Revised Third Edition, 356 pp.

Hutchinson, J.N. (1983). 'Methods of locating slip surfaces in landslides', *Bull. Assoc. Eng. Geology*, Vol. XX, pp. 235–252.

National Coal Board (1970). *Spoil Heaps and Lagoons*, NCB, London, Handbook Series, 230 pp.

Norton, P.J. (1983). 'A study of groundwater control in British surface coal mining', Ph.D Thesis, University of Nottingham.

Rao, S. (1980). 'A study of time dependent stability of excavated slopes in surface coal mining', Ph.D. thesis, University of Nottingham.

Singh, R.N., S.M. Reed and B. Denby (1985). 'The effect of groundwater re-establishment on the settlement of opencast mine backfills in the UK', 2nd *International Congress of International Mine Water Association*, Granada, Spain, pp. 802–817.

Singh, R.N., F.I. Condon and B. Denby (1986). 'Investigation into techniques of compaction of opencast mine backfill destined for development', *International Symposium on Geotechnical Stability in Surface Mining*, Calgary, pp. 285–293.

Singh, R.N. and B. Denby (1989). *Geotechnical engineering course*, University of Nottingham and The British Coal, Department of Mining Engineering, 4 volumes.

Singh, R.N., D.J. Brown, B. Denby and J.A. Croghan (1986). 'The development of new approach to slope stability assessment in UK Surface Coal Mines', *Symposium on Ground Control in Mining*, AusIMM, Wollongong, pp. 57–63.

Stimpson, B. and G. Walton (1970). 'Clay mylonites in English Coal Measures; their significance in opencast coal stability', *Proceedings of the First International Congress of International Association of Engineering Geology*, Paris, Vol. 2, pp. 1388–93.

Stead, D. and M.J. Scoble (1983). 'Rock slope stability assessment in British surface coal mines', *Surface Mining and Quarrying, IMM Symposium*, Bristol, pp. 217–224.

Stead, D. (1984). 'An evaluation of the factors governing the stability of surface coal mine slopes', Ph.D. Thesis, University of Nottingham, 487 pp.

Smith, G.N. (1973). *Elements of soil mechanics for civil and mining engineers*, Third Edition, ISBN 0 258 96948 2, Crosby Lockwood Staples, London, pp. 126–169.

Thomas, L.J. (1979). *An Introduction to Mining*, Third Edition, Methuen of Australia, ISBN 0454 00087 I, 466 pp.

Turner, K. and R.L. Schuster (Editors) (1996). *Landslides investigation and mitigation*, Special report 247, Transportation Research Board, National Research Council, Washington, D.C., pp. 337–424 and 439–501.

Woodruff, S.D. (1966). *Methods of working coal and metal mines*, Pergamon Press, London, Vol. 3, pp. 469–564.

Walton, G. and T. Atkinson (1978). 'Some geotechnical considerations in the planning of surface coal mines', *Trans IMM*, Oct. A 147–171.

Walton, and R.K. Taylor (1978). 'Likely constraints on the stability of excavated slopes due to underground coal workings', *Proceedings of Conference on Rock Engineering*, New Castle, April 1977, P.B. Atwell (Editor), University of Newcastle-upon-Tyne, pp. 329–49.

Walton, G. (1988). 'Stability implications in respect of surface mine planning in layered minerals', Ph.D. Thesis, University of Nottingham, 2 Vols., 231 pp.

Risk Analysis for Design in Geomechanics

17 Chapter

17.1 Introduction

Dealing as they do with natural materials, mining and civil engineers must take cognizance of uncertainties implicit in the material properties and other major parameters of variability in any design exercise that they undertake for a rock structure. Risk and reliability analysis are essential as uncertainty in performance of the engineered rock structure may be due to spatial variability, limitation in site exploration, and even limitations in calculation methods and most importantly because of geological "surprises". There is no way that such design efforts can lay claim for being exact, as geomechanics itself is not an exact science. Uncertainties are unavoidable and it is necessary that approaches are adopted that inform of and account for the uncertainties so as to gain some insight of the inherent risk levels. Risk and reliability considerations are thus *sine qua non* if the engineer has to propose a "probably right" design solution as design of rock structures must perforce specify, if not quantify, the risk involved in the design. In his Terzaghi lecture in 1964, Casagrande pointed out that risks are inherent in any project and that such risks should be recognized and systematic steps taken to deal with them. Harr(1987), in his seminal publication entitled "Reliability-based design in civil engineering", introduced the basic use of probability theory in civil engineering design. Even with the major strides made in the past two decades in probabilistic risk and reliability design, much remains to be done in applying the principles to practical design of engineered rock structures. Several papers in the past two decades have sought to deal with the range of uncertainties that are inherent in rock engineering problems (Hoek, 1991, Muralha, 1994, Suorinen et al, 1995, Pine and Thin, 1993, Whittlestone et al, 1995, Duzgun et al, 1998, Lacasse and Nadim, 1999) and examined the role of probabilistic approach to risk assessment. From site characterization, through the identification of likely failure modes, the determination of models and parameters that define the properties of the rock and joints, the estimation of loads, down to the choice of the methods to assess safety, uncertainty is all pervasive, which can only be addressed by expertise, empirical knowledge or by overdesign. In a Presidential Address to the South African Institute of Mining and Metallurgy, Stacey(2003) highlighted the dilemma that faces practitioners in the field of rock engineering – whether rock engineering involves good design or good judgement. While a good design can be achieved by diligently following the correct design process, including the identification of the appropriate failure or design criteria, in many situations, however, there is considerable inherent variability in the input parameters, and the estimation of appropriate input data usually involves engineering judgement.

This chapter will seek to present a brief overview of the subject in relation to design in rock in mining and civil engineering with the caveat that risk and probabilistic solutions are not panacea, but they could serve to provide a rational framework to include the relevant uncertainties. It is equally important to underscore the fact that such approaches do not improve faulty or insufficient input.

17.2 DEFINING BASIC TERMINOLOGIES

Risk analysis is all about predicting events that have not happened. Risk analysis includes the combination of the probability of an event occurring and consequences of the event should it occur. Thus,

$$\text{Risk} = P(E) \times C(E)$$

The purpose of risk analysis therefore is to provide a tool to support the decision-making process pulling together a set of relevant scenarios with the corresponding probabilities of occurrence and consequences.

In any design exercise, acceptable factors of safety are set to which computed values are compared to determine the adequacy of the design. The Factor of Safety F can be described as the ratio between **Capacity (C) and Demand (D)**, where the resistance is the **capacity C (or strength)** and the loading is the induced **demand D** imposed on the structure. Conventionally, the designer forms the well-known factor of safety as the ratio of the single-valued nominal values of C and D depicted in Fig. 17.1(a). The failure value for F is 1 in a deterministic analysis. In general, the demand function will be the resultant of the many uncertain components of the system under consideration. The capacity function likewise will depend on the variability of material parameters, testing errors, construction procedures and ambient conditions, and so on. In pillar design, for instance, capacity and demand are pillar mean strength and mean imposed stress respectively in a deterministic analysis. A schematic representation of the capacity and demand functions as probability distributions is shown in

Fig. 17.1(b). If the maximum demand (Dmax) exceeds the minimum capacity (C min), the distributions overlap (shown shaded), and there is a non-zero probability of failure. The difference between the capacity and demand functions is called the **safety margin(S)**; that is,

$$S = C - D$$

Obviously, the safety margin(S) is itself a random variable, as shown in Fig.17.1(c). Failure is associated with that portion of its probability distribution wherein it becomes negative (shown shaded); that is, that portion wherein $S = C - D \leq 0$. As the shaded area is the probability of failure p(f), we have

$$p(f) = P[(C - D) \leq 0] = P[S \leq 0]$$

The number of standard deviations that the mean value of the safety margin is beyond S = 0, Fig.17.1(c) is called the **reliability index**, β; that is,

$$\beta = \frac{\bar{S}}{\sigma[S]}$$

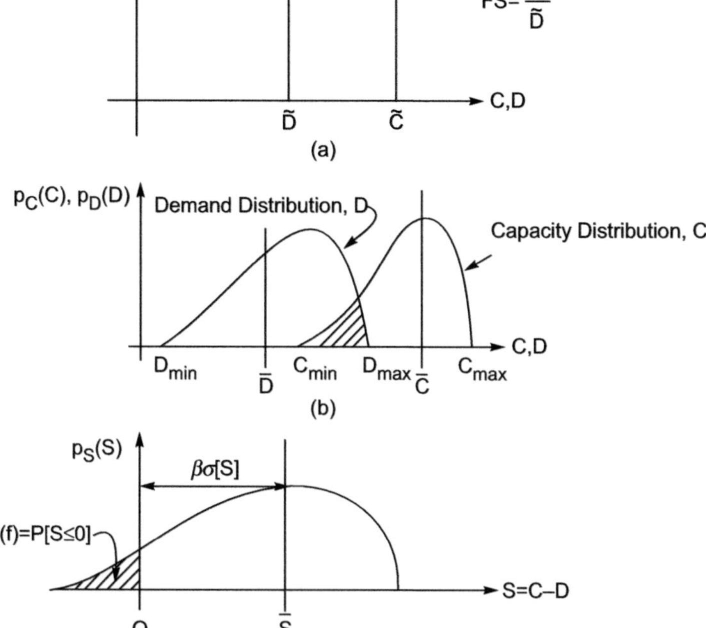

Figure 17.1

The reliability index is seen to be the reciprocal of the coefficient of variation of the safety margin. Figure 17.2 shows the concept of safety factor and probability of failure. The larger the intersection area, the higher the probability of failure.

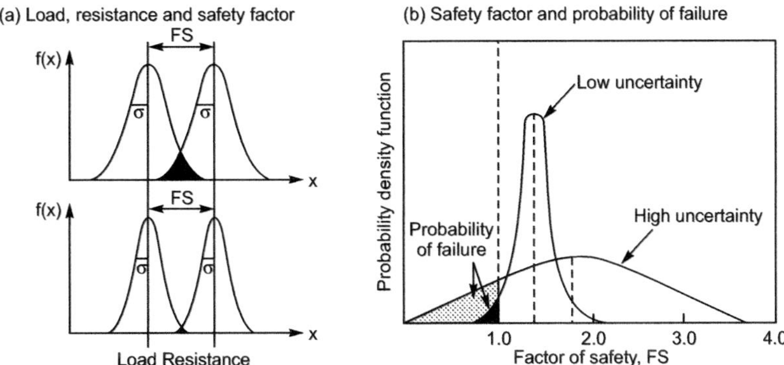

Figure 17.2 Safety factor and probability of failure, *FS = factor of safety*, *σ = standard deviation*

Figure 17.3 explains the essential differences between conventional deterministic analysis and probability analysis accounting for uncertainties.

CONVENTIONAL ANALYSIS

Input parameters → $\Sigma F_x = 0$ → Output of analysis

Load(s)
Resistance
strength, unit weight, ...
Shear strength

Factor of safety
Factor of safety

ANALYSIS ACCOUNTING FOR UNCERTAINTIES

Input parameters

Load(s) →
Resistance
strength, unit weight, ...
Model uncertainty

$G_x = \Sigma F_x = 0$
$P_f = P[G_x \leq 0]$
(P_f = P [Load > Resistance])

→ Output of analysis
Probability of failure
Reliability index
Parameter(s) which cause failure

Density function
2,0
1,0
0,0
0 1 2 3 4
Shear strength/100

Density function
1,0
0.5
0.0
−1 0 1 2 3 4
Factor of safety

Figure 17.3 Conventional analysis and analysis accounting for uncertainties, F_x = equilibrium equation, G_x = limit state function, P_f = probability of failure

17.3 UNCERTAINTIES IN DESIGN IN ROCK

Uncertainties associated with any rock engineering problem can be divided into two groups: aleatory (inherent) and epistemic (lack of knowledge) uncertainty:

- Aleatory uncertainty represents the quintessence of natural randomness of a property, such as the variation in rock strength in a horizontal direction or depth, depending upon lithology. Very little can be done to reduce this type of uncertainty.
- Epistemic uncertainty represents the uncertainty due to lack of knowledge, for example within a geologic layer. Measurement uncertainty and model uncertainty are epistemic uncertainties. Increasing the number of tests, or improving the measurement method could reduce this type of uncertainty. Within the epistemic class of uncertainties, one finds data uncertainty (due to instrument or human limitations), model uncertainty which could lead to bias, and statistical uncertainty, due to limited information and limited observations. In fact in any geologic envelope/volume, the characteristics will vary spatially, as they are controlled by a random pattern of variations.

To obtain the statistics of geotechnical parameters, traditional procedures can be used or when a lot of data exist, use of geostatistics with stochastic interpolation can be adapted. Accounting for uncertainties, even if the process requires judgment, is by far preferable to ignoring them.

17.4 PROBABILISTIC ANALYSIS TOOLS

Probabilistic analyses provide the following outcomes:

- probability of failure (or non-performance)
- reliability index
- sensitivity of result to any change in parameters.

One probabilistic analysis gives the same insight as a large number of parametric analyses. The essential input data for a probabilistic analysis are: (1) the equation defining failure and (2) the mean and distribution function for each uncertain parameter in the analysis. Defining probability distribution function can be problematic but as a simplification a normal or log-normal law is often specified as most geological processes fit such distributions. A panoply of methods exists to quantify the effects of uncertainties in a geotechnical response and its probability of non-performance. These are:

FORM (First-Order Reliability Method): seeks to approximate the limit state function by a first order tangent. The method works well over a wide range of probabilities and is simple to implement when one has an explicit limit state formulation.

FOSM (First-Order Second Moment) approximation: is suitable for a probabilistic sensitivity analysis of complex models. The method is based on determining the variance of output based on means and variances of inputs assuming all variables to be normally distributed.

MCS (Monte-Carlo Simulation): calls for repeated simulation of problem solution with randomly selected values of variables. Despite the universality of its application, it usually calls for a large number of simulations. Latin Hypercube simulation is a refinement of MCS incorporating a form of stratified sampling.

DT (Decision Tree): lists all events that can lead to failure or non-performance, the inter-relationships among events and the consequence of the non-performance.

The determination of acceptable risk levels in a given design situation is a difficult decision area. To a large extent, this difficulty is responsible for the current limited application of the probabilistic method. A determination of acceptable design risk should take into account both economic and safety aspects of the consequences of failure. Acceptance design criteria for slopes, based on experience and literature, originally suggested by Priest and Brown (1983) are given in Table 17.1. The table gives the probabilistic slope design criteria and inferences in terms of slope performance.

Table 17.1a Probabilistic slope design criteria

Category of slope	*Consequences of failure*	*Examples*	*Acceptable values*		
			Minimum	*Maximum*	
			Mean F	*P(F<1.0)*	*P(F<1.5)*
1.	Not serious	Individual benches, small (height<50m) temporary slopes not adjacent to haulage roads.	1.3	0.1	0.2
2.	Moderately serious	Any slopes of permanent or semi-permanent nature.	1.6	0.01	0.1
3.	Very serious	Medium sized (50m<height <150m) and high (height>150m) slopes carrying major haulage roads or underlying permanent mine installations.	2.0	0.0003	0.05

Table 17.1b Interpretation of probabilistic design criteria (after Priest and Brown, 1983).

Satisfaction of above criteria	*Interpretation*
Satisfies all three criteria	Stable slope
Exceeds minimum mean F but violates one or both probabilistic criteria	Operation of the slope presents a risk which may or may not be acceptable. The level of risk can be assessed by a comprehensive monitoring programme.
Falls below minimum mean F but satisfies both probabilistic criteria	Marginal slope. Minor modifications of slope geometry are required to raise the mean F to a satisfactory level.
Falls below minimum mean F and does not satisfy one or both probabilistic criteria	Unstable slope. Major modifications of slope geometry are required. Rock improvement and slope monitoring may be necessary.

The application of probabilistic analysis to geomechanics, and especially to design of rock structures in mining and civil construction, has thus far been limited and some problems of slope stability, design of pillars and design of supports have been attempted. A simple recommended approach would comprehend the following:

1. Using a point estimate method (PEM), or an equally valid probabilistic formulation, obtain the expected values and standard deviations of the capacity and demand functions: $E(C), E(D), \sigma(C), \sigma(D)$.
2. Calculate the expected value and standard deviation of the safety margin, $E(S), \sigma(S)$.
3. Fit a normal distribution to the safety margin, using appropriate upper and lower bounds.
4. Obtain the probability of failure, $p(f) = P[S \leq 0]$, the reliability, central factor of safety, and reliability index, as appropriate.

In geomechanics, the determination of the acceptable risk levels in a given situation is an intractable issue. To a great extent, this difficulty is largely responsible for the limited application thus far of the probabilistic approach. The lack of understanding of the language of probability is also a stumbling block.

17.5 CONCLUSIONS

In the design of engineered structures in rock, it is increasingly important today to adopt rational and consistent approaches that inform of and account for the uncertainties in design. The applicability of risk and reliability concepts is now emerging after a long development period. Only reliability analyses can provide the designer with insight, if not the "feel", on the risk level and quantify the effect of uncertainties on the predicted performance of an "engineered" rock structure.

References

Duzgun, H.S.B., Yucemen, M.S., and Karpuz, C. (2003), A Methodology for reliability-based design of rock slopes, Rock Mechanics and Rock Engineering, 36(2), pp. 95-120.

Harr, M.E. (1987). Reliability-Based Design in Civil Engineering. McGraw-Hill, New York.

Hoek, E. (1991). "When is a design in rock engineering acceptable?", Proc. 7th ISRM Congress on Rock Mechanics, Vol. 3, pp. 1485-1497.

Lacasse, S., and F. Nadin (1999), Risk Analysis in Geo-Engineering, ROCKSITE-99, Oxford & IBH, pp. 1-15.

Muralha, J. (1994), Uncertainty, reliability and risk in rock block stability, Rock Mechanics (eds. Nelson & Laubach), A.A. Balkema, pp. 539-564.

Pine, R.J., and I.G.T. Thin (1993). "Probabilistic risk assessment in mine pillar design", Proc. Innovative Mine Design for the 21st Century (Eds. Bawden & Archibald), Balkema, pp. 363-373.

Priest, S.D. and Brown, E.T. (1983). "Probabilistic stability analysis of variable rock slopes." Trans. Inst. Min. Metall., 92, pp. 1-12.

Stacey, T.R. (2003), Rock Engineering–good design or good judgement?, Jl.S.African Institute of Mining and Metallurgy, Sept., pp. 411-421.

Suorineni, F.T., M.B. Dusseault, and R.K.Brummer (1995), "Probabilistic risk and reliability evaluation of ground control practices in underground mining: the safety and economic implications", in Rock Mechanics (Eds. Daemen & Schultz), Balkema, pp. 579-584.

Index

D

E

F

G

H

I

J

K

L

M

N

O

P

Q

R

S

T - #0188 - 251019 - C532 - 234/156/23 - PB - 9780367391386